Advances in Intelligent Systems and Computing

Volume 632

Series editor

Janusz Kacprzyk, Polish Academy of Sciences, Warsaw, Poland
e-mail: kacprzyk@ibspan.waw.pl

The series "Advances in Intelligent Systems and Computing" contains publications on theory, applications, and design methods of Intelligent Systems and Intelligent Computing. Virtually all disciplines such as engineering, natural sciences, computer and information science, ICT, economics, business, e-commerce, environment, healthcare, life science are covered. The list of topics spans all the areas of modern intelligent systems and computing.

The publications within "Advances in Intelligent Systems and Computing" are primarily textbooks and proceedings of important conferences, symposia and congresses. They cover significant recent developments in the field, both of a foundational and applicable character. An important characteristic feature of the series is the short publication time and world-wide distribution. This permits a rapid and broad dissemination of research results.

More information about this series at http://www.springer.com/series/11156

Subhransu Sekhar Dash · Swagatam Das
Bijaya Ketan Panigrahi
Editors

International Conference on Intelligent Computing and Applications

ICICA 2016

 Springer

Editors
Subhransu Sekhar Dash
Department of Electrical and Electronics
  Engineering
SRM University
Chennai, Tamil Nadu
India

Bijaya Ketan Panigrahi
Department of Electrical Engineering
Indian Institute of Technology Delhi
New Delhi, Delhi
India

Swagatam Das
Electronics and Communication Sciences
  Unit
Indian Statistical Institute
Kolkata, West Bengal
India

ISSN 2194-5357 ISSN 2194-5365 (electronic)
Advances in Intelligent Systems and Computing
ISBN 978-981-10-5519-5 ISBN 978-981-10-5520-1 (eBook)
https://doi.org/10.1007/978-981-10-5520-1

Library of Congress Control Number: 2017946041

Printed on acid-free paper

This Springer imprint is published by Springer Nature
The registered company is Springer Nature Singapore Pte Ltd.
The registered company address is: 152 Beach Road, #21-01/04 Gateway East, Singapore 189721, Singapore

Preface

This AISC volume contains selected papers presented at the third International Conference on Intelligent Computing and Applications (ICICA 2016) held during December 21–22, 2016, at D.Y. Patil College of Engineering, Akurdi, Pune, India. ICICA 2016 is the third international conference aiming at bringing together the researchers from academia and industry to report and review the latest progresses in the cutting-edge research on various research areas of electronic circuits, power systems, renewable energy applications, image processing, computer vision and pattern recognition, machine learning, data mining and computational life sciences, management of data including big data and analytics, distributed and mobile systems including grid and cloud infrastructure, information security and privacy, VLSI, antenna, computational fluid dynamics and heat transfer, intelligent manufacturing, signal processing, intelligent computing, soft computing, web security, privacy and e-commerce, e-governance, optimization, communications, smart wireless and sensor networks, networking and information security, mobile computing and applications, industrial automation and MES, cloud computing, green IT and finally to create awareness about these domains to a wider audience of practitioners.

ICICA 2016 received 196 paper submissions including papers from two foreign countries. All the papers were peer-reviewed by the experts in India and abroad, and comments have been sent to the authors of accepted papers. Finally, 59 papers were accepted for oral presentation in the conference. This corresponds to an acceptance rate of 34% and is intended to maintain the high standards of the conference proceedings. The papers included in this AISC volume cover a wide range of topics in intelligent computing and algorithms and their real-time applications in problems from diverse domains of science and engineering.

The conference was inaugurated by Lt. Gen. Dr. S.P. Kochhar, CEO TSSC on December 21, 2016. The conference featured distinguished keynote speakers: Dr. P.N. Suganthan, NTU, Singapore; Dr. Arun Bhaskar, South Africa; Dr. B.K. Panigrahi, IIT Delhi, India; Dr. Swagatam Das, ISI, Kolkata, India; and Dr. S.S. Dash, SRM University, Chennai, India.

We take this opportunity to thank the authors of the submitted papers for their hard work, adherence to the deadlines, and patience with the review process. The quality of a refereed volume depends mainly on the expertise and dedication of the reviewers. We are indebted to the Technical Committee members, who produced excellent reviews in short time frames. First, we are indebted to the Hon'ble Dr. D.Y. Patil, Founder President, Hon'ble Dr. Sanjay D. Patil, President, Shri Satej D. Patil, Chairman, Dr. D.Y. Patil Educational Complex, Akurdi, Pune, India, for supporting our cause and encouraging us to organize the conference in their institute. In particular, we would like to express our heartfelt thanks for providing us with the necessary financial support and infrastructural assistance to hold the conference. Our sincere thanks also go to Dr. B.S. Balapgol, Principal, Dr. P. Malathi, Vice Principal, D.Y. Patil College of Engineering, Pune, for their continuous support and guidance. We specially thank Dr. Neeta Deshpande, Convener, and Dr. Kailash Shaw, Co-Convener, D.Y. Patil College of Engineering, Akurdi, Pune, for their excellent support and arrangements. We thank the International Advisory Committee Members for providing valuable guidelines and inspiration to overcome various difficulties in the process of organizing this conference. We would also like to thank the participants of this conference and IRD, India, for sponsoring. The members of faculty and students of D.Y. Patil College of Engineering, Pune, deserve special thanks because without their involvement, we would not have been able to face the challenges of our responsibilities. Finally, we thank all the volunteers who made great efforts in meeting the deadlines and arranging every detail to make sure that the conference could run smoothly. We hope the readers of these proceedings find the papers inspiring and enjoyable.

Chennai, India	Subhransu Sekhar Dash
Kolkata, India	Swagatam Das
New Delhi, India	Bijaya Ketan Panigrahi
December 2016	

Contents

About the Editors

Dr. Subhransu Sekhar Dash is presently Professor in the Department of Electrical and Electronics Engineering, SRM Engineering College, SRM University, Chennai, India. He received his Ph.D. degree from College of Engineering, Guindy, Anna University, Chennai, India. He has more than 20 years of research and teaching experience. His research areas are power electronics and drives, modeling of FACTS controller, power quality, power system stability, and smart grid. He is a visiting professor at Francois Rabelais University, POLYTECH, France. He is the Chief Editor of International Journal of Advanced Electrical and Computer Engineering and Associate Editor of IJRER. He has published more than 220 research articles in peer-reviewed international journals and conferences.

Dr. Swagatam Das received the B.E. Tel.E., M.E. Tel.E (control engineering specialization), and Ph.D. degrees, all from Jadavpur University, India, in 2003, 2005, and 2009, respectively. He is currently serving as Assistant Professor at the Electronics and Communication Sciences Unit of the Indian Statistical Institute, Kolkata, India. His research interests include evolutionary computing, pattern recognition, multi-agent systems, and wireless communication. He has published one research monograph, one edited volume, and more than 200 research articles in peer-reviewed journals and international conferences.

Dr. Bijaya Ketan Panigrahi is Professor of Electrical and Electronics Engineering Department in Indian Institute of Technology, Delhi, India. He received his Ph.D. degree from Sambalpur University. He is Chief Editor in International Journal of Power and Energy Conversion. His interests focus on power quality, FACTS devices, power system protection, and AI application to power system.

Exploration of Mars Using Augmented Reality

Manan Arya(ID), **Saad Hassan**(ID), **Swastik Binjola**(ID) **and Poonam Verma**(ID)

Abstract We describe our preliminary work on the scope for augmented reality being used in the exploration of Mars. This would be the potential use case of augmented reality and can also help the better understanding of the Mars surface and its environment. The application could be a practical use case for education and research on Mars. We also focus on providing this system to mass people which could help in evolving and bringing out efficient ways for exploration and colonization of Mars.

Keywords Augmented reality · Mars exploration · Augmenting Mars

1 Introduction

The augmented reality is the dilution of the virtual information to the real world to make it more interactive and understandable, as we define, can be used for the exploration of Mars. The exploration of Mars has long been subject of interest to the scientist and researchers. Various probes and rovers have been sent from Earth to gather information of the surface as well as the environment of Mars. With the commendable increase in the knowledge and data available about Martian system, the exploration of Mars could be made more interactive using augmented reality. With the data sets, for example, elevation of Mars surface and position of craters

M. Arya (✉) · S. Hassan · S. Binjola · P. Verma
Computer Science and Engineering Department, Bharati Vidyapeeth's College
of Engineering, BVCOE, New Delhi, India
e-mail: mananarya22@gmail.com

S. Hassan
e-mail: hassan.saad.mail@gmail.com

S. Binjola
e-mail: swastik.binjola2561@gmail.com

P. Verma
e-mail: poonamverma267@gmail.com

© Springer Nature Singapore Pte Ltd. 2018
S.S. Dash et al. (eds.), *International Conference on Intelligent Computing and Applications*, Advances in Intelligent Systems and Computing 632,
https://doi.org/10.1007/978-981-10-5520-1_1

could help us to augment the surface of Mars and understand it better and more efficiently. Making the data sets available to the public, through augmenting it can help with various innovative and efficient ideas for the exploration of Mars which could fuel up for human motivation for colonization of Mars.

As scientist's studies show possibility of human colonization of Mars [1], augmented reality could enable in viewing the three-dimensional surface and exploring the environment of Mars with least number of requirements and least amount of money by saving high amount of fuel, nuclear energy, magnetic energy, or antimatter which is consumed by a spacecraft to travel from Earth to Mars.

1.1 Background and Related Work

With promising researches on augmented reality technology in recent years and NASA's curiosity mission, it is possible to expand the augmented reality applications for the exploration of Mars with the data sets available till now. The main goal of curiosity mission was to determine whether Mars can ever support life, as well as understand the role of water, and to study the climate and geology of Mars [2].

A new technology was developed by NASA and Microsoft in 2015 which will use the data provided by real rovers to enhance the curiosity mission's exploring tools by augmenting a 3D simulation of the Martian environment with the help of Microsoft HoloLens [3].

2 Technologies Enabling the Project

2.1 Head-Worn Devices

Earlier in augmented systems, the processing of the applications took place on stationary computers and the result was displayed onto the various HMDs (head-mounted displays), which were bulky and heavy. But now the HMDs are so developed that the complex processing is done on the device itself. Head-mounted smart glasses like Microsoft HoloLens are able to process augmented applications on the device itself and are less bulky and much more precise in overlaying virtual objects onto the real environment but are still in the development stage. There are surplus of sensors that track or sense the movement of the user inside the room, and the information captured is used along with layers of colored glasses to create images you can interact with or investigate from different angles. Google Glass, another type of HMD, is designed in the shape of a pair of eyeglasses that process augmented reality applications and project it onto the glass in such a way that allows user to perceive the overlaid virtual objects while still being able to see the real world through it.

Fig. 1 Playing tennis game on Google Glass with virtual tennis ball (*Source* http://www.developers.google.com/glass/samples/mini-games)

Google Glass can perform various functions like remind the wearer of appointments and calendar events, act as a navigator by giving turn-by-turn directions, alert the wearer about the various traveling options, and much more. The explorer version of Google Glass consists of various components of hardware such as an ARM (Advanced RISC Machine)-based microprocessor for complex processing, video processors for visual processing, and a memory interface. This makes it powerful enough to handle complex augmented applications.

Google Glass is able to determine its location via the satellite signals received, just like Global Positioning System using the microchip, called the SirFstarIV [4]. Figure 1 shows how a virtual tennis ball is augmented in the real environment inside the "Tennis" game by using Google Glass.

Meta Glass is also an optical head-mounted display that too blends the virtual objects onto the real environment. Various functionalities of Meta Glass are 720p front-facing camera, sensor array for hand interaction and positional tracking and much more [5].

2.2 Handheld Devices

The increasing processing capability of personal handheld devices has helped in the transition of AR systems from the stationary computers of the mid-nineties to tablet PCs (personal computers) [6], PDAs (personal digital assistants) [7], and then mobile phones [8] in the twenty-first century.

Due to increased usage of the smartphones, it is possible to achieve "ANYWHERE AUGMENTATION." Complex augmented applications can now run on smartphones due to the massive improvements in the hardware of the smartphones such as their processor, RAM (random access memory). But the major problem that arises while running augmented applications on smartphones is that they are still unable to register the virtual objects onto the real environment precisely. There are different techniques that are being worked upon such as accelerometer-based head tracking [9] that might further reduce the problem of registration, but may not be able to resolve it to the extent where the virtual objects seamlessly blend into the real environment. The use of AR to play games on handheld smartphones is shown in Fig. 2.

The project called Tango by Google is focusing on creating devices (also known as Tango devices) with extra sensors [10]. These sensors will provide the Tango devices with the ability of motion tracking (relative motion tracking without using GPS or other external signals), area learning (by using the concept of area description files), and depth perception (by using specialized 3D sensors), i.e., the devices will be able to understand the relative distance between the objects, and the distance between the device and objects or walls in the real world. This project will further reduce the registration problem, making devices capable enough to handle complex augmented applications with ease.

Fig. 2 Playing Pokemon Go game on handheld smartphone with a virtual Pokemon augmented in the real environment (*Source* https://play.google.com/store/apps/details?id=com.nianticlabs.pokemongo)

2.3 Tracking Sensors and Approaches

To display the virtual objects like the pinpoints, flags, or rovers inside the augmented reality system, the tracking device or the device creating the augmented environment must be able to sense the external environment and then track the viewer's relative movement with all six degrees of freedom including the three-dimensional axis (x-axis, y-axis, and z-axis) and the three angles (yaw, pitch, and roll) for position and orientation, respectively, as shown in Fig. 3. Some environment models need to be used for tracking and correct AR registration. Before an AR system can detect all the six degrees of freedom movements, the environment has to be prepared. There is no single best solution available for determining the orientation of the user, and different tracking techniques work in different environment.

2.4 User Movement Tracking

In comparison with the virtual environments, the AR tracking devices must have higher accuracy, a wider input variety and bandwidth, and longer ranges [11]. The accuracy of the registration is calculated by the geometrical model along with the distance of the object to be annotated. The distance of the object from the tracking device is directly proportional to the impact errors in the orientation tracking and inversely proportional to the impact errors in position tracking which contributes

Fig. 3 Six degrees of freedom (*Source* https://developers.google.com/tango/)

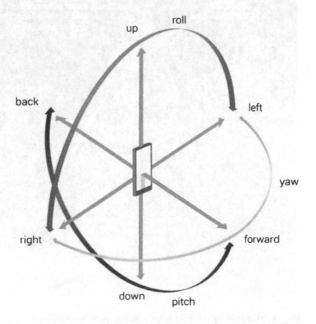

toward the misregistration of an augmented object. The tracking is usually easier in indoor settings, but unprepared outdoor environment still poses tracking problems and different ways have to be used in different environments.

The various tracking techniques include:

(1) Global positioning system (GPS)
(2) Mechanical, ultrasonic and magnetic
(3) Radio
(4) Inertial
(5) Optical
(6) Hybrid

Out of all the tracking techniques available, global positioning system (GPS) plays a major role in the AR applications based on mobile phones, which have an in-built GPS receiver for the navigation purpose.

2.5 Global Positioning System

The GPS plays a significant role in outdoor tracking by providing location and precise time information using the concept of trilateration and atomic clocks, respectively. Figure 4 shows how the GPS works and helps in determining the position of the device or the GPS receiver.

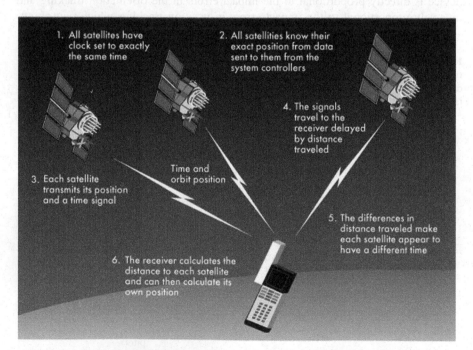

Fig. 4 Working of a GPS (*Source* http://www.bestforhunting.com/2011/06/how-does-gps-work/)

The American 24-satellite Navstar GPS, Russian Glonass, European Galileo, China's BeiDou and the seven-satellite Indian GPS system, NavIC of the Indian Regional Navigation Satellite System (IRNSS) constellation currently being launched by the Indian Space Research Organization (ISRO) in 2016 [12], will support the outdoor tracking by GPS.

The wide area augmentation system (WAAS) technology provides great accuracy of 3–4 m compared to the plain GPS with accuracy about 10–15 m. By preparing the environments using the local base station to send a differential error correction signal to the corresponding roaming unit, the differential GPS can be made accurate up to 1–3 m.

The GPS technology can be used widely in mobile-based AR applications and the Mars exploration to get the accurate location of the mobile and then augmenting the corresponding location of the Martian surface. By using the various latest technologies to convert 2D image data sets to the 3D model, the 3D Martian surface can be augmented with all the elevation and depression levels and further interaction with the surface could be done.

3 Applications

3.1 Surface Exploration

With the help of augmented reality, we can augment surface of the Mars as 3D model. NASA's Mars Rover Photos API (application program interface) [13] provides the images of Mars surface which can be converted to 3D model based on its elevation data. Augmenting will help us to analyze the Mars surface and explore it by walking on it with accurate longitude and latitude information of each point available. Based on the geological and geochemical data available for that area, we could augment essential information like chemical, isotopic, and mineralogical composition to know the detailed information about the place of minerals and water availability suitable for plant growth and mining. The processes that are involved in mineralogical composition can be available through searching and fetching real-time information on search engines.

Figure 5 shows a part of Mars surface being augmented in the real environment with the future applications like bookmarking a particular location, getting the temperature information of the respective Mars surface area, and getting information about various craters and rocks.

Markers could be set at various positions on the augmented Mars surface, and the distance between any two positions could be calculated. By creating or marking the points or pins, the sun's angle from those points could be known. By drawing the freehand lines or polylines, we could mark some particular ways to be followed on the augmented Martian surface. The landing site of various spacecraft and rovers could be visualized along with the presence of those spacecraft and rovers.

Fig. 5 Proposed view 1: illustration of Mars surfaces with various virtual information

3.2 Colonial Visualization

The main purpose of this project is to make the data available for the public use. These data sets when used in augmented systems such as smartphones with applications would help the people to realize and visualize that how their colony or their residing place at Mars would look like by adding their own innovative ideas, if they were to settle on Mars in future. The available APIs and data sets by NASA can provide us with accurate weather conditions and surface structure including the surface elevation profiles, sun angle calculation [14] that may affect the colony in one or the other ways. Hence, creating a basic visual structure for the colony would empower the people with a way of exploring the Mars just like the Earth.

In recent years, there have been many projects which successfully implemented AR in Civil Engineering so by using that technology we can enable colonial visualization of Mars. "A Touring Machine" [15] project, prototyped mobile augmented reality application that shows outdoor navigation approaches and information-seeking in that area. This Webpage [16] describes various works which could help us understand how we could visualize colonization of Mars on Earth itself using augmented reality. Figure 6 shows how virtual Mars surface can be augmented inside the real environment along with the important information about the landing sites of various rovers and their path traversed on the Martian surface.

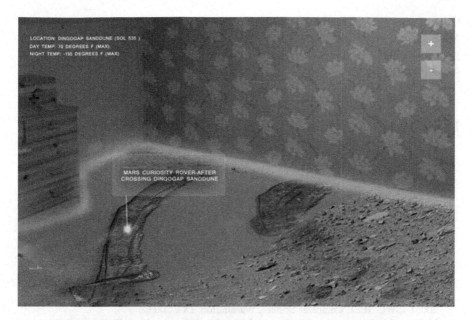

Fig. 6 Proposed view 2: illustration of Mars surface with various virtual information

4 Conclusion

AR is an emerging field, and the scientists and public in general have a vast scope for exploring its new dimensions. It can enhance the current developing scenario in the exploration of Mars and its surface. The project will initiate new ideas among the public to explore Mars by using the computer vision and the interaction used in advanced AR technology. It will enhance the understanding of students and the public members with the augmented surface features and mission activities, and then they can conduct their own exploration of Mars by visualizing the augmented NASA data in an easy-to-access and interactive way.

The project can make a great difference in the mission of colonization of Mars by viewing the whole Martian surface along with the accurate details of the Mars' environment and atmosphere. It can help in solving many challenges currently being faced in the space exploration. The success in project can lead to marking or pining the points on the augmented Mars surface with the elevations and depressions, which can be used by the scientists to decide the landing sites of rovers and spacecraft for their further missions. The advantage of drawing freehand lines or polylines can be used to decide the most suitable way rovers, or the NASA Martian landers need to follow to collect the data, on Mars surface. Not only the scientists and astronauts will be benefited by the project, but this will also create a public interest in the field of exploration of Mars and the colonization of Mars with the help of simple GPS-enabled smartphones. In future, during the colonization of Mars, this project will help as the navigation system for the people settling on Mars, giving them the in-depth knowledge of availability of resources like minerals, heat, air, liquid to sustain the lives.

References

1. Norbert Kraft, MD; Dr. James R. Kass; Dr. Raye Kass, "Mars One: Humanity's Next Great Adventure".
2. http://mars.nasa.gov/msl/mission/mars-rover-curiosity-mission-updates/.
3. http://www.jpl.nasa.gov/news/news.php?feature=4451.
4. http://www.catwig.com/google-glass-teardown/.
5. https://www.metavision.com.
6. Newman J., Schall G., Barakonyi I., Schürzinger A., and Schmalstieg D., Wide-Area Tracking Tools for Augmented Reality. In Proceedings of the 4th International Conference on Pervasive Computing, 2006.
7. Wagner D. and Schmalstieg D., First Steps towards Handheld Augmented Reality. In the Proceedings of the 7th IEEE international Symposium on Wearable Computers (Oct 21–23, 2003). ISWC. IEEE Computer Society, Washington, DC, 2003.
8. Mohring, M., Lessig, C. and Bimber, C. Video See-Through AR on Consumer Cell Phones. In the Proceedings of the International Symposium on Augmented and Mixed Reality (ISMAR'04), pp. 252–253, 2004.
9. M. S. Keir, C. E. Hann, J. G. Chase, and X.Q. Chen, A New Approach To Accelerometer-based Head Tracking for Augmented Reality & Other Applications, Department of Mechanical Engineering, Human Interface Technology Lab NZ University of Canterbury.
10. https://developers.google.com/tango/.
11. D.W.F. van Krevelen and R. Poelman, "A Survey of Augmented Reality Technologies, Applications and Limitation".
12. http://www.isro.gov.in/irnss-programme.
13. https://api.nasa.gov/api.html.
14. http://mars.nasa.gov/maps/explore-mars-map/fullscreen/.
15. Feiner, S., MacIntyre, B., Hollerer, T., & Webster, T. (1997). A touring machine: Prototyping 3D mobile augmented reality systems for exploring the urban environment. Personal Technologies, 1, 208–217.
16. http://www.engineering.com/BIM/ArticleID/12233/Augmented-Reality-for-Architects-and-Civil-%20Engineers.aspx.

Block Compressive Sampling and Wiener Curvelet Denoising Approach for Satellite Images

R. Monika, A. Anilet Bala and A. Suvarnamma

Abstract Satellite optical system produces high-resolution images which deal with large volume of data. This imposes strain on embedded resources which require more memory and computing capacity. In classical satellite imaging system, conventional compression algorithms like JPEG were used. However, they are not very efficient in reducing the data rate. In order to overcome this, block compressive sensing (BCS) technique, reweighted sampling (RWS) are used. This technique provides block-by-block sampling continuously at a rate which is very much less than the Nyquist rate. Due to the interference with high frequency signal in the environment, noise is induced in the compressed data from the satellite while transmitting them to the ground station. Curvelet transform with Wiener filtering technique (CTWF) is used for significant denoising of the BCS data. Experimental results show that BCS along with denoising technique reproduces images with better PSNR values.

Keywords Block compressive sensing · Reweighted sampling
Curvelet transform · Wiener filtering · OMP · Sparse binary random matrix

1 Introduction

Satellite image processing has been the focus of work in recent years. With the advancement in satellite imaging systems, the resolution of the image captured is very high. Transmitting or storing such a high-resolution image becomes a serious problem because of the energy and bandwidth constraints. In order to overcome

R. Monika (✉) · A. Anilet Bala · A. Suvarnamma
ECE Department, SRM University, Chennai, India
e-mail: moni.rajendran@gmail.com

A. Anilet Bala
e-mail: aniletbala.a@ktr.srmuniv.ac.in

A. Suvarnamma
e-mail: suvarnahudson@gmail.com

© Springer Nature Singapore Pte Ltd. 2018
S.S. Dash et al. (eds.), *International Conference on Intelligent Computing and Applications*, Advances in Intelligent Systems and Computing 632,
https://doi.org/10.1007/978-981-10-5520-1_2

these problems, compression of data is essential. Conventional compression techniques are not so efficient for satellite image applications.

Compressed sensing (CS) is an effective alternative, which performs compression at a rate lesser than Nyquist sampling rate [1]. But CS cannot be applied for real-time sensing of images, as the entire image is sampled at a time. Therefore, BCS is used in which the entire image is divided into small blocks [2] and CS is applied to each block independently. Since same measurement matrix is used memory requirement is less. Blocks are processed independently; therefore, the initial solution can be easily obtained and reconstruction process is speeded up.

Compressed images from the satellite are more prone to noise, which can degrade the quality of the image. It is essential to remove the noise and to improve the image quality. Image denoising can be considered as recovering a signal from inaccurately measured samples which is partially accomplished by CS. However in order to remove the noise added in the environment, denoising methods are essential. Many transform-based denoising techniques like Fourier, wavelet, ridgelets are available. However, most of them have certain shortcomings in terms of image quality and computational efficiency. In order to achieve better PSNR, CTWF technique is used.

In this paper, combination of both BCS and CTWF technique is used to retain good quality image with low data rate.

2 Related Works

Donoho [3] explained that signals or images can be recovered from fewer measurements or samples than the one described in Nyquist sampling theorem. In CS, sampling and compression are performed simultaneously to speed up the process. He suggested that CS can be applied to the signals only if it is compressible and sparse.

Gan [4] discussed the acquisition of images in block-by-block manner. This technique is simpler and efficient than normal compressed sensing technique and can effectively capture complicated structures of the image. Same measurement matrix is applied to all the blocks. Therefore, this technique requires less storage space.

Yang et al. [5] introduced a new weighting process into the conventional CS framework. Weight values are calculated for all frequency components. Signal components with larger magnitude will have large weight value and can be reconstructed more precisely. As a result, enhanced reconstructed image quality can be obtained.

Starck et al. [6] describe about the implementation of Curvelet transform for denoising. Images are reconstructed with low computational complexity. Curvelet reconstruction offers higher quality recovery of edges thereby improving the perceptual quality when compared to that of wavelet-based image reconstructions.

Ansari et al. [7] compared denoising techniques using wavelet, curvelet and contourlet transforms for remote-sensed images with additive Gaussian noise. The Curvelet-based denoising technique preserves the sharpness of the boundaries. The geometrical structure of the image can be effectively captured by the curvelets.

The rest of the paper is organized as follows: overview of compressed sensing is provided in Sects. 3 and 4 discusses about the satellite image processing system, experimental results are provided in Sects. 5 and 6 concludes the paper.

3 Overview of CS

Let x denotes a real-time finite length signal to be acquired. As per the hypothesis of CS, there exists a basis ψ where s is sparse up to sufficient level. s can be row or column vector. The equation is given by

$$x = \psi s \tag{1}$$

This means there exist k non-zero elements such that $k \ll n$. ψ can be any transform like discrete cosine transform or discrete wavelet transform. Let the matrix y represents a set of m linear combinations of x. These linear combinations can be represented as matrix ϕ with size $m \times n$ and are called measurements. The CS is represented as

$$y = \phi x = \phi \psi s \tag{2}$$

The measurements matrix ϕ used in this paper is sparse binary random matrix. This matrix has only binary values 0 and 1's, and it satisfies restricted isometry property (RIP). If the measurement matrix used satisfies RIP [3], then the length of all sufficiently sparse vectors are approximately preserved under transformation by the matrix. The other measurement matrices generally used are Gaussian matrix, Bernoulli random matrix and Sub-Gaussian random matrices. Orthogonal matching pursuit (OMP) is used as reconstruction algorithm [8].

4 Satellite Image Processing System

Figure 1 shows the block diagram of satellite image processing system. The satellite imaging system captures the image of the target. The images are high-resolution images. In order to store or process such a large volume of data, more energy will be consumed by the processor. To reduce the amount of data, BCS technique is used. This compressed data is then transmitted to the ground station. Noises in the environment get added to the data. To remove the noise, CTWF technique is used. The various techniques used are described briefly in this section.

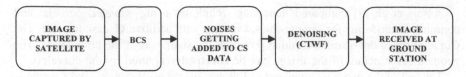

Fig. 1 General block diagram of satellite image processing system

4.1 Block Compressed Sensing

BCS adopts an adaptive projection representation, i.e. direction of projection is along the direction where the signal components have larger magnitude. As a result, it can efficiently capture the geometric structures of natural images. Also the computational complexity is reduced as same measurement operator is applied to all the blocks and can be easily stored. Each block is processed independently; therefore, the reconstruction process is speeded up. In this paper BCS technique, RWS is used.

(A) *Reweighted sampling* [5]

Reweighted Compressive Sampling for image compression introduces a weighting process to extract low-frequency components of the image. Weight values are assigned for all frequency components. Low-frequency components have large weight values. Therefore, this scheme shows discrimination to various components of the image. RWS sampling is given by the following equation,

$$y = \phi \psi W s \tag{3}$$

where W is a diagonal weighting matrix with weighting coefficients $\{w_1, w_2, \ldots w_n\}$ corresponding to different frequency component. Weight is calculated as sum of the mean and square root of variance of DC coefficients. By introducing the weighting matrix, the signal components with large magnitudes are effectively captured, which improves recovery precision.

4.2 Denoising

Noises in image degrade the quality of the image. The conventional spatial filtering technique reduces noise, but the edges of the images are blurred. Here, Curvelet transform along with Wiener filtering technique overcomes this disadvantage.

(A) *Curvelet transform* [6]

In Curvelets, multiscale ridgelets transform is combined with a spatial bandpass filtering to isolate different scale. Curvelets occur at all scales, locations, and orientations, but they have varying widths and length. Hence, they have variable anisotropy, whereas ridgelets have only global length and variable widths. Therefore, curvelets are used in this paper.

One of the important advantages of using Curvelet transform is that it analyses the image with different block sizes using a single transform. Initially the image is decomposed into a set of wavelet bands, and then ridgelet transform is applied to each band. The block size can be changed at each scale level.

(B) *Wiener filtering*

Compare to all other filtering technique, the most commonly used is Wiener filter. This is because only a few computational steps are required for execution and are very fast to process. Linear equations are used to calculate the filter weight which reduces the noise level of the signal. Curvelet transform itself provides better denoising. In order to increase the PSNR values, Wiener filtering is combined with Curvelet transform.

5 Experimental Results

Matlab R2012a is used for software simulation. The test images are taken by satellite available in image database [9]. In the BCS technique, block size of 8×8 is chosen. Number of pixel values chosen from each block is only 10 out of 64.

Table 1 shows the PSNR values obtained for various images using BCS technique. The number of pixel values chosen from each block is only 10 out of 64.

Table 1 PSNR values for various images

Technique = BCS number of measurements = 10	
Image	PSNR
Mountain image	24.2312
Airport image	25.9982

Table 2 PSNR values for various noise levels (mountain image)

Sigma value (noise in dB)	CTWF PSNR (dB)	Wiener filter PSNR (dB)
10	26.6820	26.7325
20	25.5570	25.2649
30	24.3212	23.5304
40	23.1547	21.9023
50	22.0576	20.4918

Table 3 PSNR values for various noise levels (airport image)

Sigma value (noise in dB)	CTWF PSNR (dB)	Wiener filter PSNR (dB)
10	28.4436	28.2892
20	26.9750	26.4231
30	25.3600	24.2592
40	24.0161	22.4060
50	22.7080	20.8062

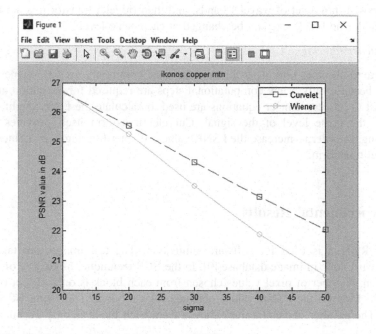

Fig. 2 Graphical representation of sigma versus PSNR

Additive White Gaussian noise is added to the BCS data by using Matlab inbuilt function. Now in order to remove the noise, denoising technique CTWF is used. The PSNR value comparison for mountain and airport images at various amounts of noise levels varying from $\sigma = 10$ to 50 dB is shown in Tables 2 and 3.

It can be seen from the table that CTWF has achieved better PSNR values than normal Wiener filtering for various noise levels. If noise level is high, then CTWF is the most efficient technique in achieving better image quality. In order to check the consistency of the technique, it is tested with various satellite images. Table 3 shows the PSNR value comparison for airport image at various noise levels as considered before.

The graphical representation of various noise levels and their corresponding PSNR values is shown below. The graph is plotted for mountain image (Fig. 2).

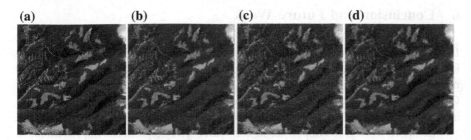

Fig. 3 **a** Original Ikonos image **b** BCS compressed image **c** noisy image **d** denoised using CTWF

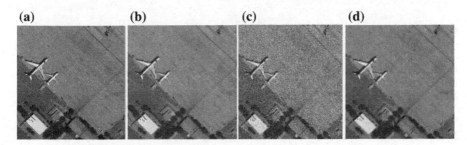

Fig. 4 **a** Original airport image **b** BCS compressed image **c** noisy image **d** denoised using CTWF

It can be seen from the graph that as the noise level increases, both CTWF and Wiener filtering have decrease in PSNR values. But however CTWF has improved PSNR values than normal Wiener filtering as the noise level increases. Hence CTWF is considered to be the best denoising technique in achieving good PSNR values.

Visual quality comparison is shown below. The quality of images after applying both techniques is found to be good with better PSNR values. Figure 3a shows the original Ikono's mountain image taken from the database available in [9]. Figure 3b–d, show the image after applying BCS, noise, CTWF, respectively. Similarly, Fig. 4a shows the original airport image taken from the database available in [9]. Figure 4b–d show the image after applying BCS, noise, CTWF, respectively.

Visual quality comparison shows that even with lesser number of measurements, BCS is very efficient in reconstructing the image. Also CTWF also helps in significant noise removal and gives better PSNR values when compared to that of normal Wiener filtering. Hence BCS with CTWF is considered to be the most efficient method in achieving good quality images with better PSNR values.

6 Conclusion and Future Work

Image reconstruction using BCS and denoising using CTWF for satellite images is investigated in this paper. BCS can effectively reconstruct images with fewer measurement values. Noise in the BCS data can be effectively removed by using CTWF technique. The final image obtained by applying BCS–CTWF is found to have better PSNR values with good image quality.

The future work is to adopt a technique which gives significant improvement in the reconstructed image quality.

References

1. E. J. Candes., M. B. Wakin.: An introduction to compressive sampling. IEEE Signal Processing Magazine, vol. 25, no. 2, pp. 21–30 (2008).
2. Z. Gao., C. Xiong., L. Ding., C. Zhou.: Image representation using block compressive sensing for compression applications. Journal of Visual Communication and Image Representation, vol. 24, no. 7, pp. 885–894, (2013).
3. D. L. Donoho.: Compressed sensing. IEEE Transactions on Information Theory, vol. 52, no. 4, pp. 1289–1306, (2006).
4. L. Gan.: Block compressed sensing of natural images. Proc. IEEE 15th International Conference on Digital Signal Processing, pp. 403–406, (2007).
5. Y. Yang., O. C. Au., L. Fang., X. Wen., W. Tang.: Reweighted Compressive Sampling for image compression. Proc. IEEE Picture Coding Symposium, pp. 1–4, (2009).
6. Starck., Jean-Luc., Emmanuel J. Candès., David L. Donoho.: The curvelet transform for image denoising. IEEE Transactions on image processing 11.6, pp. 670–684, (2002).
7. Ansari., Rizwan Ahmed., Kirshna Mohan Budhhiraju.: A Comparative Evaluation of Denoising of Remotely Sensed Images Using Wavelet, Curvelet and Contourlet Transforms. Journal of the Indian Society of Remote Sensing, pp. 1–11, (2016).
8. J. A. Tropp., A. C. Gilbert.: Signal recovery from random measurements via orthogonal matching pursuit. IEEE Transactions on Information Theory, vol. 53, no. 12, pp. 4655–4666, (2007).
9. http://www.satimagingcorp.com/services/image-processing/

Neural Deployment Algorithm for WSN: A Concept

Umesh M. Kulkarni, Harish H. Kenchannavar
and Umakant P. Kulkarni

Abstract Today's world wireless sensor networks (WSNs) are gaining more and more importance in any of the domains of industry. Plenty of research and development has happened, and it is going on in the field of WSN. The importance of conservation of energy in WSN in most of its applications will provide the best scope for WSN development. Sufficient literatures are already carried for deployment of nodes leading to conservation of energy. One of the important application areas of WSN is area monitoring where the nodes have to move forward to sense the area and come back to the original positions. This paper provides solution for this problem by the concept of change in the positions of nodes using a simple neural network concept. This concept considers the energy, distance, and sensing of nodes as the basic parameters in moving a single node or all nodes. The main concept in this paper is to propose a deployment strategy that changes the positions of the nodes according to the activation function decision.

Keywords Neural network (NN) · Wireless sensor network (WSN) · Moving rate

1 Introduction

WSNs are gaining more and more importance in today's market. Application areas of WSN range from home automation to any kinds of area monitoring systems.

U.M. Kulkarni (✉) · H.H. Kenchannavar
Department of CSE, KLS's Gogte Institute of Technology, Belagavi, India
e-mail: umesh_k@git.edu

H.H. Kenchannavar
e-mail: harishhk@git.edu

U.P. Kulkarni
Department of CSE, SDM College of Engineering and Technology,
Dharwad, India
e-mail: upkulkarni@yahoo.com

© Springer Nature Singapore Pte Ltd. 2018
S.S. Dash et al. (eds.), *International Conference on Intelligent Computing
and Applications*, Advances in Intelligent Systems and Computing 632,
https://doi.org/10.1007/978-981-10-5520-1_3

19

1.1 Deployment of Nodes

In simple terms, it is positioning of the nodes in the area of sensing. Generally, the node deployment is classified as static and dynamic deployment [1, 2]. Static node deployment is classified as deterministic and random deployment techniques [1, 2]. In deterministic node deployment, the node position is known before deployment, and every time, they are deployed at a fixed position. In random deployment, the nodes are placed at different places during every iteration. Dynamic deployment is characterized by the concept of random deployment of nodes and moving nodes. This work considers the combination of both in the sense that the nodes are pre-deployed at fixed positions, then give a fixed movement near the sensing area, and then occupy the original positions.

1.2 Neural Networks

Machine learning is an old but presently an upcoming area of research. One of the interesting and promising topics of research in machine learning is artificial neural network (ANN) commonly known as NN [1]. An NN is a simple mimic of the biological neural network that shows an interconnection of a large number of neural networks. An NN is characterized by three building blocks: neural structure, learning, and activation functions used. Neural structure means the arrangement of neurons to form the network. Learning means the knowledge gained or taken from the training process. This NN learning can be supervised, un-supervised or reinforced. Activation functions are used to arrive at particular decision for taking further actions in the neural network. In this paper, attempt is made for mapping the typical structure of NN with a WSN. In addition, the advantage of the decision-making and NN techniques is used for moving the nodes in the WSN for better sensing. The process of NN to WSN mapping is explained in the further sections.

1.3 Related Work

In [1], the meanings of ANN and its basics have been discussed. An extensive study of the types of NNs and their applicability to different types of deployments have been provided. The problems addressed are Localization [3], QoS routing, and security in WSN. This paper also provides for the mapping of application domains of WSN and their design using an NN. This paper also gives some guidelines for implementing a WSN using different types of learning methods like supervised, un-supervised, and reinforcement learning.

In [2], various deployment strategies and their meaning have been discussed. The work discusses basic terminologies of WSN like coverage, network lifetime, and connectivity. The paper gives a clear idea of deployment and different types in that. The paper provides a theoretical model for energy consumption.

In [4], it is clear that the neural network can be modeled as WSN. This paper describes that NN for WSN cannot be a solution for energy but can be used for the purpose of energy utilization. This paper even provides a mapping of NN structures that can be used for different types of WSN energy conservation techniques. This paper also demonstrates the use of Kohonen's self-organizing map for ANN to WSN implementation using MATLAB.

In [5, 6], it is very clear that how programming can be considered as the method of implementation in WSN. Different NN-based algorithms like backpropagation, feedforward concept, and their energy consumption have discussed showing the importance of energy.

In [7], a hint given for using the WSN for the next level of implementation by developing efficient algorithm for distributed cooperative learning is based on zero-gradient-sum (ZGS) optimization in a network setting.

The research work in [8–10] concentrate on specific applications like controlling the actuators, better localization schemes, medical applications, and in even distributed computing examples.

Overall, good amount of work has been proposed and implemented for converting an NN to a WSN. Plenty of these papers propose a tabular mapping of NN structures, NN algorithms, etc., and their significance in the design of a WSN. Most of these works concentrate on localization coverage and routing [11]. Most of these works try to propose a structure- or neural-based arrangement for addressing these issues [12]. The works also show an indication of the importance of energy in any of NN to WSN mapping. Some of the works like [13–15] even show the theoretical energy efficient implementation. From all these works, it is been found that there is a lack of concrete energy utilization scheme and acceptable node deployment methods. Even an NN-based acceptable mathematical model for energy utilization and dynamic node deployment models are not found. This paper attempts to propose an idea about mapping a typical NN-based mathematical calculation as applicable to WSN and shows that how an effective sensing achieved in the existing fixed deployment structure.

2 System Model

The basic reason for similarity between NN and WSN is both perform distributed computing. The meaning of distributed computing is every node and every neuron can perform the required computation.

2.1 Topology

The topology followed in this concept is like a single-layer neuron consisting of an input layer and an output layer. In this concept, all input layer neurons are mapped as input layer sensors with sensing capability. These sensors will sense the data from a sensing area as shown in Fig. 1 with a rectangle box. Figure 1 shows the model of NN to WSN design.

In Fig. 1,

x1, **x2**, **x3** (lower case) are external inputs (from sensing area).
X1, **X2**, **X3** (Uppercase) are input sensors getting the inputs as x1, x2, x3, respectively.
d1, **d2**, **d3** are the distances/weights connecting the X1, X2, X3 to a base station/output neuron.

Following are some of the topology-related assumptions for this concept:

1. The number of nodes is three, and deployment is according to Fig. 1.
2. All 3 nodes are mobile wireless sensor nodes.

2.2 Algorithm

This work proposes an NN-based algorithm for WSN for efficient energy utilization. Following are some of the terminologies used for the purpose of this algorithm:

E1, **E2**, **E3** Residual energy at any point of time
α   Moving rate of sensors
Y   Output obtained by applying some activation function on the net input yin

Fig. 1 Typical WSN designed like a single-layer NN

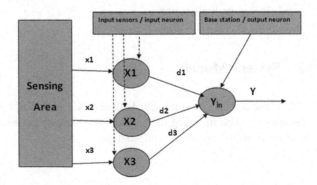

The net input y_{in} is calculated as

$$y_{in} = x1d1 + x2d2 + x3d3 \tag{1}$$

Algorithm:

Step 0 initialize the parameters α, $d1$, $d2$, $d3$.

Step 1 Deploy all the nodes at fixed places in the sensing area. The initial deployment is like a single-layer NN.

Step 2 All the sensors in the input layer of the WSN will sense the sensing area.

Step 3 All the sensors will send the sensed data to the base station along with their residual energy and distance.

Step 4 Depending on the energy levels received, the base station (output neuron) will now calculate the net input as

$$y_{in} = x1d1E1 + x2d2E2 + x3d3E3 \tag{2}$$

Step 5 Apply the activation function, and resultant value will be taken for further decision making.

$$F(y_{in}) = \begin{cases} -1 & \text{if } y_{in} > \theta \text{ (theta)}, \\ 0 & \text{if } y_{in} = \theta \text{ and} \\ 1 & \text{if } y_{in} < \theta \text{ (theta)} \end{cases} \tag{3}$$

The value θ (theta) is calculated as a threshold by previous knowledge of the basic product of the distance and residual energy, for example, Table 1.

Step 6 Perform the weight adjustment by the following logic

$$d_{(new)}i = d_{(old)}i + \alpha F(y_{in}) \tag{4}$$

Step 7 The process continues until there is no change in the new weights and old weights. Even some iteration can be fixed up for this.

This process continues until all the nodes lose their energy and converge themselves near the base station from where they have started with.

Table 1 Threshold values considered

S. No.	Distance from sensing node (in m)	Sensing (1 bit)	Transmitting (1 bit)	Receiving (1 bit)	Movement (0.1 m)	Theta θ	Energy consumed
1	1	0.05 J	0.1 J	0.05 J	0.1 J	3	0.3 J
2	5	0.25 J	0.5 J	0.25 J	0.5 J	1.5	1.5 J
3	10	0.5 J	1.0 J	0.5 J	1.0 J	0.30	3.0 J

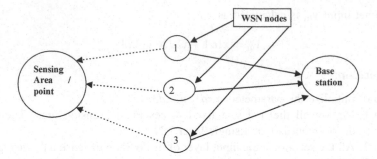

Fig. 2 Initial positions of the wireless sensor nodes before sensing

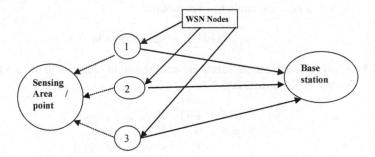

Fig. 3 Changed positions of the wireless sensor nodes during the sensing

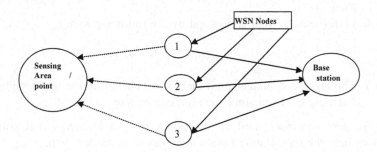

Fig. 4 Original positions of the wireless sensor nodes after sensing

Figures 2, 3, and 4 demonstrate the different stages of this algorithm for visual understanding. Basic achievement in this concept is that the optimal sensing is achieved with respect to the sensing area as an example. The whole concept is dependent on the calculation and consideration of the θ. The θ was calculated as fixed mathematical calculation of average energy consumed by a node for sensing, transmitting, receiving, and distance.

3 Demonstration

In this work, theoretical values assumed with practical viewpoint. The practical implementation the values may change. However, the pattern of change of values remains almost same with little difference in theoretical and practical values. For applying this concept for any example, one need to consider some fixed values for threshold (θ). Table 1 shows the use of theta and its significance.

The values considered for the following example may be theoretical but are not randomly assumed, modified, and arrived. The values obtained are with all mathematical logics presented in the algorithm only. Except initialization, from Table 1, all values in Tables 2 and 3 are written with theoretical calculations.

Example 1 This example works according to the theoretical concept discussed in the previous section.
   Initialization

1. Three mobile wireless sensor nodes are deployed: s1, s2, s3.
2. Initial D1 = D2 = D3 = 1 m, E1 = E2 = E3 = 3 J, α = 0.1.

   Tables 2 and 3 indicate the values considered for this example.
   The dark (bold) values in Table 3 represent that there is no change in the distances between the nodes-to-sensing point/area and nodes-to-base station; hence, no efficient utilization of energy happens. However, in case of Table 2, efficient utilization of energy can be seen. The nodes moving near the sensing area provide better sensing data than the normal sensing methods.

Table 2 Sample values for neural logic deployment

For neural logic deployment													
Iteration No.	S1	S2	S3	D1	D2	D3	E1	E2	E3	yin	Θ	Compare yin and θ	$F(y_{in})$	α
1	1	1	1	1	1	1	3	3	3	9	3	yin > θ	−1	0.1
2	1	1	1	0.9	0.9	0.9	2.7	2.7	2.7	7.29	3	yin > θ	−1	0.1
3	1	1	0	0.8	0.8	0.8	2.4	2.4	2.4	5.76	3	yin > θ	−1	0.1
4	1	0	0	0.7	0.7	0.7	2.1	2.1	2.1	4.41	3	yin > θ	−1	0.1
5	1	0	1	0.6	0.6	0.6	1.8	1.8	1.8	3.24	3	yin > θ	−1	0.1
6	1	0	1	0.5	0.5	0.5	1.5	1.5	1.5	2.7	3	yin < θ	1	0.1
7	1	1	0	0.6	0.6	0.6	1.2	1.2	1.2	2.16	3	yin < θ	1	0.1
8	1	1	0	0.7	0.7	0.7	0.9	0.9	0.9	1.89	3	yin < θ	1	0.1
9	1	1	1	0.8	0.8	0.8	0.6	0.6	0.6	1.44	3	yin < θ	1	0.1
10	1	0	1	0.9	0.9	0.9	0.3	0.3	0.3	0.81	3	yin < θ	1	0.1
11	1	0	0	1	1	1	0.0	0.0	0.0	0.0	3	yin < θ	1	0.1

Table 3 Sample values for normal deployment

For normal deployment									
Iteration No.	S1	S2	S3	D1	D2	D3	E1	E2	E3
1	1	1	1	1	1	1	3	3	3
2	1	1	1	1	1	1	2.7	2.7	2.7
3	1	1	0	1	1	1	2.4	2.4	2.4
4	1	0	0	1	1	1	2.1	2.1	2.1
5	1	0	1	1	1	1	1.8	1.8	1.8
6	1	0	1	1	1	1	1.5	1.5	1.5
7	1	1	0	1	1	1	1.2	1.2	1.2
8	1	1	0	1	1	1	0.9	0.9	0.9
9	1	1	1	1	1	1	0.6	0.6	0.6
10	1	0	1	1	1	1	0.3	0.3	0.3
11	1	0	0	1	1	1	0.0	0.0	0.0

4 Observations and Discussion

The outcomes or observations from this concept are as follows:

1. Comparison performed with respect to traditional approach in that it has an impact on correct deployment rather than just deployment at any place. In traditional deployment of nodes, the nodes will remain in the fixed place or they may move in different directions. When nodes move in any direction would make nodes to drains out their energy at any point without returning to the original position. However, the proposed approach makes the nodes to deploy at a particular position, move them with some NN logic in the direction of the sensing area, sense the area, send it to the base station and return to the original position.
2. The process defined works with some assumed initial values and may vary with respect to practical values. It is definitely true that there is a difference in the practical implementation and theory, but pattern of variation should never change.
3. The given deployment can work fine with energy utilization beginning with some initial energy making the wireless sensor node to go near the sensing area and come back to the original position before complete energy drains out.
4. This node deployment strategy presently makes some changes uniformly to all the nodes; however, this is extended for different node distances and different node residual energy.

5 Conclusion and Future Scope

Neural networks and WSN possess some similarities between each other as they possess distributed processing. Combination of NN and WSN can make sensing more appealing. NN based decision making can make the sensors in WSN to logically move towards the sensing area and come back to the original position. The base station and moving rate α (learning rate of NN) play an important role in giving the movement to the sensing nodes as compared to ad hoc deployment and ad hoc movement. Better sensing can be obtained as compared to the fixed node deployment and sensing. This work shows that how an NN concept can be mathematically mapped to WSN for better utilization of available energy and providing better sensing effect. This idea is the best suitable for area monitoring applications where nodes have to move near the sensing area come back to the original place. This work can be extended to work for heterogeneous WSN, for different initial energy levels and even to change different distances for different nodes.

References

1. Nauman Ahad A,N, Junaid Qadir A,B, Nasir Ahsan "Neural networks in wireless networks: Techniques, applications and guidelines", Elsevier Journal of Network and Computer Applications, http://dx.doi.org/10.1016/j.jnca.2016.04.006 1084-8045/& 2016 Elsevier Ltd.
2. Haitao Zhang and Cuiping Liu, "A Review on Node Deployment of Wireless Sensor Network", IJCSI International Journal of Computer Science Issues, Vol. 9, Issue 6, No 3, November 2012
3. Hanen Ahmadi (1,2), Ridha Bouallegue, "Comparative study of learning-based localization algorithms for Wireless Sensor Networks: Support Vector Regression, Neural Network and Naïve Bayes", 978-1-4799-5344-8/15/$31.00 ©2015 IEEE 1554
4. Neda Enami1, Reza Askari Moghadam 1, Kourosh Dadashtabar & Mojtaba Hoseini 3, "NEURAL NETWORK BASED ENERGY EFFICIENCY IN WIRELESS SENSOR NETWORKS: A SURVEY", International Journal of Computer Science & Engineering Survey (IJCSES) Vol. 1, No. 1, August 2010
5. DAnieal, EDUARDO et. al. "Energy Efficient reprogramming in WSN using constructive neural networks", International journal of Innovative computing, Information and control vol 8, Nov 11, 2012
6. Jabal Raval, Bhushan Jagyasi TCS Innovation Labs Mumbai Tata Consultancy Services, India "Distributed Detection in Neural Network based Multihop Wireless Sensor Network", 978-1-4799-2179-9/14/$31.00 ©2014 IEEE
7. Wu Ai A, C, Weisheng Chen b, *, Jin Xie A, "A zero-gradient-sum algorithm for distributed cooperative learning using a feedforward neural network with random weights", Elsevier Information Sciences, Available online 7 September 2016 0020-0255/© 2016 Elsevier Inc.
8. Gursel Serpen*, Jiakai Li, and Linqian Liu, Conference Organized by Missouri University of Science and Technology 2013-Baltimore, MD, AI-WSN: Adaptive and Intelligent Wireless Sensor Network Procedia Computer Science 20 (2013) 406–413, 1877-0509 © 2013 The Authors. Published by Elsevier B.V. doi:10.1016/j.procs.2013.09.294
9. Duhart Clement*†, Bertelle Cyrille, "Methodology for Artificial Neural Controllers on Wireless Sensor Network", 2014 IEEE Conference on Wireless Sensors (ICWiSE), October, 26–28 2014, Subang, Malaysia

10. Xiaobo Xie, Junqi Guo*, Hongyang Zhang, Tao Jiang, Rongfang Bie, Yunchuan Sun "Neural-Network based Structural Health Monitoring with Wireless Sensor Networks", 978-1-4673-4714-3/13/$31.00 ©2013 IEEE 163
11. Chenxian Xiao and Ning Yu Chenxian Xiao and Ning Yu, "A Predictive Localization Algorithm Based on RBF Neural Network for Wireless Sensor Networks, 978-1-4673-7687-7/15/$31.00 ©2015 IEEE
12. Text book: S.N. Sivanandam, S.N. Deepa, Principles of Soft Computing, 2nd Edition Wiley Publisher
13. Harish H Kenchannavar, U.P. Kulkarni, "Optimized Overlapping Based Energy-Aware Node Collaboration in Visual Sensor Networks", International Conference on Wireless Technologies for Humanitarian Relief (ACWR2011) sponsored by ACM SIGCOMM, SIGMM, SIGMOBILE and IEEE held at Amrita University, Amritapuri, Kerala, on 19-21st Dec 2011. doi>10.1145/2185216.2185248, ISBN: 978-1-4503-1011-6. {50% Acceptance Ratio}
14. Harish H. Kenchannavar, Umakant P. Kulkarni, "Energy Efficient Data Processing in Visual Sensor network", International Journal of Computer Science and IT Vol. 2, Number 5, October 2010, ISSN: 0975-4660. airccse.org/journal/jcsit/1010ijcsit11.pdf
15. Gowri. K, Dr. M.K. Chandrasekaran, Kousalya. K, "A Survey on Energy Conservation for Mobile-Sink in WSN", (IJCSIT) International Journal of Computer Science and Information Technologies, Vol. 5 (6), 2014, 7122–7125

Intelligent System Approach to Real-Time Video Denoising Using Lifting-Based Wavelet Technique

P. Veeranath, D. Nagendra Rao and S. Vathsal

Abstract Videos are frequently corrupted by various types of noise either during their process of capturing or during their transmission from one point to another point. We propose a new method for an efficient and improved quality real-time video denoising using a temporal video slicing (TVS) technique, lifting-based dual-tree complex wavelet transform (L-DTCWT), and adaptive thresholding approach. The TVS technique extracts discrete frames from a noisy video which will undergo a process by L-DTCWT framework, and the resulting complex wavelet coefficients are threshold by the denoising unit (which includes appropriate threshold value estimation based on noise variance and soft thresholding framework). The denoised frames are passed to the output instantaneously. This process is continued until all the frames are denoised. All the denoised frames are suitably buffered to display the denoised video. Here the major core processing blocks are TVS, L-DTCWT framework, and denoising unit. The denoised video quality will be judged through quality metrics such as PSNR, GCF, NIQE, and SSIM.

Keywords Temporal video slicing · Wavelets · Dual-tree complex wavelet transform · Lifting scheme · Thresholding · Denoising

P. Veeranath (✉) · D. Nagendra Rao
JBREC, Hyderabad, India
e-mail: veeru96@gmail.com

D. Nagendra Rao
e-mail: hrowrowv@gmail.com

S. Vathsal
IARE, Dundigal, Hyderabad, India
e-mail: svathsal@gmail.com

© Springer Nature Singapore Pte Ltd. 2018   29
S.S. Dash et al. (eds.), *International Conference on Intelligent Computing
and Applications*, Advances in Intelligent Systems and Computing 632,
https://doi.org/10.1007/978-981-10-5520-1_4

1 Introduction

Video quality has significant importance in live telecasting, navigation, video broadcasting, MRI, satellite, tomography, medical applications, etc. Videos are very often corrupted by noise either during their capture by the video sensors or during their transmission from one point to another point due to many of the reasons like internal imperfections of the capturing sensors, threats with the data acquisition process, poor illumination, atmospheric noises, and interfering natural phenomena [1]. Noise is basically classified into two types in video/image processing as independent noise and dependent noise. The independent noise shows serious effects on video because it consists of the additive white Gaussian noise (AWGN) and impulse noise. where as dependent noise consists of speckle or poison noises, thermal noises occur in the communication medium. The additive white Gaussian noise shows a serious impact on visual perception of image by its intensive degradation whereas impulse noise is caused by sharp sudden disturbances in the video/image, and hence, in our research we are focusing on removal of AWGN on real-time video processing. Denoising is a process of removing the various noises imposed on video [1, 2]. The aim of video denoising is to improve the quality and features of the video/image such as edges by suppressing the random noise.

1.1 Problem Statement

In the literature [3–6], the researchers have done enough work in image denoising, which does not remove real-time video disturbances like flickering, and it is not desirable for real-time video processing. This paper is focusing on denoising the video frames using the TVS techniques for frame synchronization, L-DTCWT for computationally efficient, and adaptive thresholding approach to preserve the fine details of video-like edges.

This paper is organized as follows: Sect. 2 gives details of frame synchronization to remove synchronization errors, Sect. 3 deals with L-DTCWT, Sect. 4 dedicates to denoising, Sect. 5 explains the implementation for proposed work, Sect. 6 discusses results analysis, Sect. 7 concludes the proposed work, and Sect. 8 devotes to future extension of our work.

2 Frame Synchronization

2.1 TVS Method

The proposed TVS technique samples the noisy video to obtain the discrete video slices or frames. All the sampled video slices or frames are numbered in order and stored in memory to synchronization and preserve the information. The main advantages of TVS approach are a real-time processing, single-level processing, and faster than existing methods [1]. The basic concept of TVS is given by Eq. (1) [7].

$$M_f = M\left(\frac{m}{f_r}\right) \quad \text{for} \quad m = 0 \text{ to } N - 1, \tag{1}$$

where 'N' is the length of the video, 'M_f' is the discrete video frames, and 'f_r' is the frame rate.

2.2 Frame Pipelining

For efficient use of the processor and to speed up the process, frame pipelining mechanism is implemented. The frame Pipelining mechanism describes that the video frames are processed in sequence without any latency; buffering capacity is given by Eq. (2) [1]:

$$B_{index} = M_{f index}, \tag{2}$$

where B_{index} is the buffer index and $M_{f index}$ is the video frame index.

3 L-DTCWT Framework

Wavelets denote better performance in video denoising due to their positive sparseness properties and multi-resolution features. DTCWT gives better noise removal and preserves quality of image compared to DWT. In the literature, many researchers have done enough work on the wavelets [8–10].

DTCWT is a powerful tool for real-time video denoising, but incorporates a numerous complex computations, which will consume a large amount of power, large amount of time for computation. To overcome this, lifting scheme-based DTCWT is proposed [2, 8].

3.1 Lifting Scheme

Lifting scheme gives the lifted wavelets, which do not necessarily translate and dilate of one fixed function [2, 11–15]. This construction is entirely spatial and

Fig. 1 Block diagram of
lifting mechanism scheme

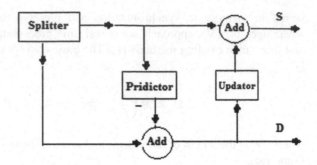

therefore be ideally suited for building second-generation wavelets that are more general in the sense that all the classical wavelets can be generated by the lifting scheme [2]. For faster implementation of wavelet transform, the lifting scheme makes an optimal use of the similarities between the high-pass and the low-pass filters [2, 16]. The inverse wavelet transform with lifting scheme can be immediately found by undoing the operations of the forward transform [2]. The lifting scheme operation is shown in Fig. 1.

Where Predictor predict the future value, and Updator update the current value [2, 18]. The lifting scheme then generally builds a new wavelet, with improved properties by adding in a several new basis functions [2].

4 Denoising Framework

The proposed denoising framework implements an adaptive thresholding approach, and it is dynamic thresholding which has two hierarchical phases such as threshold designing phase and soft thresholding phase. Threshold designing phase computes an appropriate threshold value for denoising the video frames, whereas the soft thresholding phase performs the denoising process of the noisy video frames.

4.1 Threshold Designing Phase

The threshold designing phase interpolates the complex wavelet coefficients of the noisy video frame and computes the noise signal variance of the coefficients using a robust median estimator. Depending on these noise signal variances, the depth of noise corruption will be computed instantaneously. As shown in Eq. (3) [19].

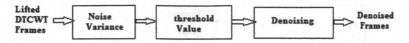

Fig. 2 Block diagram of denoising process

$$\sigma^2_{\text{noise}} = \frac{\text{Median}(|\psi_i|)}{0.6745} \qquad (3)$$

σ^2_{noise}—noise variance, $|\psi_i|$—wavelet coefficient.

Based on the depth of noise corruption and noise signal strength, an appropriate threshold value will be computed. The threshold value is calculated with the help of below given Eqs. (4) and (5).

$$\gamma = p \times \sigma^2_{\text{noise}}/\sigma_{\text{Signal}} \qquad (4)$$

γ—threshold Value, p—constant equal to $\sqrt{2}$, σ_{noise}—noise variance, and σ_{Signal}—signal variance.

$$\sigma^2_{\text{Signal}} = \text{Var}(\psi) - \sigma^2_{\text{Noise}} \qquad (5)$$

the threshold γ is chosen according to the signal energy and the noise variance (σ^2) [19]. This appropriate threshold value will be used in the soft thresholding phase to denoise the noisy video frames.

4.2 Denoising Video Frames/Soft Thresholding

In this process, we are using soft thresholding to denoise the noisy video frames; the denoised video frames are passed in a synchronized sequence to the output as shown in Fig. 2. The soft thresholding function performs the operation based in Eq. (6) as shown below.

$$
\begin{aligned}
T_h &= c - \gamma \quad \text{if } c > \gamma \\
&= 0 \qquad \text{if } c = \gamma \\
&= c + \gamma \quad \text{if } c < \gamma
\end{aligned}
\qquad (6)
$$

T_h—threshold/denoised output, c—input signal/coefficient, and γ—threshold value.

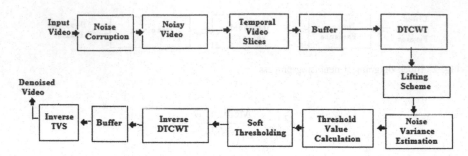

Fig. 3 Proposed system block diagram

5 Implementation

The proposed system first considers a test video file and corrupts it by Gaussian/impulse noise. The noise-corrupted video file is forwarded to the video-to-frame conversion unit, which employs the TVS technique to convert the noisy video file into discrete noisy frames, which are numbered in sequence to preserve the synchronization. The corrupted video frames are subjected to the L-DTCWT process, which will be initiated with the selection of an appropriate mother wavelet type. The selected mother wavelet is quite lazy in behavior due to its complex, irregular structure, and hence, it is lifted with the help of lifting scheme to get a linear and regular wavelet structure which is computationally efficient and faster than mother wavelet. Based on the lifted wavelet structure, analysis and synthesis filter pairs are designed for real tree, and by introducing a half sample delay into these real tree filter coefficients, the filter coefficients for analysis and synthesis filter pairs of the imaginary tree are obtained. Figure 3 shows the complete operation of proposed work.

With the help of these filters, the proposed L-DTCWT framework decomposes the video frame in real and imaginary trees parallel and provides the complex wavelet coefficients. These complex wavelet coefficients are processed for an appropriate threshold design by the threshold design block in the denoising unit. By using an appropriate threshold value supplied by the threshold design block, the soft thresholding block performs an intended denoising process of the noisy complex wavelet coefficients of the noisy video frame. The denoised complex wavelet coefficients of the denoised frame are subjected to an inverse L-DTCWT process to get the spatial domain representation of the denoised frame. This process is iterated in pipelining approach for all noisy frames by preserving the synchronization among the video frames. All the denoised video frames are buffered or combined by an inverse TVS block to get a final denoised video file.

Fig. 4 Flowchart of
proposed algorithm

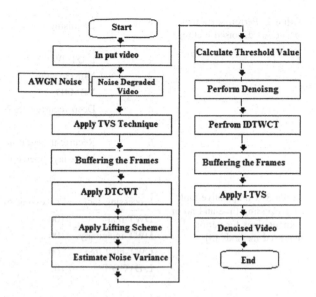

The algorithm of proposed approach is shown in Fig. 4.
Algorithm:

6 Results and Discussion

The proposed system is designed, coded, implemented, and tested in MATLAB
software. The Simulation results of the proposed system are presented as follows.
The given input Video Adds AWGN/Impulse noise, this noisy video converts into
frames and noise, processed through the denoised block. The denoised video
quality assessment matrices are shown in Table 1.

Table 2 shows comparison of PSNR value for various denoising techniques, and
from this table, we can say our proposed approach gives maximum PSNR value
compared to existing methods. From Table 3 the processing time and number of
computation reduce in L-DTCWT when compared to first-generation wavelets.

Table 1 Denoised video
quality assessment metrics

PSNR (dB)	SSIM	GCF	NIQE
35.1328	0.4911	1.6664	0.5795
35.1032	0.4932	1.6617	0.5943
35.0608	0.4917	1.6696	0.5764
35.0074	0.4905	1.6690	0.6398
34.9731	0.4970	1.6646	0.6855
34.9951	0.4950	1.6675	0.7802
35.1415	0.4796	1.7014	0.9132
35.3211	0.4817	1.7078	1.0306
35.4591	0.4805	1.7520	1.1372
35.5310	0.4630	1.7433	1.2398

Table 2 Performance table of various denoised methods

S. No.	Denoising method	PSNR
		$\sigma^2 = 25$
1	L-DTCWT	35.640
2	NLM3d non-local mean	33.268
3	VBM4D	32.48
4	Deep recurrent N-N (DRNN)	31.78
5	R-NL	31.61
6	Recurrent neural network (RNN)	30.83
7	Multi-layer perception (MLP)	29.87

Table 3 Comparative table of processing time and no of computations for various denoised methods [8]

Processing tool	Processing time (%)	No of computations
DWT-real	100	$4(N1 + N2) + 2$
DWT-imaginary	100	$4(N1 + N2) + 2$
CWT-dual tree	200	$8(N1 + N2) + 4$
L-DTCWT	100	$4(N1 + N2) + 4$

Hence, proposed approach is the best method which gives better results compared to existing methods.

7 Conclusion

In our paper we designed, coded, implemented, and tested a novel intelligent approach for an efficient and improved quality real-time video denoising of AWGN and impulse noise using TVS, L-DTCWT technique, and an adaptive thresholding approach. The extracted noisy video frames are processed by L-DTCWT framework. Resulting complex wavelet coefficients are thresholded by the denoising unit. The proposed system is simulated in the MATLAB tool. The obtained MATLAB results are good, and the proposed approach is the best option for denoising the real-time videos. The denoised video quality has been judged through quality metrics such as PSNR, GCF, NIQE, and SSIM.

8 Future Scope

The future work for this paper is improving the quality of denoising block and enhancing fine details of video like sharpening the images and extraction of edges and deblurring for better psycho-visual appearance.

References

1. P. Veeranath, Dr. D.N. Rao, Dr. S. Vathsal, "A New Approach To Real Time Video Desnoising Using Temporal Video Slicing Frame Synchronization Technique. *IOSR-JVSP Journal, Volume 6, Issue 5, Ver. II (Sep.–Oct. 2016).*
2. WWW.slideshare.net.
3. N. Kalyani 1, A. Velayudham 2, Analysis of Image Denoising Methods Using Various Wavelet Transform, *(IJARECE) Volume 4, Issue 1, January 2015.*
4. D. Suresh 1, P. Suseendhar 2, T.S. Krishna Priya 3, An Efficient Fixed Noise Removal in Images Using Dual Tree Complex Wavelet Transform and DTCWT-OSP, International Journal of Innovative Research in Computer and Communication Engineering *(An ISO 3297: 2007 Certified Organization)* Vol. 2, Issue 3, March 2014.
5. SK. Umar Faruq 1, Dr. K.V. Ramanaiah 2, Dr. K. Soundara Rajan 3, An Application of Second Generation Wavelets for Image Denoising using Dual Tree Complex Wavelet Transform, ACEEE Int. J. on Signal and Image Processing, Vol. 4, No. 3, Sept 2013.
6. Anil Dudy, 2 Kanwaljit Singh, A New Approach For Denoising Ultrasonographic Images Using DTCWT, *International Journal of Latest Research in Science and Technology* ISSN (Online): 2278–5299 *Vol. 1, Issue 2:Page No. 106–108, July–August (2012).*
7. Chong-Wah Ngo, Tinjg-Chuen pong, and Rolad T. Chin, "Video partitioning by Temporal Slice Coherency", IEEE- Transaction Circuits and Systems for Video Technology, Vol. 11, No. 8, August-2001.
8 SK. Umar Faruq, Dr. K.V. Ramanaiah, Dr. K. Soundara Rajan "An Application of Second Generation Wavelets for Image Denoising using Dual Tree Complex Wavelet Transform", ACEEE Int. J. on Signal and Image Processing, Vol. 4, No. 3, Sept 2013.
9 R. Calderbank, I. Daubechies, W. Sweldens, and B.-L. Yeo. Wavelet transforms that map integers to integers. *Appl. Comput. Harmon. Anal.*, 5(3):332.369, 1998.
10 Adeel Abbas and Trac D. Tran, "Multiplier less Design of Biorthogonal Dual-Tree Complex c Wavelet transform using Lifting Scheme. 1-4244-0481-9/06,2006-IEEE.
11 Li, Jain Ping, Jing Zhao, Victor Wickerhauser, Yuan Yan Tang, John Dauman, and Lizhong Peng, "Biorthogonal Wavelets Research Based on Lifting Scheme", wavelet analysis and its Applications, 2003.
12 W. Sweldens. The lifting scheme: A custom-design construction of biorthogonal wavelets. *Appl. Comput. Harmon. Anal.*, 3(2):186.200, 1996.
13 W. Sweldens. The lifting scheme: A construction of second generation wavelets. *SIAM J. Math. Anal.*, 29(2):511.546, 1997.
14 J. M. Combes, A. Grossmann, and Ph. Tchamitchian, editors. *Wavelets: Time-Frequency Methods and Phase Space.* Inverse problems and theoretical imaging. Springer-, New York, 1989 [13].
15 A. Grossmann and J. Morlet. Decomposing of Hardy functions into square integral wavelets of constant shape. *SIAM J. Math. Anal.*, 15(4):723.736, 1984.
16 Master.donntu.edu.ua.
17 A. Harten. Multiresolution representation of data: A general framework. *SIAM J. Numer. Anal.*, 33(3):1205.1256, 1996.
18 W. Sweldens and P. Schröder. Building your own wavelets at home. In *Wavelets in Computer Graphics*, pages 15.87. ACM SIGGRAPH Course notes, 1996.
19 M. Ramanjaneya Rao, P. Ramakrishna, "Image Denoising based on MAP Estimation using Dual Tree Complex Wavelet Transform", International Journal of Computer Applications Vol. 110-N0. 2, January-2015, ISSN0975-8887.

Data Loss Prevention Scheme Using ADCN with Effective Tour Strategy in Wireless Sensor Network

Sachin V. Todkari, Vivek S. Deshpande and A.B. Bagwan

Abstract Wireless sensor network is spatially distributed sensors with limited resources such as energy and buffer. Data loss prevention is challenging task in wireless sensor network (WSN), and hence, reliable data dissemination is desirable. In WSN, event detection ratio should be very high. In such condition, huge data are generated by event node. But there is always limitation of buffer size. As a result, it gets overflow. The data which are important may get lost. The proposed data loss prevention scheme will help to prevent the data loss. The effective tour strategy of mobile data collector node will improve QoS parameters of WSN.

Keywords Wireless sensor network · Data collector node · Adaptive data collector node · Collect data packet · Quality of service

1 Introduction

Wireless sensor network is an emerging technology that supports number of unprecedented application, and we are facing several challenges to improve reliability due to limited resources mainly buffer capacity and energy.

Reliable data collection using mobile data collector with path constraint mobile sink is one of the approaches for data collection [1] called maximum amount shortest path (MSAP), the overall amount of data which is collected by mobile sink in its single journey from all subnodes in the network. Two-phase communication protocol is used for data collection discover phase and data collection phase; in

S.V. Todkari (✉) · A.B. Bagwan
Kalinga University, Naya Raipur, Chhattisgarh, India
e-mail: sachintodkari@gmail.com

A.B. Bagwan
e-mail: aliakbar.bagwan@gmail.com

V.S. Deshpande
VIT, Pune, India
e-mail: vsd.deshpande@gmail.com

© Springer Nature Singapore Pte Ltd. 2018
S.S. Dash et al. (eds.), *International Conference on Intelligent Computing and Applications*, Advances in Intelligent Systems and Computing 632,
https://doi.org/10.1007/978-981-10-5520-1_5

discover phase, mobile sink collects the shortest hop information, and in data collection phase, it formally collects the data packets.

One of the challenging tasks in wireless sensor network is to reduce the delay in data collection and mobile data collector with multi-rate CSS [2], hereby combining the data collection region for nearby data sources and then skipping and substituting nearby sites and then applying multi-rate CSS for finding the difference between communication time and data collection time.

Some of the techniques and methods use mobile data collector node for reliable data collection in wireless sensor network [3]. Basically, sensor node senses the environmental conditions and transmits this information hop-by-hop to the destination, but in such scenario, there are more chances of bottleneck near to the sink so due to bottleneck problem huge chances of data loss, so mobile data collector nodes will handle such type of situations in reliable way.

In wireless sensor network, data collection in reliable way with maintaining QoS parameter is challenging task [4], mobile node collects the data from sensor nodes by using appropriate method called tour planning; in this method, mobile node will finalize its tour before starting its journey. Sensor node will collect the information and store in buffer and broadcast the data collection message in the network. Mobile data collector will receive the messages from multiple sensor nodes in the network after that it will finalize the route to data collection; MDC will notify all the nodes in its route and start its tour; when MDC will reach to the transmission range of the sensor node, they will hand over the data packets to the MDC. In such situation, MDC must have to predict the possible tours which will complete within standard time; otherwise, there are chances of buffer overflow at local region sensor nodes and important data loss; here, lack of some loss recovery scheme comes in picture.

Another important issue in wireless sensor network is to reduce the data loss, to reduce this some mechanism exists namely in-network storage, data collection method based on data output filtering mechanism. The method is designed aiming to achieve reliable data collection in situations of disasters and extreme environments, and sink node decoding efficiency is improved by the aid of a local configuration network and a data filtering mechanism so as to improve network data collection efficiency. A local configuration network collection model, a predecoding mechanism, a data monitoring caching mechanism, and the like are provided on the basis of an incremental network coding method, cache nodes are additionally arranged by the aid of a local configuration network model and form a single-hop data collection tree with proxy nodes, and when normal nodes perform data coding and exchange, the cache nodes store effective coded data by the aid of a monitoring and filtering function, so that decoding efficiency of sink nodes can be improved, and entire network data collection efficiency is improved. The method is capable of collecting network data at a speed approaching to the ideal collection speed, has excellent robustness, and is applicable to reliable and efficient data collection in disasters and extreme environments.

The reliable data collection using tour strategy with adaptive data collector nodes (ADCN) in highly sensitive wireless sensor network, which reduce the data loss happens due to congestion and improve the QoS of network. Normal sensor node used broadcast CDP method when burst data generated by particular event and sensor node buffer reaches peak point, i.e., threshold value. By using broadcast CDP method, it calls DCN (data collector node) within the network to collect the data packet. After receiving broadcast message, DCN (data collector node) uses the path verification method, it checks the source node id in its route database to check whether particular node previously visited or not, and DCN follows the tour strategy which is calculated using MST for every tour. If the source node is in the list, it follows the same path; otherwise, it does the respective tasks. ADCN identification method is used in the network when DCN is not capable of collecting data due to buffer overflow; at that time, ADCN method is initiated by DCN for selecting another DCN in the network adaptively.

The rest of this paper is organized as follows. Section 2 describes the related work. Section 3 covers proposed methodology. Section 4 elaborates experimental result and analysis followed by the conclusion and future work in Sect. 5.

2 Related Work

Various approaches are used for mobile nodes as data collector node [5], and one of the methods is to visit every sensor node and collect the data packets. Another approach is to use rendezvous sensor nodes with cluster structure, and for such type of methodology, MobiCluster protocol is used which comprises of different phases such as clustering, RN selection, CH attachment to RNS, data aggregation, and forwarding to RNS. Communication between RNS and mobile sink by using rendezvous node maximizes the connectivity and data throughput, and overall energy is balanced.

One of the approaches to handle the mobility for the group of mobile sink in WSN is the data transmission using virtual line concept; in this method, sensor region is consider as a circle, and the center point and radius are the location of line structure. Here, mobile sink and sensor node are aware of their current location by using locating device, and when one of the sink in a group wants to collect the data from source node, the actual group region will be detected by group registration process. For data collection, sink has to send query packet to virtual line structure and it will resolve that query and reply with data packet. A mobile sink group has two mobility type, one is group mobility which it uses when sink changes the group and another is individual mobility which means sink moves within the group; this methodology consumes more energy when flooding but by using VLDD enhances energy consumption.

The traditional way of data dissemination is to transmit data packet hop-by-hop to the sink [6]; there are some disadvantages of this methodology if amount of data are huge; there is absence of predefined topology to transfer the data packets form common sensor node to the sink. The cluster topology gives better result than flat topology, and even it increases the data collection time, but by applying some loss recovery or loss prevention scheme data delivery ratio is enhanced, another method for utilize the available energy of sensor node in efficient way is the link aware cluster tree which consists of two phases, first one is the setup phase which is used for cluster formation and another phase is for data collection in reliable way.

The steady-state phase is invoked to disseminate the data from common sensor node within the cluster to the sink; the energy-efficient link state method is network management architecture for wireless sensor network. It provides stable link, better data delivery ratio, and proper energy utilization by reducing network overhead.

In WSN, some of the sensor node utilize as in-network data collection point with relay hop constraint [7]; in traditional method, mobile node will travel to every sensor node for data collection directly, but if the speed of data collection is low, then there is delay in data collection process. This methodology is not applicable in time-sensitive WSN zone, therefore data collection point are place at proper position according to the node density and mobile data collector nodes to reduce the overall delay in data collection.

The performance of the WSN is analyzed by some of QoS parameters, and one of the important parameters is packet delivery ratio [8], which is needed to be improved in large-scale network; to improve the delivery ratio, first we have to minimize some root level issues which will be the root cause to decrease the delivery ratio. Some of the important causes of data loss which are categories as, first one is sink side failure where sink node fail or controlling system get full, second category is corruption this will happen either in-network corruption or sink side corruption, and last buffer overflow drop which will be either local level node buffer overflow or the node which are one hop away from sink node at this level buffer get full and packet drops.

3 Proposed Work

Adaptive data collector node for reliable data collection using tour strategy in highly sensitive wireless sensor network which provide solution for reducing data loss due to congestion and improve the QoS of network. Normal sensor node used broadcast CDP method when burst data generated by particular event and sensor node buffer reaches peak point, i.e., threshold value. By using broadcast CDP method, it calls DCN (data collector node) within the network to collect the data packet. After receiving broadcast message, DCN (data collector node) uses the path verification method, it checks the source node id in its route database to check whether particular node previously visited or not, and DCN follows the tour strategy which is calculated using MST for every tour. If the source node is in the list, it follows the same path; otherwise, it does the respective tasks. ADCN identification method is used in the network when DCN is not capable of collecting data due to buffer overflow; at that time, ADCN method is initiated by DCN for selecting another DCN in the network adaptively.

3.1 Buffer Occupancy Prediction by Fuzzy Logic

A set A is said to be a fuzzy set of the universal set X if each element of set A has a membership function or the degree of belongingness in X. Here, we denote the membership function of a fuzzy set by μ_A that is,

$$\mu_A: X \rightarrow [0, 1] \tag{1}$$

where μ_A is membership function of fuzzy set A

$$A: X \rightarrow [0, 1]$$

Buffer occupancy of every node is given as (2). Here, the threshold value for buffer is considered as 70%, and set element x is the set of elements of buffer occupancy status. Set A is the set of sensor node for which buffer occupancy crosses the threshold value. The membership function for set A is denoted as

$$\mu_{\breve{a}} = (x) = \begin{cases} x/70 & \text{for } 0 < x < 70 \\ 1 & \text{for } 0 < x < 100 \end{cases} \tag{2}$$

If a fuzzy number $\breve{a}$ is fuzzy set A on R, it must possess at least the following three properties:

(i) $\mu_{\breve{a}}(x) = 1$,
(ii) $\{x \in R / \mu_{\breve{a}}(x) > \propto\}$ is a closed interval for every $\propto \in (0, 1]$,
(iii) $\{x \in R / \mu_{\breve{a}}(x) > 0\}$ Bounded and it is denoted by $\left|a_\lambda^L, a_\lambda^R\right|$.

Mobile data collector starts its tour after receiving the data collection request from set A sensor node; among all sensor node in the network, those nodes which possess at least three properties as mentioned above are consider as a member of set A. First property states that those sensor node whose membership function values is 1 (buffer > 70%), these nodes are consider as critical node. Second property states that closer interval value is also considered as member of fuzzy set, and third property indicates the bounded values.

Figure 1 shows the membership function of fuzzy set of uncertainty of buffer; here, for the buffers which have membership function value 1, these nodes are

Fig. 1 Membership function for ADCN

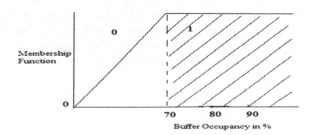

considered as critical node. After receiving request message from sensor node, mobile data collector node will calculate the shortest path for these nodes and execute the tour strategy, and for other members whose membership function value is 0, these nodes are considered as normal node because there buffer occupancy status is normal.

3.2   Broadcast CDP

Collect data packet method is used by normal sensor node to announcing request message in the network for DCN (data collector node) based on the decision. When sensor node buffer reaches to set threshold value or multiple events detected in surrounding sensing area.

3.3   Path Verification

Path verification method is used by DCN to check the source node in its database and to verify the shortest path. If the source node is in the database, then verify the route and follow same, otherwise do the following, create spanning tree and respective local region where multiple event detected, consequently tour strategy with shortest path is finalized and data collected from source node in reliable way.

3.4   ADCN Identification

ADCN method is activated by DCn when source node buffer reaches to the set threshold and data loss occurs, ADCN identification process is as follows.

First find unutilized node in the network which is in the transmission range of source, and check its buffer occupancy; if it has sufficient space, then make it as adaptive data collector node (ADCN) and broadcast to notify other DCNs in the network; new ADCN collects data from source node and hand overs the data to mobile data collector node and after that it will work as normal N.

ADCN algorithm [9] will show process and methodology for reliable data collection using tour strategy with adaptive data collector nodes (ADCNs) in highly sensitive wireless sensor network for reducing data loss due to congestion and improve the QoS of the network. The adaptive data collector algorithm mainly comprises of the following steps:

Step 1: Detect the event
Step 2: Sensor nodes start sensing
Step 3: Create Data collector nodes DCN
Step 4: Get sensed data and store in buffer
Step 5: If size (buffer) >80%
        Broadcast Collect Data Packet CDP If CDP received by DCN
              i. If Notify
                      Check source node in list If exist remove
              ii. else
                      Create spanning tree Create local regions Share unique path
                      Notify to source Travel shortest path
                      Collect data from Node Handover data to sink
              iii. End If
        End if
Step 6: Else if size (buffer) == 100% && Notify packet not received

        a.   Find buffer occupancy of nearby node & minimum buffer occupancy   node
        b.   select as ADCN
        c.   Make it as ADCN
        d.   Broadcast notify to other ADCN New ADCN collect data Disseminate the data to the DCN
        e.   Work as normal N

        End if
Step 7: If Data Dissemination time > time
        Step 4
        Else
        Stop

Sensor node uses broadcast CDP method when burst data generated by event and sensor node buffer reaches peak point, i.e., threshold value. By using broadcast CDP method, it calls DCN (data collector node) within the network to collect the data packet. After receiving broadcast message, DCN (data collector node) uses the path verification method, it checks the source node id in its route database to check whether particular node previously visited or not, and DCN follows the tour strategy which is calculated using MST for every tour. If the source node is in the list, it follows the same path; otherwise, it does the following task.

DCN creates spanning tree to find intermediate node to reach source node in a shortest way, then creates local regions, and identifies request to finalize the complete tour and after that shares unique path with subordinates and notifies to source node for data collection. And at last start the tour using travel shortest path collect data from source node. ADCN identification method is used in the network when DCN is not capable of collecting data due to buffer overflow; at that time, ADCN method is initiated by DCN for selecting another DCN in the network adaptively. For this, it performs the following.

Source node search the idle nodes in its transmission range which have sufficient energy and make them adaptive data collector node (ADCN), after that broadcast the network id of new ADCN in the network, meanwhile new ADCN start the data collection process. When MDCN will come in transmission range, then both source node and ADCN will hand over the data to MDCN. And last after situation is under control, ADCN will work as normal sensor node.

4 Experimental Result and Analysis

In this section, the performance analysis of the network will be carried out by analyzing QoS parameter such as packet delivery ratio, end-to-end delay, throughput, energy, and packet drop with different scenarios. Experiment is carried out using NS3 simulator with 700 * 700 areas with AODV as routing protocol; here, ADCN model is tested for different cases like by varying node density, with and without ADCN.

Figure 2a exhibits the PDR as function of node density without ADCN method, in traditional method the Avg PDR is 70% with node density 50 nodes, but as network density increases relatively event occurrence ratio also increases, but such situation cant handle by source node in traditional network due to absence of ADCN huge congestion occurs at source node and PDR decreases, but after applying ADCN method PDR is above 90% with same node density.

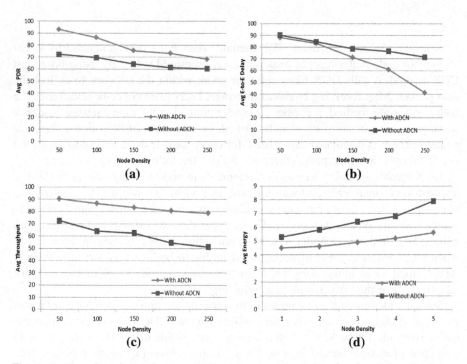

Fig. 2 Quality of service parameter **a** PDR as function of node density, **b** end-to-end delay as function of node density, **c** throughput comparison, **d** energy consumption

Figure 2b exhibits end-to-end delay comparison with and without ADCN model, here at starting both model delays require 90 m s, but as the node density increased for ADCN model delay decreases and it reaches up 40 m s, but for without ADCN model, it is near about 60 m s. By applying proper tour strategy delay requires less data dissemination in ADCN model.

Figure 2c describes throughput comparison with and without ADCN model, here throughput of ADCN model is above 90% with node density is 50 nodes, but in traditional network without ADCN method the throughput is 72% but it reduces continuously because source node cant handle huge data.

Figure 2d shows average energy consumption of the nodes within the region; here, with ADCN model, energy consumption of the nodes is in between 4 and 5 J after varying the node density because the adaptive node will avoid the overflow of data packets and DCN will travel with appropriate tour plan, but in case of without ADCN model, the energy consumption will increase randomly because nodes are not able to handle congestion at local level.

Figure 3 shows packet drop due to buffer overflow at node level, as huge amount of data generated when multiple events occurs, in such situations it very hard to manage buffer to maintain reliability as it had very small size, here the ADCN network is tested with reporting rate 0.01, the packet drop count is 150 packets with node density 50, and it will increases up to 250. But when same scenario is executed in traditional network, the packet drop count increases drastically, at node density it is 250 and when density is 250 the packet drop count is 400. The ADCN model has temporary storage in adaptive mode so the overall packet drop ratio is less.

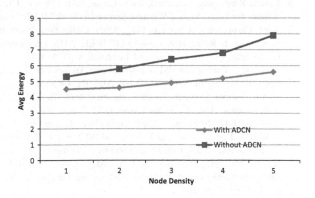

Fig. 3 Packet drop due to buffer overflow at node level

5 Conclusion

In this paper, the adaptive data collector node model has given better result in highly sensitive area where event occurrence ratio is very high. Using proposed ADCN technique, the result shows that QoS parameters improve drastically, and in burst data environment, PDR is improved by 23%, end-to-end delay is minimized by average 25 m s, throughput will be improved by 15%, energy consumption will be reduced by applying proper tour planning methodology, it will be reduced by 30%, and finally, the packet drop count is also minimized as compare to without ADCN model, so by using ADCN methodology data loss at local level is reduced and the reliability is improve in highly sensitive networks, in future the focus is to improve the selective reliability using packet priority and adaptive scheduling.

References

1. Shuai Gao, Hongke Zhang, and Sajal K. Das "Efficient Data Collection in Wireless Sensor Networks with Path- Constrained Mobile Sinks" IEEE Transactions on mobile computing, VOL. 10, NO. 5, APRIL 2011.
2. Liang He, Jianping Pan, "A Progressive Approach to Reducing Data Collection Latency in Wireless Sensor Networks with Mobile Elements" IEEE Transactions on mobile computing 2012.
3. US Patent 72422994 B2 "System and method for using mobile collector s for accessing sensor network".
4. US Patent US 20030202479 "method and system for data in a collection and route discovery communication network".
5. Charalampos Konstantopoulos "A Rendezvous-Based Approach Enabling Energy-Efficient Sensory Data Collection with Mobile Sink" IEEE Transaction on parallel and distributed system Vol 23 No 5 May 2012.
6 Elmani Ramasamy, Kaarthick Balkirishnan "An Efficient Cluster-Tree based Data Collection Scheme for Large Mobile Wireless Sensor Network" IEEE Sensor Journal/JSEN 2014.2377200.
7 Srijit Chowdhury, Chandan Giri "Data Collection Point Based Mobile Data Gathering Scheme with Relay Hop Constraint" IEEE 2013 International Conference on Advance in Computing, Communication and Informatics (ICACCI).
8 Wei Dong, Yunhao Liu "Measurement and Analysis on the Packet Delivery Performance in A Large Scale Sensor Network" IEEE 2013 INFOCOM.
9 Vivek Deshpande, Sachin Todkari Indian Patent "Reliable data collection using Tour Strategy with Adaptive Data Collector Nodes (ADCN) in Wireless sensor network" Application No.: 201621016074 Publication Date: 10/06/2016 Journal No 24/2016.
10 J. Kim, J. Lee, J. Kim, and J. Yun, "M2M Service Platforms: Survey, Issues, and Enabling Technologies," IEEE Common. Surveys & Tutorials, vol. PP, no. 99, pp. 1–16, 2013.
11 K. Ota, M. Dong, X. Chen, A. Liu, and Z. Chen, "Cross layer optimal design for wireless sensor networks under Rayleigh fast fading channels," in Proc. IEEE 10th HPCC/EUC, pp. 183189. Nov. 2013.
12 Mohamed Amine Kafi, Djamel Djenouri "Congestion Control Protocols in Wireless Sensor Networks: A Survey" IEEE communications surveys & tutorials, vol. 16, no. 3, third quarter 2014.

13 C.-Y. Wan, S. B. Eisenman, and A. T. Campbell, "Energy-efficient congestion detection and avoidance in sensor networks," ACM Trans. Sen. Netw., vol. 7, no. 4, pp. 1– 31, Feb 2011.
14 Petcharat Suriyachai, Utz Roedig, and Andrew Scott "A Survey of MAC Protocols for Mission-Critical Applications in Wireless Sensor Networks" IEEE communications surveys & tutorials, vol. 14, no. 2, second quarter 2012.
15 Wei Dong†, Yunhao Liu "Measurement and Analysis on the Packet Delivery Performance in A Large Scale Sensor Network" ©2013 IEEE.
16 Xiaoyan Yin, Yuguang Fang "A Fairness-Aware Congestion Control Scheme in Wireless Sensor Networks" IEEE transactions on vehicular technology, vol. 58, no. 9, November 2012.

Network Information Security Model Based on Web Data Mining

M. Shankar Lingam, A. Arun Kumar and A.M. Sudhakara

Abstract Recent developments in network and information security have made researchers look into the possibility of developing techniques for network economy. One of the most common techniques that are used to enhance the performance of information security in networks is 'web data mining.' This technology is found to be effective in providing information security in a network. This paper presents a meta-analytic report on information security by using data mining techniques.

Keywords Information security · Data mining · Internet

1 Introduction

With the advent of the World Wide Web, there has been an immense growth in the usage of Internet in the global level. It is evident that Internet acts as a medium that not only provides users a wide range of information but also enables them to transact information. Information given in the WWW is accessible to all individuals, groups, and organizations, and such information is obtained with the help of search engines and browsers [1, 2]. In short, WWW has become an indispensable tool for individuals and organizations as it aids in acquiring required data and in making critical decisions on business transactions. Commercial and social transactions have become order of the day, and recent initiative taken by the

M. Shankar Lingam (✉) · A.M. Sudhakara
University of Mysore, Mysore, India
e-mail: shankumacharla@gmail.com

A.M. Sudhakara
e-mail: sudhakara.mysore@gmail.com

A. Arun Kumar
BITS, Hyderabad, Telangana, India
e-mail: arun.arigala@gmail.com

© Springer Nature Singapore Pte Ltd. 2018
S.S. Dash et al. (eds.), *International Conference on Intelligent Computing and Applications*, Advances in Intelligent Systems and Computing 632,
https://doi.org/10.1007/978-981-10-5520-1_6

Government of India on the notion of 'Digital India' has enhanced the use of Internet by almost every citizen. But, along with this rapid growth of IT-enabled society, various problems associated with network information security have started posing threat to information security. In this context, an effort was made to look into the possibilities of using Web data mining techniques to enhance the network information security with the idea of providing a basic feasible solution to security-related issues [3]. To begin with, basic details of Web data mining are outlined in the next section.

2 Web Data Mining

Data mining is generally used for extracting required information from a large data, be it raw data or formatted data. Extracting information from an ultra-large data, be it raw data or formatted data, leads to what is known today as 'Big Data Analytics.' Data mining techniques help extract wanted data from random data that are large, incomplete, noisy, and possibly fuzzy [4]. Most of the traditional data mining methods are used for data analysis, only for homogeneous type of data, that comprises a huge volume of heterogeneous text information over the Internet, hyperlinks, and log information. Web data mining was created by those people who faced these issues by combining traditional data mining techniques to Web, thereby producing a new mining technology altogether. Discovery and extraction of potentially useful patterns of interest and hidden information from various Web activities and Web sources are some of the issues that this new technology can handle. It mines the Web for useful information as a primary goal by implementing data mining, document mining, multimedia mining along with taking the help of databases, computer networks, data warehouses, artificial intelligence, information retrieval, natural language in understanding the technology, and visualization that are passed by combining Web with data mining [4]. The basic process involved in Web data mining is shown in Fig. 1.

Resource Discovery process is used to acquire and return text resources from Web. Web pages, Web databases, Web architecture, user records, and other similar information include in their objects of treatment.

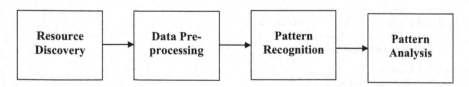

Fig. 1 Basic process of data mining

3 Information Security Model—A Cursory Analysis

Network information security model based on Web data mining is shown in Fig. 2.

There are three main areas of a network security one has to consider while analyzing a network security model. Filtering is the first process that extracts data from the database and uses ambiguous methods of analysis than the right to check for consistency. Next is the mining synthesizer, that is, an engine-driven digging method. It uses appropriate mining systems to claim and carry out excavation of data in the algorithm library. The next is the method selection expert system and knowledge base. These play a significant role in Web data mining where the data mining system not only considers the specific requirements of the user but also selects the most effective mining algorithm for extracting technical information. In addition, it assists through the development of Web content and frequently updated rules to improve the intelligent systems.

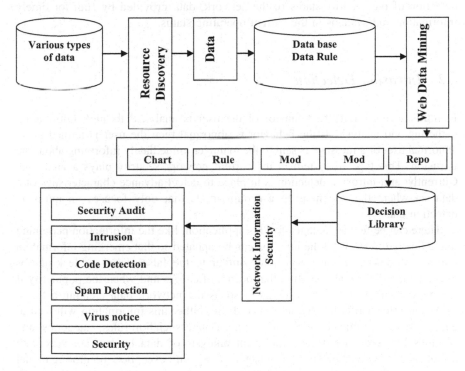

Fig. 2 Network information security model

4 Data Mining Prevention Model for Security

4.1 Security Audit

Security audit is the first system that targets security-related data and records generated on the network to analysis and statistical process. User operating system, user network activity, system applications and services, and network behavior are few of the recorded security events that are analyzed by security audit which can be also assisted by Web technology. Its main agenda is to dig in the normal data to normal network traffic patterns those in turn would, after a number of attacks linked to the associated rule base analysis, will be detected in the system during the post-analysis of the various harmful vulnerabilities that are found in the program, and then an appropriate action is taken to resolve all of these threats. Using the technical and safety audit system integration, Web data mining can turn on the HF firewall and the IDS intrusion systems to protect information along with the timely detection of the security status of the network, data provided by staff for timely information and systems in the current operating status.

4.2 Intrusion Detection

In intrusion detection, the behavior of the user is analyzed through information gathering and analysis. If the behavior is abnormal then the user is termed as an abnormal user or a message is sent to the manager immediately informing about the intrusion. This type of system in the current network security plays a vital role. Currently, anti-intrusion detection is in place to detect advance characteristics of a data set where the system is in a certain mode that only focuses on intrusion detection data.

There can be few advantages for this application, like the information pertaining to the timely discovery of the invasion can be updated so that emerging information will not recognize the invasion quite similar to the false alarms or leak alarms phenomenon during an operation that occurs often. As and when the popularity if the network in people's lives increases so as the network data creating a lot of irrelevant information for the audit record. So, either this information will lead to data overload or mitigate the detection rate. Data association rules, categories, and columns like sequence mode, intelligent analysis of data by law are very well established in Web data mining, and hence, along with exception monitoring model in intrusion detection system, it is used to maximize the reduction in dealing with audit data on a priori knowledge discovery and also reduces the false alarm rate of the system.

4.3 Malicious Code Detection

The most prominent in the anti-malware research is the signature detection technology but it comes with a weakness. It can only detect known malicious code, but malicious code for emerging it can have no impact. Web data mining will significantly improve the efficiency and quality of malicious code detection. Initially, it gathers a huge number of malicious codes, and then a library is created by adding normal code to the large volume of malicious code and then is divided into training set and test set. Next it uses a variety of algorithms such as the rules of classification algorithms or the Bayesian algorithms for training sample, and training is performed to identify malicious code and normal code accurately. Finally, the test set is used to evaluate the training set that produces effective results that are more satisfactory.

4.4 Malicious Spam Detection

With the advancement of information technology, various malicious codes based on emergence of e-mail attachments are born. A common practice is to detect the binding virus scanner through e-mail filter where the virus scanner is based on the signatures to detect malicious code. For those unknown malicious codes without any corresponding signatures cost of prevention is high and less efficient. Using Web data mining, e-mails can be scanned for virus bundling by testing a sample of the e-mail for malicious code or characteristic pattern of malicious e-mail, and then by the use of naïve Bayes classifier and enhanced methods of machine learning techniques, a mail filter will filter out all the malicious e-mails and only useful messages go into the inbox.

4.5 Virus Warning

The rise of Internet's popularity has seen significant increase and prevalence of broadband networks along with computer viruses in the network-oriented development known as Worms. Generally, traditional antivirus technologies detect known virus signatures and when it comes to this type of emerging virus they are powerless. By connecting behavioral anomaly detecting network in real time with the help of Web data mining, an early warning system is implemented to provide virus warning and thus enables to find traces of worms and allowing network administrators to take appropriate measures to avoid big loses before the worm outbreaks.

4.6 Security Assessment

Information security protection management is an important part of risk management. Web data mining uses meta-search engine to retrieve information on the structure of resulting process that allows users to large volumes of semi-structured risk assessment information by selecting Web, mining, integration, documentation, network information security risk assessment data, and risk assessment. In accordance with the characteristics of different information on the retrieved information in order to classify risk, establishment of information security, risk assessment information database is created which is constantly updated and expanded that will help in the establishment of a risk decision support system in order to provide tremendous support for the network information security risk management.

5 Conclusion

From the meta-analysis of network information security using Web data mining techniques, one may conclude that enhancement in the performance of network information security should be feasible. The Web mining prevention model is used here as a guideline to achieve this. Data mining can give accurate and valuable results. Priority has to be given to sensitive data. Hackers could use data mining to access these sensitive data, and the model discussed in this paper could be used to avoid such scenarios.

References

1. Research; China highlights four individual network security market development trends.
2. H. Vernon Leighton and J. Srivastava. Precision Among WWW Search Services (Search Engines): Alta Vista, Excite, Hotbot, Infoseek, Lycos. http://www.winona.msus.edu/is-f/libraryf/webind2/webind2.htm, 1997.
3. R. Cooley, B. Mobasher and J. Srivsatava. Web Mining: Information and Pattern Discovery on the Word Wide Web. Technical Report TR 97-027, University of Minnesota, Dept. of Computer Science, Minneapolis, 1997.
4. SEIFERT JW. Data mining and the search for security challenges for connecting the dots and data bases. Government Information Quarterly, 2004 (21): 461–480.

Diagnosis of Disease Using Feature Decimation with Multiple Classifier System

Rupali R. Tajanpure and Sudarson Jena

Abstract Nowadays, due to change in lifestyle, the problem of heart disease has become very common. When patients visit hospitals for nominal reason, they have to undergo different tests suggested by a doctor which create lot of stress in patients leading to loss of money as well as time. Since doctors suggest number of tests to patients to identify the problem, there are chances that few tests may not be required at preliminary stage. Also poor clinical decisions may lead to some disastrous conditions. In existing systems, all the features are tested at a time by the classifier in order to detect whether patient is suffering from that particular disease or not. The entire feature testing consumes a lot of time. Also if system is testing all attributes of a healthy person, then it is wastage of time. So the proposed idea is that the attributes/features are decimated into groups according to their importance. These groups of features are then input to different stages of multiple classifier system. If output of stage I is showing risk of disease, then only system will go for second stage of classifier with second-level attribute set as input and so on. Thus, for healthy person system will stop at first stage with conclusion no risk of disease. In this way, multiple classifier system utilizes time efficiently. Simultaneously, patient is also relieved from unnecessary stress as well as fatigue. Calculations for basic architecture of neural network show that time complexity in terms of number of additions and multiplication is reduced by 58 and 30%, respectively, for assumed case.

Keywords Multiple classifier system (MCS) · Single classifier system (SCS) Neural network (NN) · Feature decimation

R.R. Tajanpure (✉) · S. Jena
GITAM University, Hyderabad, India
e-mail: rupalidixit1@yahoo.co.in

S. Jena
e-mail: sudarsonjena@gitam.edu

© Springer Nature Singapore Pte Ltd. 2018
S.S. Dash et al. (eds.), *International Conference on Intelligent Computing and Applications*, Advances in Intelligent Systems and Computing 632,
https://doi.org/10.1007/978-981-10-5520-1_7

1 Introduction

Many people lose their lives since they are not familiar with the symptoms associated with the diseases. According to the fact sheet of WHO, there are 10 leading causes of death in the world during the year 2000–2012, where ischaemic heart disease, stroke, lower respiratory infections and chronic obstructive lung disease are major killers during the past decade. Also it is mentioned that 7.4 million deaths are due to ischaemic heart disease which is at first position of causes of death [1].

Today, the concept of smart hospitalization has become buzz word. Many hospitals are maintaining their patient's database online such as the records related to tests suggested, their results and the prescriptions suggested. This generates huge data which could be in any form such as text, numbers as well as images and videos. In fact, all this data is important in making clinical decisions. In order to handle such a large data efficiently, usage of multistage classifier has become necessary. The proposed system uses three-stage neural network classifier to check whether patient is suffering from the disease or not. At every stage as per level, some attributes are tested to decide whether to move to next level or stop. This system optimizes resources very efficiently. Also it aims to identify the problem in the very preliminary stage and suggest reliable solution to problems increasing life of patients.

If we take example of heart disease, then the tests for detecting possibility of heart disease can be categorized as level-1, level-2 and level-3 tests. Level-1 tests include blood pressure, heart rate/pulse rate, ECG, lipid profile (HDL, LDL, triglycerides, and total cholesterol), sodium, potassium tests. Level-2 tests include 2D/3D echo, stress test, angina symptoms. Level-3 includes angiography. As a routine check-up, if patients perform level-1 tests, then according to possibility of heart disease either next level test will be suggested or some preventive measures will be given by the system. So our software will help doctor to make proper diagnosis. This system will help doctor to think about next treatment to be given. The patients can also refer to system by doing level-1 test as routine activity and will decide whether to visit the doctor or not.

Since for evaluation of heart disease, if we consider 10 tests/attributes/features to be used at a time, we have to test each and every feature. If we think of 50% chances of being positive for heart disease, then this means that for those without having any risk of heart disease, we have to test 10 attributes compulsorily. Now if we have three-stage classification strategy, then by testing only level-1 attributes, we can conclude whether to go for further level testing. This reduces complexity in checking all patient verses all attributes. Also features/attributes can be grouped according to different levels as per importance of the test. Patients can be diagnosed and relieved at the level-1 itself, if they do not have any symptoms of heart disease. Also if it is necessary, doctor can decide to go to next level for some critical cases.

Primary aim of the proposed system is to apply data mining algorithm to predict the chances and level of heart-related problems. By use of this model along with routine check-up, patients will be able to decide to visit doctor or not for further treatment.

Section 2 of this paper presents a literature survey. In Sect. 3 proposed system of multiple classifier is discussed. Section 4 describes working principle of proposed system. Finally, Sect. 5 of this paper presents concluding remark.

2 Literature Survey

From early nineties, the research started on multiple classifier system due to the fact that multiple classifier system will work well since it will be able to avoid limitations coming from single classifier system. Some classifiers work well for one dataset while shows poor performance on other dataset. So if we have set of classifiers showing good performance on problem data, then instead of choosing only one classifier which may be inadequate, we can choose number of classifiers and the average result of them will be near to best classifier of that problem.

Dietterich [2] suggested three reasons (statistical, computational and representational) for analysis of how classifier ensemble may prove better over single classifier. Different strategies are used to combine multiple classifiers.

The different criteria used for combinations of classifiers are discussed in Ludmila I. Kuncheva's book [3]. This book elaborates four approaches for building ensembles of diverse classifiers. In first approach, classifiers can be used based on different combiner designs and is known as combination level approach. Second approach uses different base classifiers, while third approach uses different feature subsets. Last approach is based on different data subsets.

Lim et al. [4] have studied multiple classifier system (MCS) in medical diagnosis. They have used integration of fuzzy ARTMAP (FAM) and probabilistic neural network (PNN) as base of classifier system. They concluded effectiveness of MCS over individual classifier. Since multistage classifier system may have combination of multiple classifiers in cascade, parallel or hybrid form. Depending on which the MCS can be ensemble-based or voting-based system. Such systems are discussed next.

Yang et al. [5] in their paper used neural network ensembles as effective technique to improve generalization of MCS system with neural network. In this paper, stack of neural network (NN) is used. One of the NN is used as combiner of outputs of well-trained neural networks. Now, second layer of NN is trained with results of first-layer NNs. They concluded that multistage neural network ensembles show improved performance as compared to majority voting as an ensemble combination method over wide range of datasets.

Rokach [6] has made survey of different existing ensemble techniques. He discussed variants of AdaBoost algorithm as well. He mentioned ensemble framework as consisting basic blocks such as training set, base inducer, diversity generator and combiner. Kuncheva et al. [7] have observed that Rotation Forest with independently trained decision trees for building classifier ensembles was proved to be more accurate than bagging, AdaBoost and Random Forest ensembles across a collection of benchmark data sets. So they worked to find the factors due to which Rotation Forest is proved to be better.

In 2011, Khemphila et al. [8] introduced classification of presence of heart disease using multilayer perceptron with back propagation. For feature selection, they used Information Gain to choose number of attributes. With reduction in number of attributes, they studied change in accuracy which was minimal.

Amato et al. [9] discussed the survey of artificial neural network in medical diagnosis. In this paper, he discussed the philosophy, capability and limitations of neural networks. Fundamental steps for using ANN in different disease detection are given along with conclusion that ANN are having potential for their use in medical field for disease diagnosis due to their ability to process large data in less time. He also discussed the details of inputs and outputs related with medical diagnosis by NN. Gargiulo et al. [10] have done the survey on multiple classifier system (MCS) covering theory like different strategies of combining multiple classifiers as cascade, parallel and hierarchical combination. Different topologies such as conditional, serial and parallel for combining classifiers are also discussed. Also their applications in important fields as biometric, medical, document analysis, remote sensing data analysis, computer and network security etc. are presented along with details. At the end, important tools for implementation of MCS such as Weka, KNIME, PR Tools (MATLAB) are discussed.

Sumana et al. [11] focused on cascaded combination of clustering and classification for prediction of diseases. Here, K-means is used as preprocessing algorithm as well as for clustering. Finally, clustered samples are used as input to classifier model and tested with 10-fold cross-validation. Twelve different classifiers are used to classify the data. Here, on an average accuracy obtained is 95% for all classifiers. Jabbar et al. [12] in his work used Naïve Bayes classifier. Here, discretization and genetic search is used for optimization of features. This prediction model is designed for early detection of heart disease.

Toshniwal et al. [13] proposed multistage classification of cardiovascular disease. The first stage of classification classifies normal and abnormal ECG beats and second stage is used to reduce number of false negatives so as to refine the results of first stage. The result shows that the proposed technique is better.

Niranjan Murthy [14] used multilayer perceptron NN with back propagation algorithm to classify heart stroke. The results have shown 85.55% of accuracy.

Weng et al. [15] have worked to find the performance of different classifiers present in ensemble classifier as well as in their individual performance. Also the performance of classifier is investigated with real-life datasets. The study concludes that ensemble classifier performs well than single classifier in the ensemble. But the performance of single classifier is not so worse than ensemble classifier. Rau et al. [16] used artificial neural network (ANN) for the prediction of liver cancer in diabetes patients. He mentioned that ANN is effective classifier for disease prediction. Prieto et al. [17] presented overview of modelling, simulation and implementation of neural network for improvement of our understanding of nervous system and to find improvement in its application areas for efficient use of them.

Liu et al. [18] proposed a new method of feature selection which is based upon association analysis which takes into consideration an association between features. This method helps to make syndrome diagnosis of coronary heart disease more standard and objective.

3 Proposed MCS System

The basic idea of proposed system is to design a Decision Support System which helps in analyzing database of different patients in a simple and convenient way. The system is based on three different stage data levels, viz. level-1, level-2 and level-3. This system will automatically suggest different way out based on severity of disease and will suggest preventive and corrective measures accordingly. The system will first categorize the input features according to their level of importance [19]. For such feature decimation, it is necessary to take guidance of doctor having specialization in that area. Feature decimation is the important part of our proposed system. The overall flow of system will be as follows.

Figure 2 shows the architecture of MCS which is part of Fig. 1.

Where, F1, F2, F3 are the level-1 attributes/features/tests whose values will evaluate risk of heart disease at first stage. For heart disease, the level-1 attributes

Fig. 1 Workflow of the system

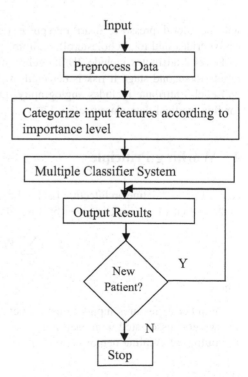

Fig. 2 Architecture of
proposed multiple classifier
system

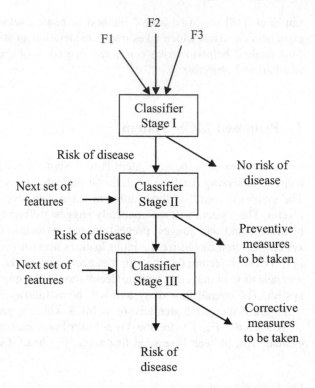

will be blood pressure, heart rate/pulse rate, ECG, lipid profile (HDL, LDL, triglycerides and total cholesterol), sodium, potassium tests.

Level-2 attributes include 2D/3D echo, stress test, angina symptoms which are inputs to second stage if risk is detected in first stage.

Level-3 attribute includes angiography. Due to its advantages, NN is preferred for this medical data classification [20].

4 Working Principle

According to Adaline architecture [21]—the simplest architecture of neural network —the output of network is given by formula,

$$Y = \sum_{i=0}^{n} Xi * Wi$$

where

X number of features inputs to neural network
W weights associated with each input
Y output of Adaline neural network [21]

Here, following calculations are done by considering Adaline neural network as base classifier.

For single classifier, if we want to evaluate the data of 100 patients with total 10 attributes associated with each patient with Adaline NN classifier, then output at any node is given by

$$Yj(1{:}100) = \sum_{i=0}^{n=10} Xi * Wi$$

where

i is numeric factor indicating number of attributes (Let total number of attributes per patient is 10)

j is total number of patients (Let total number of patients = 100)

From above equation,

Total number of multiplications will be 10 * 100 = 1000
Total number of additions will be 9 * 100 = 900

But if we go for multistage classifier

1. Consider Classifier Stage I with only 5 basic features as first-level input,

$$Yj(1{:}100) = \sum_{i=0}^{n=5} Xi * Wi$$

Hence, number of multiplications for 5 attributes and 100 patients will be 5 * 100, and number of additions will be 4 * 100

2. For Classifier Stage II, say only 50% patients are qualified and input features are 3 then

$$Yj(1{:}50) = \sum_{i=0}^{n=3} Xi * Wi$$

So the number of multiplications for 3 attributes and 50 patients will be 3 * 50 and number of additions will be 2 * 50

3. Similarly, for last stage with remaining 2 features and 25% patients,

$$Yj(1{:}25) = \sum_{i=0}^{n=2} Xi * Wi$$

number of multiplications will be 2 * 25 and number of additions will be 1 * 25.

So with MCS, total numbers of multiplications are reduced from 1000 to 700 and additions from 900 to 525.

Table 1 Summary of calculations for MCS and single-stage classifier

Stage number	Number of patients	Number of attribute	Number of multiplications	Number of additions
Stage I	100	5	500	400
Stage II	50	3	150	100
Stage III	25	2	50	25
Total with MCS	100	10	**700**	**525**
Total with single classifier	100	10	1000	900

Fig. 3 Comparison of time complexity of multiple classifier system and single classifier system with Adaline NN as base classifier

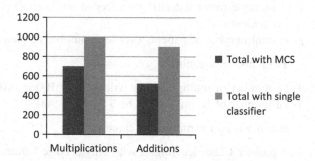

Above calculations are summarized in Table 1.

The graphical representation of above data in Fig. 3.

Above calculations are done by considering Adaline architecture of NN as classifier. Table 1 shows that with MCS, number of calculation (additions and multiplications) is reduced. Also it indicates that 100% processing of attributes at single stage is not needed, thereby reducing time and space complexity and improving effectiveness of the system. With this system, genuine result is expected leading to high accuracy level.

Same system could be used for diagnosing other diseases by changing attributes and deciding levels accordingly.

Looking at criticality of heart disease, the system has to work under guidance of the doctor. Here, the test levels should be defined very specifically. Also wrong inputs will lead to wrong decisions.

5 Conclusion

The proposed system will work effectively with respect to accuracy and time complexity as per the calculations done. The system will be beneficial to society for early detection of heart disease if used with routine health check-up. Also the headache of money and stress will be somewhat released for the patients.

References

1. "The Top 10 Causes Of Death". *World Health Organization*. Fact sheet N°310 Updated May 2014.
2. T. G. Dietterich,: Ensemble methods in machine learning. Multiple Classifier Systems., vol. 1857, pp. 1–15 (2000).
3. Ludmila I. Kuncheva.: Combining Pattern Classifiers: Methods and Algorithms (2004).
4. Chee Peng Lim, Phaik Yean Goay, Poh Suan Teoh, R.F. Harrison et al,: Combination of decisions from a multiple Neural Network classifier system., pp 191–194 (1999).
5. Shuang Yang, Anthony Browne and Philip Picton,: Multistage Neural network ensembles. Multiple Classifier System (2002).
6. Rokach L.: Ensemble based classifiers. Vol. 33, Issue 1, pp 1–39 ArtifIntell Rev, pp 1–39 (2010)
7. Ludmila I. Kuncheva and Juan J. Rodriguez,: An Experimental Study on Rotation Forest Ensembles (2007).
8. Archana Khempilla and Veera Bhoonjing.: Heart disease classification using neural network and feature selection (2011).
9. Filippo Amato et al,: Artificial neural networks in medical diagnosis. Journal of applied Biomedicine (2013).
10. Francesco Gargiulo, Claudio Mazzariello, and Carlo Sansone.: Multiple Classifier Systems: Theory, Applications and Tools. pp 335–378 (2013).
11. B V Sumana, T Santhanam.: Prediction of disease by cascading clustering and classification (2014).
12. M A Jabbar, B L Deekshatulu and Priti Chandra.: Computational intelligence technique for early diagnosis of heart disease (2015).
13. Durga Toshniwal, Bharat Goel and Hina Sharma.: Multistage classification for cardiovascular disease risk prediction. pp 258–266 (2015).
14. H S Niranjana Murthy and M Meenakshi.: Comparison between ANN based heart stroke classifiers using varied folds data set cross validation. pp 693–699 (2015).
15. Cheng-Hsiung Weng, Tony Cheng-Kui Huang, Ruo-Ping Han.: Disease prediction with different types of neural network classifiers (2015).
16. Hsiao Hsien Rau et al.: Development of a web based liver cancer prediction model for type II diabetes patients by using an artificial neural network (2015).
17. Alberto Prieto n, Beatriz Prieto, Eva Martinez Ortigosa, Eduardo Ros, Francisco Pelayo.: Neural networks: An overview of early research, current frameworks and new challenges. pp 1–20 (2016)
18. Xi Liu et-al.: A new method for modeling of inquiry diagnosis for coronary heart disease in traditional disease medicine. BMEI (2011).
19. Jiawei Han and Micheline Kamber: Data Mining: Concepts and Techniques. Morgan Kaufmann Publishers, 3rd Edition (2006).
20. Balasaheb Tarle, Rupali Tajanpure, Suderson Jena.: Medical Data Classification Using Different Optimization Techniques: A Survey, IJRET, Volume 05, Special Issue 05, pp 101–108, ICIAC 2016.
21. S N Sivanandam, S Sumathi, S N Deepa.: Introduction to Neural Networks using Matlab 6.0. Tata McGraw Hill Publication (2007).

Adaptive Filter Algorithms Based Noise Cancellation Using Neural Network in Mobile Applications

A.M. Prasanna Kumar and K. Ramesha

Abstract Noise-free output is a desired characteristic of any mobile communication system. Adaptive noise cancellation is achieved by subtracting unwanted noise signal from the corrupted signal. We propose signal extraction using artificial neural network hybrid back propagation adaptive for mobile systems. The performance analysis of the proposed hybrid adaptive algorithms is carried out based on the error convergence and correlation coefficient. By taking into consideration of the existing algorithms, the proposed algorithms require small neural training sets and it gives good results. Noise cancellation operation is established through adaptive control with the goal of achieving minimum noise error level at output. This paper focuses on the analysis of noise cancellation using least mean square algorithms, gradient adaptive lattice algorithms, and hybrid adaptive algorithms. From computed output, we observed that the hybrid adaptive algorithms perform better.

Keywords Adaptive filter · Neural network · Least mean square
Adaptive noise cancellation · Adaptive algorithms

1 Introduction

Mobile telecommunication system comprises transmitter, channel, and receiver. Normally, channel experiences with noise and interference of other channel signals. The concept of adaptive noise elimination [1, 2] obtain by using adaptive filter interfering signal estimation to subtract it from the corrupted signal. We have taken

A.M. Prasanna Kumar (✉)
Department of Electronics & Communication Engineering,
ACS College of Engineering, Bengaluru 560074, India
e-mail: amprasanna64@gmail.com

K. Ramesha
Department of Electronics & Communication Engineering,
Dr. Ambedkar Institute of Technology, Bengaluru 560056, India
e-mail: kramesha13@gmail.com

© Springer Nature Singapore Pte Ltd. 2018   67
S.S. Dash et al. (eds.), *International Conference on Intelligent Computing
and Applications*, Advances in Intelligent Systems and Computing 632,
https://doi.org/10.1007/978-981-10-5520-1_8

simulation results of adaptive algorithms for noise removal process. Efforts are made use to emphasize electrical and audio signals of practical use. The analysis of the result gives guidelines to further analyses. Performance analysis has been carried out among the adaptive algorithms of their parameters and also discussed effect of epochs on converging error level. Adaptive algorithms proved highly stable and potential for noise cancellation. In this paper, we articulate the adaptive noise cancellation system using different algorithms and artificial neural network used for noise cancellation. We conclude the paper with performance analysis along with complete set of computational results. The choice of adaptive algorithms to be applied is always a compromise between computational complexity and faster error convergence. Performance analysis of hybrid adaptive filter simulation results is found satisfactory.

2 Literature Review

Widrow et al. [1] presented experimental analysis of adaptive noise cancellation, proposed concepts of adaptive noise cancelling principles, and narrated different field of applications such as robotics, automobile control systems, on selected medical field. Experiments are carried out on both random and periodic signal. Benefits are low output noise, distortion, and adaptability compared with conventional filter configurations. The adaptive noise cancellation using Wiener solutions are given. Dixit and Nagaria [2] proposed neural network-based performance study on the recursive least squares algorithms. Authors recommended a novel method in which analog coefficients are varied for satisfactory performance. Analysis carried out on performance factors and depicted results showed that as the filter order increases, execution time of the recursive least square algorithms decreases. Reduction in processing time is achieved. This method works better for speech signal. Ashok et al. [3] proposed Fast Haar wavelet transform employing Neural Network Back Propagation algorithm without convolution for the analysis of signals and obtained reduction in mathematical complexity and dynamic power. Hadei and Lotfizad [4] presented a novel method to evaluate noise cancellation using adaptive algorithms for enhancement of speech. They have developed two new algorithms named fast Euclidean direction search algorithms and fast affine projection algorithms for attenuating noise.

Vijay Kumar et al. [5] presented multi-view classification considering humans speech to convey intelligent audio signal with each other at a bandwidth of 4 kHz. Audio frequency spectrum is periodic in time. The quality of the speech signal difficulty decreases due to noise interference. Modified adaptive algorithms proved superior with white noise by 2 dB. Ferdouse et al. [6] presented noise interference in signals with output free from noise in various electronic and telecommunication systems. Comparison between different adaptive algorithms performance is carried out. Anand et al. [7] presented a system for cancellation of noise using adaptive intelligent filtering method. Adaptive neuro-fuzzy inference system method is applied for corrupted voice signals for noise removal. This algorithm is easy to implement, and it takes less time for convergence.

3 Adaptive Noise Cancellation System

During recent times, hybrid adaptive algorithms are a smart method for noise cancellation using neural networks. Noise elimination is a common phenomenon existing in mobile communication systems. Some of the standard algorithms are gradient adaptive lattice (GAL) and least mean square (LMS) algorithms. For accomplishing learning of multilayer perceptron network, optimized hybrid back propagation learning algorithms are proposed. The analysis of adaptive noise cancellation is carried out using proposed artificial neural network hybrid back propagation adaptive algorithms (ANNHBPAA). Hybrid adaptive algorithms comprising of least mean square algorithms and gradient adaptive lattice algorithms.

The basic principle of artificial neural network hybrid back propagation adaptive algorithms is to get signal cancellation from reference signal and noisy output signal, and by subtracting these two components, noise signal is removed from original signal. Adaptive noise cancellation system with adaptive control and weight adjustment using neural networks could effectively restore the original baseband signal from the noise.

The proposed block diagram of hybrid adaptive filter configuration is given in Fig. 1. Input signal is $x(k)$, adaptive filter output is $y(k)$, and $d(k)$ is the reference signal, where k is the iteration number. The error $e(k)$ is computed by difference of $d(k) - y(k)$. Adaptation algorithms use error signal to form execute function which computes the suitable updating coefficients of filter. The decreasing of the performance function conveys output signal and is same as the expected signal.

To determine error rate of each neuron, back propagation algorithms are used to get desired output. A typical structure of back propagation network (BPN) is given in Fig. 2. Input layer neurons are totally connected to the hidden layer. Hidden layer output is completely connected to the output layer. Entire network gets affected even by one neuron error. In back propagation, audio signal is allowed to propagate through the neural network to produce an output. Output layer error results are obtained after subtracting desired output with reference. Until the output reaches expected value, the error transmitted backward toward input layer via hidden layers. For reducing its error signal, small weight adjustments are carried out for every neuron. Procedure is repeated for each and every value of input. In this hybrid

Fig. 1 Hybrid adaptive filter configuration

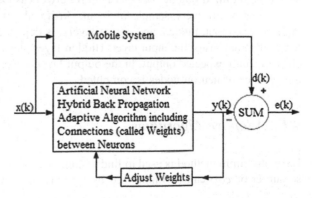

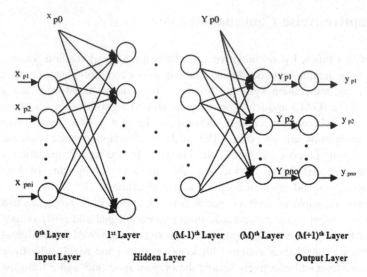

Fig. 2 Neural network back propagation learning model

learning algorithms, least square method is combined with the back propagation method. An adaptive filtering procedure used in the application of the neural network techniques are taken into account, and a control system adjusts parameters of the adaptive filter by means of a multilayered neural network. Hybrid algorithms are introduced to overcome slow convergence.

4 Artificial Neural Network

Connections between elements can be trained using artificial neural network by adjusting the appropriate weights. The output of artificial neuron is summation of product of both inputs with corresponding weights. Every input link has an individual weight accompanied. Output value is equal to one if weighted input sum is greater than or equal to the threshold value, otherwise output value is zero. In order to get the required output, artificial neural network is capable of adjusting weights corresponding to every sample of input. Artificial neural network formed using three layers, input, hidden, and output layers, is depicted in Fig. 3. Audio signal is fed to the neurons of the input layer. Hidden layer plays a vital role in reducing the error to attain expected output. In the output layer, according to the required output the number of neuron nodes is computed.

4.1 Adaptive Algorithms

LMS algorithm method is used to find instantaneous value of gradient vector. Mean square error $e(n)$ can be minimized by varying the weights of the filter. Optimal

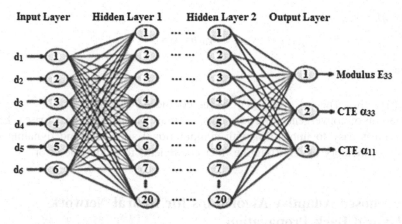

Fig. 3 Artificial neural network layers

Weiner solution [2] is obtain for every iteration of the adaptive filter weights given in Eq. (1). $\mu(n)$ represents step size, n is the time, $W(n)$ gives the coefficients of adaptive filter, and $X(n)$ represents the input vector.

Optimal value of μ is chosen to avoid large convergence time, instability, and divergence of output. Negative gradient of the function is taken into account in the following algorithm to minimize error.

$$W(n+1) = W(n) + \mu(n)e(n)X(n). \tag{1}$$

Mean square error $e(n)$ is given in Eq. (2)

$$e(n) = d(n) - W^T(n)X(n). \tag{2}$$

Input signal vector $X(n)$ is given by

$$X(n) = [x(n)x(n-1)\ldots x(n-L+1)]^T.$$

LMS algorithms are given as

$$y(n) = \sum_{i=0}^{M-1} w_i(n)^* x(n-i) \tag{3}$$

$$e(n) = d(n) - y(n) \tag{4}$$

$$w_i(n+1) = w_i(n) + \mu^* e(n)^* x(n-i). \tag{5}$$

For $M = 1$,

$$y(n) = w_0(n)^* x(n) \tag{6}$$

where

$W(n) = [w_0(n)w_1(n)...w_{L-1}(n)]^T$ is the coefficient vector.

The rate of convergence of error signal increases as the value of μ rises. LMS algorithm is easy to implement with computational simplicity and dynamic utilization of memory by adjusting the filter coefficients to reduce the error.

5   Proposed Adaptive Algorithms for Neural Network Hybrid Back Propagation

Step 1
Normalize inputs and outputs for maximum values [3]. Neural network performs better if input and output values varies within 0 to 1

For every training pair, L inputs and n output are given by

$$(I)_1 \div (I \times 1)$$
$$(O)_0 \div (n \times 1).$$

Step 2
Assess neurons present in hidden layer

$$1 < m < 2I.$$

Step 3
[V] gives synapses weights into input neurons along with neurons of hidden layer, and [W] gives synapses weights linking hidden layer neurons with output neurons. Weights initialized between -1 and 1. Threshold value taken as zero. λ assumed as 1.

$$[W]^0 = [\text{random sampling weights}]$$
$$[V]^0 = [\text{random sampling weights}]$$
$$[\lambda V]^0 = [\lambda W]^0 = [O].$$

Step 4
Output of the input layer can be evaluated for one set of input and output as

$$[O]_1 = [I]_1.$$

Step 5
By computing, multiplying corresponding weights of inputs to the hidden layer is given by

$$[I]_H = [V]^T [O]_I.$$

Step 6
Determine the comparable weights of synapses as

$$[I]_0 = [W]^T [O]_H.$$

Step 7
Using sigmoidal function, enumerate network output

$$[O]_0 = 1 \div \left(1 + e^{-I_{oj}}\right).$$

Step 8
Quantify error and subtract desired output from network output for ith training set

$$E^P = \sqrt{\sum (T_j - O_{oj})^2} \div n.$$

Step 9
Evaluate $[d]$ as

$$[d] = (T_k - O_{ok})O_{ok}(1 - O_{ok}).$$

Step 10
Assess $[Y]$ matrix

$$[Y] = [O]_h(d).$$

Step 11

$$\lambda W = \alpha[\lambda W]^t + \eta[Y].$$

Step 12
Enumerate

$$[\lambda V]^{t+1} = \alpha[\lambda V]^t + \eta[X].$$

Step 13
Calculate

$$[V] = [V]^t + [\lambda V]^{t+1}$$

$$[W]^{t+1} = [W]^t + [\lambda W]^{t+1}$$

Step 14

$$\text{Error rate} = \sum Ep \div n \text{ set.}$$

Repeat 4–14 steps until error rate converges within tolerance.

6 Results and Performance Analysis

Neural network-based hybrid adaptive algorithms analysis on four signals mixed with noise and its computation are carried out. Signals are chirp signal, sinusoidal periodic signal, saw-tooth signal, and audio signals. These signals have been subjected to noise. Filter parameters like correlation coefficient, error recovery rate, and time were enumerated. Convergence performance and behavior of our algorithms found better with existing algorithms. Analysis of these results offered useful perception on behavior of the hybrid adaptive algorithms. The task has been accomplished using the new hybrid adaptive algorithms for adaptive noise cancellation process. The work tries to generate correlations coefficient of least mean square algorithms, gradient adaptive lattice algorithms, and hybrid adaptive algorithms. Using LMS, signal-to-noise ratio of 30 dB is applied as input for six hidden layers. Simulated results of correlations coefficient least mean square algorithms, gradient adaptive lattice algorithms, and hybrid adaptive algorithms are given. Figure 4 gives the comparison of LMS, GAL, and hybrid correlation coefficient for 5,000 and 10,000 iterations. Calculated data is given in respective predicted column, shown in Tables 1 and 2. Difference between actual and predicted is given in corresponding error column. Figure 5 is the MATLAB (R2013a version Intel® Core™ i3-4130 CPU @ 3.40 GHz, 4.00 GB RAM, 64-bit Operating System) response for convergence error for 5,000 and 10,000 iterations.

The proposed ANNHBPAA algorithm is compared with existing LMS and GAL on convergence time which is given in Table 3. ANNHBPAA gives least convergence time in seconds.

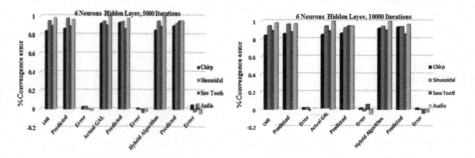

Fig. 4 Comparison of LMS, GAL, hybrid correlation coefficient for 5,000 and 10,000 iterations

Table 1 Six neurons hidden layer, 5,000 iterations

Signal type	Actual LMS	Predicted	Error	Actual GAL [6]	Predicted	Error	Hybrid algorithms	Predicted	Error
Chirp	0.8401	0.8651	0.0250	0.9218	0.9305	0.0087	0.8501	0.8663	0.0162
Sinusoidal	0.9464	0.9736	0.0272	0.9422	0.9378	−0.0044	0.9459	0.9501	0.0042
Saw tooth	0.8935	0.8911	−0.0024	0.9021	0.8651	−0.0370	0.8909	0.9477	0.0568
Audio	0.9798	0.9601	−0.0197	0.9989	0.9736	−0.0253	0.9988	0.9477	−0.0511
	Total error: 0.0301			Total error: −0.0580			Total error: 0.0261		

Table 2 Six neurons hidden layer, 10,000 iterations

Signal type	Actual LMS	Predicted	Error	Actual GAL [6]	Predicted	Error	Hybrid algorithms	Predicted	Error
Chirp	0.8401	0.8591	0.0190	0.9218	0.9361	0.0143	0.8501	0.8657	0.0156
Sinusoidal	0.9464	0.9663	0.0199	0.9422	0.9403	−0.0019	0.9459	0.9470	0.0011
Saw tooth	0.8935	0.8796	−0.0139	0.9021	0.8591	−0.0430	0.8909	0.9473	0.0564
Audio	0.9798	0.9711	0.0087	0.9989	0.9663	−0.0326	0.9988	0.9473	−0.0564
	Total error: 0.0337			Total error: −0.0632			Total error: 0.0167		

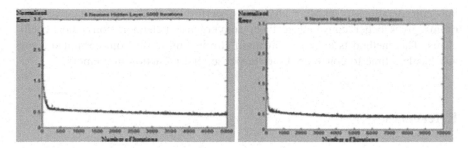

Fig. 5 MATLAB response for convergence error variations

Table 3 Comparison of convergence time (s)

Signal type	LMS	GAL [6]	ANNHBPAA
Chirp	0.4530	0.6720	0.0156
Sinusoidal	0.6410	0.8440	0.0011
Saw tooth	0.4220	1.4530	0.0564

Fig. 6 Comparison of convergence time (s)

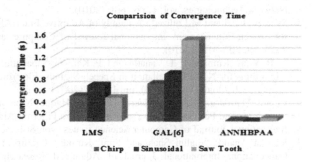

Simulated results of convergence time of ANNHBPAA algorithm compared with LMS algorithm and GAL algorithm and its comparison are given in Fig. 6.

7 Conclusion

The proposed hybrid adaptive algorithms input signals are deterministic. LMS and GAL algorithms are stochastic. Adaptive noise cancellation using hybrid adaptive algorithms is implemented. Compared with conventional algorithms, the hybrid adaptive algorithms exhibit extremely fast convergence. With persistent improvement of the adaptive hybrid algorithm and the rapid development of signal processing chip, it will be more widely used in mobile telecommunication system, and signal processing fields. The simulation perception analysis of hybrid adaptive algorithms is carried out on the convergence behavior, correlation coefficient, and convergence time. After comparing, simulated results were tabulated. By taking into considerations of existing algorithms, performance of hybrid adaptive algorithms gives better convergence time, convergence behavior, correlation coefficients. This method is more systematic in eliminating noise from corrupted signal and has less time to converge, faster response, and reduction in memory.

References

1. Bernard Widrow., John R Glover., John M McCool., John Kaunitz., Charles S Williams., Robert H Hearn., James R Zeidler., Eugene Dong Jr and Robert C Goodlin.: Adaptive Noise Cancelling Principles and Applications. Proceedings of IEEE Spectrum, vol. 63, issue 12, pp. 1692–1716, (1975).
2. Shubhra Dixit and Deepak Nagaria.: Neural Network Implementation of Least Mean Square Adaptive Noise Cancellation. International Conference on Issues and Challenges in Intelligent Computing Techniques, vol. 1, pp. 134–139, (2014).
3. V Ashok., T Balakumaran., A Nirmalkumar, P Ravikumar.: Spatially Resolved Laser Doppler Based Diffused Beam Reflectance Measurement of Blood Glucose Diagnosis in Real Time by Noninvasive Technique. International Conference on Computing, Communication and Network Technologies, vol. 1, pp. 1–8, (2010).
4. Syed A Hadei and M Lotfizad.: A Family of Adaptive Filter Algorithms in Noise Cancellation for Speech Enhancement. International Journal of Computer and Electrical Engineering, vol. 2 (2), pp. 307–315, (2010).
5. V R Vijay Kumar., P T Vanathi and P Kanaga Sapabathi.: Modified Adaptive Filtering Algorithms for Noise Cancellation in Speech Signals. Electronics and Electrical Engineering, Kanus Technologija, vol. 2 (74), pp. 17–20, (2007).
6. Lilatul Ferdouse., Nasrin Akhter., Tamanna Haque Nipa and Fariha Tasmin Jaigirdar.: Simulation and Performance Analysis of Adaptive Filtering Algorithms in Noise Cancellation. International Journal of Computer Science Issues, vol. 8, issue 1, pp. 185–192, (2011).
7. Vartika Anand., Shalini Shah and Sunil Kumar.: Intelligent Adaptive Filtering for Noise Cancellation. International Journal of Advanced Research in Electrical, Electronics & Instrumentation Engineering, vol. 2, issue 5, pp. 2029–2039, (2013).

A Novel Approach to Database Confidentiality in Online Voting System

P. Sanyasi Naidu and Reena Kharat

Abstract In a democratic country, public voting is used to choose the government. The eligible citizens of that country are allowed to vote. In order to increase public voting and reduce the cost involved in an election, we need an online voting system. Public will be motivated to vote if they believe in the privacy and security of the system. This paper addresses issues in identification and authentication in the online voting system. Password and six-digit key are used for authentication. Confidentiality of password and key is preserved using our proposed algorithm. The integrity of data from user and database is verified using hash code. Brute force attack is impossible. The output generated by our algorithm makes cryptanalysis impossible.

Keywords Authentication · Online voting · Steganography
Visual cryptography · Hash

1 Introduction

Country will be truly democratic if 100% population of the country participates in voting. In today's fast life, we want everything at our doorstep. For voting, citizen has to go through voting booth, wait in line till his/her turn comes, and then vote. Many people think this as hectic and time-consuming process, so they don't go for voting. In order to motivate more people to vote, we need to provide online voting

P. Sanyasi Naidu (✉) · R. Kharat
Department of Computer Science and Engineering, GITAM Institute
of Technology, GITAM University, Visakhapatnam, India
e-mail: snpasala@yahoo.com

R. Kharat
e-mail: reenakharat@gmail.com

R. Kharat
Department of Computer Engineering, Pimpri Chinchwad College
of Engineering, Pune, India

© Springer Nature Singapore Pte Ltd. 2018
S.S. Dash et al. (eds.), *International Conference on Intelligent Computing
and Applications*, Advances in Intelligent Systems and Computing 632,
https://doi.org/10.1007/978-981-10-5520-1_9

system through which people can finish their voting within 5–10 min without waiting and travelling. People will be motivated enough to vote online if they believe in confidentiality of the system and secrecy of their vote. Security requirement for a secure e-voting system is given in [1].

Following are security considerations given by NIST in [2] for online voting system.

- Identification and Authentication—a unique identity is provided to the voter. During registration, system will provide credentials to the voter. These credentials are verified in authentication phase in order to establish trust in voter's identity.
- Confidentiality—system must store registration data such that even someone get access to it cannot understand the data. It is very important component of the system. If confidentiality of registration database is broken, then one gets access to credentials used during authentication. Once unauthorized person gets credentials of another valid voter, unauthorized person can vote instead of authentic voter.
- Integrity—there is no tampering happened with registration database is checked using integrity check.

Our system uses password and six-digit key for authentication. Confidentiality of password and six-digit key is provided using SHA-512 algorithm, RSA signature [3], steganography, and visual cryptography. Breaking password and six-digit key are impossible.

2 Related Work

We have studied different authentication methods of the online voting system. Use of login ID and PIN/password is specified in [4, 5]. Confidentiality of PIN/password is not at all addressed. In [6, 7], fingerprint is used for authentication which is stored as it is in voter's smart card. In [8], live fingerprint is used but no discussion on confidentiality. In [9], confidentiality to fingerprint is provided by visual cryptography. The problem here is if a voter has a crack on finger then he will not be authenticated as a legitimate voter. Therefore, multiple biometric factors like fingerprint and face is suggested in [10]. Here, confidentiality is provided to the database using recursive XOR scheme. In [11], authors have given a scheme to provide confidentiality to password. In [9–11], integrity of database and VIC is not checked. Use of multiple biometric features will increase cost of implementation. So, in [12], choice of one biometric along with password is provided for authentication. Integrity to database is also provided through hash code.

For use of online voting system by all common people, we need authentication system without new purchase of biometric device. So we have used two factors: PIN and password. Confidentiality of PIN and password is provided using steganography, cryptography, and visual cryptography.

3 Background

In [13], cryptography and steganography are used together to provide confidentiality to embedded data. In [14], authors used LSB of the cover image to hide data. In [15], cryptography and steganography are used to provide secure authentication in an online voting system. The fingerprint is used as a secret key to store PIN. The live fingerprint of a voter is not taken during authentication. In 1994, Naor and Shamir have introduced visual cryptography [16]. Reconstructed image using visual cryptography has reduced image resolution and contrast [17]. In [18], XOR-based visual cryptography is used to reconstruct lossless image. In (2, 2) scheme, for a white pixel, same shares are selected and, for black pixel, two inverse shares are selected. Recursive XVC introduced in [19] divides shares created in the first step recursively into sub-shares.

4 Proposed Algorithm for Registration and Authentication

The main purpose of authentication is that only eligible voter should be allowed to vote. Each voter is allowed to cast only one vote at a time. For correct authentication, we need to use non-transferable credentials such as biometric features.

4.1 Registration Phase

Eligibility of an individual for voting is checked first before allowing him to go for registration. This is done by checking original identity card issued by the government, original address proof, etc. After verification, an eligible individual is allowed to register. We have used cryptographic hash algorithm SHA-512 to generate hash code.

Following are the steps in registration phase:

1. After verification of identity card issued by government, a person is allowed to register to online voting system.
2. System will ask voter to enter four-digit number which acts as key for embedding algorithm. This 32-bit key is repeated over the bits to generate expanded key which is half the image size.
3. Voting system will ask voter to enter password (PW) which is greater than six characters.
4. Voting system will issue unique identity number (IDN) for each voter which will identify particular voter uniquely from the database.

5. System takes timestamp value (TSV) when voter registers.
6. Password is signed using PWS = E(PR$_{AS}$, (HASH(PW ‖ IDN ‖ TSV) ‖ TSV)). Here RSA encryption algorithm is used with key pair (PR$_{AS}$, PU$_{AS}$) as private and public key of authentication server. SHA-512 is used as HASH algorithm to generate hash code.
7. Encrypted password PWS is stored in cover image using LSB technique. The key (4-digit number) is entered by user in step-2 as starting position for embedding PWS. User's fingerprint will act as cover image.
8. The key (32-bit) is expanded to half the size of image by repeating key over the bits.
9. The stego image is divided into two halves. Left half of image is XORed with expanded key. This output is expanded with right half of image. Now left half and right half are exchanged. This step is repeated eight times to make cryptanalysis harder. Finally, left half and right half are exchanged. It is shown in Fig. 1.
10. Now system will generate two shares of cryptographed stego image using XOR-based visual cryptography. Hash of shares is calculated for integrity check.
11. Share$_1$ and Hash (Share$_2$) are stored in voters registration database. Share$_2$ and Hash (Share$_1$) are sent to voter's mail along with IDN.

4.2   Authentication Phase

On Election Day, a voter will be allowed for authentication if he has a valid Voter Identification Card (VIC). The following are the steps for authentication phase:

1. The voting system will direct voter to enter his/her unique identity number (IDN).
2. This IDN is used to fetch record from database to retrieve Share$_1$ and Hash (Share$_2$) corresponding to that voter.
3. Voter is directed to upload Share$_2$ and Hash (Share$_1$).
4. Share$_1$ and Share$_2$ are XORed to get cryptographed stego image.
5. Now system will direct voter to enter four-digit number which is acting as key. This 32-bit key is repeated over the bits to generate expanded key which is half the image size.
6. The cryptographed stego image is divided into two halves. Left half of image is XORed with expanded key. This output is expanded with right half of image. Now left half and right half are exchanged. This step is repeated eight times to make cryptanalysis harder. Finally, left half and right half are exchanged. These steps take out the effect of cryptography on image and give us a stego image. It is shown in Fig. 2.

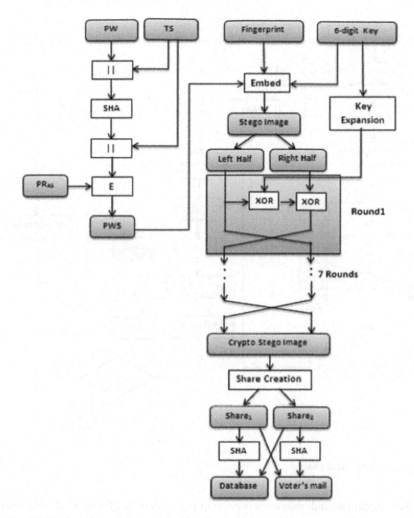

Fig. 1 Registration phase

7. Encrypted password PWS is retrieved from stego image by reading LSB bits with key as start position.
8. Now system will direct voter to enter a password. Let's call it as entered_PW.
9. The PWS is decrypted using $D(PU_{AS}, PWS)$. This will give HASH(PW || IDN || TSV) and TSV as timestamp value. Let received_PW_HashCode = HASH (PW || IDN || TSV).
10. System will generate hash using generated_PW_HashCode = HASH (entered_PW || IDN || TSV).

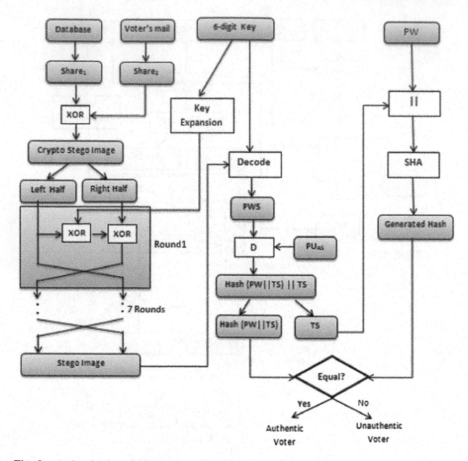

Fig. 2 Authentication phase

11. For signature and password verification, system will check if generated_ PW_HashCode is equal to received_PW_HashCode. If it is equal, then voter is authentic voter. If it is not equal, then either the person is unauthentic voter or there is coercion attack.

5 Security Discussion

Hash algorithm is one way. It is used to get hash code of (PW‖TS). From output hash code, we cannot go to (PW‖TS). Someone may gain access to database and change the database in order to get through authentication. In order to avoid this, hash code is signed by private key of authentication server. So, to change database one needs private key of authentication server as well. Fingerprint is used as cover image as it is unique biometric feature of voter. The six-digit key is used to embed

PWS into cover image. The six-digit key is not at all stored in any form, but it is used such that if the key is correct then only PWS will be correctly retrieved. The key expansion and Fiestel structure are used in order to break the relation between the bits. Due to this, cryptanalysis becomes complex. Brute force attack is also impossible. For brute force, one needs all $2(256 \times 256) = 265,536$ combinations of share. Suppose system takes maximum 12 characters for PW and 32 bits for TS. Then, attacker also needs $(212 + 32) = 4096$ combinations for PW and 26 for key. In total, brute force requires $265,536 \times 4096 \times 26$ combinations. It is infeasible for single vote as well.

In order to avoid voters being coerced into casting their ballot differently, we allow voter to vote multiple times. Online voting system will maintain two types of tokens as valid and invalid tokens. For voter who passes authentication step, a token from valid token database is given with ballot. For voter who does not pass authentication step, a token from invalid token database is given with ballot. When a voter is forced to give vote in front of party, a voter can enter invalid key and password and can submit vote. As a person has entered invalid key and password, system will generate invalid token and it will be linked to ballot submitted. All submitted ballot will be classified into valid vote and invalid vote as per the token attached with it. The party will not gain any knowledge if the vote is valid or not though person vote in front of the party. So the person can give his valid vote when no one is around. Once valid vote is casted, the status in the database is changed to TRUE. Once valid vote is casted, system generates invalid tokens for that person though he gives correct credentials in authentication phase.

Allowing multiple votes by voting system makes system prone to availability. So threshold is set per system. Number of votes per system is also restricted to some limit say not more than 100. So, only one valid vote per voter is guaranteed. We block the client system for 10 min after every vote. So to cast 100 votes completely more than 16 h are required. For coerced voting, party people have to wait for whole Election Day which is impossible.

6 Correctness of Algorithm

In authentication module, XOR-based visual cryptography is used. As XOR is invertible, we get reconstructed image and crypto stego image as same. XOR operation is also used in Fiestel structure, giving us back the same stego image as that used in the registration phase. The six-digit key is used to decode stego image and get PWS. The PWS is decrypted using public key of authentication server to get hash value which is same as that generated at the time of registration.

7 Conclusion

The proposed system is cost-effective. Credentials used in authentication are password, six-digit key, and share. The six-digit key is not stored. Password is stored using our novel approach which uses key, fiestel structure, and visual cryptography. Due to this, cryptanalysis and brute force are infeasible. Signature verification helps us to know if the password stored in stego image is authentic or not. The coerced attack and availability problem are well addressed in our paper.

References

1. Gritzalis, D. A.: Principles and requirements for a secure e-voting system. Computers & Security, Elsevier, Vol. 21, No. 6, (2002): pp. 539–556. doi:10.1016/S0167-4048(02)01014-3.
2. Hastings, N., Peralta, R., Popoveniuc, S., Regenscheid, A.: Security Considerations for Remote Electronic UOCAVA Voting. National Institute of Standards and Technology, Feb 2011. doi:10.1.1.204.5483.
3. Rivest, R., A. Shamir, L. Adleman. "A Method for Obtaining Digital Signatures and Public-Key Cryptosystems", Communications of the ACM, Volume-21, Issue-2, Feb-1978, pp. 120–126. doi:10.1145/359340.359342.
4. Abd-alrazzq, H.K., Ibrahim, M.S., Dawood, O.A.: Secure internet voting system based on public key kerberos. IJCSI Int. J. Comput. Sci. Issues 9(2), 428–435 (2012). No 3.
5. Al-Anie, H.K., Alia, M.A., Hnaif, A.A.: e-Voting protocol based on public key cryptography. Int. J. Netw. Secur. Appl. (IJNSA) 3(4), 87–98 (2011). doi:10.5121/ijnsa.2011.3408.
6. Sridharan, S.: Implementation of authenticated and secure online voting system. In: 4th ICCCNT 2013. IEEE, 4–6 July 2013. doi:10.1109/ICCCNT.2013.6726801.
7. Vermani, S., Sardana, N.: Innovative way of internet voting: secure on-line vote (SOLV). IJCSI Int. J. Comput. Sci. Issues 9(6), 73–78 (2012). No 3.
8. Khasawneh, M., Malkawi, M., Al-Jarrah, O., Hayajneh, T.S., Ebaid, M.S.: A biometric-secure e-Voting system for election processes. In: Proceeding of the 5th International Symposium on Mechatronics and its Applications (ISMA08), Amman, Jordan, 27–29 May 2008. doi:10.1109/ISMA.2008.4648818.
9. Sanyasi Naidu, P., Kharat, R., Tekade, R., Mendhe, P., Magade, V.: E-Voting System Using Visual Cryptography & Secure Multi-party Computation. 2nd ICCUBEA 2016, IEEE. doi:10.1109/ICCUBEA.2016.7860062.
10. Sanyasi Naidu, P., Kharat, R.: Multi-factor Authentication using Recursive XOR-based Visual Cryptography in Online Voting System. Security in Computing and Communications: 4th International Symposium, SSCC 2016, Springer, pp. 52–62. doi:10.1007/978-981-10-2738-3_5.
11. Sanyasi Naidu, P., Kharat, R.: Secure Authentication in Online Voting System Using Multiple Image Secret Sharing. Security in Computing and Communications: 4th International Symposium, SSCC 2016, Springer, pp. 336–343. doi:10.1007/978-981-10-2738-3_29.
12. Sanyasi Naidu, P., Kharat, R.: Secure Authentication and Tamper Detection in Remote Voting System. International Journal of Control Theory and Applications, 2016, pp. 199–206.
13. Abdulzahra, H., Ahmad, R., Noor, N.M.: Combining cryptography and steganography for data hiding in images. In: Applied Computational Science, pp. 128–135 (2014). ISBN: 978-960-474- 368-1.

14. Wayner, P.: Disappearing Cryptography. Boston: AP Professional Books, 3rd Edition, ISBN: 9780080922706, 2008.
15. Katiyar, S., Meka, K.R., Barbhuiya, F.A., Nandi, S.: Online voting system powered by biometric security using steganography. In: 2011 Second International Conference on Emerging Applications of Information Technology. IEEE (2011). doi:10.1109/EAIT.2011.70.
16. Naor, M., Shamir, A.: Visual cryptography. In: De Santis, A. (ed.) EUROCRYPT 1994. LNCS, vol. 950, pp. 1–12. Springer, Heidelberg (1995). doi:10.1007/BFb0053419.
17. Tuyls, P., Kevenaar, T., Schrijen, G., Staring, T., Dijk, M. V.: Visual Crypto Displays Enabling Secure Communications. Security in Pervasive Computing, Volume 2802 of the series Lecture Notes in Computer Science pp. 271–284. doi:10.1007/978-3-540-39881-3_23.
18. Tuyls, P., Hollmann, H. D. L., Lint, J. H. V., Tolhuizen, L.: A polarisation based Visual Crypto System and its Secret Sharing Schemes. Available at the IACR Cryptology ePrint Archive, http://eprint.iacr.org/2002/194/. doi:10.1.1.12.812.
19. Thomas, M., Anto, P. B.: Recursive Visual Cryptography Using Random Basis Column Pixel Expansion. ICIT 2007, IEEE, pp. 41–43. doi:10.1109/ICIT.2007.32.

Attending Prominent Face in the Set of Multiple Faces Through Relative Visual Saliency

Ravi Kant Kumar, Jogendra Garain, Dakshina Ranjan Kisku and Goutam Sanyal

Abstract Visual saliency determines the extent of attentiveness of a region in a scene. In the context of attending faces in the crowd, face components and its dominance features decide the focus on attention. Attention boosts up the recognition and identification process in a crowd and hence plays an excelling role in the area of visual surveillance and robotic vision. Using different computer vision-based techniques, enormous researches have been carried out on attention, recognition, and identification of the human face in context of different applications. This paper proposes a novel technique to analyze and explore the prominent face in the set of multiple faces (crowd). The proposed method stretched out the solution, using the concept of relative visual saliency, which has been evaluated on the various parameters of face as a whole and its componentwise too. These parameters are face area, spatial location, intensity, hue, RGB values, etc. The proposed work furnishes satisfactory results. The assessment made with this approach shows quite encouraging results which may lead to a future model for robotic vision and intelligent decision-making system.

Keywords Face attention · Hue · Intensity · RGB values · Spatial distance
Visual saliency

R.K. Kumar (✉) · J. Garain · D.R. Kisku · G. Sanyal
Department of Computer Science and Engineering,
National Institute of Technology Durgapur, Durgapur, India
e-mail: vit.ravikant@gmail.com

J. Garain
e-mail: jogs.cse@gmail.com

D.R. Kisku
e-mail: drkisku@gmail.com

G. Sanyal
e-mail: nitgsanyal@gmail.com

© Springer Nature Singapore Pte Ltd. 2018 89
S.S. Dash et al. (eds.), *International Conference on Intelligent Computing
and Applications*, Advances in Intelligent Systems and Computing 632,
https://doi.org/10.1007/978-981-10-5520-1_10

1 Introduction

The human vision system does not visit all the faces in the crowd with the same focus. The faces having the dominating features automatically drag our attention [1]. These dominating features may occur due to overlook color, texture, geometry, intensity, etc. In other terms, these dominating features of an object (here face) make it standalone with respect to surrounding objects in a scene is called visual saliency [2]. In context of attending a crowd (set of multiple faces), some faces may have salient because of perceiving faces that can be in the different expression and mood [3]. In this paper, our aim is to locate the most prominent or salient face in the set of multiple faces. A face may look attentive because of dominating features of its components. Therefore, cumulative saliency values of all the belonging faces in the set of multiple faces have been calculated. For this, saliency values have been determined in two steps: first, based on low-level feature and size differences of the corresponding components of faces (i.e., locally), next, on the basis of average features differences and the respective whole face area (i.e., globally). Attention of a face is also affected by its proximate faces [4]. Therefore, it is exponentially modulated with the spatial distances of the corresponding faces. Here, *RGB* (Red, Blue, and Green color components) values and hue have been considered as the vital features. In the aspects of finding the attentiveness of faces, researchers and scientists introduced the visual saliency concepts in the area of computer vision system. But, in this area researches have not been accomplished much work. Some important literatures are: Saliency in crowd [5], To predict where people look in natural scenes [6, 7], Anomaly detection in crowded scenes [8], Context Aware Saliency [9, 10], Attention capture by faces [11], Enlighten the effect of neighbor faces in the crowd [12], estimating normalized attention of faces [13].

Rest of this paper is organized as follows: In Sect. 2, proposed mathematical formulation for computing saliency score of the faces in the crowd (a set of multiple faces) has been discussed. The overall proposed procedure (Algorithm) and step-wise intermediate outcome is depicted in Sect. 3. Experimental validation, description of database, and result comparison of the proposed technique have been presented in Sect. 4. Finally, in Sect. 5, follow up the concluding remark and future work.

2 Proposed Mathematical Formulation

The mathematical formulation of our proposed work for calculating saliency has been inspired by some models [14, 15]. Finally, an improved mathematical formulation for obtaining the salient faces has been framed based on face component features, average face features, face area, and their spatial location. Here, intensity, hue, and *RGB* values have been considered as the features.

In the general terminology, attention toward any object or location of a visual scene is determined by its feature difference (contrast) with the surrounding locations or nearby objects. For the two faces say 'i' and 'j,' saliency formula has been established based on their important face components (left eye, right eye, nose area, and mouth area) features, as well as whole face features, their sizes, and corresponding Cartesian or spatial distances. If 'k' is the subsequent face components of faces 'i' and 'j', feature difference (FD$_c$) between the kth component (C_k) of face 'i' and 'j' is:

$$\mathrm{FD}_c = |f_i C_k - f_j C_k|. \tag{1}$$

Saliency (Sal$_{ij}$) of face 'i' with 'j' due to their face is:

$$\mathrm{Sal}_{ij} = \left(|f_i C_{k_i} - f_j C_{k_j}|\right) e^{-D^2_{C_{ij}}/2\sigma^2}. \tag{2}$$

The saliency of face 'i' with respect to all other faces ('j') based on their face components is obtained as:

$$S_{ij} = \sum_j \sum_k \left(|f_i C_k - f_j C_k|\right) e^{-D^2_{C_{ij}}/2\sigma^2}. \tag{3}$$

The cumulative saliency of face 'i' due to all the other faces 'j' is obtained as:

$$S_{ij} = \sum_j \sum_k \left[\left(|f_i C_k - f_j C_k|\right) + (S_i - S_j)\right] e^{-D^2_{C_{ij}}/2\sigma^2} \tag{4}$$

where (f_i, f_j) indicates the feature values and (S_i, S_j) are the face area of 'i' and 'j'. $D_{C_{ij}}$ denotes their subsequent Cartesian distance in the crowd. 'σ' denotes the standard deviation.

In a special case, if feature difference values as well as the size of faces become zero, Eq. (4) grants the overall saliency values as zero by ignoring the contributions of saliency values due to Cartesian distances among the faces. Therefore, the generalized formulation has been done as:

$$S_{ij} = \sum_j \sum_k \left[\left(|f_i C_k - f_j C_k|\right) + (S_i - S_j) + 1/\sigma\sqrt{2\pi}\right] e^{-D^2_{C_{ij}}/2\sigma^2}. \tag{5}$$

3 Proposed Algorithm

The proposed approach to get saliency values of all the faces in the input image and to determine the most salient face involves the following steps:

Algorithm 1: Calculating Saliency Value

Input: Crowd image having faces.

Output: Salient Face based on saliency values of faces.
1. Detect all the faces of the image using Voila Jones Algorithm [16] and Locate the Centre position of every face in the image and store in the Centre Vector $(C_i) = [C_1, C_2,....,C_n]$
2. From every face 'i' of the input image do
Calculate the Euclidean Distance between face i and the all other faces j and keep the consequent values in a Spatial Distance Vector $(D_{Sij}) = [D_{S1}, D_{S2},....,D_{Sn}]$
3. From the input image, extract all the faces and do
   3.1 Calculate area of every faces in the area vector Area Vector $(A_i) = [A_1, A_2,....,A_n]$
   3.2 Measure the Hue, Intensity and overall R, G, B values of every faces and store these values in the respective vectors as shown below:
   Hue, $(H_i) = [H_1, H_2,....,H_n]$; Red component $(R_i) = [R_1, R_2,....,R_n]$; Green component $(G_i) = [G_1, G_2,....,G_n]$; Blue component $(B_i) = [B_1, B_2,....,B_n]$ and Intensity $(I_i) = [I_1, I_2,....,I_n]$.
4. Detect the important face components (Left Eye, Right Eye, Nose area and Mouth area from each of the faces using [16] and calculate Area, Hue, Intensity and overall R, G, B values of corresponding face components (C_j): Area $(A_j) = [A_1, A_2,....,A_n]$ Hue, $(H_j) = [H_1, H_2,....,H_n]$; Red component $(R_j) = [R_1, R_2,....,R_n]$; Green component $(G_j) = [G_1, G_2,....,G_n]$; Blue component $(B_j) = [B_1, B_2,....,B_n]$ and Intensity $(I_j) = [I_1, I_2,....,I_n]$.
5. For each face in the input image
       If equal face size and same feature values
   Apply Eq. (4) to determine Saliency (S_{ij}) by using Centre, Area, Hue, Intensity and RGB Vectors described in Steps 2 and 3.
Else
   Apply Eq. (5) to determine Saliency (S_{ij}) by using Centre, Area, Hue, Intensity and RGB Vectors described in Steps 2 and 4.
6. Find the highest Relative saliency score after normalizing the saliency values with maximum normalized value as 1.
7. Find the most attentive face having the highest Saliency score (i.e. 1).

Dataset of 'set of multiple faces' has been created by assembling the individual faces taken from FEI database [17] and World Wide Web (WWW). Detailed description of database is explained in the Sect. 4.

Viola Jones face detection [16] has been applied in the given input crowed image for detecting the faces as well as its important components (left eye, right eye, nose, and mouth). Saliency of faces can be distracted due to non-face part of the images. At this beginning level of the experiment, background distraction has not been considered. Therefore, detected faces have been preprocessed (cropped and rearranged) with no background (Fig. 1).

Next, center coordinates of each of the faces of the input image have been calculated, which is described in Table 1 and shown in Fig. 2.

After getting center coordinates, spatial distance (Cartesian distance) among the faces has been calculated. Detected faces are extracted, and area of each faces is obtained. Average values of intensity, hue, and *RGB* components are calculated using image analysis tool [18], for each of the faces without much focusing on finer details of its components (WFC). The average/normalized feature values of all the faces and their components have been obtained using [18], and depicted in Table 2.

Fig. 1 *First row left* input image, *first row right* face detection using [16], *second row left* face components detection using [16], *second row right* corresponding face numbering

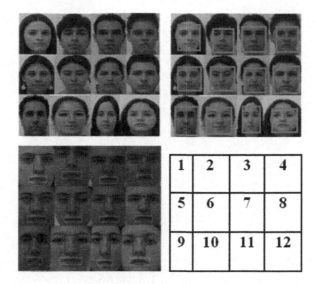

Table 1 Center coordinate of each face

(54,244)	(135,253)	(227,248)	(320,251)
(48,154)	(142,152)	(231,153)	(319,154)
(43,50)	(136,53)	(220,51)	(307,56)

Fig. 2 Center of each face in the input image

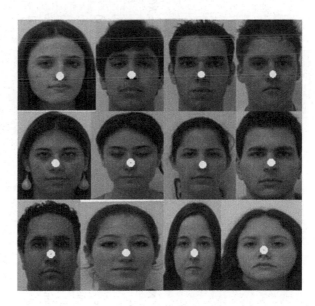

Table 2 Normalized area, average hue, average intensity, and average *RGB* values of each face in the input crowd

Face No.	1	2	3	4	5	6	7	8	9	10	11	12
Area	0.78	0.73	0.76	0.76	0.79	0.70	0.74	0.72	0.79	0.94	0.76	1.00
Avg hue	0.81	0.83	0.75	0.78	0.83	0.76	0.75	0.77	0.80	1	0.78	0.97
Avg intensity	22.0	15.1	13.2	12.6	21.0	19.9	16.4	23.9	14.6	21.1	22.8	22.2
Avg red	0.42	0.33	0.30	0.31	0.33	0.34	0.34	0.35	0.27	0.38	0.39	0.45
Avg green	0.54	0.45	0.39	0.43	0.46	0.45	0.46	0.45	0.36	0.49	0.51	0.57
Avg blue	0.40	0.29	0.27	0.27	0.31	0.31	0.30	0.33	0.25	0.36	0.38	0.43

Table 3 Normalized area, average hue, average intensity, and average *RGB* values of left eye of each face in the input image

Face No.	1	2	3	4	5	6	7	8	9	10	11	12
Area	1.00	0.63	0.69	0.69	0.97	0.47	0.54	0.72	0.57	0.96	0.82	0.87
Avg hue	22.0	12.3	11.6	11.0	16.7	15.9	13.7	22.5	10.5	23.7	24.6	21.5
Avg intensity	0.24	0.17	0.13	0.16	0.20	0.16	0.16	0.18	0.13	0.18	0.24	0.26
Avg red	0.34	0.23	0.17	0.23	0.27	0.22	0.22	0.24	0.16	0.23	0.31	0.33
Avg green	0.23	0.15	0.12	0.13	0.18	0.15	0.14	0.17	0.12	0.17	0.23	0.23
Avg blue	0.16	0.13	0.11	0.11	0.14	0.12	0.11	0.13	0.11	0.13	0.17	0.17

Table 4 Normalized area, average hue, average intensity, and average *RGB* values of right eye of each face in the input crowd

Face No.	1	2	3	4	5	6	7	8	9	10	11	12
Area	1.00	0.69	0.68	0.82	0.88	0.49	0.51	0.63	0.56	0.94	0.82	0.95
Avg hue	22.0	9.9	11.4	10.9	16.5	14.2	12.2	20.6	10.6	25.0	22.8	22.0
Avg intensity	0.19	0.14	0.11	0.15	0.18	0.13	0.13	0.13	0.10	0.15	0.20	0.24
Avg red	0.27	0.19	0.14	0.21	0.24	0.18	0.18	0.18	0.11	0.20	0.27	0.32
Avg green	0.18	0.12	0.10	0.13	0.16	0.12	0.11	0.13	0.09	0.15	0.19	0.23
Avg blue	0.13	0.11	0.09	0.11	0.13	0.10	0.09	0.10	0.09	0.11	0.15	0.17

Table 5 Normalized area, average hue, average intensity, and average *RGB* values of nose of each face in the input crowd

Face No.	1	2	3	4	5	6	7	8	9	10	11	12
Area	0.76	0.70	0.63	0.72	0.84	0.69	0.62	0.59	0.73	1.00	0.79	0.81
Avg hue	23.4	17.9	12.4	12.2	22.3	21.3	18.8	23.6	16.3	23.0	25.5	25.6
Avg intensity	0.44	0.33	0.29	0.31	0.35	0.33	0.36	0.35	0.28	0.39	0.42	0.45
Avg red	0.57	0.46	0.41	0.45	0.49	0.47	0.49	0.47	0.38	0.51	0.54	0.59
Avg green	0.43	0.30	0.26	0.26	0.33	0.31	0.33	0.33	0.25	0.38	0.41	0.44
Avg blue	0.33	0.23	0.21	0.21	0.24	0.22	0.25	0.24	0.20	0.29	0.30	0.33

Table 6 Normalized area, average hue, average intensity, and average *RGB* values of mouth of each face in the input image

Face No.	1	2	3	4	5	6	7	8	9	10	11	12
Area	0.76	0.70	0.63	0.72	0.84	0.69	0.62	0.59	0.73	1.00	0.79	0.81
Avg hue	23.4	17.9	12.4	12.2	22.3	21.3	18.8	23.6	16.3	23.0	25.5	25.6
Avg intensity	0.44	0.33	0.29	0.31	0.35	0.33	0.36	0.35	0.28	0.39	0.42	0.45
Avg red	0.57	0.46	0.41	0.45	0.49	0.47	0.49	0.47	0.38	0.51	0.54	0.59
Avg green	0.43	0.30	0.26	0.26	0.33	0.31	0.33	0.33	0.25	0.38	0.41	0.44
Avg blue	0.33	0.23	0.21	0.21	0.24	0.22	0.25	0.24	0.20	0.29	0.30	0.33

Similarly, normalized area, average intensity, average hue, and average color components (normalized R, G, B values) have been computed for important components (left and right eyes, nose, mouth, etc.) of each of the faces of the input image in Tables 3, 4, 5, and 6, respectively.

After obtaining all the feature values, saliency score of the faces without spotlighting the finer details of its components (FWC) has been computed based on intensity, Red, Green, and Blue components using Step 5 of the proposed algorithm (Sect. 3). Saliency scores of important face components (left eye, right eye, nose, and mouth) are also computed due to various low-level features like intensity, Red, Blue, and the Green component using Step 5 of the proposed algorithm (Sect. 3).

Table 7 Obtained saliency score faces (without finer details of face called FWC) and their components (finer details of face) based on different low-level features

Face No.	1	2	3	4	5	6	7	8	9	10	11	12
FWC (intensity)	3.56	3.96	6.03	6.69	2.47	1.27	2.54	5.63	4.50	2.58	4.43	3.78
FWC (red)	0.75	0.02	0.05	0.04	0.02	0.01	0.01	0.01	0.08	0.03	0.04	0.10
FWC (green)	0.08	0.01	0.08	0.03	0.01	0.01	0.01	0.01	0.11	0.02	0.05	0.11
FWC (blue)	0.08	0.03	0.06	0.06	0.01	0.01	0.02	0.01	0.08	0.03	0.06	0.11
LEYE (intensity)	0.26	0.04	0.20	0.11	0.05	0.08	0.12	0.01	0.24	0.01	0.23	0.27
LEYE (red)	0.10	0.01	0.08	0.01	0.02	0.02	0.02	0.01	0.09	0.01	0.07	0.09
LEYE (green)	0.06	0.02	0.05	0.04	0.01	0.02	0.03	0.01	0.05	0.01	0.06	0.06
LEYE (blue)	0.03	0.01	0.02	0.02	0.01	0.01	0.02	0.01	0.02	0.01	0.04	0.04
REYE (intensity)	0.31	0.01	0.15	0.06	0.10	0.03	0.07	0.04	0.19	0.03	0.28	0.32
REYE (red)	0.08	0.01	0.06	0.01	0.01	0.01	0.03	0.04	0.09	0.04	0.01	0.12
REYE (green)	0.09	0.02	0.04	0.05	0.02	0.01	0.05	0.01	0.05	0.05	0.01	0.11
REYE (blue)	0.06	0.01	0.02	0.03	0.04	0.01	0.05	0.03	0.03	0.04	0.01	0.08
NOSE (intensity)	0.08	0.03	0.07	0.05	0.01	0.03	0.01	0.01	0.07	0.03	0.06	0.10
NOSE (red)	0.09	0.02	0.08	0.03	0.01	0.01	0.01	0.01	0.11	0.02	0.05	0.10
NOSE (green)	0.10	0.03	0.08	0.08	0.01	0.02	0.01	0.01	0.09	0.04	0.08	0.11
NOSE (blue)	0.08	0.02	0.04	0.04	0.01	0.03	0.01	0.01	0.05	0.03	0.05	0.08
MOUTH (intensity)	0.08	0.01	0.04	0.02	0.03	0.01	0.05	0.02	0.09	0.04	0.02	0.08
MOUTH (red)	0.08	0.01	0.06	0.01	0.01	0.01	0.03	0.04	0.09	0.04	0.01	0.12
MOUTH (green)	0.09	0.02	0.04	0.05	0.02	0.01	0.05	0.01	0.05	0.05	0.01	0.11
MOUTH (blue)	0.06	0.01	0.02	0.03	0.04	0.01	0.05	0.03	0.03	0.04	0.01	0.08
FWC (intensity)	3.56	3.96	6.03	6.69	2.47	1.27	2.54	5.63	4.50	2.58	4.43	3.78
FWC (red)	0.75	0.02	0.05	0.04	0.02	0.01	0.01	0.01	0.08	0.03	0.04	0.10
FWC (green)	0.08	0.01	0.08	0.03	0.01	0.01	0.01	0.01	0.11	0.02	0.05	0.11
FWC (blue)	0.08	0.03	0.06	0.06	0.01	0.01	0.02	0.01	0.08	0.03	0.06	0.11

Table 8 Faces covering the most salient components based on intensity, *RGB* values, and hue properties using proposed algorithm

Face No.	1	2	3	4	5	6	7	8	9	10	11	12
Intensity saliency				FWC					MOUTH			LEYE REYE NOSE
Red saliency	LEYE								NOSE			FWC REYE MOUTH
Green saliency	LEYE										LEYE	FWC LEYE REYE NOSE MOUTH
Blue saliency	NOSE										LEYE	FWC LEYE REYE NOSE MOUTH
Hue saliency								REYE MOUTH		FWC	LEYE	NOSE

The saliency score of the faces and its components based on low-level features (intensity, Red, Blue, and the Green) has been depicted in Tables 7 and 8.

Cumulative saliency score of every face is calculated based on saliency values of all the components, estimated by low-level features, area, and respective spatial proximities. The overall saliency score (cumulative saliency of faces) has been depicted in Table 9. The saliency values have been normalized with respect to the maximum value.

Based on saliency score, we have obtained attention ranking of faces to find out most attending face, next attending face, and so on (Table 9). In the case where there is minor feature dissimilarity among the faces, it is very difficult to attending the most salient face through human visual system. But, by using this proposed technique, computer vision system can be trained for handling such situations. The most salient face has been shown in Fig. 3.

Table 9 Overall obtained normalized saliency score and corresponding attention ranking using proposed algorithm

Face No.	1	2	3	4	5	6	7	8	9	10	11	12
Saliency	0.75	0.54	0.40	0.36	0.45	0.29	0.35	0.37	0.55	0.80	0.63	1.00
Attention ranking	3	6	8	10	7	12	11	9	5	2	4	1

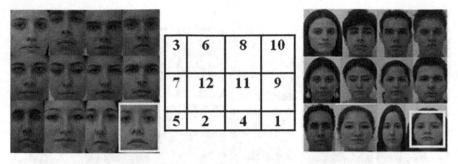

Fig. 3 *Left* most salient face (*inside yellow box*), *middle* attention ranking of corresponding faces, *right* most salient face in the input image (*inside yellow box*)

4 Experiment Validation and Result Comparison

Our experimental results have been well verified with the existing saliency map, generated on the basis of low-level features, i.e., intensity, RGB values, etc. In this proposed approach, saliency of faces has been computed based on low-level features. For calculating saliency score, face area and feature differences have been modulated with the positional proximities of the faces. Hence, it is clear that the proposed technique also incorporates the concept of context (saliency dependency due to the nearby faces). Therefore, our result has been compared with the state-of-the-art technique named 'context aware saliency' [9].

4.1 Database Description

This algorithm has been applied on 55 sets of face images having 4–25 faces in every set. Face image set has been prepared from collecting faces from World Wide Web (WWW) and FEI dataset. Human's focus goes toward a face due to various low-level and high-level features along with different facial emotions and expressions. In this experiment, parameters to measure saliency due to variation of facial expression have not been taken. Therefore, database has been prepared in such a way that all the faces may have with the same mood and expressions. Here, the objective is to validate the experiment based on intensity, size, and spatial distance variations among faces. At this stage, impact on saliency due to face expressions and background distraction has also been ignored. So, database has been prepared with same face expression, without any background but it may vary in terms of low-level features like intensity (for gray scale images), RGB (for color images), face area, and the spatial locations. Some sample database of input image in gray scale and in color has been shown in Fig. 4a, b respectively. These images are framed by cropping and reassembling the faces of the crowd detected by Viola Jones face detection algorithm [16].

Some sample images of the framed database are depicted Fig. 4.

(a)   (b)

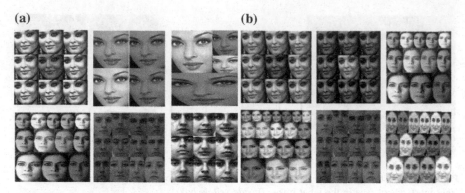

Fig. 4 **a** Set of multiple faces in gray scale, **b** set of multiple faces in color

4.2   Comparison and Analysis

For comparing the saliency of faces in the crowd image, no such benchmark ground is available. Because attention toward faces in a scene depends on enormous factors like low-level features, high-level features, (choices, likeness, mood, and behavior) of attendee, supervised and unsupervised attendee. Also, choices, likeness, mood, and behavior of same attendee may vary from time to time. Therefore, preparing ground truth for finding salient faces in the crowd is itself a big challenge till date. However, some good techniques are available for dealing with the object saliency in a scene. Our result has been compared with the existing state-of-the-art technique for object saliency, where we have treated every face as an object. Context aware saliency [9] has been focused on four basic principal. It works on low-level features (color and contrast), global feature consideration and salient object location, and visual visualization rule which states that visual forms may possess one or several centers of gravity about which the form is organized. Proposed method is also influencing by low-level features. In addition to this, in visualization rule, face area and hue property and positional proximity have also been considered. Saliency map for all the input images is generating by using [9]. The most salient regions are clearly visible in the saliency map. Now, most salient face (face having the highest saliency score) found by our proposed method is compared with technique [9].

Experiment has been conducted on gray scale and color input images. For gray scale, 'intensity' is considered as the prominent feature whereas for color images other low-level features like *RGB* values and hue properties are furthermore taken. Face area and its proximity among the faces (spatial distances) have also been considered for computing saliency of faces and its components. In Fig. 5a, b row wise: Input image (left), saliency map using method [9] (first middle), most salient face (inside yellow boundary) using our proposed method (second middle) and face numbering table (right) are shown in Fig. 5.

(a)   (b)

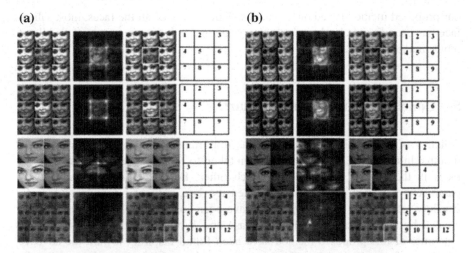

Fig. 5 **a** Input image (*left, row wise*), saliency map using method [9] (*first middle*), most salient face (inside *yellow* boundary) using our proposed method (*second middle*), and face numbering table (*right*), **b** input image (*left, row wise*), saliency map using method [9] (*first middle*), most salient face (*inside yellow boundary*) using our proposed method (*second middle*) and face numbering table (*right*)

In Fig. 5a, b one can observe that in the most of the cases, the maximum salient face using our proposed method is matching with the state-of-the-art method for object saliency [9].

4.2.1 Observation 1

Saliency map of input image 1, 2, and 3 in color samples (Fig. 5b: first middle) is clearer than respective gray scale sample (Fig. 5a: first middle) for visualizing the salient locations. The most salient locations are found at the same face number 3, 3, and 5, respectively. The proposed approach also locates the most salient faces at the same number 3,3, and 5 for both corresponding gray scale and color image samples. For the input image 4 (Fig. 5a, b), different salient locations have been found in the saliency map, whereas our proposed method found the most salient face at the same place (face number 12) for both gray scale and color images.

4.2.2 Observation 2

In the situation where corresponding features of the faces are not differing significantly (Input Image 4 between Fig. 5a, b), it is difficult to find out the most salient face in the saliency map. In the saliency map (first middle of input image 4 of Fig. 5a, b) salient location does not give much attention toward any face. But using our proposed method based on obtained saliency score of all the faces, most salient face has been found clearly in both gray scale and color images (second middle of input image 4 of Fig. 5a, b).

5 Conclusion and Future Remark

Saliency of a face highly depends on its important components like eyes, nose, mouths. In this paper, an attempt has been taken to estimate the relative saliency of faces in the crowd, with respect to each other. For calculating saliency, many parameters like area, spatial location, intensity, hue, and color components (R, G, B values) of face and their components have been considered. The dataset has been created by taking faces from FEI database and World Wide Web (WWW). The outcomes of this proposed method have been found inspiring toward the advance modeling of computer vision. Therefore, the work can be further enhancing by the inclusion of other high-level features of faces. This approach can give a novel dimension of revelation to the computer vision system.

References

1. Itti L., and Koch C.: Computational modelling of visual attention.: Nature Reviews Neuroscience. vol. 2, no. 3, pp. 194–203, (2001).
2. Itti L., and Koch C.: A Saliency-based search mechanism for overt and covert shifts of visual attention. Vision Research, 40, pp. 1489–1506, (2001).
3. Niedenthal P. M., Halberstadt J. B., Margolin J. and Innes-Ker A. H.: Emotional state and the detection of change in facial expression of emotion. European Journal of Social Psychology, vol. 30, no. 2, pp. 211–22, (2000).
4. Koch C. and Ullman S.: Shifts in selective visual attention: towards the underlying neural circuitry. In Matters of intelligence, Springer Netherlands, pp. 115–141, (1987).
5. Jiang M., Xu J. and Zhao Q.: Saliency in crowd, European Conference on Computer Vision, Springer International Publishing, pp. 17–32, September (2014).
6. Jiang M., Xu J. and Zhao Q.: Where Do People Look at in Crowded Natural Scenes? Journal of Vision, vol. 14, no. 10, pp. 1052–1052, (2014).
7. Das, A., Agrawal H., Zitnick C. L., Parikh D., and Batra D.: Human Attention in Visual Question Answering: Do Humans and Deep Networks Look at the Same Regions? arXiv preprint arXiv:1606.03556, (2016).
8. Li V., Mahadevan W., Bhalodia V. and Vasconcelos N.: Anomaly detection in crowded scenes. In CVPR, pp. 249–250, June (2010).
9. Goferman S., Zelnik-Manor L., and Tal A.: Context-aware saliency detection. IEEE Transactions on Pattern Analysis and Machine Intelligence, vol. 34, no. 10, pp. 1915–1926, (2012).
10. Jiang M., Huang S., Duan J., and Q. Zhao.: SALICON: Saliency in context. In 2015 IEEE Conference on Computer Vision and Pattern Recognition (CVPR), pp. 1072–1080, June (2015).
11. Langton S. R., Law A. S., Burton A. M. and Schweinberger S. R.: Attention capture by faces, Cognition, vol. 107, no. 1, pp. 330–342, (2008).

12. Kumar R. K., Garain J., Sanyal G., and Kisku D. R.: A novel approach to enlighten the effect of neighbor faces during attending a face in the crowd. In: Proceedings of TENCON, IEEE, pp. 1–4, (2015).
13. Kumar R. K., Garain J., Sanyal G., and Kisku D. R.: Estimating Normalized Attention of Viewers on Account of Relative Visual Saliency of Faces (NRVS). International Journal of Software Engineering and Its Applications, vol. 9, no. 7, pp. 85–92, (2015).
14. Harel J., Koch C., and Perona P.: Graph-based Visual Saliency. In: Proceedings of Annual conference on Neural Information Processing Systems (NIPS), pp. 545–552, (2006).
15. Pal R., Mukherjee., Mitra A. P., and Mukherjee J.: Modelling visual saliency using degree centrality. Computer Vision, IET CV, vol. 4, no. 3, pp. 218–229, (2010).
16. Viola P., and Jones M. J.: Robust real-time face detection. International journal of computer vision., vol. 57, no. 2, pp. 137–154, (2004).
17. Thomaz C. E., and Giraldi G. A.: A new ranking method for principal components analysis and its application to face image analysis. Image and Vision Computing., vol. 28, no. 6, pp. 902–913, (2010).
18. https://www.digimizer.com.

Parallel Palm Print Identification Using Fractional Coefficients of Palm Edge Transformed Images on GPU

Santwana S. Gudadhe, A.D. Thakare and C.A. Dhote

Abstract This paper gives the performance analysis of palm print identification techniques based on fractional coefficients concept of palm edge transformed images for Haar transform, Cosine and Kekre transform on the GPU-based system running on CUDA platform. In this, three edge detection methods are applied on palm image to obtain the palm edge images. Transform palm edge images are obtained using three different image transforms. The characteristics of image transform to concentrate the energy of image toward low-frequency region are used to obtain feature vector of palm images. Seven different feature vectors are generated by the low frequency coefficient in transformed palm edge image. Matching and feature extraction is performed on GPU. Time required to match the image is calculated on CPU and GPU. Experimentation is ended with test bed of 2000 images (1000 left and 1000 right). Performance comparison is done using GAR. Experimental result shows the performance improvement in terms of GAR value using fractional coefficient concept for all transforms. In all transform, Cosine Transform gives the best performance. Time required for matching on GPU shows two times faster performance than CPU.

Keywords Palm print identification · Biometrics · Genuine acceptance ratio (GAR) · Fractional coefficient · Discrete Cosine Transform (DCT) Compute unified device architecture (CUDA) · Graphical processing unit (GPU) Mean square error (MSE)

S.S. Gudadhe (✉) · A.D. Thakare
Computer Engineering Department, PCCOE, Savitribai Phule
Pune University, Pune, India
e-mail: s.santwana20@gmail.com

A.D. Thakare
e-mail: Adthakare2014@gmail.com

C.A. Dhote
Computer Engineering Department, PRMIT&R,
Amravati University, Badnera, India
e-mail: cadhote@mitra.ac.in

© Springer Nature Singapore Pte Ltd. 2018
S.S. Dash et al. (eds.), *International Conference on Intelligent Computing and Applications*, Advances in Intelligent Systems and Computing 632,
https://doi.org/10.1007/978-981-10-5520-1_11

107

1 Introduction

Biometrics is an inevitable tool used for secured person authentication and identification of a person. Palm print identification acts as one of the widely used biometric authentication as compared to other biometrics like finger, voice, iris in terms of their universality, collectivity, uniqueness, and cost-effectiveness in terms of installation. Palm print patterns are especially reliable and have many feature points like wrinkles, ridges, principle lines.

The amount of time required by the palm print identification device is considerably long if the number of images in the database is more. It has been observed that it is susceptible to acceptance of incorrect input image leading to unauthorized access to fraudulent one or providing no access to the designated person.

The proposed system improves the existing system performance and gives better efficiency to the existing system. The use of GPU considerably reduces the time involved in the matching and feature extraction.

2 Palm Print Basic Identification System

The existing palm print identification system runs on the sequential execution in matching which increases the response time. The system involves the following steps such as preprocessing, feature extraction, and feature-matching computations. The palm print basic identification system has shown in Fig. 1 [1, 2].

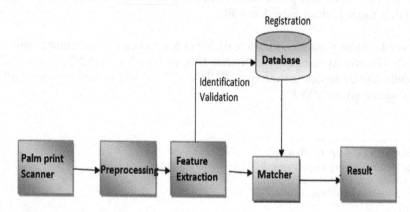

Fig. 1 Palm print basic identification system

The accuracy of the palm print identification system highly depends on distinctive and robust features. These features can be accurately extracted by shift and rotation techniques via masking. Also other preprocessing techniques can also be used to extract the features accurately. In basic palm print identification system, query palm image is matched with the database images. As compared to finger print, feature available in palm print is more distinct and unique for matching which improves the accuracy but it has the time complexity. In proposed system, the concept of fractional coefficient is used to obtain reduced feature vector. Matching and feature extraction is performed parallel on GPU to speed up the execution. Genuine acceptance ratio [3] is considered for performance comparison, and MSE [1] is used for matching.

3   Fractional Coefficient

The concept of fractional coefficient [1] can be seen in terms of energy which is concentrated toward an end. The fractional coefficient reduces the no of pixel considerably. It enables us to analyze the results with reduced feature vector. In this paper, eight different fractional coefficient [4] sets are considered as feature vectors which are 100, 25, 12.5, 6.25, 1.56, 0.39, 0.097, 0.024%. Figure 2 shows the extracted different fractional coefficient feature vectors.

Fig. 2 Extracted different fractional coefficient feature vectors

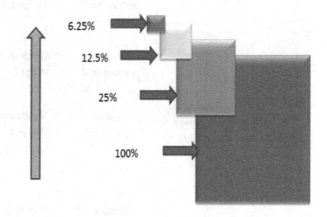

4 Proposed System

4.1 Proposed Architecture

In palm print identification, palm image has to be identified with greater accuracy and with less amount of processing time.

In proposed identification system, system works with first stage resizing the palm image to standard size; edges are extracted using different edge detection methods such as Laplacian, Sobel, and Robert. After getting edge image, transformed edge image is generated by applying Cosine, Haar, and Kekre transforms on palm edge image matrix. Finally, features are extracted from the palm edge transformed images with the concept of fractional coefficient [1], and these extracted features are stored in database as a feature vectors. During identification, query image feature vector is generated and it is compared with feature vector stored in database. In this proposed system, feature extraction and matching is done

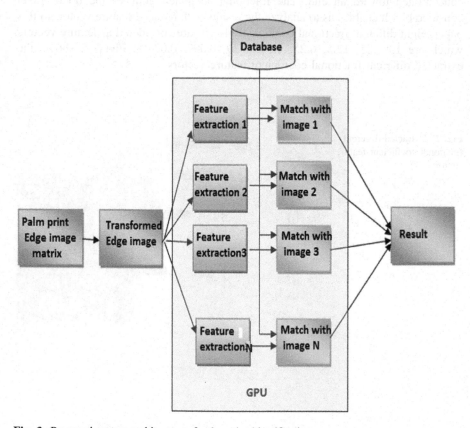

Fig. 3 Proposed system architecture of palm print identification

parallelly on GPU. Final result is sent back from GPU to CPU. Implementation on GPU is discussed in section (B).

Figure 3 shows the proposed system architecture of palm print identification [2].

4.2 GPU

Graphics processing units (GPUs) have proven to be a typical computing resource which has number (thousands) of cores. On computing platform, providing large scale parallelism is the main aspect of GPU. It is enormously extensible which

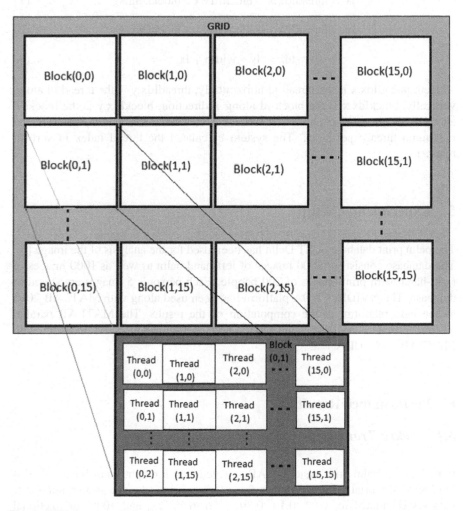

Fig. 4 GPU threads and blocks

provides the flexibility to user to do necessary changes according to computational complexities [2, 5].

Figure 4 shows the structure of GPU grid composed of blocks of threads for 2D computation.

The GPU executes set of computations by assigning pixel, size, and indexing of the image to the threads which resides inside the blocks of GPU. The system is allocated 256 blocks where each block consists of 256 threads. So during the matching, the system uses $256 \times 265 = 65,536$ threads which have been assigned the pixels of the image.

The system calculate the thread index through the below given code in CUDA language.

$$\text{Ix} = \text{threadIdx.x} + \text{blockIdx.x} * \text{blockDim.x}; \qquad (1)$$

$$\text{Iy} = \text{threadIdx.y} + \text{blockIdx.y} * \text{blockDim.y}; \qquad (2)$$

$$\text{Idx} = \text{Iy} * \text{width} + \text{Ix}; \qquad (3)$$

Here threadIdx.x is the thread id horizontally, threadIdx.y is the thread id along vertically, blockIdx.x is the block id along x-direction, blockIdx.y is the block id along y-direction, blockDim is the block dimension, and width indicates the maximum threads per block. The system calculated the thread index in vertical manner

5 Experimental Setup

The palm print database of IIT Delhi has been used for the analysis of the image [6]. The database consists of 1000 images of left-hand palm as well as 1000 images of right-hand palm print images of 200 people. Each person's 5 images are stored in database. The NVIDIA CUDA platform has been used along with MATLAB 2013 as the base platform for the computation of the results. The MATLAB parallel toolbox provides the basic connectivity and operational tools for working with MATLAB and CUDA.

6 Performance Results

6.1 Kekre Transform

For proposed palm print system, GAR is calculated for right palm image as well left-hand palm print using Kekre transform for a test image given as an input. GAR value is calculated for 0.02, 0.10, 0.39, 1.56, 6.25, 25, and 100% of fractional coefficient on applying Sobel, Robert, and Laplace edge detection algorithms [1]. The GAR values change with respect to the feature vectors of palm print image. GAR Plot using Kekre transform on right-hand palm based on proposed palm print identification method is shown by Fig. 5.

GAR Plot using Kekre transform on left-hand palm based on proposed palm print identification method is shown by Fig. 6.

From the result, it is observed that the Sobel edge detection technique shows better GAR value is with 96% in right-hand palm with 1.56 and 6.25% of fractional coefficient. The GAR is 87% for the left-hand palm with 6.25% as well as 1.56% of fraction.

Table 1 shows the time required for matching the query image with database on GPU and CPU for left and right palm print with different edge detection techniques for Kekre transform.

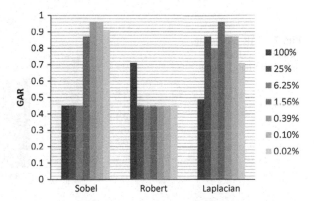

Fig. 5 GAR plot using Kekre transform for right-hand palm image

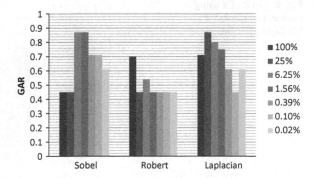

Fig. 6 GAR plot using Kekre transform for left-hand palm image

Table 1 Kekre transform

		Sobel	Robert	Laplacian
GPU	Left	9.05	8.53	8.22
	Right	9.31	8.71	8.03
CPU	Left	18.91	18.60	18.24
	Right	18.50	18.43	17.57

6.2   Discrete Cosine Transform

For proposed palm print system GAR is calculated for right hand as well left-hand palm print using discrete Cosine Transform [7] for a test image given as an input.

GAR value is calculated for 0.02, 0.10, 0.39, 1.56, 6.25, 25 and 100% of fractional coefficient by applying Sobel, Robert, and Laplace edge detection algorithms [1]. The GAR values change with respect to the feature vectors of palm print image. GAR Plot using Cosine Transform on right-hand palm is shown by Fig. 7.

GAR Plot using Cosine Transform on left-hand palm is shown by Fig. 8

From the experimental results, it is observed that the Laplace edge detection technique shows better GAR value is with 100% in right-hand palm with 1.56 and 0.39% of fractional coefficient. The GAR is 96% for the left-hand palm with 25, 6.25, 1.56, 0.39% as well as 0.10% of fraction.

Fig. 7 GAR plot using Cosine Transform for right-hand palm image

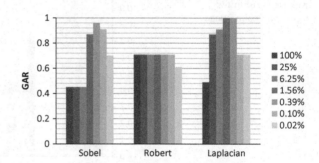

Fig. 8 GAR plot using Cosine Transform for left-hand palm image

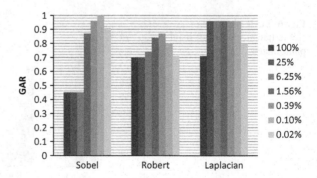

Table 2 Cosine Transform

		Sobel	Robert	Laplacian
GPU	Left	8.91	9.08	9.37
	Right	8.85	9.34	8.99
CPU	Left	18.05	18.51	18.44
	Right	18.31	17.67	18.81

Table 2 shows the time required for matching the query image with database on GPU and CPU for left and right palm print with different edge detection techniques for Cosine Transform.

6.3 Haar Transform

For proposed palm print system, GAR is calculated for right-hand as well as left-hand palm print using Haar transform [3] for a test image given as an input. GAR value is calculated 0.02, 0.10, 0.39, 1.56, 6.25, 25, and 100% of fractional coefficient by applying Sobel, Robert, and Laplace edge detection algorithms [1]. The GAR values change with respect to the feature vectors of palm print image.

GAR Plot using Haar transform on left-hand palm is shown by Fig. 9.

GAR Plot using Haar transform on right-hand palm is shown by Fig. 10.

From the experimental result, it is observed that the Laplacian edge detection technique shows better GAR value is with 100% in right-hand palm with 0.1% of fractional coefficient and 0.96% for the left-hand palm with 1.56% of fractional coefficient.

Table 3 shows the time required for matching the query image with database on GPU and CPU for left and right palm print with different edge detection techniques for Haar transform.

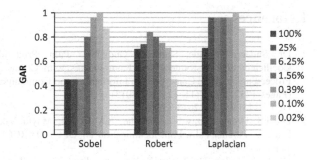

Fig. 9 GAR plot using Haar transform for left-hand palm image

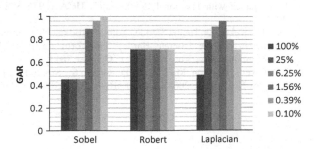

Fig. 10 GAR plot using Haar transform for right-hand palm image

Table 3 Haar transform

		Sobel	Robert	Laplacian
GPU	Left	9.03	8.43	8.35
	Right	9.02	9.11	9.12
CPU	Left	17.31	18.63	18.85
	Right	19.07	18.08	18.27

7 Conclusion

Palm print identification is a most prominent system used for personal identification technique. Palm images contain unique information such as principal lines, ridges, and wrinkles. In recent work, palm print identification system uses the frequency domain approach. It involves use of transform, and wavelets transform techniques. This is proven to be efficient in implementation of palm print identification. This paper presents the parallel palm print identification on GPU with concept of fractional coefficient.

For experimentation, it is observed that, as compare to CPU, processing on GPU is much faster. The use of GPU for parallel computation in feature-matching techniques is capable of improving the performance and efficiency of the system. It has been observed through the analysis that Sobel and Laplacian edge detection algorithms give better results in terms of GAR as compared to Robert.

References

1. Dr. Sudeep Thapade, Santwana Gudadhe, "Palm Print Identification using Fractional Coefficient of Transformed Edge Palm Images with Cosine, Haar and Kekre Transform", Proceedings in IEEE Conference on Information and Communication Technologies, ICT-2013. (Pages 1232–1236) (11–12 April 2013) DOI 10.1109/CICT.2013.6558289.
2. Dr. J.S. Umale, Santwana Gudadhe, N.V. Rohit, Mayur Vikas Patil, Yogesh Patil, Kiran Muske, "Parallel Palm Print identification on GPU" IJARCCE (2278–1021)Vol. 4, Issue 1, January 2015.
3. Dr. H. B. Kekre, Dr. Tanuja K. Sarode, Aditya A. Tirodkar, "An Evaluation of Palm Print Recognition Techniques using DCT, Haar Transform and DCT Wavelets and their Performance with Fractional Coefficients", IJCA (0975–8887) Volume 32–No. 1, October 2011.
4. Dr. H.B Kekre, Dr.Tanuja Sarode, Aditya Tirodkar, "Performance Enhancement of Fractional Coefficients of the Discrete Hartley Transform for Palm Print Recognition", Journal of Telecommunication, Vol. 11, Issue 1, October 2011.
5. J.D. Owens, Houston, M.; Luebke, D.; Green, S.; Stone, J.E.; Phillips, J.C. "GPU Computing" Vol. 96, No. 5, May 2008 DOI: 10.1109/JPROC.2008.917757.

6. "CUDA by Example" Addison Wesley Jul 2010 (Last Referred on 30th march 2016).
7. Dr. Kekre. H, Thepade. S; Maloo. A (2005): "Image Retrieval using Fractional Coefficients of Transformed Image using DCT and Walsh Transform", International Journal of Engineering Science and Technology (IJEST), Volume 2, Number 4, 2010, pp. 362– 371. (ISSN: 0975-5462).
8. IIT Delhi Touchless Palmprint Database (Version 1.0), http://web.iitd.ac.in/~ ajaykr/Database_ Palm.htm.

A Secure Spatial Domain Image Steganography Using Genetic Algorithm and Linear Congruential Generator

Pratik D. Shah and R.S. Bichkar

Abstract Significant increase in data traffic over the Internet has given rise to many data security issues. Steganography is a technique which is used to hide the existence of secret communication. Hence, it is extensively used to solve the issues related to data security. In this paper, a secure and lossless spatial domain image steganography technique is proposed. Stream of secret data is hidden in quarter part of image by identifying suitable locations to hide 2 bits of secret data in each pixel, resulting in generation of coefficients corresponding to the location of match. These coefficients are hidden in remaining part of image using LSB replacement steganography. Genetic algorithm is used to find best possible location to hide these coefficients in the image, making the proposed technique very secure and almost impossible to extract secret data from it. The result of the proposed technique is compared with LSB replacement steganography where in same amount of secret data is embedded. It is observed that the proposed technique is much superior as compared to LSB steganography. It provides improvement in MSE and PSNR values; in addition, the degradation in histogram is also minimal thus eliminating histogram attack. Average PSNR value of stego-image obtained from proposed technique is 53.11 dB at two bits per pixel data embedding rate compared to 52.21 obtained by LSB technique.

Keywords Genetic algorithm (GA) · LSB steganography · Steganalysis

P.D. Shah (✉) · R.S. Bichkar
G. H. Raisoni College of Engineering and Management, Pune, Maharashtra, India
e-mail: shahpratik219@gmail.com

R.S. Bichkar
e-mail: rajankumar.bichkar@raisoni.net

© Springer Nature Singapore Pte Ltd. 2018
S.S. Dash et al. (eds.), *International Conference on Intelligent Computing
and Applications*, Advances in Intelligent Systems and Computing 632,
https://doi.org/10.1007/978-981-10-5520-1_12

119

1 Introduction

Steganography is an art of hiding the existence of secret communication [1]. In steganography, huge amount of secret data is hidden in cover media to conceal it from the attack of eavesdropper's and unauthorized persons [2]. Steganography can be done in many digital file formats e.g. video, text, image, audio, etc., but the formats with high degree of redundancy are more suitable. In image steganography, secret data is hidden inside a normal image. The secret information can be in any file format such as text, image, excel file [3]. The image used for hiding the data is called as cover image, and the image obtained after embedding the secret data onto the cover image is called as stego-image. The main aim of steganography is to reduce the difference between stego-image and cover image so that it can conceal the existence of any secret communication. The performance of an image steganography technique is mostly evaluated using four parameters viz. imperceptibility, payload capacity, robustness and security [1]. Imperceptibility is the ability of a steganography technique to be undetected by visual inspection. Payload capacity is the amount of secret data that can be hidden inside the cover image. Robustness is the resistance of steganography technique against image manipulation attacks like cropping, scaling, rotation, compression. Security is the ability of steganography system to resist the attacks of steganalysis system. Steganalysis is study of detecting messages hidden using steganography [4].

Image steganography can be categorized as spatial domain steganography and transform domain steganography. In spatial domain image steganography technique, the data hiding is performed directly on the pixel values of the cover image. Spatial domain techniques include methods which operate at bit level such as bit insertion and noise manipulation [5]. Transform domain techniques utilize the domain-specific characteristics of image to embed data on it. The image is first transformed to frequency domain using numerous transforms like DCT, DFT, DWT, curvelet transform, contourlet transform [6]. In these techniques, the data is embedded on the coefficients of transformed image instead of direct pixels and then the image is retransformed to spatial domain.

In last decade, enormous amount of work is carried out in the field of image steganography but very few studies have explored the use of metaheuristic and stochastic optimization operators in improving the result of steganography. Kanan and Nazeri [7] proposed genetic algorithm-based image steganography technique in which they used genetic algorithm to find the proper locations in cover image to hide secret data. The data was hidden using LSB replacement steganography, GA was used to find out starting location and direction for data embedding. The output was tunable, i.e. it generated various different stego-images, any one of which can be selected based on desired results and application. Average PSNR value of 45.12 dB was obtained during various experiments. Wang et al. [8] proposed a secure steganographic method to bypass RS steganalysis attack. Data is hidden in first LSB bit of the image, and second LSB bit is modified so that image by passes RS attack. Genetic algorithm is used to search for best adjustment matrix which is used to modify second LSB of image so that the stego-image bypasses RS analysis and also provide a better PSNR value. Average PSNR value of 41.2 dB was obtained during various experiments. Nosrati et al. [9] proposed a before embedding steganographic scheme. In this technique, the secret information is hidden in image segments of cover image. Genetic algorithm is used to find the suitable locations in cover image to embed the secret information.

The rest of the paper is organized as follows: Sect. 2 briefly explains genetic algorithm. The proposed algorithm is explained in detail in Sect. 3. Experimental results and discussion are presented in Sect. 4. Section 5 concludes the paper.

2 Genetic Algorithm

Genetic algorithm is biologically motivated metaheuristic technique used to solve both constrained and unconstrained optimization problems. It is population-based approach driven by the principle of natural evolution based on Darwin's theory [10]. Each population is potential solution for a given search and optimization problem. Fitness function is used to assess the quality of each possible solution. The solution with high fitness value will survive and form a new population of the next generation. Genetic algorithm operators like reproduction, crossover and mutation are used to obtain new generations. It is an iterative process which is carried out till the desired result is obtained or till the number of predetermined iterations is reached. Over successive iterations the population evolves towards a near-optimal solution.

3 Proposed Technique: Optimally Mapped Least Significant Bit Replacement Steganography

Proposed technique is a modified variant of LSB replacement steganography. In this technique, secret data is not hidden using LSB replacement method but instead a new approach called data mapping is proposed. In data mapping, two bits of secret data are embedded in every pixel. In grey scale image, each pixel consists of

Fig. 1 Co-efficient generation ('**' indicates the location at which the secret data is embedded)

Embedding position	Coefficient
010001**	0
01000**0	1
0100**00	2
010**100	3
01**0100	4
0**00100	5
**000100	6
100010	7

eight bits, so there are eight possible locations for mapping the secret data bits on each pixel of cover image. Based on location of match, we generate corresponding coefficients, as illustrated in Fig. 1.

If there is no match between secret data bits and cover image pixel, then these bits are embedded in LSBs of cover image pixel. It is mostly possible to embed two bits of secret data in each pixel without changing its value. However to successfully recover the secret data, we should know the value of coefficients. We propose to hide these coefficients in the unused part of image using genetic algorithm to find the most optimal embedding locations.

Figure 2 illustrates the proposed mapping process with the help of an example in which 4 bytes of secret data is embedded in 4 × 4 cover image. Figure 2a shows four bytes of secret data and its binary representation. Figure 2b shows 4 × 4 cover image and its binary representation along with the position of match between secret data and cover image pixel highlighted. Figure 2c shows the coefficients generated by mapping.

In proposed technique, secret data is embedded in 1/4th part of the image and rest part of the image is used to save the coefficients generated from data mapping. For a grey scale image with 256 × 256 resolution, we can use 128 × 128 pixels to hide the secret data; hence, the resultant data embedding capacity is 2 × 128 × 128 = 32,768 bits. In this case, a matrix of size 128 × 128 is generated for the coefficients. From Fig. 2c, it can be observed that value of coefficients is in the range of 0–7; hence, each coefficient requires 3 bits for embedding. The coefficients matrix is split into three-bit planes. These three-bit planes are optimally embedded in LSBs of

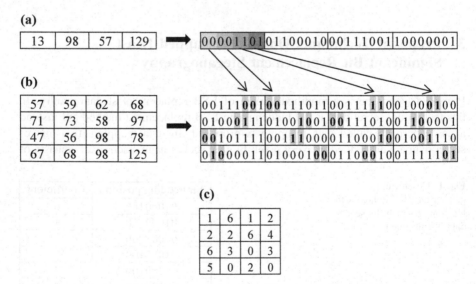

Fig. 2 Proposed mapping process to hide secret data in cover image. **a** Secret data and its binary representation. **b** Cover image (4 × 4 pixels) and its binary representation. **c** Generated coefficient matrix

remaining 3/4th part of image using genetic algorithm. The task of finding an optimal match between coefficient matrix and LSBs of cover image is modelled as search and optimization problem. Genetic algorithm is employed to solve this search and optimization problem. Each bit plane of coefficient matrix is divided into blocks of size 8×8 yielding 256 blocks. An optimal sequence of 256 numbers is searched to embed these blocks of coefficients in the rest of image. Linear congruential generator (LCG) is used to generate a pseudo-random sequence, and genetic algorithm is used to modify the parameters which control the result of LCG. LCG requires initial seed value, a multiplying factor and an offset to generate sequence of m numbers; it is illustrated in Eq. 1. $X_n + 1$ is the value of next integer in the sequence, X_n is the value of present integer, a is the multiplying factor, c is the offset value and m is the length of sequence.

$$X_n + 1 = (a \cdot X_n + c) \bmod m \tag{1}$$

Various sequences can be generated with different values of a, X_n and c. The proposed algorithm uses the sequence generated by LCG to embed the coefficients on cover image. The sequence generated may not be the best sequence to embed coefficients on cover image; hence, we use genetic algorithm to control the parameters and generate new solutions which might be more suitable.

Genetic algorithm is a population-based method; hence, every population is a possible solution to the search and optimization problem. The proposed chromosome consists of three genes as depicted in Fig. 3.

Each gene in the proposed chromosome is of 8 bits. These three genes are used to control the sequence generated by LCG. The first gene of the chromosome is multiplier which is of eight-bit length, second is offset and third one is initial seed X_0, both of them are of eight-bit length. Initial seed X_0 decides the initial value of the pseudo-random sequence.

Initial population is selected randomly. The population size is set to hundred, and these population will fight for their existence and also to seed populations in next generation. The efficiency of each individual is evaluated using fitness function. In this case, fitness function is defined as the number of optimally mapped coefficients values on the LSBs of cover image. More the similarity between the coefficients value and the LSBs of cover image more the fitness of a population. We randomly choose five populations and based on fitness of these populations two are selected to seed next generation. We use crossover and mutation operators to obtain next generation populations. This process is repeated until all the populations are chosen to compete with each other in tournament selection competition. The process of generating new solutions is repeated till 50 iterations.

Fig. 3 Chromosome structure

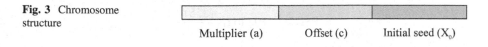

Multiplier (a)   Offset (c)   Initial seed (X₀)

Algorithm for embedding data in cover image

1. Map two bits of secret data on each cover image pixel and generate coefficient based on position of match.
2. Hide the data in 1/4th part of image and generate coefficients for the same.
3. Divide the coefficients matrix in three-bit plane.
4. Divide each bit plane into blocks of size 8 × 8.
5. Generate pseudo-random sequence using LCG.
6. Use GA to optimize parameters of LCG to generate a sequence which produces optimal match between coefficients and LSBs of cover image.
7. Obtain the final chromosome after 50 iterations and hide it as a secret key.

Algorithm for extracting secret data from stego-image

1. Obtain secret key which is chromosome of GA.
2. Generate pseudo-random sequence using the secret key and LCG.
3. Extract data from LSBs of cover image in the sequence obtained by LCG to get bit planes coefficient matrix.
4. Obtain all the three-bit planes and combine them to obtain coefficient matrix.
5. Extract secret data from the 1/4th part of image using coefficient matrix.

The proposed method is a lossless method because it does not have any bit error rate. The security of algorithm is very high because it is difficult for anyone to extract secret data from stego-image as knowledge of secret key is necessary to get the sequence used to hide the coefficients.

4 Experimental Results and Discussion

To perform experiment, we have selected standard grey scale test images. The resolution of cover image is 256 × 256, and the secret data is also a grey scale image with resolution 64 × 64. The performance of proposed technique is compared with 2-bit LSB replacement steganography with same data embedding capacity. To perform experimentation, Matlab 8.1 version was used. MSE and PSNR of stego-images are obtained by Eqs. 2 and 3, respectively, where M and N are total number of rows and columns in the image, respectively. X_{ij} and Y_{ij} are pixel values of ijth location of original image and stego-image, respectively.

$$MSE = \frac{1}{MN} \sum_{i=1}^{M} \sum_{j=1}^{N} \left(X_{ij} - Y_{ij} \right)^2 \tag{2}$$

$$PSNR = 10 \cdot \log_{10} \frac{(255)^2}{MSE} \tag{3}$$

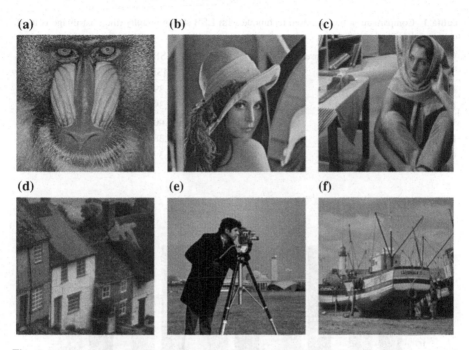

Fig. 4 Test images used in experiment. **a–e** Cover images (baboon, lena, barbara, house and cameraman). **f** Secret data image (boat)

Subjective analysis and histogram analysis are also performed on the results obtained from the both techniques. The test images used for experimentation are shown in Fig. 4. Five natural images are chosen as cover images; they are Baboon, Lena, Barbara, House and Cameraman. Boat image is used as secret data image.

4.1 Imperceptibility Analysis

Imperceptibility analysis is done to measure the amount of change in original image during the process of data embedding. In imperceptibility analysis, the stego-image obtained from LSB replacement steganography and proposed technique are compared with respect to MSE and PSNR parameters. MSE is the cumulative squared error between original image, i.e. cover image and stego-image. MSE values should be as less as possible, lesser the value of MSE better the imperceptibility. On the other hand, PSNR is used to obtain peak signal to noise ratio. PSNR value of stego-image should be as greater as possible, since high PSNR value will ensure a better visual quality of stego-image.

Table 1 shows a comparison of MSE and PSNR values of proposed technique and LSB replacement technique at 2 bits per pixel data embedding rate. In this experiment, boat image is used as secret data image for all five cover images.

Table 1 Comparison of the proposed technique with LSB steganography (the boat image is used as secret data)

Cover image	Proposed technique		LSB replacement	
	MSE	PSNR	MSE	PSNR
Baboon	0.23	54.43	0.39	52.18
Lena	0.38	52.33	0.39	52.20
Barbara	0.27	53.80	0.40	52.17
House	0.35	52.64	0.38	52.27
Cameramen	0.37	52.36	0.39	52.23

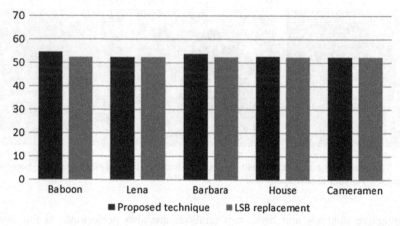

Fig. 5 PSNR values of stego-images

Fig. 6 Stego-images:
a Baboon stego-image
obtained from LSB
steganography, **b** Baboon
stego-image obtained from
proposed technique, **c** Lena
stego-image obtained from
LSB steganography, **d** Lena
stego-image obtained from
proposed technique

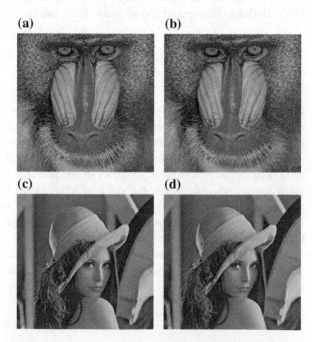

The results clearly suggest the superiority of proposed optimally mapped LSB technique over LSB replacement technique. The PSNR values of stego-image obtained from proposed techniques are higher compared to LSB steganography for various images which prove the efficiency of proposed technique. The MSE value of all stego-images obtained from proposed technique is less than 0.38 which undoubtedly indicates that there is very less amount of deviation between cover image and stego-image. Hence, it can be said that proposed technique is highly imperceptible. Figure 5 illustrates the comparison between PSNR values of stego-image obtained from proposed technique and LSB replacement steganography for various images. Figure 6 shows the stego-images obtained from LSB steganography and proposed technique for Baboon and Lena image.

4.2 Histogram Analysis

In histogram analysis, the distortion between histogram of cover image, stego-image obtained from LSB replacement steganography and proposed optimally mapped LSB technique is compared. Baboon image is selected to perform histogram analysis. The histogram of cover image, stego-image obtained from proposed method and LSB technique is shown in Figs. 7, 8 and 9, respectively.

It can be observed that there is a huge amount of variation in histogram of stego-image obtained from LSB replacement steganography and histogram of original image, whereas the proposed technique produced very less amount of distortions in the histogram of stego-image. Histogram analysis of proposed technique would not generate any suspicion regarding existence of secret message in the stego-image since it does not generate any typical pattern in histogram. However on the other hand, histogram analysis of LSB steganography makes the existence of

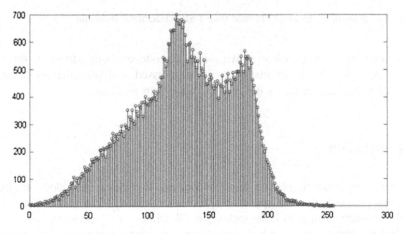

Fig. 7 Histogram of original cover image

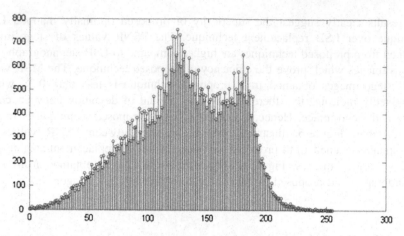

Fig. 8 Histogram of stego-image obtained from proposed technique

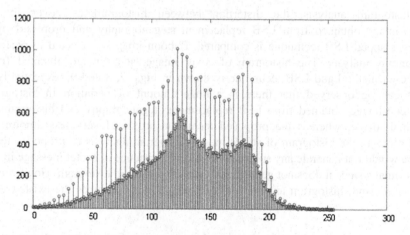

Fig. 9 Histogram of stego-image obtained from LSB replacement technique

secret data very evident because its stego-image produces a step pattern. In this step pattern, one pixel value of histogram is suppressed and the adjacent value is increased which can be easily detected by histogram analysis.

5 Conclusion

This paper proposes a genetic algorithm-based lossless spatial domain image steganography scheme with special emphasis on security. The result of the proposed optimally mapped LSB technique illustrates its superiority over LSB replacement steganography in both subjective and objective analysis. The proposed technique is highly imperceptible since it causes very less changes in stego-image making it extremely difficult to detect the existence of secret data by visual inspection. The proposed technique is extremely secure since the secret data embedded in stego-image cannot be extracted by without the knowledge of secret key. Even if the secret key is obtained the sequence generation is challenging and further more challenging is to use the sequence to extract secret data from image.

In this paper, histogram analysis of stego-images obtained from both techniques is also performed. Histogram analysis reveals the weakness of LSB replacement technique against histogram attack as it generates a particular pattern which may arouse suspicion to the eavesdropper. On the other hand, the proposed technique produces very less changes in histogram of stego-image; hence, it is robust against histogram attack. Future work will be focused on improving the time complexity of proposed technique and also on exploring possibility of using other sequence generators which provide more control over the generation of sequence.

References

1. Cheddad A., Condell J., Curran K. and Mc Kevitt P.: Digital Image Steganography: Survey and Analysis of Current Methods. Signal Processing, vol. 90, no. 3, pp. 727–752, (2010).
2. Chugh G., Yadav R. and Saini R.: A New Image Steganographic Approach Based on Mod Factor for RGB Images. International Journal of signal processing, image processing, and pattern recognition, vol. 7, no. 3, pp. 27–44, (2014).
3. Subhedar M.S. and Mankar V.H.: Current status and key issues in image steganography: A survey. Computer science review, vol. 13, pp. 95–113, (2014).
4. Fridrich J., Goljan M. and Du R.: Reliable detection of LSB steganography in color and grayscale images. In Proceedings of the 2001 workshop on Multimedia and security: new challenges, pp. 27–30, (2001).
5. Li B., He J., Huang J. and Shi Y.Q.: A survey on image steganography and steganalysis. Journal of Information Hiding and Multimedia Signal Processing, vol. 2, no. 2, pp. 142–172, (2011).
6. Chanu Y. J., Tuithung T. and Singh K.M.: A short survey on image steganography and steganalysis techniques. In Proceedings of 3rd National Conference on Emerging Trends and Applications in Computer Science, pp. 52–55, (2012).
7. Kanan H.R. and Nazeri B.: A novel image steganography scheme with high embedding capacity and tunable visual image quality based on a genetic algorithm. Expert Systems with Applications, vol. 41, no. 14, pp. 6123–6130, (2014).
8. Wang S., Yang B. and Niu X.: A secure steganography method based on genetic algorithm. Journal of Information Hiding and Multimedia Signal Processing, vol. 1, no. 1, pp. 28–35, (2010).
9. Nosrati M., Hanani A. and Karimi R.: Steganography in Image Segments using Genetic Algorithm. In proceedings of Fifth International Conference on Advanced Computing & Communication Technologies, pp. 102–107, (2015).
10. Goldberg and David E.: Genetic algorithms. Pearson Education India, (2006).

Three-Dimensional MRI Brain Image Analysis on Hadoop Platform

Jyoti S. Patil and G. Pradeepini

Abstract Tremendous medical data generation takes place in hospitals due to advanced image capturing techniques. Real-time analysis of that large knowledge becomes essential for medical practitioners. This paper focuses on implementation of automatic diagnosis of three-dimensional data using the MapReduce framework. After flourishing completion of two-dimensional image analysis of magnetic resonance imaging (MRI) brain pictures for detecting tumor, we have introduced 3D image analysis of brain MRI for tumor detection. Every image in diagnosis is analyzed by well-known Scale Invariant Feature Transform (SIFT) rule to spot keypoints at varied scales, whereas Hadoop's MapReduce is employed to match extracted feature vector of every inputted image, and HDFS supports storing options of existing medical pictures within the system. This application will certainly cut back diagnosis time and facilitate doctors for higher decision making.

Keywords Medicinal image diagnosis · Content-based medical image retrieval (CBMIR) · Scale Invariant Feature Transform (SIFT) · Hadoop HDFS · MapReduce

1 Introduction

In today's era of big data and because of advanced camera capturing techniques, tremendous 3D data is generated per second. Simply by manual observation with the assistance of eyes, it's tough for doctors to create correct diagnosing of disease.

Custom DICOM file format (Digital Imaging and COmmunication in Medicine) [1] is employed to just accept the 3D pictures, and then it regenerated into JPEG format and kept into HDFS [2]. On every image, standard SIFT algorithm [3]

J.S. Patil (✉) · G. Pradeepini
CSE Department, KL University, Vaddeswaram, Andhra Pradesh, India
e-mail: jyotipatilnba@gmail.com

G. Pradeepini
e-mail: pradeepini_cse@kluniversity.in

© Springer Nature Singapore Pte Ltd. 2018
S.S. Dash et al. (eds.), *International Conference on Intelligent Computing and Applications*, Advances in Intelligent Systems and Computing 632,
https://doi.org/10.1007/978-981-10-5520-1_13

is used to observe keypoints options. These keypoints are kept in HDFS as feature library. Then, this feature library is employed to match and retrieve pictures at the time of diagnosing. The content-based medical image retrieval (CBMIR) [4] has the benefit of quick retrieval speed and great preciseness. In this technology, visual characteristics of images are mined and matched for image retrieval. It measures Euclidean distance measure between the extracted features of the image in question and the feature contents saved in feature library. Our previous method, retrieval of medical images using MapReduce [5], is based on the model of medical term referred as evidence-based practice (EBP) [6]. In this method, doctors give treatment to patients on the basis of medications given to similar type of patients in the past. Doctors have to refer the historical data of previous cases, and current patient information is matched with database. Our application greatly reduces query images matching and diagnosis time. This model will be the best tool used by doctors for quick investigating diseases.

The paper organization is as follows. Section 2 explains the related work, and proposed architecture is covered in Sect. 3. Further, Sect. 4 elaborates design and implementation details, while Sect. 5 focuses on the results obtained. Section 6 is performance analysis followed by conclusion.

2 Related Work

2.1 Diagnosis of 2D MRI Brain Images

In this paper published for two-dimensional MRI brain images diagnosis using MapReduce [6], idea of image matching using distance measure is described. During implementation, author got with success enforced MapReduce for two-dimensional medical image information. Every tumor image from pathology laboratory is taken as a sample and kept in HDFS using its feature vector, then the Sobel edge detection algorithm [7] is employed to detect edges of the image and its histogram is additionally calculated. Figure 1 shows the output when edge detection is performed. The image in question, i.e., MRI brain tumor image, is queried to the application, and appropriate identical images are fetched from the application based on measure of Euclidean distance. The job of MapReduce is continuously working within the system at back, to compare and fetch relevant images. So the employment of MapReduce framework leads to reduced retrieval time and fast diagnosing for two-dimensional data.

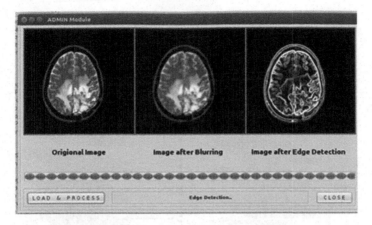

Fig. 1 Output of Sobel edge detection

2.2 Scale Invariant Feature Transform (SIFT)

'Scale-invariant feature transform (SIFT)' is developed by D. Lowe in 2004, University of British Columbia [3]. SIFT algorithm extracts keypoints of image and calculates its descriptors. Steps involved in SIFT are as follows:

Step 1 Detection of Scale-Space Extrema

Laplacian of Gaussian is observed in every picture with numerous σ values. σ acts as a scaling parameter. SIFT set of rules is based on difference of Gaussians that is an approximation of the Laplacian theorem of Gaussian. Subsequently, the difference of Gaussian (DOG) is computed because of the distinction of Gaussian blurs of a picture with 2 totally different parameters σ; let its σ and $k\sigma$. Stated method is recurrent for all offered octaves of given image in Gaussian Pyramid. Figure 2 shows keypoints best drawn in that scale.

After finding, this DOG per image all pictures is searched counting on native extrema over scale and space.

$$\text{DOG}(x, y', s + \Delta s) - L(x, y', s) \approx \frac{\Delta s}{2} \nabla^2 L(x, y', s)$$

For example, one constituent from a picture is compared to its eight neighboring pixels also as all nine pixels in next higher scale and every one of nine pixels in its previous lower scales. If it's an area extrema, then it's a possible keypoint. It primarily implies that the keypoint is that the best illustration of that scale.

Step 2 Identification of location of keypoints

In this step, location and scale of every candidate location are decided by supported stability, as well as with low distinction points which results in rejection of points lying on the edge.

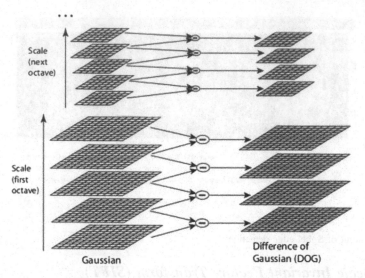

Fig. 2 Difference of Gaussian (DOG) in SIFT

Step 3   Assignment of orientation to each keypoint

For reaching invariance to picture rotation at every keypoint orientation is assigned. A neighborhood region is taken into consideration around each keypoint vicinity depending on its scale; its gradient is magnitude and direction. Then, orientation histogram is generated with 36 bins protecting all 360°. The highest height inside the histogram is located, and numbers of peaks which are above 80% of the best top also are taken into account to determine the orientation. It produces keypoints having identical location and scale, but having distinctive directions.

Step 4   Creation of Descriptor for each keypoint

Keypoint descriptor is formed. A section of 16 × 16 neighborhood constituent round the keypoint is taken into thought. The 4 × 4 size sub-blocks are made from each sixteen block. Eight-bin orientation histogram is formed for every single value of the sub-block. So set of 128 total bin values is available consistent with sub-block. It's far represented as a vector to shape keypoint descriptor. Illumination modifications and rotation factors are also taken into consideration, and unique measures are taken.

Step 5   Matching of keypoints with its neighbors

Keypoints between two different images are matched by calculating their adjacent neighbors' values. In few exceptional cases, the second closest match is also very around the primary. It happens attributable to noise or different reasons. In these cases, magnitude relation is taken between initial closest distance and second closest distance.

2.3 Analysis of 3D Images Using MapReduce Framework [8]

This paper has proposed the idea of the usage of Hadoop's framework for massive scale image processing. Hadoop is a Linux-based open source software platform especially designed for storage and retrieval of files used in big data. Hadoop made up of two essential components, (1) MapReduce for computational abilities and (2) HDFS for storing records. MapReduce runs on distributed framework for information processing, specially handles large records. The MapReduce method of Hadoop consists of levels Map and Reduce. In Map segment, saved split information is inputted to map function a good way to generate an intermediate key-value pair. Afterward Reduce phase takes delivery of key-value pair generated as its inputs and combines all intermediate generated values related to equal intermediate key. Figure 3 shows the structure of MapReduce.

Hadoop distributed file system (HDFS) is a sub-part of the Apache Hadoop Project. This was implemented with the intension to offer fault tolerance to file system and its execution on commodity hardware. HDFS can create, pass, delete, or rename the documents like traditional database management system; however, it varies in the technique of storage. It consists of two working modules first Namenode and second Datanode. This Datanode saves files in Hadoop, and Namenode is the controller of HDFS. HDFS provides reliable information storage potential even within the occurrences of failure, along with failure of Namenode, failure of Datanode, and failure of network partitions. Usually, HDFS follows the master/slave structure, wherein one master device controls more than one slaves.

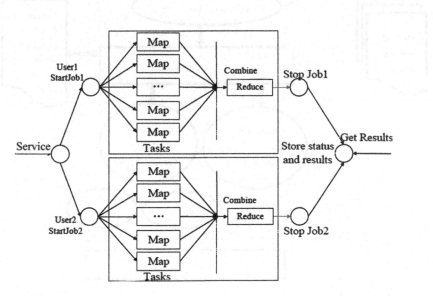

Fig. 3 Processing cycle of MapReduce

3 Proposed Architecture of the System

As shown in Fig. 4, the proposed system comprises a single Hadoop cluster with one master node and three slave nodes (multinode Hadoop cluster). User can input any image in question to the application. This image query is absorbed via Hadoop master node and stored in HDFS (Hadoop distributed file system). JobTracker running on the master node will initiate a MapReduce activity for extracting features from image query. Then, the JobTracker will divide the MapReduce activity into several tasks and later those responsibilities are distributed among slave nodes (Map Phase). The TaskTracker on slave node will carry out the function extraction, and then these extracted features are back to the JobTracker (Reduce phase). Afterward, this JobTracker will initiate another MapReduce job to match those extracted capabilities with present characteristic vectors, previously present in the feature library stored in HDFS. TaskTrackers carry out the task of extracted feature matching with existing one. After assessment, resulting matching vectors are sent to the JobTracker. Then, JobTracker generates final MapReduce job to recollect matching images primarily based on the result of previous MapReduce job. The matched image set is retrieved and given to the user. MapReduce jobs are carried

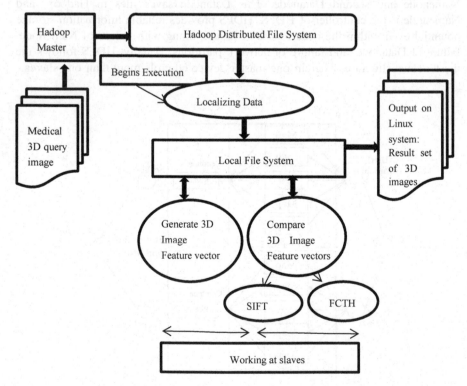

Fig. 4 Proposed architecture

out via TaskTrackers parallelly at every slave node; due to this, the said time of execution for the system is a comparatively less to single-node system. Figure 4 shows proposed architecture of the system.

4 Design and Implementation

Step 1 Load query image into HDFS

Sample DICOM files [1] for 3D images of brain MRI are saved and converted into JPEG layout. For single pattern of brain picture, nearly 27 to 277 layers exist.

Figure 5 shows some of the samples brain tumor files. The actual 277 image layers are taken into consideration to capture single patient's document in 3D.

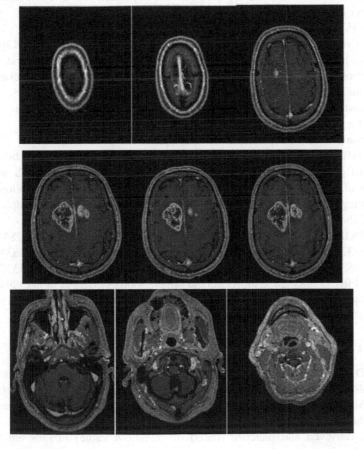

Fig. 5 DICOM files 3D brain images

Step 2 **Feature extraction for 3D image using SIFT** [3]

Feature extraction is a method of mining relevant features from the group of features of particular image. These mined characteristics are predicted to include appropriate information for additional processing. For the reason that it's very complicated to process the whole picture (along with evaluation) while it's very easy to process selected functions of the image. The SIFT algorithm is implemented on every photograph to stumble on keypoints, and it's far saved as the feature vector of every photograph.

Step 3 **3D vector generation**

Three-dimensional vector is an extension of 2D vector. Three-dimensional vector is generated for mapping of 3D images. Each vector is particular for every different individual vector. While any 3D picture of eight layers is furnished to the system, it goes via the gray-scale transformation. It generates coloration histogram for every layer. Those eight histograms are then blended, separated via '$' symbol, to shape eight-layer histogram as shown as follows

'41439$38945$43252$41439$38945$43252$45321$45673'

Above vector generated is for just single 3D image of eight layers. This approach uses Bucketing method to generate vector for each layer. In this method, pixel's color values are divided into various buckets. Let us consider, for 8 buckets, segment color value through eight and then allocate that value to the bucket which having cost equal to the reminder of the modulo division operation. Observe the method for every layer.

For example, color representation of the particular pixel is 138.

So 138% 8 = 2. Hence, we tend to assign 138 values to a bucket number 2. Then, this generated vector is stored in the vector library, i.e., HDFS with the direction of original photograph and its diagnosis. Equal process is followed for each photograph in database, for you to generate function library. Sample vector report is shown below:

41439$38945$43252$41439$38945$43252$45321$456731680$4227$41439
$38945$43252$41439$38945$432521087$3790$12376$45326$41321$$38945
$43252$45321$456731443$3121$38945$43252$45321$4567338945$43252
$45321$45673$3985$2460$45632$325434679$2060$$43252$41439$38945
$43252$45321$456733410$1682$2389$1476$1495$1177$1199$1054911$913
$624$707$512$643$510$448/home/hduser/1111Tumor_Images

Step 4 **3D images mapping**

When 3D query image is given to the system, it generates eight layer vectors for it. Then, this vector is given to the MapReduce for mapping with images from feature library. MapReduce is calculating the Euclidean distance of the question image with every other image from feature library.

These retrieved and matched (based on threshold value) images are kept in ascending order with reference to their geometrician distance. Currently, analyze those pictures for identification. If max threshold numbers of images are signifying brain tumor of stage 1 and suggesting surgical procedure as treatment, then display final analysis as

1. Type: brain tumor
2. Stage: stage 1
3. Remedy: surgical operation

Further, last outputted image should be the image measuring minimal Euclidean distance.

5 Results and Discussion

DICOM files are converted into JPEG and then loaded into HDFS. Figure 6 shows a snapshot before loading a 3D image into HDFS.

A typical 3D image consists of number of layers to represent z-axis. For a single brain, image consists of layers varying from 12 to 256. So, while loading 3D images, we have to load each and every layer captured by camera in 3D. Figure 7 shows loading of Layer 0.

Similarly, Fig. 8 shows loading of Layer 3 of the same image.

Accordingly, each and every available layer is loaded into HDFS. Figure 9 depicts results and suggested treatment for image in question.

Each image in question is compared with every indexed file. Depending on the corresponding measured threshold value of the image, stage of explicit disease also can be diagnosed. Consistent with previous experiences, what sort of treatment is given to the patient is recommended through the system.

Fig. 6 Snapshot before loading 3D image

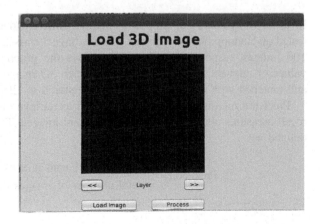

Fig. 7 Snapshot of loading
3D image layer 0

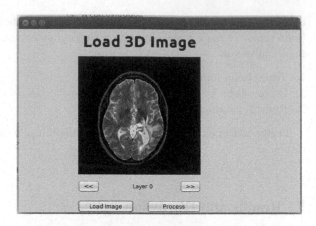

Fig. 8 Snapshot of loading
3D image layer 3

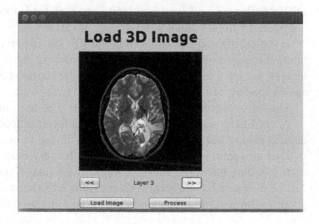

6 Performance Analysis

In order to evaluate our proposed method, different datasets of MRI images are tested on Hadoop platform. Image datasets have 3 classes which contain 20, 60, and 100 images, respectively. Figure 10 shows the graphs for precision and recall values of relevant retrieved images. For each 3D image, 22 layers are considered and matched with each layer of 3D image stored in HDFS.

Precision measures ability of the model to retrieve only relevant images, while recall measures ability to retrieve all relevant images. The precision and recall are defined as

$$\text{Precision} = \frac{\text{number of relevant images retrieved}}{\text{total number of images retrieved}}$$

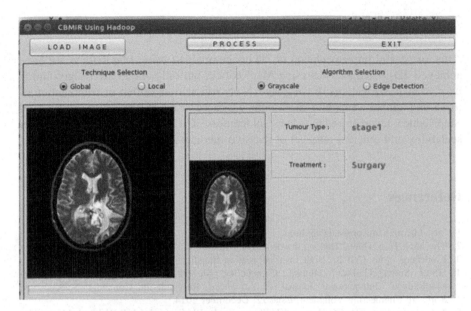

Fig. 9 Results showing suggested treatment

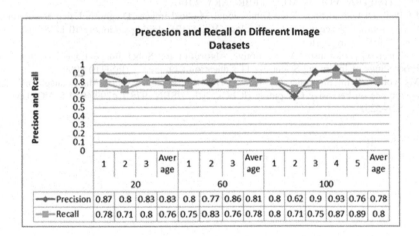

Fig. 10 Graphs for precision and recall

$$\text{Recall} = \frac{\text{number of relevant images retrieved}}{\text{total number of relevant images}}$$

Average values are tested which gives precision up to 80%. And recall results in 78%. Proposed method will give accuracy up to 80% in the relevant images retrieved. As Hadoop platform is used, its accuracy is not compared with other traditional methods. In future, comparative analysis will be done with 2D and 3D image retrieval using Hadoop with traditional techniques available.

7   Conclusion

After successful implementation of 2D image analysis using MapReduce, we have achieved 3D image analysis using MapReduce, but only single SIFT algorithm is used to map 3D features of the image. More efficient algorithm or combination of local and global descriptor algorithms must be used in the future along with MapReduce. It is required to manage all features of 3D image and to achieve greater scalability and efficient retrieval of massive medical images.

References

1. http://dicom.nema.org/standard.html.
2. The book,Tom white, "Hadoop the definitive guide".
3. Lindeberg, Tony (2012). "Scale invariant feature transform". Scholarpedia 7 (5): 10491.
4. Hinge Smita, Gaikwad Monika, Chincholkar Shraddha "Retrieval of Images Using MapReduce" International Journal of Advanced Research in Computer Science and Software Engineering Volume 4, Issue 12, December 2014.
5. YAO Qing-An 1, ZHENG Hong 1, XU Zhong-Yu 1, WU Qiong 2, LI Zi-Wei 2, and Yun Lifen 3 "Massive Medical Images Retrieval System Based on Hadoop" In: JOURNAL OF MULTIMEDIA, VOL. 9, NO. 2, FEBRUARY 2014.
6. J.S.Patil, Dr G.Predeepini,"Two Dimensional Medical Images Diagnosis Using MapReduce" Indian Journal of science and Technology Vol 9(17), doi:https://doi.org/10.17485/ijst/2016/v9i17/93014, May 2016.
7. O.Vincet, O. Folorunso "A Descriptive Algorithm for Sobel Image Edge Detection" In: Proceedings of Informing Science & IT Education Conference (InSITE) 2009.
8. Ms.Jyoti S. Patil, Sunayana A. Mane "3-D Image Analysis Using MapReduce" In: IEEE conference on Pervasive Computing (ICPC), 2015. ISBN: 978-1-4799-6054-5, VOLUME-1, Pg-521–525.

Performance Evaluation of Fingerprint Trait Authentication System

Rupali S. Patil, Sonali D. Patil and Sudeep D. Thepade

Abstract Biometric authentication is a way of verifying the person which he claims. Biometric authentication verifies the identity of the person by comparing the biometric trait with the stored template. Fingerprint is the most popularly used for authentication due to the simplicity in capturing the fingerprint image. Human biometric traits such as fingerprint, iris, and palmprint can be used to identify the person due to its unique texture pattern. In the proposed work, texture feature extraction method using discrete cosine transform (DCT) is implemented for fingerprint. Benefit of the transformation in the energy compaction and reduction in the feature vector size by considering fractional coefficients is taken. The test bed of 500 fingerprint samples of 100 persons is used for experimentation. Various size feature sets are taken for the performance evaluation. GAR and FRR are used as performance measures. Experimental results shown that feature vector size 4×4 gives the highest accuracy of 91.75% as compared to the accuracy with feature vector sizes taken as 8×8 or 16×16 or 32×32.

Keywords Biometric authentication · Discrete cosine transform (DCT)
Feature vector · FRR · GAR · Secret sharing

R.S. Patil (✉)
Computer Engineering Department, RMD Sinhgad School of Engineering,
Pune, India
e-mail: patil18rupali@gmail.com

S.D. Patil (✉) · S.D. Thepade (✉)
Computer Engineering Department, Pimpri-Chinchwad College of Engineering,
Pune, India
e-mail: sonalimpatil@gmail.com

S.D. Thepade
e-mail: sudeepthepade@gmail.com

© Springer Nature Singapore Pte Ltd. 2018   143
S.S. Dash et al. (eds.), *International Conference on Intelligent Computing
and Applications*, Advances in Intelligent Systems and Computing 632,
https://doi.org/10.1007/978-981-10-5520-1_14

1 Introduction

Digitization has increased the interest over personal information and created security threats. Identification and recognition of the person are essential to limit the crimes in the cyberspace. Textual passwords or tokens can easily be stolen or shared. Hence, the biometric system is becoming essential part of human society with the increasing need of security [1]. Biometrics is the way of recognizing the person with the characteristics he or she possessed, and biometrics cannot be stolen or shared.

In the biometric systems, recognition is based on physical and behavioral characteristics of the person [2]. Physical characteristics of a person include the structural features like fingerprint structure, iris patterns, palmprint structure, facial feature and behavior of the person includes behavioral characteristics such as signature of a person, gesture, voice, gait. Fingerprint is popularly accepted trait due to its distinctiveness, permanence, and simplicity in scanning [3].

1.1 Secret Sharing [4, 5]

It is a method which splits the picture or image into the multiple parts known as shares. Sufficient number of shares are taken to reproduce the original picture or image. The secret sharing concept is shown for fingerprint trait in Fig. 1.

Security of defensive information in applications like banking system, voting system, home locking system is the critical concern. Hence, biometrics is best suited for such applications where the person needs to authenticate. This paper takes fingerprint trait for authentication. Secret sharing scheme is used in the proposed method to split the fingerprint into the two shares. While authenticating the person, biometric trait is reconstructed and compared with the live trait presented at the time of authentication.

The paper is further explicated Literature Survey, Proposed method, Experimentation Environment, Result and discussion, and Conclusion.

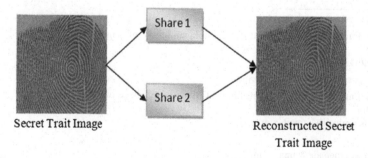

Secret Trait Image   Reconstructed Secret
                                                                  Trait Image

Fig. 1 Sharing of fingerprint trait secretly

1.2 Literature Survey

Biometric-based authentication using visual cryptography is elaborated [6–9]. The problem with such scheme is large size of shares. Secret sharing is useful in keeping the size of the created shares as same or less in comparison with the original secret. In this section, review of previous work done in the area of secret sharing and discrete cosine transform is given.

1.3 Secret Sharing Techniques [10, 11]

$$f(p) = t_0 + t_1 p + t_2 p^2 + \cdots + a p^{k-1} (\text{mod } d) \tag{1}$$

where d is any prime number and t_0 is the secret number which is to be shared.

Share values are taken as (p_i, q_i) where $q_i = f(p_i)$, $1 \leq i \leq n$ and $0 < p_1 < p_2$ ... $< p_n \leq d - 1$.

After calculating share values, the function $f(p)$ is destroyed and each participant will receive a share (p_i, q_i). No participants alone is able to reproduce the original secret. No information about secret is revealed by of $k - 1$ or lesser shares. When available participants are k or more than k, then only secret can be revealed by generating k linear equations using pair of value (p_i, q_i) such as $q_i = f(p_i)$ [12].

Linear equations can be solved using Lagrange's interpolation formula by Eq. (2) given below.

$$t_0 = \sum_{i=1}^{k} q_i \prod_{1 \leq j < k, j \neq i} \frac{p_j}{p_j - p_i} \tag{2}$$

1.4 Discrete Cosine Transform [13]

For image compression, discrete cosine transform (DCT) is popularly used. DCT used with the image is known as two-dimensional DCT. Fundamental equation of DCT is given below.

$$B_{pq} = \alpha_p \alpha_q \sum_{m=0}^{X-1} \sum_{n=0}^{Y-1} A_{mn} \cos \frac{\pi(2m+1)p}{2X} \cos \frac{\pi(2n+1)q}{2Y} \tag{3}$$

The values of α_p and α_q are obtained from Eqs. (4) and (5), respectively, as given below.

$$\alpha_p = \begin{cases} \frac{1}{\sqrt{X}} & \text{if } p = 0 \\ \sqrt{\frac{2}{X}} & \text{if } 1 \leq p \leq X - 1 \end{cases} \tag{4}$$

$$\alpha_q = \begin{cases} \frac{1}{\sqrt{Y}} & \text{if } q = 0 \\ \sqrt{\frac{2}{Y}} & \text{if } 1 \leq q \leq Y - 1 \end{cases} \tag{5}$$

2 Proposed Biometric Authentication

The proposed method is based on fingerprint as a trait for authentication. Complete biometric trait is not important for identification. Only structural information such as principle lines, minutiae are used for distinguishing one person from the other. Hence, the advantage of the transform domain for energy compaction and for database size reduction is taken in the proposed methods. Only few coefficients are taken from the transformed trait for feature vector creation. In the proposed methods there are two phases.

2.1 Enrollment Process

Figure 2 shows the process of enrollment. The steps of enrollment are given below.

Step 1 Take input as fingerprint image.
Step 2 Apply DCT to fingerprint image.
Step 3 Take the fractional coefficients of various sizes. Part of transformed image is taken as 4×4 pixels, 8×8 pixels, 16×16 pixels, and 32×32 pixels.

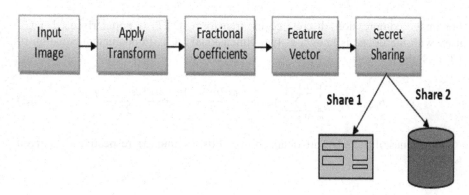

Fig. 2 Enrollment process

Step 4   Create feature vector from the selected fractional coefficients in step 3.

Step 5   Apply secret sharing scheme to generate two shares.

Step 6   Store one share in database and assign other share to the user.

2.2   Authentication Process

Process of authentication is presented in Fig. 3, which involves following steps.

Step 1   User provides the share which is stored in ID card.

Step 2   Database share corresponding to user's share is retrieved from the database.

Step 3   Secret reconstruction algorithm is applied to reconstruct the feature vector of the fingerprint trait.

Step 4   Live fingerprint is captured using sensor.

Step 5   Apply DCT to the live fingerprint trait.

Step 6   Take the fractional coefficients of various sizes. Part of transformed image is taken as 4×4 pixels, 8×8 pixels, 16×16 pixels, and 32×32 pixels.

Step 7   Feature vector of live fingerprint trait is created.

Step 8   Compare the feature vector of live fingerprint with the feature vector generated in step 3.

Step 9   Matching score of root-mean-square-error (RMSE) is verified against the threshold value, and the decision is taken accordingly.

Step 10   If matching score is below threshold, then the user will be accepted as genuine, otherwise he or she will be rejected as if fraudulent user.

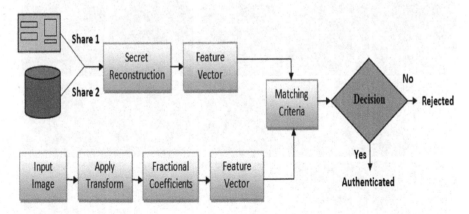

Fig. 3 Authentication process

3 Experimentation Environment

The proposed idea is experimented using MATLAB. A system with specifications such as Intel Core2Duo (2.00 GHz) processor and 2 GB RAM is used for experimentation.

3.1 Dataset of Fingerprint Trait

Experimentation has been done on total 500 images per person five images of fingerprint. Hongkong PolyU [14] fingerprint dataset is used for experimentation. Figure 4 given below shows the sample of fingerprint dataset of 3 persons.

4 Experimentation Results and Discussion

The proposed system is implemented for fingerprint trait as a secret where secret sharing scheme used is (2, 2). That is, the numbers of participants are 2 and both the participants are required for secret reconstruction. The main concern is to achieve highly secure authentication using secret sharing. The proposed method not only provides the security to the fingerprint trait database but also provides the reduction in the storage space required to store the feature template.

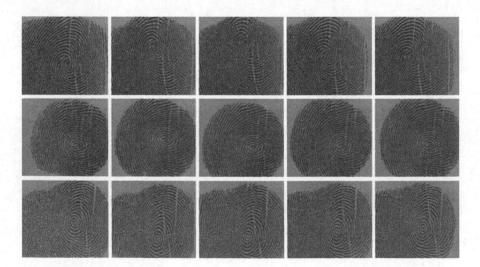

Fig. 4 Fingerprint samples of 3 persons

Table 1 Comparison of various size feature vectors using GAR and FRR for different thresholds

Feature vector size	Threshold	Min + SD	Min + 2 * SD	Min + 0.75 * (Max − Min)	Min + 3 * SD	Min + 4 * SD	Min + 5 * SD
4 × 4	GAR	23.25	52.25	**91.25**	71.5	78.25	**91.75**
	FRR	76.75	47.75	8.75	27.5	21.75	8.25
8 × 8	GAR	20.5	48.75	**89.25**	71.0	83.75	**90.5**
	FRR	79.5	51.25	10.75	29.0	16.25	9.5
16 × 16	GAR	15.0	44.25	**88.75**	66.5	81.75	**89.25**
	FRR	85.0	55.75	11.25	33.5	18.25	10.75
32 × 32	GAR	10.25	37.25	**88.25**	62.5	78.5	**88.75**
	FRR	89.75	62.75	11.75	37.5	21.5	11.25

RMSE similarity measure is used to calculate difference between live fingerprint trait and regenerated fingerprint trait feature vector template. RMSE is computed mathematically by the Eq. (6) given below.

$$\text{RMSE} = \sqrt{\frac{1}{n}\sum_{i=1}^{n}(P_i - Q_i)^2} \tag{6}$$

Here, genuine acceptance ratio (GAR) [15] and false rejection ratio (FRR) are used as performance measures. The system GAR and FRR are computed with the Eqs. (7) and (8), respectively.

$$\text{GAR} = \frac{\text{No.of persons correctly matched}}{\text{Total number of persons in the database}} \tag{7}$$

$$\text{FRR} = \frac{\text{No.of persons rejected}}{\text{Total number of persons in the database}} \tag{8}$$

Table 1 shows the results for accuracy in authentication for various sizes of feature vectors. From Fig. 5, it is clearly seen that 4 × 4 feature vector size has given the highest accuracy. That is, 91.75% of GAR is achieved in 4 × 4 feature vector size using threshold as Min + 5 * SD followed by 8 × 8 size feature vector, 16 × 16 size feature vector, and 32 × 32 size feature vector. Hence, in the proposed system, the storage saving is achieved by using feature vector of size 4 × 4, i.e., by considering only 4 × 4 coefficients from the transformed image.

5 Conclusion

Secure biometric authentication system is proposed using secret sharing scheme. Experimentation has been done on total 500 fingerprint images by applying the discrete cosine transform. The accuracy in the authentication is measured using the fractional coefficients of the DCT for various size feature vectors. The percentage

Fig. 5 Percentage GAR for
various size feature vectors

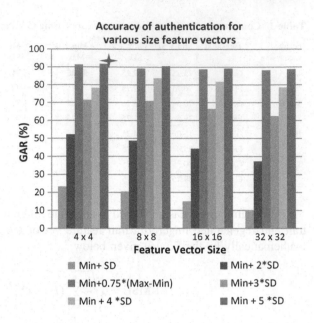

accuracy achieved in the authentication is 91.75 for feature vector size 4 × 4.
Hence, feature vector size is reduced enormously as compared to the total image
size. Only 0.0244% storage is required for storing the feature vector when image
size is 256 × 256.

References

1. Vinayak Bharadi, Bhavesh Pandya, Bhushan Nemade, "Multimodal Biometric Recognition
   using Iris and Fingerprint – By Texture Feature Extraction using Hybrid Wavelets",
   Confluence The Next Generation Information Technology Summit, International Conference,
   pp. 697–702, IEEE, 2014.
2. Sudeep Thepade, Rupali Bhondave, "Biomodal Biometric Identification with Palmprint and
   Iris Traits using Fractional coefficients of Walsh, Haar and Kekre Transforms", International
   Conference on Communication, Information and Computing Technology (ICCICT), pp. 1–4,
   IEEE, 2015.
3. R. Mukesh, V. J. Subashini, "Fingerprint based Authentication System using Threshold
   Visual Cryptographic Technique", International Conference on Advances in Engineering,
   Science and Management, pp. 16–19, IEEE, 2012.
4. Adi Shamir, "How to share a Secret", Communications of the ACM, vol. 22-No. 11, pp. 612–
   613, 1979.
5. Sonali Patil, Prashant Deshmukh, "An Explication of Multifarious Secret Sharing Schemes",
   International Journal of Computer Applications, vol. 46, no-19, pp. 6–10, 2012.
6. J. Sirdeshpande, Sonali Patil, "Amended Biometric Authentication using Secret Sharing",
   International Journal of Computer Applications, vol. 98, No. 21, 2014.

7. J. Sirdeshpande, K. Tajne, Sonali Patil, "Secret Sharing Schemes for Secure Biometric Authentication", International Journal of Scientific & Engineering Research, Volume 4, Issue 6, pp. 2890–2895, 2013.
8. Sonali Patil, Komal Bhagat, Susmita Bhosale, Madhura Deshmukh, "Intensification of security in 2-factor biometric authentication system", International Conference on Pervasive Computing, pp. 1–4, IEEE 2015.
9. J. Sirdeshpande, K. Tajne, Sonali Patil, "Enhancing Security and Privacy in Biometrics Based Authentication System Using Multiple Secret Sharing", International Conference on Computing Communication Control and Automation (ICCUBEA), pp. 190–194, IEEE, 2015.
10. Li Bai, S. Biswas, A. Ortiz, D. Dalessandro, "An Image Secret Sharing Method", IEEE, International Conference on Information Fusion (ICIF), pp. 1–6, 2006.
11. C. C. Thien, J. C. Lin, "Secret image sharing", Computers & Graphics, volume 26, pp. 765–770, 2002.
12. Sudeep Thepade, Rupali Patil, Sonali Patil, "Novel Reversible Image Secret Sharing Based on Thien and Lin's Scheme using Discrete Haar wavelet Transform", International Conference on Pervasive Computing (ICPC), pp. 1–4, IEEE, 2015.
13. H. B. Kekre, Tanuja K. Sarode, Aditya A. Tirodkar, "An Evaluation of Palmprint Recognition Techniques using DCT, Haar Transform and DCT Wavelets and their performance with Fractional Coefficients", International Journal of Computer Applications, vol. 32, No. 1, pp. 31–38, 2011.
14. The Hong Kong Polytechnic University, http://www.comp.polyu.edu.hk/~biometrics/HRF/HRF.htm.
15. Manisha Dale, Madhuri Joshi, Neena Gilda, "Texture based palmprint identification using DCT features", International Conference on Advances in Pattern Recognition, pp. 221–224, 2009.

Graphical Password Using an Intuitive Approach

**Rajat Mahey(iD), Nimish Singh(iD), Chandan Kumar(iD),
Nitin Bhagwat(iD) and Poonam Verma(iD)**

Abstract In this modern age of increasing interconnectivity, information security has come to occupy a pivotal position in recent research works. The vast expanse of Internet and human–computer interaction has created new threats of unauthorized access and data exploitation. Traditional methods of authentication have been text-based password schemes. However, the major drawback with them has been the fact that their strength varies inversely with user convenience. Users generally tend to select passwords that are easier to recall and shorter in length. This, though, makes them vulnerable to cracking attempts. A graphical password is a confirmation framework that works by having the client select pictures, in a particular order. Graphical passwords have inherent advantages over conventional textual password schemes. In this paper, we propose one such graphical password methodology which makes use of the distinct shape, color, and type of image a user chooses, for the purpose of authentication.

Keywords Authentication · Graphical password · Security

R. Mahey (✉) · N. Singh · C. Kumar · N. Bhagwat · P. Verma
Computer Science and Engineering Department,
Bharati Vidyapeeth College of Engineering, New Delhi, India
e-mail: rjtmhy@gmail.com

N. Singh
e-mail: nimishoc@gmail.com

C. Kumar
e-mail: chandantapu30@gmail.com

N. Bhagwat
e-mail: n.bhagwat123@gmail.com

P. Verma
e-mail: poonamverma267@gmail.com

© Springer Nature Singapore Pte Ltd. 2018
S.S. Dash et al. (eds.), *International Conference on Intelligent Computing
and Applications*, Advances in Intelligent Systems and Computing 632,
https://doi.org/10.1007/978-981-10-5520-1_15

1   Introduction

Recent years have seen a huge growth in computer interconnectivity. It is almost impossible for any organization to exist without networking capabilities. Huge amount of sensitive and non-sensitive data are flowing around the world every second. Information sensitivity and constant data exchange have led to the creation of a number of authentication schemes. User authentication has come to be of prime importance in securing data repositories. Authentication has become the central idea in providing a safeguard between data and its unintended access. A number of people have been interested in exploring and creating new, stronger, and user-friendly authentication schemes.

One of the earliest methods has been to use textual passwords for authentication. Users are required to recreate a text pattern that they had originally selected, to gain access to data. Textual passwords are easy to implement and provide a fair deal of security. The strength of a text-based password depends on its length and randomness of its characters: A longer password with more randomly generated characters is safer against cracking attempts. However, this very fact poses a problem for the user. The user is usually tempted to select a pattern that is easier to recall, which means that more often than not, the user will either choose an insufficiently long pattern or will opt for a more predictable and commonly occurring pattern. Such passwords are easier to crack using brute force or dictionary-based attacks, which are the primitive forms of password cracking techniques employed by malign persons. As a result, password becomes weak, and security is compromised.

The next stepping stone in this domain was graphical passwords. Graphical passwords authenticate a user based on his capability to reproduce a visual pattern or discover a combination of interrelated images. Humans have been known to have a better visual memory and find it easier to memorize images than plaintext characters. It is this innate tendency that is made use of, in graphical passwords.

A graphical password is easy to remember and is less prone to keylogger-based attacks that target the normal text password-based authentication system. This makes the graphical password-based system a better alternative to standard text password-based system.

Several techniques have been proposed to minimize the limitations of alphanumerical password. Mulwani et al. [1] used a grid of characters from which the password is selected by the user according to a set of rules; the characters of the text-based password are used by the system to direct the user on how the rules are to be applied to get access to the system. The first four characters of the password are arranged in such a way that they form the vertices of a square, and the user has to select a character from the square according to some rules as determined by the system.

Most of the current implementation of graphical passwords is based on cued click points (CCP) [2] which uses the concept of viewport and shuffle. In this, the user, at the time of registration, selects a certain part of the image and identifies the points in that selected part as click points. The main disadvantage of this scheme is that, it again requires the user to remember exactly, his selected click points.

Here, we propose a graphical password authentication system that aims to eliminate most of the drawbacks of common graphical password schemes. In Sect. 1.1, we give a brief review of graphical passwords. The proposed system is described in Sect. 2. In Sect. 3, we briefly discuss implementation and highlight some aspects about the proposed system.

1.1 Graphical Password

In 1996, Blonder first introduced graphical passwords [3]. He presented it as the task of clicking on a few selected regions, when an image is shown on the screen. If the correct regions were clicked, the user would be authenticated.

Graphical passwords techniques are broadly classified into two main categories: recognition-based graphical techniques and recall-based graphical techniques [4]. In recognition-based techniques, a user is authenticated by making him recognize one or more images that are interrelated in some way. In recall-based techniques, a user is asked to reproduce a pattern or an image that he created or selected earlier during the registration stage. Some of the reviewed graphical password techniques include DAS [5], character-based grids [1], and click-point authentication [3] (Figs. 1, 2 and 3).

While these techniques offer excellent advantages over textual techniques, each of them suffers from one drawback or the other. Though DAS is language

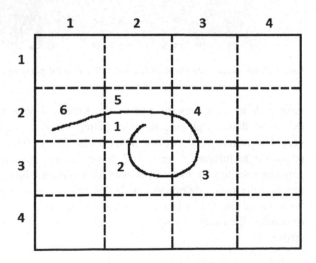

Fig. 1 DAS technique (input of a graphical password on a 4 × 4 grid) [6]

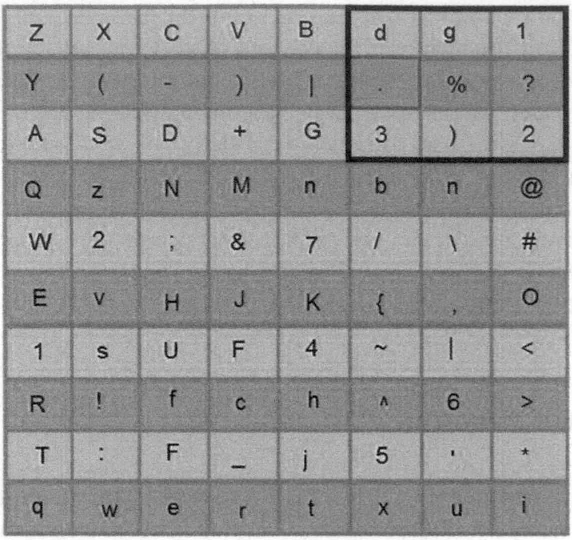

Z	X	C	V	B	d	g	1
Y	(	-	)	|	.	%	?
A	S	D	+	G	3	)	2
Q	z	N	M	n	b	n	@
W	2	;	&	7	/	\	#
E	v	H	J	K	{	,	O
1	s	U	F	4	~	|	<
R	!	f	c	h	^	6	>
T	:	F	_	j	5	'	*
q	w	e	r	t	x	u	i

Fig. 2 Character-based grid [1]

Fig. 3 Click-point authentication (cued-recall graphical passwords) [7]

independent, it requires the user to remember the pattern exactly. Nali and Thorpe [8] found that a high percentage of patterns had symmetry, and many others had very few numbers of strokes. Thus, even though DAS is an effective graphical password technique, there is still some element of predictability in the general patterns selected by users as password. Character-based grid, on the other hand, requires the user's familiarity with the characters of a particular language that is being used in the scheme. Click-point technique again requires the user to remember the exact points of authentication and offers a low level of tolerance for error.

2 Proposed System

The proposed system works as follows. At the time of registration, a user has to select three images from a list of images, as follows. The user has to select an image type (I) (say, from the set of images of animals) and also a shape (S) from a set of "n" shapes (e.g., triangle, square, circle, etc.). Lastly, the user is needed to choose a color (C).

The combination of the chosen shape, color, and image (the "parameters") will be used to authenticate the user whenever he tries to login to the system. This authentication scheme has three stages. At every stage, only two images out of the generated image grid are to be selected.

At the first stage, the system generates a combination, randomly selecting two from the three parameters. Thus, the possible combinations can be either color–shape CS, shape–image type SI or color–image type CI. The user is required to select the images, bearing his selected components, pertaining to the parameter combination generated by the system. He has to select only those two images that contain both the parameters from the given set (Fig. 4).

In the second stage, the user has to select only those two images that contain a combination of the parameter that was not used in the first stage, with each one of the parameters used in the first stage. Essentially, one of the selected images will contain a combination of this particular parameter (unused in the first stage) and one of the previous two parameters; the other image shall contain the other of the previous two parameters, in conjunction with this parameter (that was unused in the first stage) (Fig. 5).

From the last set, the user has to select two such images that do not contain any of the parameters selected, but only contains the parameter left out, in the first stage (Fig. 6).

For selecting m images from a set of n images, we have ^nC_m possibilities. Having p such selection sets makes the number of possibilities as $(^nC_m)^p$. Having n and m sufficiently close, with the value of n decently large, will give a sufficiently huge number of possibilities, even for a small value of p.

Fig. 4 Images to be selected for color–shape combination generated by system and user choices: *yellow color, circle shape, lion image*

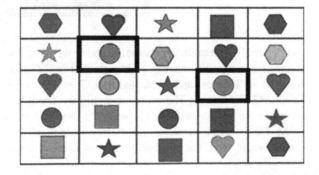

Fig. 5 Images to be selected if the user choices were: *yellow color, circle shape, lion image*

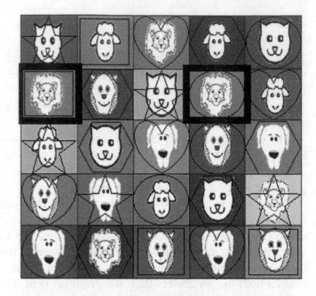

Fig. 6 Third level selection for user choices: *yellow color, circle shape, lion image*

For the aforementioned scheme, suppose we have 3 sets, and in each set, the two images from say, 30 images can be selected in $^{30}C_2$, i.e., 435 ways; and the entire group of six images can be selected in $435 \times 435 \times 435$ ways, which come out to be 82,312,875 ways. Therefore, it is highly improbable for an attacker to gain access to the system without knowing all the three parameters.

As Table 1 shows, the number of combinations becomes massive, even for a smaller value of n, the number of images randomly generated in a set, and m, the number of images to be selected by the user. This allows the algorithm to be user-friendly, requiring him to choose a small number of images, while also being manageable in terms of the memory requirement for storing and displaying the image sets. The high number of possible combinations makes it almost impervious to brute force attacks.

We believe that proposed approach is promising and unique for at least two reasons:

- It provides multi-parameter authentication in a friendly intuitive system.
- Uses three different levels for the authentication of a user.

The proposed scheme inherits the merits of any other graphical password method. Textual passwords are prone to dictionary attacks and can also be cracked via brute force attacks. The strength of a textual password is directly proportional to its length, as well as the randomness of its characters. However, in order to be able to remember passwords, a user is tempted to select a commonly occurring, preferably English, word as the password. A shorter length password makes it vulnerable to brute force attack, while a commonly used word makes it vulnerable to dictionary attacks. In addition, because such passwords are keyboard-input, they can be traced using software such as keyloggers and are also prone to shoulder surfing attacks.

In contrast, the human brain is capable of processing and remembering images, easier than text. Thus, an image-based password chosen by the user can be tougher to crack, in spite of being simpler to remember.

The proposed system, being a combination of recall- and recognition-based techniques, acquires the advantages offered by both the worlds. It is thus, more secure against keyloggers, dictionary attacks, and brute force attacks. Since the scheme does not involve any characters in the presented images, it is language independent and can be easily used even by the people who are uninitiated to characters of different languages.

Our methodology revolves around the color, shape, and type of image. These parameters are easy to remember for even a naive user and do not require him to be well-versed with any of the technical aspects of security. The involved parameters can be recalled by the user naturally, and hence provide considerably more user-friendliness.

Also, as Table 1 suggests, making the user choose more number of answer images at each level enhances the method's defense against brute force attacks. This can be achieved by having more than two images that follow the mentioned rules at each stage amidst a larger grid of random images.

Table 1 Possible combinations for fixed values of n and p (keeping $n = 20$ and $p = 3$)

m	Total possibilities, $(^nC_m)^p$
2	6,859,000
3	1,481,544,000
4	113,731,651,125

3 Discussion

Let us assume that the user has selected circle shape, yellow color, and a picture of a lion as parameters during the registration with the system. Suppose that the system-generated color–shape combination for the first level. The selections that the user has to make in order to gain access to the system in each set of images are described below:

From the first set, the user has to select only those two images that contain the circle shape and yellow color.

From the second set, the user has to select the image which contains lion's image segment and yellow color and another image which contains lion's image segment and the circle shape.

From the last set, the user can select any image that contains the lion's image segment but does not contain either of the yellow color and circle shape.

4 Conclusion

User verification is a central part in most PC and network security settings. In today's world, data security and authorization schemes have come to occupy central positions in designing any information system. The proposed system combines the ease of remembering graphical passwords with the difficulty in cracking highly parameterized password schema. It can also be easily extended to select a larger number of answer images at each stage, without having to tweak the criterion of each stage. Upon combining this methodology with some other authentication process, system security can be highly improved.

The proposed scheme is effective in itself, in providing satisfactory authentication and security. However, its strength can be further improved by making it a part of a multi-level authentication system. Several three-level authentication methods have been mentioned of, like in [3, 8, 9]. Our proposed method can be made a part of such an arrangement to heighten data security.

References

1. Mulwani, Kunal, et al. "3LAS (three level authentication scheme)." *International Journal of Emerging Technology and Advanced Engineering* 3 (2013): 103–107.
2. Chiasson, Sonia, et al. "Persuasive cued click-points: Design, implementation, and evaluation of a knowledge-based authentication mechanism." *IEEE Transactions on Dependable and Secure Computing* 9.2 (2012): 222–235.
3. Vemuri, Vamsi Krishna, and SD Vara Prasad. "A Secure Authentication System by Using Three Level security".
4. Usenix, "Input of a graphical password on a 4x4 grid, digital image", https://www.usenix.org/legacy/events/sec99/full_papers/jermyn/jermyn_html/img31.gif, accessed 18 October 2016.

5. Suo, Xiaoyuan, Ying Zhu, and G. Scott Owen. "Graphical passwords: A survey." *21st Annual Computer Security Applications Conference (ACSAC'05)*. IEEE, 2005.
6. Manjunath, M., Mr K. Ishthaq Ahamed, and Ms Suchithra. "Security Implementation of 3-Level Security System Using Image Based Authentication".
7. Cued-recall graphical passwords, digital image, http://pubs.cs.uct.ac.za/honsproj/cgi-bin/view/2013/mametja_mhlanga.zip/GASSP_Website/ClickPoints.jpg, accessed 18 October 2016.
8. Nali, Deholo, and Julie Thorpe. "Analyzing user choice in graphical passwords." *School of Computer Science, Carleton University, Tech. Rep. TR-04-01* (2004).
9. Sophia, Mughele Ese, "Three-Level Password Authentication." *European Journal of Computer Science and Information Technology* 3.5 (2015): 1–7.

A Model of Legal and Procedural Framework for Cybercrime Investigation in India Using Digital Image Forensics

Vijith T.K. Thekkekoodathil and Pramod K. Vijayaragavan

Abstract Advances in technology lead people to a new space called cyberspace. It introduces a radical change in every span of society. Rapid growth of cyberspace leverages the proliferation of cybercrimes. Use of advanced cyber forensic tools gives confidence to the law enforcement agencies for investigating cybercrimes. Digital image forgery is an easy task by using free-of-cost image editing software. So the image-based cybercrimes are also increasing. There are tools and procedures for image forgery detection in digital image forensics. In the case of image as primary evidence in cybercrime, a law enforcement agency has to adopt a legal and procedural framework for their investigation process. This paper aims to propose a model of legal and procedural framework in cybercrime investigation where a digital image is considered as primary evidence. The model is constructed based on Indian laws and regulations.

Keywords Cybercrime · Cyber law · Image forgery · Digital image forensics Cybercrime investigation

1 Introduction

Cybercrime investigation is an art as well as an intelligent act. Although there are similarities between cyber and conventional crime scene, there are eloquent differences. Conventional procedures are inadequate in cybercrime investigation process. And crime scene concept in cybercrime cases has no fixed boundary. Because of the fragile and sensitive nature of the digital evidences, an investigating

V.T.K. Thekkekoodathil (✉) · P.K. Vijayaragavan
Department of Computer Applications, Cochin University of Science and Technology,
Cochin, India
e-mail: vijithtk@gmail.com

P.K. Vijayaragavan
e-mail: pramodkv4@gmail.com

© Springer Nature Singapore Pte Ltd. 2018
S.S. Dash et al. (eds.), *International Conference on Intelligent Computing and Applications*, Advances in Intelligent Systems and Computing 632,
https://doi.org/10.1007/978-981-10-5520-1_16

team should be well aware about the legal and procedural steps in the investigating process.

In the investigation scenario, the officer follows laws enacted in that country, and legal and procedural framework prescribed by the concerned authority. In the case of cybercrime, instead of adopting a conventional method, a specialized legal and procedural framework has to be adopted.

Digital images are becoming tools for cybercrimes, especially through social media. This is a major problem before the law enforcement agencies and social media service providers. If cybercrime investigating officers have the facility to ensure the credibility of digital image by the way to test image tampering, it will help them to find out the genuinity of the image. Most of the image forensic tools attempt to identify specific or single type of manipulation. No single tool is available to detect all kinds of manipulations. Combination of tools is the solution for effective image forgery detection.

A multi-tier, hierarchical framework for guiding digital investigation was proposed [1]. It proposes a model which considers different stages in the investigation process and precautionary measures. It compares prevailing single-tier approaches. In the USA, Department of Justice framed a guideline for electronic crime scene investigation [2]. In India, a cybercrime investigation manual was prepared by data security council of India in 2011 [3]. This is the only document in this regard available as a model, but which is not yet notified by the government. For comprehensive study on cybercrime, a document is prepared by United Nations office on Drugs and crime by considering new trends and pitfalls in existing laws for cybercrime and its investigation process [4]. This venture proposes a model guideline for laws and procedures and recommends its participating states to make necessary correction in the rules and regulations.

The model discussing here considers a digital image is the vital evidence in a cybercrime. If this is primary evidence, the law enforcement agency team can take this model as a guideline. Compared to other digital evidence, digital image is more powerful talking evidence, and it is self-exploratory. This property of digital image leverages the effort of investigating team. The model describes law sections connected with image-based crimes and procedure adopted for evidence collection, digital image forensic process and reporting.

This paper is organized as follows: Sect. 2 presents a brief outline about cybercrime and its Indian perspective. Section 3 describes about the Information Technology Amendment Act 2008. Section 4 introduces the concept of digital image forensics. Section 5 presents the proposed model. Finally, Sect. 6 presents the discussion and conclusion, which mention the limitation of the study also.

2 Cybercrime: An Indian Perspective

Computer and computer networks are the backbone of the cyberspace. Increased dependency on cyberspace increases cybercrime. Cybercrime is not formally defined in any law enacted in India. In general, it can be said that cybercrime is any unlawful or unethical act performed in cyberspace [5, 6].

The increased use of digital devices increases cybercrime rate also. Cases reported and persons arrested for cybercrime in 2014 is 12248. Out of this, 5548 cases are under computer-related offenses (Sections 66 and Sections 66 A to E of ITAA 2008) [7]. The rate of increase in cyber cases registered under the ITAA 2008 in last five years is keep growing. This is alarming information about cyberspace. Cases related to cybercrimes not only come under ITAA 2008 but also under IPC and SLL (Special and Local Laws) in India. For example, cyber defamation is a cognizable offense. In law, the term defamation is defined in IPC Section 499. So it will be accounted under IPC.

3 Cyber Law

The parliament of India has passed the Information Technology Act 2000 (ITA 2000), which came in force on October 17, 2000. The act mainly focuses on handling legal validity of electronic transactions, digital signatures, and cyber-crimes [8]. Even in the presence of ITA 2000, cybercrimes rely on IPC 1860 for its legal validity. This act was inadequate to handle new developments in technology and cybercrimes. Under these circumstances, debates and discussions on ITA 2000 lead to an amendment of it, resulting in the Information Technology Amendment Act 2008 (ITAA 2008) [9].

4 Digital Image Forensics

4.1 Digital Image Formation

A digital image is formed through various processing stages which are classified as three main phases—acquisition, coding, and editing. Figure 1 shows the various stages in the digital image formation [10]. In acquisition phase, light is coming from the real-world scene which is framed by camera components starting from lens and ending at sensor and its associated processing. In coding phase, camera converts image into a compressed form. Most of the camera uses JPEG compression format with specific parameters for compression. In editing phase, images may undergo various formatting by editing software.

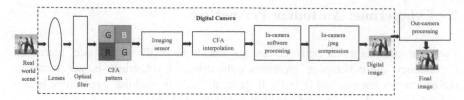

Fig. 1 Stages of image formation

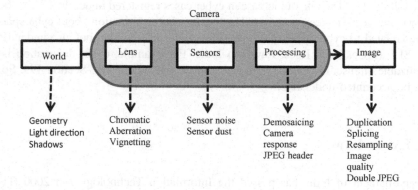

Fig. 2 Sources of image regularities

4.2   Sources of Image Regularities

Digital images have certain regularities that are disturbed by tampering. The regularities are introduced by stages of image formation. Forensic techniques are used to measure the regularities and find the difference in the measurements. Image creation is a sequential process, and each process stage leaves its traits on image called fingerprint. Image forensics procedure identify fingerprints and analyze it by testing whether any irregularities present.This can be used for finding the source of an image or its tampering detection.

Each image acquisition device has its own unique characteristics even if it is a same model of the same brand [11]. And each image has its own fingerprint which is unique. Figure 2 shows the sources of regularities in the images [12, 13]. The object and the scene in the world have physical characteristics and relationships. The things considered here are geometry, direction of light, and position of shadow. The image that has been captured keeps these relationships and characteristics. Chromatic aberration occurs due to the physical limitations of the lens. And defects during the manufacturing of a lens affect the image. The manufacturing defects of the sensors called sensor noise, and the effect of dust on the sensors leaves its own traits on image. The various in-camera processes create regularities such as demosaicing, camera response, JPEG header. Duplication, splicing, re-sampling, image quality, and double-JPEG are the image editing operations that are used to alter image regularities.

4.3 Image Tampering

Photographs are self-explanatory evidences which are widely accepted. Today, digital images completely override the conventional analog photograph. But they do not have that much credibility, because of the ease of manipulation as a result of the availability of professional and free-of-cost image editing software [14]. The forgery of images has spread every sphere of its use. Establishing authenticity of images and distinguishing the original and the forged image are a challenging task. Image tampering is defined as 'adding, changing, or deleting some important features from an image without leaving any obvious trace' [15]. Any manipulations on image like splicing, copy-move, cloning, retouching, cropping, etc., are the methods used for image tampering [16].

4.4 Digital Image Forensics

There are various tools for detecting image forgery. These all come under two different categories, active forensics and passive forensics, and the classification is shown in Fig. 3 [13]. The active forensic approach requires pre-computation of some fragile information like cryptographic hash of the media or insertion of a watermark through information hiding. But this approach is only applicable in a sophisticated domain. The passive forensic approaches are readily applicable to a wide range of forensic applications. This approach is based on two sub categories, i.e., intrinsic regularities and tamper anomalies. Intrinsic regularity uses unique intrinsic footprints for source identification or to assess the disturbances. Tamper anomalies use some specific anomalies left over by some processing operations.

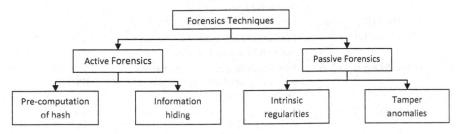

Fig. 3 Classification of digital image forensic techniques

5 Proposed Model of Legal and Procedural Framework for Cybercrime Investigation in India Using Digital Image Forensics

5.1 Legal Framework

The legal framework for cybercrime investigation here is consider what are the laws enacted by the government in this regard and what are the sections which are connected with cybercrime investigation where digital image is the primary evidence.

Information Technology Amendment Act 2008 (ITAA 2008), Indian Evidence Act 1892, Indian Penal Code 1860, Banker's Book of Evidence Act 1891 and Indian Telegraph Act 1885 are the laws enacted by government of India which are very much connected with cybercrime investigation. Table 1 describes the sections which are dealing with digital image-related cases.

5.2 Procedural Framework

5.2.1 Evidence Collection

The scene where cybercrime happened is entirely different from conventional crime scene. Since digital evidences are so sensitive and fragile, utmost care should be

Table 1 Sections present in Indian laws associated with image-related cases

S. No.	Law	Section	Offenses/statement	Other applicable laws and sections
1	The Indian Evidence Act 1872 [17]	65 B	Admissibility of electronic records	
2	ITAA 2008 [9]	66 E	Capturing, publishing, or transmitting the image of the private area without any person's consent or knowledge (violation of privacy)	Section 292 IPC
3	IPC 1860 [18]	468	Forgery for the purpose of cheating	
4	ITAA 2008 [9]	67	Publishing or transmitting obscene material in electronic form	Section 292 IPC
5	ITAA 2008 [9]	67 A	Publishing or transmitting of material containing sexually explicit act, etc., in electronic form	Section 292 IPC
6	ITAA 2008 [9]	67 B	Publishing or transmitting of material depicting children in sexually explicit act, etc., in electronic form	Section 292 IPC

taken in advance for searching and collecting evidence from crime scene. The following is the sequences of steps in cybercrime scene investigation [3].

1. Identifying and securing the crime scene
2. 'As is where is' documentation of the scene of offense
3. Collection of evidence

   a. Procedure of gathering evidences from switched-off systems
   b. Procedure of gathering evidence from live systems

4. Forensic duplication
5. Conducting interviews
6. Labeling and documenting of the evidence
7. Packaging and transportation of the evidences

5.2.2 Forensic Analysis of Digital Image

The procedure for conducting digital image forensics on evidence image for test its genuinity is depicted in Fig. 4. Depending up on the nature of the image and type of the case investigating, the process for image forensics differs. Each process is a complex task, therefore it requires considerable amount of expert manpower and time. The forensic expert and investigating officer after discussing about the case decide which approach is most suitable for the forensic process.

1. *Read Image*: For the purpose of analysis, read the evidence image using interface of the digital image forensic software
2. *Meta-data Analysis*: The first operation to perform for identifying forgery on the image is meta-data analysis. In this method, the software fetches and displays meta-information about the image. Meta-information usually holds information like file name, date of creation, date of modification, date of last access, file type, compression type, capturing device.

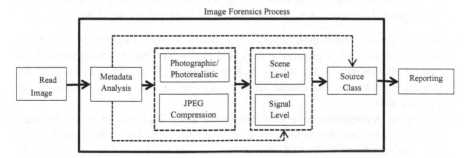

Fig. 4 Procedure for digital image forensics

The meta-information displayed by the system tells us more about the history of the image. From this point itself, analyzer can draw information about the trace of forgery. By analysing the attribute values present in the meta-information trace of forgery can be disclosed. But this is not dependable and reliable information.

3. *Photographic or Photorealistic*: Images are generally classified as photographic and photorealistic images. Images which are created by a camera is called photographic image, where images which are created by computers are called photorealistic images.

   Camera-captured images always keep certain natural statistics. Those statistics are introduced from real-world scene. Real-world objects' geometrical features and their relationships are present in images, and the light/illumination distribution is also contributed for the presence of unique statistical features to the image. Artificially creating natural image statistics is almost impossible.

   Through this process, analyzer can easily identify the genuinity of the image under consideration.

4. *Analysis of Image Coding using JPEG Compression Information*: Most of the cameras store image in JPEG format, which is a lossy compression format. For a particular camera, the parameters used for compression are camera specific. If a person who performs a manipulation on a digital photograph using a photograph editing software, it re-saves to JPEG format. Here actually double compression was performed; one by camera software and second by editing software.

   Using the trace of double compression, analyzer can identify amount of error presents in it by performing error-level analysis (ELA). It reveals the trace of JPEG compression impact and will decide the trace of the manipulation.

5. *Tampering Detection*

   (a) *Scene-Level Based*: The scene-level intrinsic regularities present in the image are altered when it was forged. Splicing type of forgery can be detected by testing what are the scene-level alterations by the way of testing inconsistencies in lighting/shadows and geometry/perspectives.

   (b) *Signal-Level Based*: The signal-level intrinsic regularities present in the image are changed when it was forged. Copy–move forgery and re-sampling type manipulations can be detected by testing what are the statistical changes present among the pixels in the photograph.

6. *Source Class Identification*: Every camera has its own fingerprint in it. Each photograph keeps this fingerprint. This information of the photo can be used for finding the camera using which it was captured. This can be achieved by identifying camera introduced fingerprints by reverse identification of its irregularities.

7. *Reporting*: A comprehensive report about the analysis generated by the software describes what are the techniques were used for identifying the manipulation and specifies the procedure adopted in each step of the process that is admissible to the court of law.

6   Discussion and Conclusion

The aim of this paper is to create a model of legal and procedural framework for cybercrime investigation based on rules and regulations framed by government of India where digital images are the primary evidence. Cyber crime is increasing in India. The crime scene of and investigation strategy for cyber crime have lot of differences when compared to convetional crimes. Law enforcement agencies handling cyber crime as convetional crime. Lack of training and awareness about the technology and underlying process, law enforcement agencies, and associate machineries could not utilize the simplified and effortless way for investigating and handling cyber-related crimes.

Apart from the conventional evidences from a crime scene, digital evidences are talking evidences. In the case of digital image is the primary evidence in a cybercrime, an investigating officer should not follow conventional procedure for the investigation of the associated case. The officer can adopt legal and procedural framework prescribed in this model. If an investigation team is aware about the proposed framework they will have an idea about the precautionary measures adopted in the investigation process like evidence collection, evidence packaging and send it to forensic analysis. Hope, these will help and improve the efficiency of investigating agencies and associated machineries. Legal validation, feasibility, and applicability of the model have to be tested.

References

1. Beebe, N. L., Clark, J. G.: A hierarchical, objectives-based framework for the digital investigations process. Digit. Investig. 2, 147–167 (2005).
2. Mukasey, M. B., Sedgwick, J. L., Hagy, D. W.: Electronic Crime Scene Investigation: A Guide for First Responders, Second Edition. Electron. Crime Scene Investig. A Guid. First Responders, Second Ed. 74 (2008).
3. Bhawan, N., Floor, rd, Tula Ram Marg, R., Bangalore Cyber Lab Karthik Chennai Cyber Lab Abhishek Kumar Haryana Cyber Lab Chaitanya J Belsare Mumbai Cyber Lab Dinesh Dalvi Mumbai Cyber Lab Sandip P Gadiya, M.R.: DATA SECURITY COUNCIL OF INDIA Cyber Crime Investigation Manual. (2011).
4. Conference Support Section, Organized Crime Branch, Division for Treaty Affairs, Unodc: Comprehensive Study on Cybercrime. United Nations Off. Drugs Crime. 1–320 (2013).
5. N Godbole, S. B.: Cyber Security Understanding Computer Forensics and Legal Perspective. Wiley-India (2011).
6. Garima, T.: Understanding Laws-Cyber laws and Cyber Crimes. Lexis Nexis. (2014).
7. NCRB: Cyber Crimes Chapter of Crimes in India. 175–180 (2008).
8. Minsitry of Law, J. and C. affairs: Information Technology Act. 1–13 (2000).
9. Law, M., Affairs, C.: The Information Technology ACT, 2008. Jyaistha. 1922, 1–38 (2008).
10. Lukáš, J., Fridrich, J., Goljan, M.: Digital camera identification from sensor pattern noise. IEEE Trans. Inf. Forensics Secur. 1, 205–214 (2006).

11. Faugeras, O.: Three-dimensional computer vision: a geometric viewpoint. (1993).
12. Piva, A.: An overview on image forensics. ISRN Signal Process. (2013).
13. Sencar, H., Memon, N.: Digital image forensics. (2013).
14. Lester, P.M.: Photojournalism: An Ethical Approach. (2015).
15. Farid, H.: Image forgery detection. IEEE Signal Process. Mag. (2009).
16. Popescu, A. C., Farid, H.: Statistical Tools for Digital Forensics. 6Th Int. Work. Inf. Hiding.
    3200, 128–147 (2004).
17. The Indian Evidence Act. 1934, (1872).
18. Government of India Government: The Indian penal code. 377. 120 (1860).

Comparative Study of Android-Based M-Apps for Farmers

Sukhpreet Kaur and Kanwalvir Singh Dhindsa

Abstract M-Apps grow to be renowned feature of mobile phone. Farmers also getting solution to their farming problems with the help of the mobile applications. This paper highlights the mobile applications for agriculture, where architecture of mobile phone with its current trends and the most useful M-Apps related to agriculture market, soil description, weather information, and government schemes have been described in detail. In this paper, Android-based M-Apps act as foremost applications, where Arduino used as an external hardware kit for sensors. Comparison is based on the features provided by the developed M-Apps to the users. The benefits and drawbacks of developed M-Apps have also been discussed. The conclusion describes the most useful M-Apps and its further improvements.

Keywords Android · E-agriculture · Mobile application · Arduino
M-App · Agriculture · GPS · ICT · IoT

1 Introduction

Mobile manufacturers are enhancing configuration of mobile phones, whereas developers are trying to solve real-life problems through the development of M-Apps. Agriculture is the backbone of Indian economy. Mobile application is an efficient approach to enhance the knowledge of farmers. Cloud computing, sensor control, GPS, image processing, data mining, speech processing and other techniques have been used to provide knowledge about farming [1, 2]. M-Apps for

S. Kaur (✉)
I.K.G. Punjab Technical University, Jalandhar, Punjab, India
e-mail: sukh5waheniwal@gmail.com; sukhpreetkaurggscmt@gmail.com

K.S. Dhindsa
Baba Banda Singh Bahadur Engineering College, Fatehgarh Sahib, Punjab, India
e-mail: kdhindsa@gmail.com

© Springer Nature Singapore Pte Ltd. 2018
S.S. Dash et al. (eds.), *International Conference on Intelligent Computing and Applications*, Advances in Intelligent Systems and Computing 632,
https://doi.org/10.1007/978-981-10-5520-1_17

173

agriculture are used for utilizing and managing agriculture resources. Farmers use M-Apps to get information about weather forecasting, soil conservation, insecticides, harvesting and market price of crops [3]. A mobile application can convey information about farming strategies in detail. Interactive voice response system (IVRS) has been used in India for farmers. IVRS gives automatic voice response regarding farming problems. Various IVRSs give facility to talk with agriculture experts. The remaining paper is structured as follows: Scct. 2 introduces the existing work related to mobile phones and M-Apps. Section 3 highlights the mobile phone architecture and trends related to mobile phone users. Section 4 describes M-Apps and its role in agriculture. Section 5 provides the details of existing agriculture M-Apps in tabular form. Section 6 provides M-Apps comparison from user point of view. Section 7 highlights the limitations of developed M-Apps. Section 8 concludes the work with future scope.

2   Related Work

Various M-Apps have been developed by researchers. Dhaliwal and Joshi [4] highlighted mobile phone as a boon for rural system, where mobile phone can help farmers to take decision about production and marketing of their crops. Mondal et al. [5] preferred to use information technology and biotechnology for agriculture than traditional agriculture system. A smartphone agricultural app was developed by Karetsos et al. [6], where a user can find information about farming and a farmer can submit his requests by following simple procedures. According to Zhu et al., cloud computing technology can play a very important role in enhancement of agriculture business, but there is a need to build data centres, enhance service quality, integrate resources and make information security [7]. Sandhu et al. analysed Kisan mobile advisory services of Punjab [8]. Mobile ICT is a better technology to increase the efficiency of farming. MahaFarm is an Android-based solution developed by Bhave et al. [9] for agriculture in which a user can get information about crop, weather, and market price. This application provides location-specific information. Narechania [10] proposed Arduino system for agriculture. Arduino system is a four-wheeled wireless robot, controlled by Android cell phone, used for farm management. Crop details, market price of crops and weather information are showcased by Patodkar et al. [11]. Gao and Yao [12] had implemented a set of moveable system to check climate of agricultural area, using Android mobile applications and IoT. The literature review on M-Apps is accessibility for farmers; all conclude that M-Apps for agriculture are not providing a user-friendly environment for agriculture. A layman should be able to understand functioning of M-Apps.

3 Mobile Phone Architecture and Current Trends

Mobile architecture comprises of hardware and software, where internal and external mobile applications are being used for task performing. The basic architecture of mobile phone is shown in Fig. 1.

Figure 1 shown describes the memory, keypad, speaker, microphone, SIM and LCD as the common hardware devices of a mobile phone, whereas the application processor and communication processor are used for internal functionalities. To enhance the capabilities of mobile phone, various applications are being developed. Figure 2 shows the growth of total wireless subscribers.

Figure 2 shows the projected growth of telephone subscribers in India. The number has increased from 1,029.34 million (430.09 + 599.25) in October 2015 to 1035.18 million (436.26 + 598.92) at the end of November 2015, with 0.57% monthly growth rate [13]. The use of mobile technology is increasing in rural areas rapidly.

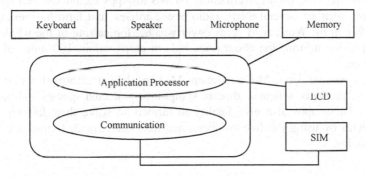

Fig. 1 Basic architecture of mobile phone device

Fig. 2 Total wireless
subscribers (in millions) [13]

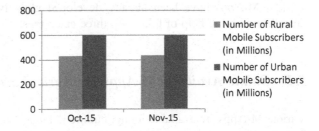

4 M-Apps Role in Agriculture

A few M-Apps are pre-installed in mobile phone devices, whereas other M-Apps can be installed with the help of application distribution platform. Owners of mobile operating systems have provided their own platform such as Google Play Store by Android, Windows Phone Store by Windows, Apple App Store by Mac and BlackBerry App World by BlackBerry. Android is a popular OS (operating system). The source code of Android operating system is available by way of open source, and M-Apps for Android are also available on Google play store. Mobile phone has been playing significant role in the enhancement of agricultural production. Various categories of agriculture M-Apps are listed as follows:

- *Agriculture Information Resource Apps*: These types of M-Apps include information about farmers, fertilizers, insecticides, pesticides, crop disease resistance rating in local and global market.
- *Agriculture Calculator Apps*: This kind of M-Apps are used to calculate the state of cultivated plant, by inspecting present and past days' information of field locale. A user can search for farm equipments also.
- *Agriculture NEWS Apps*: Agriculture NEWS M-Apps include NEWS related to market, weather, several alerts, radio news, articles, and farm business blogs.
- *Weather Apps*: Weather M-Apps cover weather forecasting of the whole world and provide information about price bids of crops within 100 miles of farmer location.
- *M-Government Apps*: M-Government M-Apps deliver information about varieties of products, schemes, disease symptoms, protected species and protected areas. These apps also give facility to farmers to solve their farming-related problems by using toll-free numbers, real-time chat, email, forum, and private messaging.

5 M-Apps Available for Agriculture

Various M-Apps have been already developed globally. Available M-Apps are described with the help of following three categories.

5.1 Online Available M-Apps for Agriculture in India

Various M-Apps are available for agriculture in India. Table 1 shows the significant features of the most beneficial M-Apps. These M-Apps are useful for Indian farming and are available online.

Table 1 Online available M-Apps for agriculture in India

S. No.	M-App name, version	Minimum OS required	Developers	Area covered, supported languages	Significant features
1	myRML, version 2.0.2	Android 2.3.3	RML AgTech.	18 states of India, languages: 9 different languages	Market price, weather forecasting, agriculture advisory, agriculture news, government schemes, unit converter
2	AgriSmart, version beta	Android 2.1	Punjab agriculture Dept.	Punjab, languages: English	Extension, crop reporting, pest warning
3	Digital Mandi, version 1.0	Android 2.2	AppKiddo	15 states of India, languages: English	Indian agricultural commodity market price list
4	Modern Kheti, version 4.0	Android 3.0	Magzter Inc. education	Northern India, languages: Punjabi, Hindi and English	Magazine of latest agriculture news.
5	Water reporter, version 1.4.2	Android 4.0	Viable industries	India, languages: English	Pollution and activity report submission
6	Kisan Yojna, version 3.0	Android 3.2	ANN India	Maharashtra, languages: local language	Government's schemes
7	CropInfo, version 2.1.1	Android 2.3	Nirantara LRPL	Global, languages: Kannada and English	Articles about horticulture crops
8	Kisan Books	Android 3.0	Kisan form Pvt. Ltd	India, languages: English, Hindi and 5 local languages of India	E-Books magazines and booklets for agriculture
9	Fertilizer calculator, version 1.02	Android 2.2	Dr. Vishwanath Koti	India, languages: English	Calculator to estimate the fertilizer
10	AgriApp, version 1.3.1	Android 4.0	Criyagen, AgriApp communications	India, languages: English, Hindi, Kannada	Information related to crop, chat, call and videos, place orders

As shown in Table 1, myRML and Digital Mandi are the most useful M-Apps to know the market price.

- myRML provides market price, weather forecasting, agriculture advisory, news, government schemes, unit converter, and calculator for plants, whereas Digital Mandi provides market price only.
- Fertilizer Calculator estimates the fertilizer based on nitrogen, phosphorous and potassium, whereas AgriSmart provides information about pest warning.
- CropInfo and Kisan Books are effective apps to access e-books, magazines and booklets, whereas Kisan Yojna gives information about government schemes only.

5.2 Online Available M-Apps for Agriculture in Foreign Countries

Various M-Apps are available for agriculture in foreign countries. Table 2 shows significant features of the most constructive M-Apps. These M-Apps are available online.

As shown in Table 2, MySoil and Fertilizer Cost Calculator both are helpful to check the value of soil.

- MySoil is an application to provide information about soil description only, whereas Fertilizer Cost Calculator, roughly calculates the value of nitrogen in soil.
- FarmEmergencyContacts contain all farm emergency contacts of the UK. Growing Organic Vegetables is a paid mobile application to get tips for growing organic vegetables.
- AgriNews Live and Farm Progress are used to get information about agriculture news, market commentary, reports from grain and livestock farmer.

5.3 M-Apps for Agriculture by Various Developers

Various M-Apps have been developed by researchers for farming, but these M-Apps are not available in play stores. Table 3 describes the significant features of the latest M-Apps.

Table 3 shown provides the details of various M-Apps developed by developers.

- Smart Agri Services is Android-based M-App for farmers. This M-App is information-based app, where a user can get knowledge about all types of farming.
- E-Agro M-App provides information regarding crop management.

Table 2 Online available M-Apps for agriculture in foreign countries

S. No	M-App name, version	Minimum OS required	Developers	Area covered, supported languages	Significant features
1	My Soil, version 1.1	Android 2.3	BGS and Centre of Ecology and Hydrology	Britain and Europe, languages: English	Soil description
2	GAF 2014, version 1.2	Android 2.3	ISuL Techno logia entertainment	Global, languages: English	Global agribusiness forum for discussion
3	Farm Emergency Contacts, version 1.1	Android 1.5	WeFarmItAppLabs Communications	UK, languages: English	Contain all UK farm emergency contacts
4	Farm progress, version 4.4	Android 2.3.3	iNet solutions group, Inc.	Global, languages: Publications in 14 languages	Agri. News, Grain and Livestock market, Weather forecasting blogs, magazine
5	Agricultural policy research, version 1.2	Android 2.2	Global development network	Sub-Saharan Africa and South Asia, languages: English	Papers regarding different policy research, policy briefs, documentaries of project, news and events
6	Crop and medicinal plants, version 1.0	Android 4.0.3	Wiki Kids Limited	Global, languages: English	Knowledge about preparation of soil and seeds, fertilizers crop rotation and harvesting, plant breeding, medicinal plants
7	Learn agricultural engineering, version 1.7	Android 2.1	WAGmob	Global, languages: English	Snack Sized chapters, Flashcards for agriculture engineering
8	AgriNews Live, version 3.0.1	Android 2.1	LoadOut	America, languages: English	Live market commentary, staff blogs, audio and video
9	Fertilizer cost calculator, version 1.0	Android 2.3.3	JWS Group LLC	Australia, languages: English	Estimates the value of nitrogen in soil
10	Growing organic vegetables, version 1.0	Android 2.2	Brook Barn, Graffham West Sussex	UK, languages: English	Providing tips for growing organic vegetables

Table 3 M-Apps for agriculture by various developers

S. No.	M-App name	Year	Developers/paper cited	Tool and techniques used	Significant features
1.	Smart Agri Services	2014	Karetsos et al. [6]	Android developer tools, Android version 2.2	Information regarding: horticulture, forestry, apiculture, sericulture
2.	MahaFarm	2014	Bhave et al. [9]	GPRS, Wi-Fi, Android, XML, Yahoo weather API, HTTP	Weather updates, market prices, news updates
3.	Arduino system	2015	Narechania [10]	Android APIs, SQLite, Google maps V 2.0, HTTP, GPRS, GPS, XML, JSON	Weather forecast, market prices, agriculture news, farmer help lines, map view of fields, farm management, plough mechanism, wireless pump operation, obstacle avoidance and indicator
4.	E-Agro	2015	Patodkar et al. [11]	MySql, GCM, JSON, XML, Android OS 4.2.2	Crop details, crop alerts, ask questions, weather info, crop rates

- MahaFarm, Ardunio System and E-Agro provide the option of market price, whereas Ardunio system is a combination of hardware and software systems. Software part includes weather forecasting, market prices, agriculture news, farmer help lines, map view of fields, farm management and plough mechanism, whereas hardware system provides obstacle avoidance and indicator and wireless pump operation management.

6   Comparison of Available M-Apps for Agriculture

Farmers install M-Apps to find solutions to their problems, so comparison of available M-Apps is based on user point of view. Table 4 describes all the facilities, which have been given to the user by the developers. All given facilities would be used to ask queries and to access information.

Table 4 shown compares the existing M-Apps based on facilities given to the users.

- myRML, AgriSmart, Water Reporter, Kisan Books and GAF 14 provide facility of uploading information using text, images and feedback options.

Table 4 Comparison of available M-Apps for agriculture

S. No.	Mobile app name	Upload information				Significant features
		Feed-back form	Text	Image	Audio and video	Group chat, IVRS, make call, download information, place orders, GPS, FAQs, email
1	myRML	Yes	Yes	Yes	–	Group chat and GPS
2	AgriSmart	Yes	Yes	Yes	–	–
3	Digital Mandi	–	–	–	–	–
4	Modern Kheti	–	–	–	–	Download information
5	Water Reporter	Yes	Yes	Yes	–	GPS
6	Kisan Yojna	–	–	–	–	–
7	CropInfo	–	–	–	–	Downloading
8	Kisan Books	Yes	Yes	Yes	–	Make call, place orders
9	Fertilizer calculator	–	Yes	–	–	–
10	AgriApp	Yes	Yes	Yes	Yes	Group Chat, IVRS, make call, download information, place orders
11	MySoil	–	–	–	–	GPS
12	GAF 14	Yes	Yes	Yes	–	Group chat, GPS
13	FarmEmergencyContacts	–	–	–	–	Make call
14	Farm Progress	–	–	–	–	Download information
15	Agricultural policy research	–	–	–	–	–
16	Crop and medicinal plants	–	–	–	–	Download information
17.	Learn agricultural engineering	Yes	–	–	–	Download information
18	AgriNews Live	–	–	–	–	–
19	Fertilizer cost calculator	–	Yes	–	–	–
20	Grow your own organic vegetables	Yes	–	–	–	–
21	Smart Agri Services	–	–	–	–	FAQs
22	MahaFarm	–	–	–	–	GPS
23	Ardunio System	–	Yes	–	–	GPS
24	E-Agro	Yes	Yes	Yes	Yes	Email, make call, IVRS

- AgriApp is the only App, which covers maximum features, whereas Agricultural Policy Research, Digital Mandi and AgriNews Live are based on static information. Neither a user can upload his query nor he can download information.
- In Fertilizer Calculator and Fertilizer Cost Calculator, user can upload information using text, whereas in Grow Your Own Organic Vegetables, user can upload information using feedback option only.
- The feature of downloading information is available through Modern Kheti, CropInfo, AgriApp, Farm Progress, and Crop Medicinal Plants, whereas the feature of GPS is provided by myRML, Water Reporter, MySoil, GAF 14, MahaFarm and Ardunio System.
- The facility of IVRS is part of AgriApp only, whereas option of making call is provided by M-Apps as in Kisan Books, AgriApp and FarmEmergencyContacts.
- myRML, AgriApp and GAF 14 include group chat option, whereas Smart Agri Services provide facility of FAQs only.
- E-Agro is the only M-App, which provides option of uploading queries through email.

7 Limitations in Available M-Apps

Existing M-Apps have their own attributes. Each app is useful for farmers, but there exist some limitations.

Two-way communication is not provided by the developers. The majority of M-Apps provide static information, where user cannot get knowledge about current affairs. There is no option for feedback provided by Agricultural Policy Research, Digital Mandi and AgriNews Live M-Apps. Consequently, there is a gap between user, M-App and provider. AgriApp and E-Agro are the only M-Apps, where user can upload audio and video regarding their agriculture difficulties. IVRS is not followed by all the developers even there is no M-App, which support toll-free numbers. Thus, farmers cannot give a call because of high charges. Advance techniques of IoT have not been used till now. Arduino-based M-Apps have used a few sensors, which can do wireless pump operations only. Language dependency is also the major drawback because all farmers are not good in English language. So, there is a need of multilingual support in M-Apps. As of today, none of the M-Apps provides facility to farmers to buy agriculture products or sell their produce in the market online. There is a need of user-friendly environment.

8 Conclusion

Various types of agriculture M-Apps such as Information Resources Apps, Weather Apps, NEWS Apps, E-Govt. Apps have been described in detail. Mobile architecture and trends provide detailed information about rapid growth of mobile phones over the year. Currently, the best M-Apps in India are myRML (agricultural market), Kisan Yojna (government schemes), Modern Kheti (literature) and AgriSmart (crop information). Farm Progress, Agricultural Policy Research, Growing Organic Vegetables and MySoil are found to be the best M-Apps in foreign countries for agricultural marketing, government schemes, literature and soil information simultaneously. Arduino system is found to be the best M-App by developers in which mobile phone is attached with combination of sensors. AgriApp is the best app in user's point of view, where user can upload his queries with the help of text, image, audio and video. IVRS (paid service) is supported by E-Agro and AgriApp. Language dependency, static information, minimum use of IoT and no online marketing are the major drawbacks of M-Apps. M-Apps with IoT have been found to be the best source for the modernization of agriculture.

References

1. Hori, M., Kawashima, E., Yamazaki, T.: Application of cloud computing to agriculture and prospects in other fields. Fujitsu Scientific & Technical Journal, vol. 46, no. 4, pp. 446–454, (2010).
2. Pocatilu, P.: Developing mobile learning applications for android using web services. Economic Data, vol. 14, no. 3, pp. 106–115, (2010). doi:10.1145/2261605.2261641.
3. Patel, B.V., Thakkar, G.R., Desai, V.H.: An android application for farmers for kharif and rabi crop diseases information. International Journal of Research in Computer Science and Software Engineering, vol. 10, issue 10, pp. 788–791, (2014).
4. Dhaliwal, K.R., Joshi, V.: Mobile phones- boon to rural social system. Literacy Information and Computer Education Journal, vol. 1, issue. 4, pp. 261–265, (2010).
5. Mondal, P., Basu, M., Bhadoria, S.B.P.: Critical review of precision agriculture technologies and its scope of adoption in india. American Journal of Experimental Agriculture, vol. 1, no. 3, pp. 49–68, (2011). doi:10.9734/AJEA/2011/155.
6. Karetsos, S., Costopoulou, C., Sideridis, A.: Developing a smartphone app for m-government in agriculture. Journal of Agriculture Informatics, vol. 5, no. 1, pp. 1–8, (2014). doi:10.17700/jai.2014.5.1.129.
7. Zhu, Y., Wu, D., Li, S.: Cloud computing and agriculture development of china: theory and practice. International Journal of Computer Science Issues, vol. 10, no. 1. pp. 7–12, (2013).
8. Sandu, S.H., Singh, G., Grover, J.: Analysis of kisan mobile advisory services in south western punjab. Journal of Krishi Vigyan, vol. 1, no. 1, pp. 1–4, (2013).
9. Bhave, A., Joshi, R., Fernandes, R.: Mahafarm-an android based solution for remunerative agriculture. International Journal of Research in Advent Technology, vol. 2, no. 4, pp. 43–47, (2014).
10. Narechania, A.: An android-arduino system to assist farmers in agricultural operations. Proceedings of IRF International Conference, New Delhi, India, pp. 19–26, (2015).
11. Patodkar, V., Simant, S., Sharma, S., Shah, C., Godse, S.: E-Agro Android Application. International Journal of Engineering Research and General Science, vol. 3, issue 3, pp. 458–465, (2015).
12. Gao, C., Yao, K.: The design and implementation of portable agricultural microclimate data acquisition system based on android platform. 8th International Symposium on Computational Intelligence and Design, IEEE Explore, vol. 1, pp. 210–213, (2016). doi:10.1109/ISCID.2015.275.
13. Telecom regulatory authority of India: Highlights of telecom subscription date as on 30th June 2015, New Delhi, India, pp. 1–19, (2015).

Feature Extraction of DICOM Images Using Canny Edge Detection Algorithm

Diptee Chikmurge and Shilpa Harnale

Abstract Generally, in the medical field early diagnosis of the disease was performed using MRI, CT scans, X-ray, and ultrasound images. These medical images were captured in Digital Imaging and Communication in Medicine (DICOM) format (Bhagat and Atique in Medical Images: Formats, Compression Techniques and Dicom Image Retrieval Survey, 2012) [1]. As per the structure of DICOM image, physicians were unable to detect strangeness or disease in the patient without any image processing. Image processing and machine learning process can be useful to identify strangeness in these images by evaluating feature extraction and boundary detection of DICOM images which aims to help experts to analyze medical images. These medical images actively engaged in the medical field to diagnose disease and give proper treatment. Nowadays due to increase in the large database of DICOM images, the classification and retrieval of images have been a critical task for diagnosis of disease. The content-based image retrieval is effectively applicable for effective treatment of disease. Canny edge detection algorithm is useful for extracting features of medical images.

Keywords Canny edge detector · Gradient magnitude · Non-maxima repression

D. Chikmurge (✉)
Computer Engineering, MIT Academy of Engineering,
Alandi, Pune, Maharashtra, India
e-mail: dvchikmurge@comp.maepune.ac.in

S. Harnale
Computer Engineering, Bheemanna Khandre Institute of Technology,
Bhalki, Karnataka, India
e-mail: shravi97@rediffmail.com

© Springer Nature Singapore Pte Ltd. 2018
S.S. Dash et al. (eds.), *International Conference on Intelligent Computing
and Applications*, Advances in Intelligent Systems and Computing 632,
https://doi.org/10.1007/978-981-10-5520-1_18

185

1 Introduction

Generally, in the medical field, early diagnosis of the disease was performed using MRI, CT scans, X-ray, and ultrasound images. In advanced medical technology, medical images were captured using DICOM. The structure of DICOM image contains two major parts:

1: Patient Information
2: Actual medical image

(Digital Imaging and Communication in Medicine) format [1]. As per structure of DICOM image, physicians were unable to detect strangeness or disease in the patient without any image processing. Image processing and machine learning process can be useful to identify strangeness in these images. The aim of feature extraction and boundary detection of DICOM images is to help experts in analyzing medical images. These medical images play a vital role in the medical field to diagnose disease and give proper treatment. Nowadays due to increase in the large database of DICOM images, the classification and retrieval of images have been a critical task for diagnosis of disease. The content-based image retrieval is effectively applicable for effective treatment of disease with the help of extracting features of the image. Using effective techniques [2] of CBIR, doctors can give the correct treatment to a patient.

2 Feature Extraction–Edge Detection Algorithm

In this paper, CBIR process is described using edge detection. The edge detection is used for removing unwanted information and extracting an interested region from images. The edge detection is a process which detects and figures out interested entity and makes the distribution of entity and background region in the image. Detection of the edge is the major task of identifying discontinuities in pixel intensity value. There are many techniques [3] for edge detection algorithms such as Prewitt method, Roberts edge detection, Sobel method, and Canny edge method. In DICOM image, the interested region and remaining parts need to be differentiated using image processing. Differentiation between these two parts was possible using edge detection algorithm. Figure 1 represents the example of DICOM image.

The medical image processing is complicated because of noise occurrences in the image. However, the noisy medical images cannot give the correct diagnosis; in the proposed work, we have presented improved Canny edge algorithm to remove noise and smooth the image along with feature extraction of DICOM image [4]. The feature extraction of images is performed by identifying the interested area along with borderlines in different sections of an image. So the image feature

Fig. 1 Sample DICOM
medical image

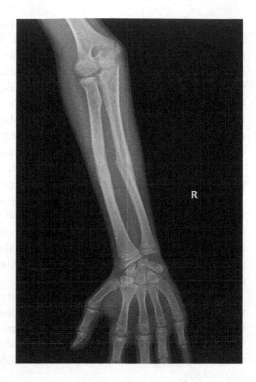

extraction is accomplished by detecting edges in an image. The Canny edge
detection algorithm is the effective technique to find enhanced edges without disturbing features of the image.

3 Canny Edge Detection Algorithm

The Canny edge detector was defined by John F. Canny in 1986. This algorithm
involving several stages is used to detect edges in images. It depends on gradients
(edges) where gray scale intensity changes the most.

The aim of Canny Edge detector algorithm is to assure following three [5]
criterion.

- Detection of the edge of the image with small delusion rate, which intends that
  the perception has detected all possible edges presented in the image.
- Recognized edge pinpoints exactly concentrate on the center of the edge.
- The third criterion removes multiple responses to an edge means there is only
  one counter to a single edge.

The traditional Canny edge detector [6] is used as the Gaussian filter to smooth
and remove noise from an image and then evaluate the gradient intensity and

direction angle of the pixel in an image then employ non-maxima on gradient intensity to suppress the weak edge and use double thresholding to find connected the edge. The traditional Canny edge is not specifying efficient result with two parameters that are a standard deviation and decided double threshold by users. The standard deviation (σ) parameter affects the degree of smoothness of the image. The Gaussian filter is concerned to smoothen the image and get rid of the disturbance from the image. It is a linear low-pass filter. Image filtering using a Gaussian filter is applied to an image to remove noise and smooth the edge in the image. This will cause failure to encounter weak edge occurrence of isolated edges in the result.

The second parameter that is a double-threshold values which are decided by the programmer, will result in loss of some edge and will be unable to remove noise. There are two threshold values decided by a programmer that are high and low threshold values. The gradient magnitude of image edge pixel is greater than the higher threshold value then that edge is considered as the strong edge. If the gradient magnitude is less than the lower threshold value, then the edge is suppressed. And if gradient magnitude lies in high and low threshold values, the edge is considered as a weak edge.

As per the predefined criteria of Canny edge algorithm, high accuracy of real-edge detection is not possible with the use of the Gaussian filter. The expected output should be smoothness which is added to noise more and less to the edge.

In proposed Canny edge detection, these two drawbacks will be removed by using adaptive filters instead of the Gaussian filter and use adaptive double thresholding.

4   Proposed Improved Canny Edge Detection Algorithm

1. Adaptive filter is applied to separate the disturbances from image and smooth the image.
2. Determine the intensity gradients and image angle of the selected image.
3. Employ non-maximum repression to the intensity gradient in the edge direction.
4. Decide threshold value by utilizing OTSU's technique and locate the associated edges.

The system design of proposed work is shown in Fig. 2.

4.1   Adaptive Filtering

In proposed Canny edge algorithm, adaptive filter is used to filter the image. This adaptive filter algorithm [7] will select flexible weight as per the properties of the gray scale value of each pixel and at the same time sharpening edges. If there is

Fig. 2 System design of
proposed system

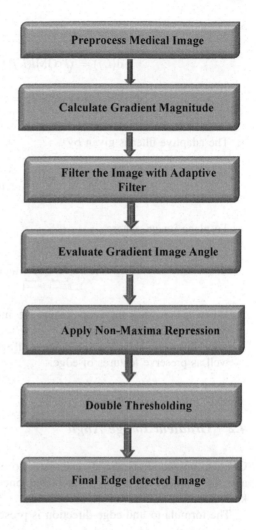

large disruption in a gray scale value of edge pixel, then a weight of adaptive filter is set to low scale, and when a gap is lower, then weight adjusted to the large scale.

Consider the two-dimensional image as $f(i,j)$, $wg(i,j)$ is weight of adaptive filter.

As per behavior of filter, the flow of adaptive filter is presented as follows [8].

1. Consider $g = 1$ set the iteration n and coefficient of edge e.
2. Compute gradient magnitude $GMi(i,j)$ and $GMj(i,j)$ are partial derivation of i and j direction.

$$GMi = \mathrm{d}f(i,j)/\mathrm{d}i \quad \text{and} \quad GMj = \mathrm{d}f(i,j)/\mathrm{d}j$$

3. Compute weight according to formula

$$d(i,j) = \sqrt{(GMi(i,j))^2 + (GMj(i,j))^2}$$

$$\text{And } wg(i,j) = \exp\left(-\frac{\sqrt{d(i,j)}}{2e^2}\right)$$

4. The adaptive filter is given by

$$f(i,j) = 1/N \sum_{a=-1}^{1} \sum_{b=1}^{-1} f(i+a,j+b)wg(i+a,j+b)$$

To plane image, where

$$N = \sum_{a=-1}^{1} \sum_{b=-1}^{1} wg(i+a,j+b)$$

5. When $g = n$, stop the loop, otherwise increase the value of g as $g = g+1$, repeat step 2 to 4.
   This approach of the adaptive filter is effectively used for removing the noise as well as preserve features of edge.

4.2   Gradient Image Angle

Gradient magnitude is already calculated to evaluate the weight of adaptive filter, and the gradient angle [8] is calculated once gradient in x and y directions are known.

The formula to find edge direction is prescribed below:

$$\text{theta} = \arctan\left(\frac{GMi}{GMj}\right)$$

With the value of edge direction angle, associate the edge direction angle with the direction that can be tracked down in the image. The four possible tracing directions such as horizontal, vertical, positive diagonal and negative diagonal that point the surrounding pixel. So edge direction which is calculated will be rounded up to one of four angles or nearest angle.

4.3   Non-Maxima Repression

After the calculation of edge directions, the non-maxima [9] is applied to satisfy the third criteria of Canny edge detection algorithm. Non-maxima repression is useful for thinning the edge and used to focus the gradient in edge orientation. The cost of the current pixel is compared with pixel values that are directed at 90°. Pixel will be repressed or stamped out if their values are lower than the gray scale of a pixel on edge; otherwise, higher pixel value is preserved at the edge.

4.4   Double Thresholding

After using non-maxima repression, edge pixels represent the real edge. But still, some edge pixels in the image are affected by some noise. So it is necessary to remove weak gradient value edge pixel and preserve high gradient value edge pixel [10]. To analyze these forms of edge pixel value, two threshold values need to set or decided by the programmer, one is a major threshold and the second one is a minor threshold value. In traditional Canny edge detection algorithm, double-threshold value is given by the user can cause some problem. To resolve this problem, in improved Canny edge detection algorithm the threshold value is determined using Otsu's method [11, 12]. It is applied on non-maxima repressed gradient magnitude to determine a major threshold value. The minor threshold value is adjusted to 1/3 of a major threshold value. With two threshold values, specified edge will be repressed or edge will be identified as a prominent edge. If the gradient value of a pixel is greater than the minor threshold and smaller than a major threshold value, then an edge is considered as the weak edge. If pixel gradient cost is greater than a major threshold, then that edge is the prominent edge. If pixel gradient cost is less than a minor threshold, then that edge is repressed or removed.

5   Experimental Result

In this part, we will discuss the experimental result of improved Canny edge detection for feature extraction of DICOM images as depicted in Fig. 3. The proposed edge detection method is implemented using MATLAB and tested with DICOM digital images. Figure 3 represents the original image, and the gradient magnitude is calculated and applied on actual image as shown in Fig. 4.

Fig. 3 Original DICOM
image

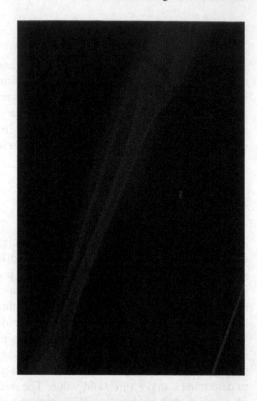

Fig. 4 Gradient magnitude
image

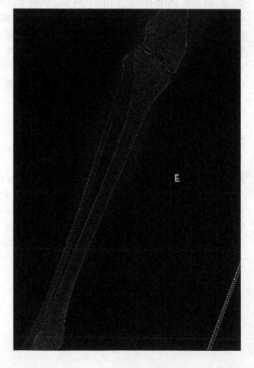

Fig. 5 Gaussian filter image

Fig. 6 Adaptive filter result

Fig. 7 Traditional Canny
edge detector

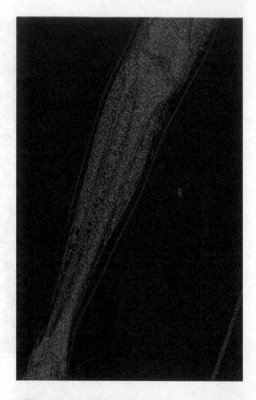

Fig. 8 Final result using
improved Canny edge
algorithm

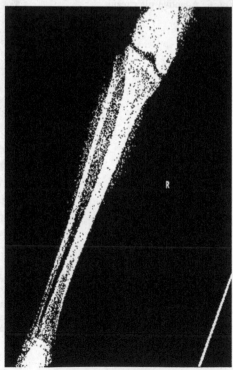

As per circulation of proposed process, the adaptive filter is employed on the image as shown in Fig. 6. After applying an adaptive filter, the result gives the better performance as compared to Gaussian filter output as shown in Fig. 5(c). Then, peak threshold value is calculated using OTSU's method and applied on filtered image and result of double thresholding as shown in Fig. 8 which gives better output than traditional Canny edge detector in Fig. 7.

The main objective of feature extraction is identifying the real edge in the image. We have implemented improved Canny edge algorithm on DICOM image with adaptive filter and adaptive thresholding method. The final output of proposed system is represented in Fig. 8 which edge detected image without disturbing features of images.

6 Conclusion

As we study about feature extraction of DICOM images using improved Canny edge algorithm, it gives effective result in terms of noise removal, extraction of required images. This approach is efficiently useful for medical image retrieval. The medical image retrieval needs high accuracy and robustness without loss of features of images. It can be seen from the experimental result that the adaptive filter and adaptive double thresholding methods give high generalization performance in improved Canny edge detection algorithm.

References

1. A. P. Bhagat, Mohammad Atique, "Medical Images: Formats, Compression Techniques And Dicom Image Retrieval Survey" IEEE international conference on 15–16 March 2012.
2. Subrahmanyam Murala, Q.M. Jonathan Wu "MRI and CT image indexing and retrieval using local mesh peak valley edge patterns" Science Direct journal December 2013.
3. Rashmi, Mukesh Kumar, and Rohini Saxena "Algorithm And Technique On Various Edge Detection: A Survey" Signal & Image Processing: An International Journal (SIPIJ) Vol. 4, No. 3, June 2013 DOI: 10.5121/sipij.2013.430665.
4. M.S. Sudhakar, K. Bhoopathy Bagan "A Novel Approach for Retrieval of Medical Image in Bit Plane Domain" in 2011 IEEE International Conference on Signal and Image Processing Applications (ICSIPA2011).
5. Bing Wang, Shao Sheng Fan, " An improved CANNY edge detection algorithm" in 2009 Second International Workshop on Computer Science and Engineering.
6. Tian-Shi Liu, Rui-Xiang Liu, Ping-Zeng and Shao-Wei Pan, "Improved Canny Algorithm for Edge Detection of Core Image, The Open Automation and Control Systems Journal, 2014, 6, 426-432 1874-4443/14 2014 Bentham Open.
7. Xun Wang, Jianqiu JIN. An Edge Detection Algorithm Based on Improved CANNY Operator. Seventh International Conference on Intelligent Systems Design and Applications, 623–628, 2007. J. Clerk Maxwell, A Treatise on Electricity and Magnetism, 3rd ed., vol. 2. Oxford: Clarendon, 1892, pp. 68–73.

8. Cai-Xia Deng, Gui-Bin Wang, Xin-Rui Yang, "Image Edge Detection Algorithm Based on Improved Canny Operator" International Conference on Wavelet Analysis and Pattern Recognition, Tianjin, 14–17 July, 2013.
9. Lalrinawma, Ramanjeet Kaur, " Edge Detection of an Image Using an Improved Canny Algorithm: A Review " IJARCCE ISSN (Online) 2278-1021 ISSN (Print) 2319 5940 International Journal of Advanced Research in Computer and Communication Engineering ISO 3297:2007 Certified Vol. 5, Issue 8, August 2016 Copyright to IJARCCE DOI 10.17148/IJARCCE.2016.58122587.
10. Yu Chen, Caixia Dengand Xiaxia Chen, "An Improved Canny Edge Detection Algorithm" in International Journal of Hybrid Information Technology Vol. 8, No. 10 (2015), pp. 359–370 http://dx.doi.org/10.14257/ijhit.2015.8.10.33 ISSN: 1738-9968 IJHIT.
11. Huang Jianling, Xiong Jia. Zou Tengbo, "Self-adaptive image edge detection method based on Canny", Computer Engineering and Applications, 2010, 47(34): 219–221.
12. Cai-Xia Deng, Gui-Bin Wang, Xin-Rui Yang, "Image Edge Detection Algorithm Based on Improved Canny Operator" International Conference on Wavelet Analysis and Pattern Recognition, Tianjin, 14–17 July, 2013.

Design of 2-Bit Parallel Asynchronous Self-timed Adder and 2-Bit Parallel Adder Using Radix Adder

Kuleen Kumar and Tripti Sharma

Abstract This paper presents design of asynchronous parallel adder by using recursive approach and comparison of various parameters such as average power, power-delay product, and number of transistors to design different adders. The parallel asynchronous self-timed adder designed using half adder along 2:1 multiplexer requires minimum interconnection. These adders have propensity to run faster than existing adders for random data. Parallel adder based on radix method provides faster computation of sum and reduces delay which is generated by carry chain. One-bit asynchronous parallel adder is designed with 24T transistor, while 1-bit radix adder is designed with 28T. In radix-based parallel adder, firstly carry is generated and then generated carry is used in sum propagation, which provides low area. Both adders are implemented using Mentor Graphics tool on tsmc018.mod process.

Keywords Asynchronous circuits · Radix-based full adders · Self-timed adders 24T · 28T

1 Introduction

The binary addition is basic operation that is performed by processor. Most of the adders have been designed with synchronous circuit, but due to no assumption of quantization of time asynchronous circuit is used for processor/circuits. The quantization of time is not assumed by asynchronous circuits [1]. So logic design with the help of asynchronous circuit is free from various problems such as less speed of operation and more power dissipation. The pipelining is established by request handshaking protocol in the absence of clocks, and also dual-rail carry is

K. Kumar · T. Sharma (✉)
Chandigarh, India
e-mail: tripsha@gmail.com

K. Kumar
e-mail: kuleen.elx@gmail.com

© Springer Nature Singapore Pte Ltd. 2018
S.S. Dash et al. (eds.), *International Conference on Intelligent Computing and Applications*, Advances in Intelligent Systems and Computing 632,
https://doi.org/10.1007/978-981-10-5520-1_19

197

managed by handshaking protocol and performing addition of bits. The acknowledgment is provided by dual-rail carry, and 1-bit adder block provides the carry output. The NULL conventional logic is used in dual-rail encoding for addition of bits. The robustness of circuit is generated by dual-rail carry representation [2]. In this paper, 2-bit self-timed adder using recursion method and 2-bit adder using radix method are designed.

Multiplexer and half adder are used to design parallel asynchronous adder which requires minimum interconnection. Thus, it is more efficient when implemented on VLSI technology. For independent-carry design, continuous parallel block and maximal rate pipelining are used to stabilize the output. The timing assumption of self-timed circuit depends upon AND/OR operation [3]. Adders based on this recursive approach have potential to run faster. The completion detection unit provides output in worst case, and carry is assumed to be zero by changing the multiplexer select lines.

1.1   Single-Rail Data Encoding for Parallel Adder

Single-rail data encoding is used for pipelined adder, i.e., enabled by req/ack handshake signal, and simultaneously carry is propagated. Bit-wise flow of carry output is provided by dual-rail carry convention. Two logic values can be used for dual-rail logic (invalid, (0, and 1)) [4]. When a bit operation is completed, bit-level ack is generated. The ack signals are received (high) when iterations are completed. The completion of iterations is detected by completion detection circuit [5]. Due to high fan-in requirement, abort logic implementation is expensive.

2   Design of 2-Bit Pasta

The general block diagram of 2-bit asynchronous parallel adder is described in this section. For half addition of each bit, adder accepts two input operands, and subsequently, iterations are taken for carry and sum to perform addition until all carry values are set to be zero [6]. Multiple binary bits addition can be performed by using parallel adders.

2.1   Depiction of 2-Bit Parallel Adder

The general block diagram of 2-bit adder is shown in Fig. 1. The multiplexer responds to input through synchronous signal and will be single 0 to 1 transformation based on select lines.

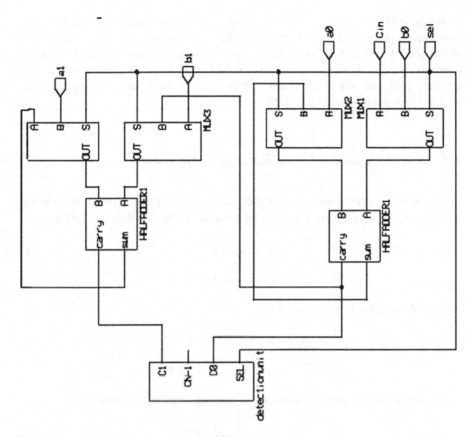

Fig. 1 General block diagram of 2-bit PASTA

Fig. 2 Initial phase

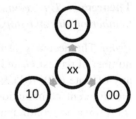

When SEL = 0 actual operands are selected. In next case when SEL = 1, feedback carry path is selected for subsequent iterations. The iterations are taken until all carry signals are fixed to zero values.

2.2 State Transition Diagram

The initial phase is shown by the state transition diagram given in Fig. 2. Each state transition is carried out by carry signal for every iteration.

By recursive approach, the circuit half adder works in fundamental mode during initial phase [6]. The (11) can be neglected because we are using half adder instead of full adder [7, 8]. In half-adder circuit, there is no provision to add carry bits from previous addition. In iteration phase when SEL = 1, the multiplexer block selects the carry path, and sum is calculated with previous carry. Iteration is stopped when all carry bits are assumed to be zero.

2.3 Binary Addition Using Recursive Approach

Let S_i and C_{i+1}^j denote the sum and carry, and sum will be produced for different combination of inputs. Condition $(j = 0)$ for addition is calculated as follows:

$$S_i^0 = a_i \oplus b_i \tag{1}$$

$$C_{i+1} = a_i\, b_i \tag{2}$$

The Ith iteration is given as

$$S_i = S_i \oplus C_i, \quad 0 \le i < n \tag{3}$$

$$C_{I+1} = Si \cdot Ci, \quad 0 \le i \le n \tag{4}$$

The iteration is terminated when the following condition is achieved.

$$c_n^k + c_{n-1}^k + c_1^k \quad 0 < k < n \tag{5}$$

Theorem 1 *The recursion is continued until it produces correct sum for any number of bits and stops within a predefined time.*

Proof The accuracy of the algorithm can be checked by introducing the required number of recursion, and completing the addition. For bit parallel adder considering any of state (00), (01) & (10); states can be selected. In $(k + 1)$th iteration among the different combinations (00), (01), (10), (11), (00) is considered. The several transitions are continuously consider in this approach to generate the actual carry and sum generation. Carry completion sensing adder and speculative completion adders are designed with parallel asynchronous technique.

2.4 2-Bit Radix Adder

An asynchronous parallel adder can also be designed using a Radix Full Adder (FA). One-bit full radix adder can be designed with 24-transistor. Therefore, radix adder is more efficient than parallel asynchronous self-timed adder. Firstly, carry is generated on its critical path and then recursively sum is generated. Two 1-bit full adders are joined for 2-bit radix adder and speed can be optimized [9]. The CLA technique is used to shorten the carry path. In Radix Full Adder, the power-delay product and energy-delay product are reduced due to carry chain propagation. We use the 24T transistor mirror FA as a base 2-bit adder. The total transistor count for 2-bit radix adder is 56, which is more than 2-bit PASTA. But the power-delay product is less than the 2-bit PASTA.

For full-adder implementation, carry and sum expression are given as follows:

$$C_{\text{out}} = (A + B)C_{\text{in}} + AB \tag{6}$$

$$\text{Sum} = AC_{\text{out}} + C_{\text{out}}B + C_{\text{out}} \cdot C_{\text{in}} + ABC_i \tag{7}$$

The Radix Full Adder is used to calculate the least significant sum. We use the CLA technique to design the circuitry for the most significant carry bit C_{out}. The sum and carry are generated explicitly. By using CMOS inverter, all inputs are complementary and least significant sum bit is calculated.

3 CMOS Implementation

A CMOS implementation is carried out for both adders. For implementation of 2-bit PASTA and 2-bit radix adder, we have used Mentor Graphics ELDO SPICE version tool on tsmc018.mod process. The 2:1 multiplexers are used to design 2-bit parallel adder. The circuit diagram 2-bit PASTA is shown in Fig. 3.

The output waveform of 2-bit PASTA is shown in Fig. 4. The delay between inputs and output is less in case of 2-bit radix adder. Power-delay product of 2-bit PASTA is more than that of 2-bit radix adder. Dual-rail logic has been used to reset the invalid state in circuit, and PASTA is designed with transmission gates which reduces number of transistor count. So on the basis of power-delay product, 2-bit radix adder is more efficient than asynchronous parallel adder. But transistor count slightly increases with radix adder. The 2-bit PASTA is designed with 48T transistor, while in case of 2-bit radix adder having great potential to reduce the total power consumption and used in high-speed processors. But transistor count slightly increased in radix adder [9]. The propagation of carry through successive bit adders and maximum carry propagation in worst case are increased for different adder such as 4-bit and 16-bit adder [10] and [11]. The first iteration is completed by selecting 2-bit adder's multiplexer select line to zero (Fig. 5).

The output waveform for 2-bit radix adder is shown in Fig. 6. As from comparison of Table 1, delay between input and generated sum is less in case of 2-bit radix adder as compared to 2-bit addition is performed parallel. So radix adder generates the output with less delay.

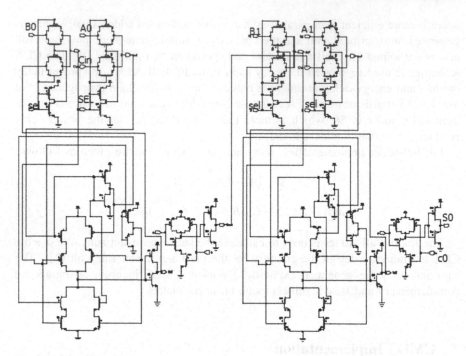

Fig. 3 2-Bit PASTA

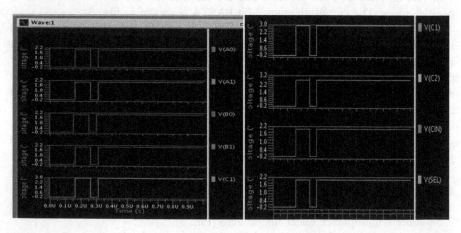

Fig. 4 Input–output waveforms of 2-bit PASTA

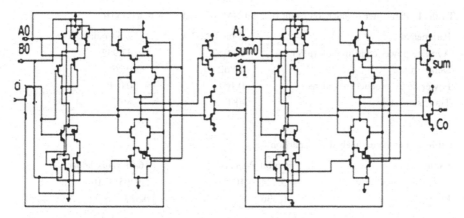

Fig. 5 2-Bit radix adder

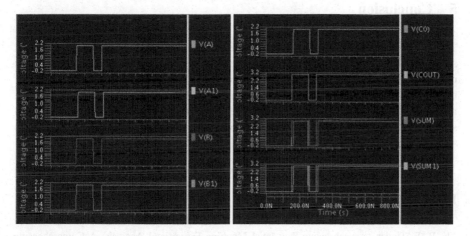

Fig. 6 Input–output waveforms of 2-bit radix adder

4 Simulation Results

In this section, we present simulation results for 2-bit PASTA and 2-bit radix adder running on Mentor Graphics tool 64-bit LINUX platform.

4.1 Result Comparison

Comparison result of parameters such as average power, delay, power-delay product, and transistor count for both adders is given as follows.

Table 2 gives comparison result of existing 16-bit PASTA and 16-bit radix adder.

Table 1 Parameter comparison result of 2-bit PASTA and 2-bit radix adder

Parameters	2-bit PASTA	2-bit radix adder
Average power consumption (μW)	55.942	37.942
Delay (n-sec)	1.862	0.265
Power-delay product (nWatt-sec)	0.104	0.010
Transistor count	48T	56T

Table 2 Parameter analysis comparison

Parameters	16-bit PASTA	16-bit radix adder
Average power (mw)	5.824×10^{-3}	5.9705×10^{-7}
PDP (μWatt-sec)	0.536	0.333

5 Conclusion

This paper describes execution of 2-bit PASTA and 2-bit radix adder. Initially, single-rail logic and dual-rail logic are used for designing parallel adder, and then recursion is used to design parallel asynchronous self-timed adder. The drawback of more power dissipation and less speed in asynchronous adder is overcome by radix method. Radix adder has potential to decrease the delay and total power consumption over parallel adder using recursive approach. Thus, radix adder is more suitable for high-performance processors.

References

1. R. F. Tinder.: Asynchronous Sequential Machine Design and Analysis: A Comprehensive Development of the Design and Analysis of Clock-Independent State Machine and Systems. San Mateo, CA, USA: Morgan 2009.
2. F.-C. Cheng, S. H. Unger, and M. Theobal.: Self-timed carry-lookahead adders, IEEE Trans. Computer. vol. 49, no. 7, pp. 659–672, Jul. 2000.
3. P. Choudhury.: implementation of basic arithmetic operations using cellular automaton.: in Proc. ICIT, 2008, pp. 79–80 "Implementation of basic arithmetic operations using cellular automaton," in Proc. ICIT, 2008, pp. 79–80.
4. M. D. Riedel.: Cyclic combinational circuits," Ph.D. dissertation, Dept. Comput. Sci. California Inst. Technol., Pasadena, CA, USA, May 2004.
5. D. Geer.: Is it time for clockless chips? [Asynchronous processor chips]," IEEE Comput., vol. 38, no. 3, pp. 18–21.
6. J. Sparsø and S. Furber.: Principles of Asynchronous Circuit Design. Boston, MA, USA: Kluwer Academic, 2001.
7. S. Nowick.: Design of a low latency asynchronous adder using speculative completion IEEE proc. Computer Digital Tech, vol. 49, no. 5, pp. 301–307, Sep 1996.
8. N. Weste and D. Harris. CMOS VLSI Design.: A Circuits and Systems Perspective. Reading, MA

9. C. Cornelius, S. Koppe, and D. Timmermann.: Dynamic circuit techniques in deep submicron technologies: Domino logic reconsidered in Proc. IEEE ICICDT, Feb. 2006, pp. 1–4.
10. W. Liu, C. T. Gray, D. Fan, and W. J. Farlow.: A 250-MHz wave pipelined adder in 2-μm CMOS," IEEE J. Solid-State Circuits, vol. 29, no. 9, pp. 1117–1128, Sep. 1994.
11. M. Anis, S. Member, M. Allam, and M. Elmasry.: Impact of technology scalingon CMOS logic styles. IEEE Trans. Circuits Syst., Analog Digital Signal Process., vol. 49, no. 8, pp. 577–588, Aug. 2002.

Efficient Image Secret Sharing Using Parallel Processing for Row-Wise Encoding and Decoding

Sonali D. Patil

Abstract Secret sharing plays a vital role in secure transmission of secret information in the form of images. Majority of the secret sharing algorithms are build using Lagrange's Interpolation due its information theoretic secure property. These image-sharing algorithms use pixel values of images for construction of shares and uses share's pixel values for reconstruction of a secret. The larger the size of an image the pixel values are more. The problem with such image secret sharing algorithm is its large computational complexity while implementing it in real-time application. A concurrent approach is proposed here for row-wise encoding and decoding. The concurrent approach helps to expedite the construction and reconstruction process. The proposed approach is implemented using UNIX-based Quadra Core system. The algorithm improves the time complexity of the construction and reconstruction process. The step-up of the concurrent algorithm is relatively even.

Keywords Secret sharing · Parallel algorithm · Cryptography · Image processing

1 Introduction

Military or commercial applications mainly consist of secret images or texts. As that is the case, how to store that secret images or text securely is the main issue. Security of these secret data in terms of images or test is the huge problem. In few decades, many techniques are developed to increase the security of secrete images by using data hiding and watermarking algorithms. But the limitation of these algorithms is single point failure if information carrier is lost or corrupted. Because of which the secrete data is not retrieved back. Here, the risk is having all secret data in one copy and with one authority. To avoid this problem, divide the secret data among group of people. We know the data hiding and conventional encryption

S.D. Patil (✉)
Department of Computer Engineering, Pimpri Chinchwad College
of Engineering, Nigdi, Pune 411044, India
e-mail: sonalimpatil@gmail.com

© Springer Nature Singapore Pte Ltd. 2018
S.S. Dash et al. (eds.), *International Conference on Intelligent Computing
and Applications*, Advances in Intelligent Systems and Computing 632,
https://doi.org/10.1007/978-981-10-5520-1_20

techniques, but the secret sharing method gets separated from these techniques as it converts the secret data into multiple (n) parts. These parts are shared among the (n) parties to avoid the loss of data accidentally or intentionally. During reconstruction of the original secret, any t $(2 \leq t \leq n)$ number of participants are sufficient. In 1979, Shamir and Blakely proposed the concept of secretly sharing the secret in terms of shares [1, 2]. After which many researchers began putting efforts on secret sharing techniques. Naor and Shamir [3] put forth simple pattern-based secret transmitting schemes for images. In [3], dealer creates shares into n shadows and transmits to n predefined participants. Share with each participant cannot disclose any information about original secret. The original secret data is reformed using any t or more shares. The original image will not get reconstructed if sufficient shares are not submitted. Secret sharing along with extended capabilities [4] also makes the schemes useful for many applications.

For sharing secret image, we need to encode and decode it pixel-by-pixel which is computationally rigorous due to large amount of data which is to be processed. As a result, such processing algorithms suffer with large computational complexities. Practical implications need fast processing of algorithms. In [4–7] proposed concurrent approaches for image processing algorithm. These implementations are based on High Computing Machines. Nowadays, almost all end user machines are with multi-core capabilities. These machines can be effectively utilized using concurrent algorithm skeleton.

A concurrent algorithm for construction and reconstruction process is proposed using row-wise encoding and decoding of secret data in the form of image. Every step in the share creation and reconstruction of image from share is parallelized. Four cores are utilized simultaneously for the implementation using POSIX library.

This paper is aligned in next five sections. Section 2 outlines literature review. Section 3 focuses on sequential strategy for creation of shares and reconstruction of secure image from shares. Section 4 elaborates the proposed concurrent algorithm approach for construction and reconstruction process. Section 5 focuses on the relative results of sequential and concurrent algorithms. Section 6 states the conclusion regarding proposed approach.

2 Literature Survey

The polynomial threshold secret sharing schemes are briefly interpreted in this section.

2.1 Secret Sharing Scheme by Shamir [1]

Shamir [1] invented the sharing scheme for (t, n) where t is threshold and $t \leq n$. A polynomial of degree one less than threshold is constructed. The constant term is a secret to be shared in the polynomial. The other coefficients in the polynomial are random values.

The share values will be computed by putting participant number in the polynomial. The generated share values are distinct. Those values are distributed among the participants.

For using Shamir's scheme for sharing secret images, the pixel value is used as a constant term in the polynomial. For each pixel, one polynomial gets formed, by keeping that pixel value as constant term. The number of total pixels defines the number of polynomials to be formed. The scheme is based on linear equations. The problem with this technique is if the size of image is large it adds the computational complexity for the construction and reconstruction process. For m size image and (t, n) scheme, m number of polynomials are derived. Thien and Lin's [8] scheme shortens computational complexity which is discussed further.

2.2 Image Secret Sharing Scheme by Thien and Lin [9]

This scheme [9] used the Shamir's technique of secret sharing effectively for images. It also uses the pixel values of an image to construct the shares. Unlike using the random values as a coefficient, it effectively uses further pixel values in sequence to form a polynomial. For m size image and (t, n) scheme, m/t number of polynomials are derived. It helps in reducing computational complexity as evaluating the number of polynomial gets reduced. Also it helps in reducing the share size, which helps in reducing bandwidth in the network. Still, the total computational complexity can be further reduced using proposed concurrent approach.

Researchers are working on parallel strategies to make the image processing algorithms to be in use of real-life applications. These algorithms require very high configuration of the system like n-core systems. The paper proposes a simple technique of parallelism in secret sharing algorithms based on regular quad core machines with open-source libraries. The proposed algorithm is elaborated in next section.

3 Image Secret Sharing: Construction and Reconstruction—A Sequential Approach

3.1 Construction of Shares

i. Initially, threshold value will be taken as t. For each row, $(t - 1)$ order polynomial will be formed using first t pixels in sequence.
ii. Polynomial value will be computed for all participants from 1, 2, ..., n.

iii. Then, next t pixel values will be taken from the row for next polynomial.
iv. Repeat the same procedure for each row till the last pixel value of a row to compute the value of polynomial for all participants from 1, 2, ..., n.
v. Different shares will get created using computed values from polynomials.

3.2   Reconstruction of Original Secret from Shares

i. Collect t shares from the interested participants.
ii. Select the first pixel from each share which is selected.
iii. Apply Lagrange's Interpolation formula to form an equation from t selected pixel values of t shares.
iv. All coefficients of derived equation will be taken as pixel values for resultant image.

4   Image Secret Sharing: Construction and Reconstruction—A Parallel Approach

The proposed algorithm uses multi-threading strategy to achieve parallelism in construction and reconstruction of secret sharing algorithm. It is implemented on multi-core system. POSIX library is used to create the number of threads. For polynomial-based threshold secret sharing [9], concurrent approach is as given below.

4.1   Construction of Shares

Parent Process:

i. The Parent process will split the image row-wise into p chunks for the respective threads.
ii. Parent process will create p child threads.
iii. Each child thread function will be passed with parameters which will indicate starting and ending row number of each portion.
iv. Each child thread will be assigned to a particular core.

Threads:

i. Each child thread will apply Thein and Lin's [4] algorithm on specified rows to form a polynomial of degree $(t - 1)$.
ii. Each child thread will compute the value of polynomial for all participants from 1, 2, ..., n.

Parent Process:

i. Parent Process will wait till all thread joins.
ii. Parent process will transmit all the created shares to the participants.

4.2 Reconstruction of Original Secret from Shares

Parent Process:

i. The Parent process will receive t shares from the interested participants.
ii. Parent process will be responsible for deciding the column-wise p chunks for the respective threads.
iii. Parent process will create p threads.
iv. Each thread function will be passed with parameters which will indicate starting and ending column number for all shares.
v. Each thread will be assigned with a particular core.

Threads:

i. Each thread will select the first pixel from each share which is selected.
ii. Each thread will apply Lagrange's Interpolation formula to form an equation from t selected pixel values of t shares.
iii. All coefficients of derived equation will be taken as pixel values for resultant image.
iv. Each thread will repeat the steps II and III for all specified rows.

Parent Process:

i. Parent Process will wait till all thread joins.
ii. Parent process will display the reconstructed secret image.

5 Experimental Results

The above concurrent approach is implemented using POSIX functions on **Quadra Core system** to create threads and join. The results for Lena image are shown in Figs. 1, 2 and 3.

The high PSNR values of original secret and reconstructed secret show that the applied concurrent approach is maintaining the accuracy of the image-sharing scheme.

The level of parallelism is checked by varying the number of threads from 1 to 4. The time results are calculated for different images.

Table 1 shows the average time required for construction of shares for these images. Table 2 shows reconstruction time results. The results are linear with respect to time and number of threads.

Fig. 1 Secret image Lena.jpg

Fig. 2 Reconstructed image Lena.jpg

Fig. 3 Reconstructed image
Lena.jpg

Table 1 Time comparison table for construction of shares using parallel versus sequential program

Secret image	Time required (in s)			
	Sequential execution	With 2 threads	With 3 threads	With 4 threads
Lena	1.011	0.596	0.417	0.333
Baboon	0.985	0.583	0.408	0.326
Barbara	0.963	0.572	0.401	0.321
Pepper	0.911	0.546	0.384	0.308

Table 2 Time comparison table for reconstruction of secret using concurrent versus sequential program

Secret image	Time required (in s)			
	sequential execution	With 2 threads	With 3 threads	With 4 threads
Lena	0.068	0.041	0.030	0.025
Baboon	0.065	0.040	0.029	0.024
Barbara	0.064	0.039	0.028	0.024
Pepper	0.061	0.038	0.027	0.023

The reconstruction time is less as compare to construction time. As in construction algorithm, more time is required to form polynomials and calculating shares for all n participants. The reconstruction algorithm accepts t shares only and applied Lagrange's Interpolation formula to reconstruct original image.

6 Analysis and Conclusion

There is a huge gap in time requirements of sequential approach and expected timings for real-time applications. The proposed concurrent approach shows even results with respect to adding threads with the total time required for construction and reconstruction. The speed-up time for construction and reconstruction is observed to be linear. Multi-threading approach keeps the proposed approach very simple and efficient. Such concurrent approaches are very useful to apply secret sharing in practical usage. In future, optimum threshold values can be decided for parallel approach with various schemes.

References

1. Shamir, "How to share a secret," Communications of the ACM, vol. 22, no. 11, pp. 612–613, Nov. 1979.
2. G. R. Blakley, "Safeguarding cryptographic keys," in Proc. the National Computer Conference, American Federation of Information Processing Societies, pp. 313–3' 7, June 1979.

3. Noar M., Shamir A., "Visual cryptography", Advances in Cryptography. Eurocrypt, Lecture Notes in Computer Science, vol. 950, Springer-Verlag. 1–12, 1994.
4. Sonali Patil, Prashant Deshmuth, "An Explication of Multifarious Secret Sharing Schemes", International Journal of Computer Applications (0975–8887), Volume 46– No. 19, May 2012
5. N. Shimizu, T. Watanabe, "High performance parallel FFT on distributed memory parallel computers", Lecture Notes in Computer Science, Vol. 1336, 1997
6. B. Chen, C. Zeng, Y. Jiang, "A Parallel FFT Scheme Based on Multi-Machines Environment", The 3rd International Conference on Grid and Pervasive Computing - Wortshops, 2008, pp. 221–226.
7. W.P. Fang, S.J. Lin, "Fast Secret Image Sharing Scheme in HPC," Proceeding on the 10th International Conference on High-Performance Computing in Asia-Pacific Region (HPC ASIA 2009) joint WortShop on PC-Grid, Grand Hi-Lai Hotel, Taohsiung, Taiwan, 2009, 3, 2–2009, 3, 5.
8. Arpita Gopal, Sonali Patil, Amresh Nikam, "A Parallel Algorithm for Image Edge Detection using Difference Chain Encoding", International Journal of Computer Science and Application Issue 2010 pp. 113–13.
9. C. C. Thein and J. C. Lin, "Secret Image Sharing", Computers & Graphics, Vol. 26, 2002, pp. 765–770.

TXGR: A Reverse Engineering Tool to Convert Design Patterns and Application Software into Graph

Arti Chaturvedi, Manjari Gupta and Sanjay Kumar Gupta

Abstract Software reusability is considered as a crucial technical condition to improve the overall software quality and also reduce production and maintenance cost. Design pattern detection is one of the important techniques, which helps to improve reusability of existing software. In this paper, we proposed a new reverse engineering tool (TXGR) with the help of two open-source tools Java NetBeans and Class Visualizer. TeXt to GRaph (TXGR) tool generates graphs for structural information of design pattern and existing software JHotDraw (application software). With the help of this tool, we apply graph application (subgraph isomorphism method) for design pattern detection.

Keywords UML · XML · Directed-labeled graph · Design pattern
JHotDraw 7.0 · Java NetBeans · Class Visualizer

1 Introduction

Gamma et al. [1] proposed 23 design patterns based on object-oriented programming which are solutions of problems that are coming again and again in software development process. Therefore, design pattern detection is considered vital and coined as part of reverse engineering process. However, reverse engineering is "the process of analyzing a subject system to (a) identify the system's component and their

A. Chaturvedi (✉) · S.K. Gupta
School of Studies in Computer Science and Applications,
Jiwaji University, Gwalior, MP, India
e-mail: arti.2408@gmail.com

S.K. Gupta
e-mail: Sanjaygupta9170@gmail.com

M. Gupta
Department of Computer Science, Faculty of Science,
Banaras Hindu University, Varanasi, UP, India
e-mail: Manjari_gupta@rediffmail.com

© Springer Nature Singapore Pte Ltd. 2018   215
S.S. Dash et al. (eds.), *International Conference on Intelligent Computing
and Applications*, Advances in Intelligent Systems and Computing 632,
https://doi.org/10.1007/978-981-10-5520-1_21

interrelationship and (b) create representation of a system in another form at a higher level of abstraction" [2]. To identify design patterns, extracting structural information from the source code of system under study and converting it into graph is one of the steps. Graph has been a powerful and popular representation formalism in pattern recognition. So, class diagram can be perfectly mapped to graph where vertices represent the classes, while edges correspond to a selected type of relationship (i.e., association, generalization, dependency, aggregation) [3]. The approaches described in [4–13] represent UML of design patterns in graph format, but unable to express their relationships as a whole in a single design pattern graph.Thus,reverse engineering process is formulated for the representation of an Object-Oriented system as a set of graphs to represent various relationships of design pattern. Furthermore this set of graphs will use in pattern detection methodology based on graph matching approach using subgraph isomorphism to identify reusable design patterns. In such a methodology, both the system under study and the design patterns to be detected are described in terms of graphs. According to [3], a directed graph $G = (V, E)$ will represent the class diagram of object-oriented system under study. The set of vertices V corresponds to the classes of the system, while the set of all edges E presents selected kind of relationship between the classes. (For example, if the association is to be represented, a directed edge $(r, s) \in E$ will indicate an association between classes, r and s with a direction from r to s). In [4–13], the proposed approaches represent number of graphs, one for each kind of relationship that makes detection of design patterns more complicated and increases the possibility of finding false-positive instances (occurrence of design pattern). Other graph-based techniques proposed by researchers and scientists for design pattern detection include similarity scoring method [14] and template matching method [15], are also used multiple graphs to represent relationships of design pattern.

Thus, TXGR tool has been developed to represent the complete structure of design patterns in graphical format so that this will overcome the problems where multiple graphs are used to represent the relationships of design patterns. Our approach covers various aspects and shows complete structural view of design patterns in a single graph that not only provides comprehensive view to expedite the detection practice but also minimizes the likelihood of discovering false-positive instances. The limitation of this tool is that it does not show two relationships between two nodes in the same direction, so that it partially draws three design patterns (i.e., decorator, interpreter, and composite) where two relationships exist in same direction between two vertices. Therefore, the purpose of this work is to provide a general tool that generates graphical format using input as data text file. Moreover, the objective of this tool is to use this graphical format for further speedy detection of design patterns using subgraph isomorphism method to improve the reusability of legacy software.

The outline of this paper is as follows. We begin with the brief overview of proposed approach to develop the TXGR tool in Sect. 2. Case study of on an open-source project JHotDraw7.0 in design pattern detection perspective is discussed in Sect. 3. TXGR tool results are illustrated as JHotDraw7.0 graph and design pattern graphs in Sect. 4. Lastly, we conclude in Sect. 5.

2 Overview of Approach

There are lots of tools given by many researchers for reverse engineering in past. In this proposed work, we present a tool which converts design pattern data file into a graph. Therefore, tool provides huge assistance for users to generate the graph of design patterns. For the development of this tool, we have used two open-source available tools, one is Java NetBeans and other is Class Visualizer (Fig. 1).

For the development of this TXGR tool, structural information of the existing application software and design pattern is extracted from XML and UML design domain, and then, text file is prepared to give the information of classes and relationship among them. Thus, TXGR will convert text information of source code into directed and labeled graph. Here, each class is represented by node (vertex), and each edge represents relationship between two classes. In this, three types of classes are taken into consideration: One is of abstract class (pink color node) which is represented by (1, 0, 0), second is concrete subclass (cyan color node) represented by (0, 1, 0), and third is concrete class (magenta color node) represented by (0, 0, 1). Furthermore, four types of relationships are considered, where label "1" shows dependency, label "2" shows generalization, label "3" shows association, and label "4" shows aggregation. TXGR tool uses these properties to convert text into graph. This has been described and interpreted using the single tuple of text file characteristic as an example (23(1,0,0) 42(0,0,1) 4 explain class ID 23 is abstract class and class ID 42 is concrete class, and edge label 4 shows aggregation relationship between these classes).

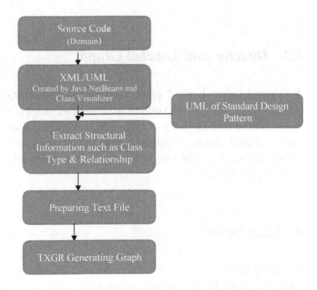

Fig. 1 Flow chart of methodology used to develop tool

Fig. 2 Example of
directed-labeled graph [17]

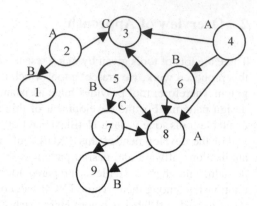

2.1 Rules for Directed Graph

According to [16], following rules are used for directed graph.

(1) If a direct association relationship is between two classes, then an edge starts from source class and targeted to destination class.
(2) If a generalization relationship is between two classes, then an edge starts from child class and targeted to parent class.
(3) If a dependency relationship is between two classes, then an edge starts from dependent class and targeted to class on which it depends.
(4) If an aggregation relationship is between two classes, then an edge starts from whole class and targeted to part class.

2.2 Directed and Labeled Graph

A directed vertex-labeled graph G is denoted as $\{V(G), E(G), L_V, L_E\}$, where $V(G)$ and $E(G)$ are set of vertices and directed edges, respectively; L_V is a set of vertex labels; and L_E is set of edge labels. Figure 2 shows an example of a directed vertex-labeled graph. Numbers inside the vertices are vertex IDs, and the letters beside the vertices are vertex labels. In our approach, edges are also labeled as described in Sect. 2.

3 Case Study

We have selected an open-source project JHotDraw 7.0 as existing application software (domain for pattern detection) and used an open-source tool Java NetBeans that converts JHotDraw source code file into XML file. From this file, all class names of JHotDraw and three relationships among these classes (where "Use a" relationship indicates dependency, "Has a" relationship shows aggregation, and "Is a" relationship shows generalization or inheritance) are extracted [16]. Java NetBeans is not able to recover some relevant information for design pattern identification. In case of Java, this XML file (Fig. 3) does not represent the Association relationship between classes and also does not show type of classes. To overcome these deficiencies of this tool, we use another open-source tool Class Visualizer to recover the missing information from the source code. Class Visualizer converts Java source code class file into UML format (Fig. 4). Using Class Visualizer, we extract classes that are of three types: One is abstract class, in which we take all abstract classes and interfaces and are shown as (1, 0, 0), second is concrete subclass, in which we take those classes which are inherited from JHotDraw 7.0 class and are shown as (0,1,0), and third is concrete class, in which we take those classes which are inherited from Java class and are shown as (0,0,1). Using these two tools, we extracted 308 classes from XML file of JHotDraw 7.0 and approximate 900 relationships among them.

Now, with the help of this information, we prepare separate text file manually for JHotDraw 7.0 as well as for 23 design patterns [1]. Each tuple of this file (Fig. 5) has two classes: One is source class and other is target class with class ID (i.e., 1, 2, ... for JHotDraw) and their types (i.e., (1,0,0), (0,1,0), (0,0,1)) and relationship label between these classes as mentioned in Sect. 2. However, for design pattern, class ID (node) is represented by 1001, 1002, 1003, etc.

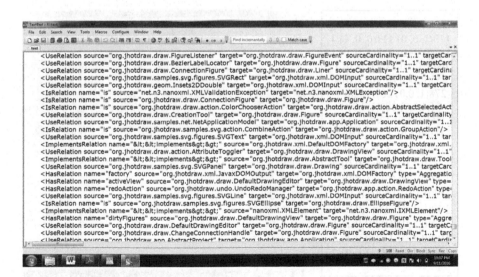

Fig. 3 Snapshot of Java NetBeans XML file

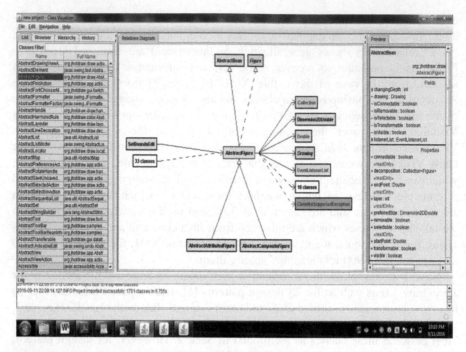

Fig. 4 Snapshot of UML representation by Class Visualizer

Source Target Relationship
110(1, 0, 0) 111(0, 0, 1) 4
183(0, 1, 0) 287(1, 0, 0) 1
68(1, 0, 0) 286(1, 0, 0) 1
255(0, 0, 1) 56(0, 1, 0) 4
278(1, 0,0) 281(1,0,0) 1
64(1,0,0) 306(1,0,0) 1
266(0,1,0) 12(0,1,0) 4
170(0,1,0) 290(1,0,0) 1
35(0,1,0) 281(1,0,0) 1
136(0,1,0) 284(1,0,0) 4
172(0,1,0) 293(1,0,0) 1
97(0,1,0) 150(0,0,1) 1
140(0,1,0) 290(1,0,0) 1
124(0,1,0) 197(0,0,1) 1
190(0,1,0) 152(0,0,1) 4
165(0,1,0) 283(1,0,0) 2

Fig. 5 Example of data base for JHotDraw 7

4 Results

4.1 Tool Components

The outcome of this tool comprises of two main components that define several classes and their relationships.

(i) **Graph Visualizer**: It shows graph of design pattern and JHotdraw 7.0.
(ii) **Help**: It is an information provider of graph where node color represents the type of class and label shows the information about relationship between classes.

4.2 JHotDraw 7.0 Graph

TXGR tool generates a directed and labeled graph in Fig. 6 (snapshot of first screen and result as graph). Here, TXGR takes input as mentioned in Fig. 5. This graph has three types of node: Pink color node represents abstract type class, cyan color node indicates concrete subclass, and magenta color node signifies concrete class as mentioned earlier in Sect. 2. Edges are well represented by relationships among these classes as mentioned in Sect. 2.

4.3 Design Pattern Graph

TXGR also generates all 23 design pattern graphs that are shown below in Table 1 with their input text file according to UML structure elaborated for each pattern in [1].

Fig. 6 JHotDraw 7.0 graph

Table 1 TXGR generated outcome as design pattern graphs from input text file

Abstract factory: text file	Builder: text file	Factory method: text file	Prototype: text file
Source target relationship	Source target relationship	Source target relationship	Source target relationship
1001(0,0,1) 1002(1,0,0) 3	1001(0,0,1) 1002(1,0,0) 4	1003(0,1,0) 1002(1,0,0) 2	1001(0,0,1) 1002(1,0,0) 3
1001(0,0,1) 1003(1,0,0) 3	1003(0,1,0) 1002(1,0,0) 2	1003(0,1,0) 1004(0,1,0) 1	1003(0,1,0) 1002(1,0,0) 2
1004(0,1,0) 1003(1,0,0) 2	1003(0,1,0) 1004(0,0,1) 1	1004(0,1,0) 1001(1,0,0) 2	
1005(0,1,0) 1002(1,0,0) 2			
1004(0,1,0) 1005(0,1,0) 1			
Abstract factory pattern graph	Builder design pattern graph	Factory method design pattern graph	Prototype design pattern graph

Singleton: text file	Adapter class: text file	Adapter object: text file	Bridge: text file
Source target relationship	Source target relationship	Source target relationship	Source target relationship
1001(0,0,1) 1001(0,0,1) 3	1001(0,0,1) 1002(1,0,0) 3	1001(0,0,1) 1002(1,0,0) 3	1002(1,0,0) 1001(1,0,0) 4
	1003(0,1,0) 1002(1,0,0) 2	1003(0,1,0) 1002(1,0,0) 2	1003(0,1,0) 1002(1,0,0) 2
	1003(0,1,0) 1004(0,0,1) 2	1003(0,1,0) 1004(0,0,1) 3	1004(0,1,0) 1001(1,0,0) 2
Singleton design pattern graph	Adapter class design pattern graph	Adapter object design pattern graph	Bridge design pattern graph

Composite: text file	Decorator: text file	Façade: text file	Flyweight: text file
Source target relationship	Source target relationship	Source target relationship	Source target relationship
1001(0,0,1) 1002(1,0,0) 3	1002(0,1,0) 1001(1,0,0) 2	1001(1,0,0) 1002(0,0,1) 3	1002(0,1,0) 1001(1,0,0) 2
1003(0,1,0) 1002(1,0,0) 2	1003(0,1,0) 1001(1,0,0) 4		1003(0,1,0) 1001(1,0,0) 2
1004(0,1,0) 1002(1,0,0) 2	1003(0,1,0) 1001(1,0,0) 2		1004(0,0,1) 1002(0,1,0) 3
1004(0,1,0) 1002(1,0,0) 4	1004(0,1,0) 1003(0,1,0) 2		1004(0,0,1) 1003(0,1,0) 3
			1004(0,0,1) 1005(0,0,1) 3
			1005(0,0,1) 1001(1,0,0) 4
Composite design pattern graph (Partial)	Decorator design pattern graph (Partial)	Façade design pattern graph	Flyweight design pattern graph

(continued)

Table 1 (continued)

Proxy: text file	Chain of responsibility: text file	Command: text file	Interpreter: text file
Source target relationship		Source target relationship	Source target relationship
1002(0,1,0) 1001(1,0,0) 2	Source target relationship	1001(0,0,1) 1002(0,0,1) 3	1001(0,0,1) 1002(1,0,0) 3
1003(0,1,0) 1001(1,0,0) 2	1001(0,0,1) 1002(1,0,0) 3	1001(0,0,1) 1003(0,1,0) 1	1001(0,0,1) 1003(0,0,1) 3
1003(0,1,0) 1002(0,1,0) 3	1002(1,0,0) 1002(1,0,0) 4	1003(0,1,0) 1002(0,0,1) 3	1004(0,1,0) 1002(1,0,0) 2
	1003(0,1,0) 1002(1,0,0) 2	1003(0,1,0) 1004(1,0,0) 2	1005(0,1,0) 1002(1,0,0) 2
		1005(1,0,0) 1004(1,0,0) 4	1005(0,1,0) 1002(1,0,0) 4
Proxy design pattern graph	Chain of responsibility design pattern graph	Command design pattern graph	Interpreter design pattern graph (Partial)

Iterator: text file	Mediator: text file	Memento: text file	Observer: text file
Source target relationship	Source target relationship	Source target relationship	Source target relationship
1003(0,1,0) 1002(1,0,0) 2	1002(1,0,0) 1001(1,0,0) 3	1001(0,0,1) 1002(1,0,0) 1	1001(1,0,0) 1002(1,0,0) 3
1003(0,1,0) 1004(0,1,0) 3	1003(0,1,0) 1002(1,0,0) 2	1003(0,0,1) 1002(1,0,0) 4	1003(0,1,0) 1002(1,0,0) 2
1004(0,1,0) 1003(0,1,0) 1	1004(0,1,0) 1003(0,1,0) 3		1003(0,1,0) 1004(0,1,0) 3
1004(0,1,0) 1001(1,0,0) 2			1004(0,1,0) 1001(1,0,0) 2
1005(0,0,1) 1001(1,0,0) 3			
1005(0,0,1) 1002(1,0,0) 3			
Iterator design pattern graph	Mediator design pattern graph	Memento design pattern graph	Observer design pattern graph

State: text file	Strategy: text file	Template: text file	Visitor: text file
Source target relationship	Source target relationship	Source target relationship	Source target relationship
1001(0,0,1) 1002(1,0,0) 4	1001(0,0,1) 1002(1,0,0) 4	1002(0,1,0) 1001(1,0,0) 2	1001(0, 0, 1) 1002(0, 0, 1) 3
1003(0,1,0) 1002(1,0,0) 2	1003(0,1,0) 1002(1,0,0) 2		1001(0, 1, 0) 1003(1, 0, 0) 3
			1004(0, 1, 0) 1003(1, 0, 0) 2
			1002(0, 0, 1) 1005(1, 0, 0) 3
			1006(0, 1, 0) 1005(1, 0, 0) 2
State design pattern graph	Strategy design pattern graph	Template design pattern graph	Visitor design pattern graph

5   Conclusion

In this paper, we have developed a reverse engineering tool (TXGR) that converts data text file into graph with the help of two open-source tools Java NetBeans and Class Visualizer. Earlier developed tools are not quite efficient to represent the complete flow of specific design pattern. This work may provide great ease to researchers, scientists, and developers to analyze and represent available relationships between classes of design patterns graphically. This may results in fast design pattern detection and therefore may be used frequently by the user to enhance the software reusability. In future, using TXGR graphical tool, we will apply design pattern mining techniques based on graph theory for improving the software reusability.

References

1. Gamma E., Helm R., Johnson R., Vlissides J. (1995) Design Patterns Elements of Reusable Object-Oriented Software, Addison-Wesley.
2. Keller, Rudolf K. (1999) Pattern-based reverse-engineering of design components, Proceedings of the 21st international conference on Software engineering. ACM.
3. Chatzigeorgiou A., Tsantalis N., Stephanides G. (2006) Application of Graph Theory to OO Software Engineering, WISER'o6, May 20, 2006, Shanghai, China.
4. Gupta, M., Rao, R. S., Pande, A., & Tripathi, A. K. (2011, January). Design pattern mining using state space representation of graph matching. In International Conference on Computer Science and Information Technology, Springer Berlin Heidelberg. (pp. 318–328).
5. Gupta, M., Rao, R. S., & Tripathi, A. K. (2010, December). Design pattern detection using inexact graph matching. In Communication and Computational Intelligence (INCOCCI), 2010 International Conference on IEEE. (pp. 211–217).
6. Gupta M. (2010) Inexact Graph Matching for Design Pattern Detection using Genetic Algorithm, International Conference on Computer Engineering and Technology, Nov, Jodhpur.
7. Gupta M., Akshara Pande, Rajwant Singh Rao, A.K. Tripathi (2010) Design Pattern Detection by Normalized Cross Correlation, International Conference on Methods and Models in Computer Sciences (ICM2CS-2010), December 13–14, JNU,.
8. Pande A., Gupta M., Tripathi A. K. (2010) DNIT—A New Approach for Design Pattern Detection, International Conference on Computer and Communication Technology, MNNIT-Allahabad, proceeding published by the IEEE.
9. Pande A., Gupta M., Tripathi A. K. (2010) A Decision Tree Approach for Design Patterns Detection by Subgraph Isomorphism, International Conference on Advances in Information and Communication Technologies, ICT 2010, Kochi, Kerala, LNCS-CCIS, Springer.
10. Pande A., Gupta M., Tripathi A. K. (2010) A New Approach for Detecting Design Patterns by Graph Decomposition and Graph Isomorphism", International Conference on Contemporary Computing, Jaypee Noida, CCIS, Springer.
11. Pande A. & Gupta M. (2010) Design Pattern Mining for GIS Application using Graph Matching Techniques, 3rd IEEE International Conference on Computer Science and Information Technology. pp. 09–11, Chengdu, China.
12. Pande A., Gupta M. (2010) Design Pattern Detection Using Graph Matching, International Journal of Computer Engineering and Information Technology (IJCEIT), Vol. 15, No. 20, Special Edition, pp. 59–64.

13. Gupta, M. (2011). Design Pattern Mining Using Greedy Algorithm for Multi-labeled Graphs, International Journal of Information and Communication Technology 3.4 (2011): 314–323
14. Tsantalis N., Chatzigeorgiou A., Stephanides G., Halkidis S. (2006) Design Pattern Detection Using Similarity Scoring IEEE transaction on software engineering, 32(11).
15. Dong J., Sun Y., Zhao Y. (2008), Design Pattern Detection by Template Matching, the Proceedings of The 23rd Annual ACM Symposium on Applied Computing (SAC), pages 765–769, Ceará, Brazil.
16. Booch G., Rumbaugh J., Jacobson I. (1998) Unified Modeling Language User Guide, Addison Wesley.
17. Peng, Peng. (2011) Subgraph search over massive disk resident graphs, International Conference on Scientific and Statistical Database Management. Springer Berlin Heidelberg.

Comparative Analysis of Image Fusion Using DCT, DST, DWT, Walsh Transform and Kekre's Wavelet Transform

Jyoti S. Kulkarni and Rajankumar S. Bichkar

Abstract Image fusion focuses on combining information from different images of a scene to obtain more useful information for various applications. A large number of transform techniques have been used for image fusion process that include DCT, DST, DWT, Kekre's wavelet transform and Walsh transform. This paper compares the quality of image fusion obtained using these transforms. These techniques have been compared with each other in the past using different quality indices that include mean, variance, standard deviation, RMSE, PSNR and SF. This paper attempts to compare these fusion techniques using the various quality indices so as to get a clear picture about relative performance of these techniques. The experimental results on some standard test images demonstrate that DWT and DCT are better compared to other.

Keywords Multisensor · Multitemporal · Multiresolution · Multifocus Visual sensor network

1 Introduction

Image fusion is a process of combining the images from different sensors. The various types of image fusion include multifocus, multisensor, multiresolution and multitemporal image fusion. The image fusion technique has been applied in various fields that include medical imaging, geographical analysis, change detection, wireless sensor network.

J.S. Kulkarni (✉) · R.S. Bichkar
E & TC Department, G.H. Raisoni College of Engineering and Management,
Pune, India
e-mail: sanjyot_8@rediffmail.com

R.S. Bichkar
e-mail: rajan.bichkar@raisoni.net

J.S. Kulkarni
E & TC Department, Pimpri Chinchwad College of Engineering, Pune, India

© Springer Nature Singapore Pte Ltd. 2018
S.S. Dash et al. (eds.), *International Conference on Intelligent Computing and Applications*, Advances in Intelligent Systems and Computing 632,
https://doi.org/10.1007/978-981-10-5520-1_22

For example, in medical imaging applications, the images obtained by PET and CT convey different information about internal organs in human body. These images can be fused to obtain a single image which provides more information to a doctor for effective diagnosis.

In multiresolution image fusion, pyramid and wavelet analyses are used. The edge and margin of image is extracted by using wavelet transformation. The region-based multifocus image fusion assesses the quality of images in spatial domain. Additionally, feature aspects can be added through genetic algorithm. Genetic algorithm is used to set the optimal block size in image [1]. A multiresolution image fusion can be implemented using least mean square error method. The average brightness of the image is calculated through mean, whereas intensity is calculated through relative variance [2].

Urban development is observed through the change detection reflected from multitemporal image fusion. The change is the difference in reflectance spectra of the object. The change detection is useful for various applications such as urban growth, land use, forest and vegetation dynamics, disaster monitoring. But a specific algorithm is not applicable for all these applications. Thus, researchers are motivated to find algorithms to process remotely sensed multitemporal images and find change without any prior knowledge [3]. Dampher Shafer evidence theory elaborates the change detection from multispectral imagery [4].

Several transform techniques have been employed for image fusion. These include DCT, DST, DWT, Kekre's wavelet transform and Walsh transform. This paper compares the quality of fused images using these transforms for multispectral image fusion. The rest of the paper is organized as follows: Sect. 2 describes the different aspects of image fusion. Section 3 briefly describes the various transforms for image fusion used in this paper. Section 4 demonstrates experimental results and discussions, and Sect. 5 gives conclusion and future scope.

2 Different Image Fusion Aspects

Image fusion method is applicable in different areas for variety of applications as stated below.

1. Medical image fusion: Multimodal images are fused to find useful information. Arpita et al. proposed a comparative study of fuzzy fusion approach. In multiresolution analysis, the image is decomposed using Haar wavelet transforms into high- and low-frequency bands. The high-frequency component can be selected using fuzzy clustering or genetic-based selection technique and low-frequency component using averaging method [1].
2. Different resolution image fusion: Tao et al. proposed least mean square error-based fusion method. Here, the information through pixel of coarser image is used to modify the finer image [2].

3. Multitemporal image fusion: Multitemporal image captured after specific time period which is useful to find the change detection in urban development. Du et al. proposed fusion technique to merge multiple difference images to find spectral changes [3, 4].
4. Intelligent service robot: Luo et al. used simultaneous localization and mapping for intelligent service robot. Fusion of images from multiple sensors is done to perform self-localization with mapping. This robot also detects the moving objects in service area [5].
5. Visual sensor networks: Phamila et al. designed a multifocus image fusion method using discrete cosine transform. This is for wireless visual sensor network to increase the capacity of node where images are stored [6]. Discrete cosine transform is more useful in real-time systems [7–9].

Depending on the input images, the types of image fusion are decided. If images are taken from same sensor with different time intervals, then it is called as multitemporal image fusion. If input images are from different sensors, then it is called as multisensor image fusion. If input images are taken from single sensor with a different focus, then it is called multifocus image fusion. Different fields are using the image fusion such as computer vision, automatic object detection, medical imaging, remote sensing and robotics. Image fusion is useful in a different area for multiple objectives. It is used to find the pose of object, to find the change in area, for flood monitoring, for disaster monitoring, land use, agriculture and forestry, to observe the urban growth etc. [10, 11].

3 Image Fusion Using Different Transforms

Image fusion is done at three different levels. These levels are pixel level, feature level and decision level. As the name suggests pixel-level image fusion deals with pixel values in the image, feature level uses the features of images before fusion. In decision level, the appropriate decision about the algorithm to integrate the information is taken before the fusion of images.

The image fusion techniques are divided into spatial domain and frequency domain. The spatial-domain image fusion uses the images as it is (i.e. they directly operate on pixels), whereas frequency domain image fusion uses transformed images for image fusion. Spatial-domain image fusion methods are averaging method, principal component analysis (PCA), intensity hue saturation (IHS), high pass filtering and Brovey transform. The limitation of spatial-domain image fusion is spectral degradation. Due to this, transform-domain image fusion is preferred.

Transform-domain image fusion method uses different transforms that include DCT, DST, DWT, Kekre's wavelet transform and Walsh transform. The performance of these transforms is compared using performance parameters.

DST is represented by symmetrical, real and orthogonal matrix with time complexity of $N \log N$. It has very good energy compression. DCT is fast transform represented by real and orthogonal matrix with time complexity of $N \log N$ that provides very good energy compaction for highly correlated data.

Discrete wavelet transform uses wavelet expansion functions. The wavelet expansion function gives time–frequency localization and energy of the signal. Walsh Transform is represented by separable and symmetrical matrix. The maximum energy is available near origin and goes on decreasing away from origin. The energy compaction is less than DCT. Kekre's Wavelet Transform represented by a generic matrix need not have power of 2. In this, values above diagonal are 1 and lower diagonal except below diagonal line is zero.

4 Experimental Results and Discussions

Simulation was carried out with "image processing toolbox" in MATLAB. Image fusion is performed on a different set of input images taken from different sensors as multispectral and panchromatic sensors. Multispectral image gives the spectral information, and panchromatic image gives the spatial information. The set of images is taken from http://datatang.com. The input images are applied for experimentation using different transforms. Figure 1 shows the result generated for test set 1. Figure 1a, b shows multispectral and panchromatic images used as input images. Figure 1c shows standard image which is subsequently used for visual comparison as well as to calculate various quality indices. Figure 1d–h shows the fused images obtained by DCT, Kekre's wavelet transform, DST, DWT and Walsh transform, respectively. It is observed that the fused image is very similar to standard image using DCT, whereas better spectral quality is given by DST and Walsh transform-based technique. Also, better spatial quality is given by discrete wavelet transform.

The second experiment is conducted on test set 2. Figure 2a gives the spectral information, and Fig. 2b has more clarity of objects present in the image. The result in Fig. 2d–h when compared with standard image shows that all transforms except discrete wavelet transform give better spectral information and discrete wavelet transform reflects the spatial information.

The fused images are compared with standard image to find the quality of image fusion using different transforms based on quality indices parameters that include mean, variance, entropy, standard deviation, root-mean-square error, peak signal-to-noise ratio and spatial frequency.

These quality indices are described below where x is the vector representing input image and F is the vector resenting fused image with dimension N. The mean of image is the average pixel value indicating the brightness of image. The mean is given as follows:

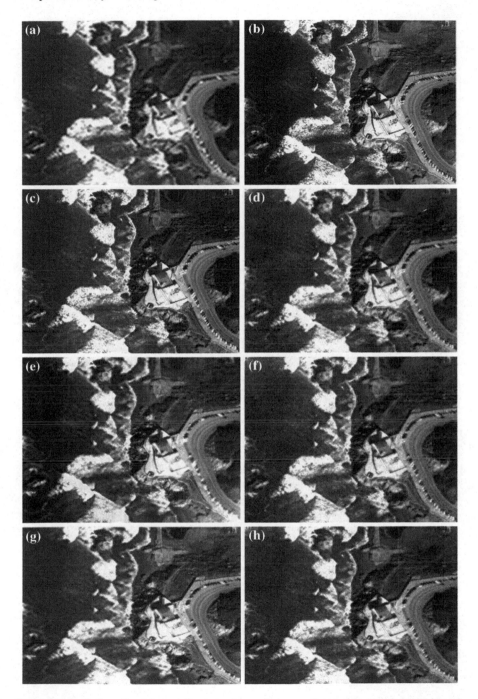

Fig. 1 Image fusion of multispectral and panchromatic images from set 1 **a** multispectral image, **b** panchromatic image, **c** standard image, **d–h** fused images using **d** discrete cosine transform, **e** Kekre's wavelet transform, **f** discrete sine transform, **g** discrete wavelet transform, **h** Walsh transform

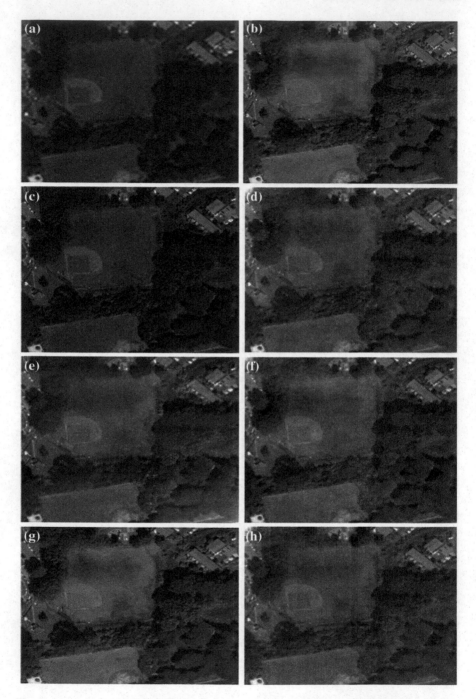

Fig. 2 Image fusion of multispectral and panchromatic images from set 2 **a** multispectral image, **b** panchromatic image, **c** standard image, **d–h** fused images using **d** discrete cosine transform, **e** Kekre's wavelet transform, **f** discrete sine transform, **g** discrete wavelet transform, **h** Walsh transform

$$\mu = \frac{\sum_{i=1}^{N} xi}{N} \tag{1}$$

Variance is the change in intensity of image. For smooth image, variance is less. The variance is given as:

$$\sigma^2 = \frac{\sum_{i=1}^{N} (xi - \mu)^2}{N} \tag{2}$$

Standard deviation is the amount of variation with respect to mean. The standard deviation is denoted as σ.

Entropy describes the information contents in the image. The entropy is given as:

$$E = \sum_{i=1}^{N} xi\log(xi) \tag{3}$$

The root-mean-square error is the measure of accuracy giving information about the difference between standard values and observed or calculated value. This is given as:

$$RMSE = \sqrt{\frac{1}{N} \sum_{i=1}^{N} (xi - Fi)^2} \tag{4}$$

The peak signal-to-noise ratio describes the signal strength with respect to noise in image. This is given as:

$$PSNR = 10 \log\left[\frac{max}{RMSE}\right] \tag{5}$$

where max is the maximum fluctuation in input image data type.

Spatial frequency gives the level of fused image. The spatial frequency is given as:

$$SF = \sqrt{CF^2 + RF^2} \tag{6}$$

where CF is the column frequency and RF is the row frequency of an image.

Tables 1 and 2 give the quality indices using different transform techniques. From the quality indices calculated for image set 1, DCT, DST and Walsh transform are giving better performance for signal quality (PSNR) and root-mean-square error. The brightness of image can be maintained by discrete sine transform. The spatial frequency is good by using discrete wavelet transform. In image set 2,

Table 1 Quality indices for test set 1 using different image transform techniques

	Mean	Variance	Std. dev.	RMSE	PSNR	SF
Input images						
MS image	69.80	4355	65.99	18.83	95.41	10.67
PAN image	32.70	5102	71.43	20.64	94.50	29.57
Standard image	53.91	5102	68.46	0	∞	27.87
Output images						
DCT	76.72	4344	65.91	15.00	97.68	19.25
Kekre's WT	36.96	5036	70.97	22.70	93.54	19.83
DST	81.35	4350	65.96	14.94	97.73	19.09
DWT	71.18	4619	67.96	19.09	95.28	22.45
Walsh transform	74.78	4365	66.07	14.93	97.73	20.09

Table 2 Quality indices for test set 2 using different image transform techniques

	Mean	Variance	Std. dev.	RMSE	PSNR	SF
Input images						
MS image	65.47	436	20.89	9.59	102.16	4.35
PAN image	68.02	1299	36.04	32.61	89.92	15.36
Standard image	49.30	1299	22.89	0	∞	13.97
Output images						
DCT	100.64	752	27.43	29.64	90.87	9.67
Kekre's WT	67.99	703	26.51	30.32	90.65	10.19
DST	84.00	883	29.72	31.16	90.38	9.72
DWT	76.26	1223	34.97	33.05	89.79	13.48
Walsh transform	99.48	802	28.32	29.97	90.76	10.35

discrete cosine transform and Walsh transform are giving better performance for signal quality. The brightness of image is maintained by discrete cosine transform, and the spatial frequency component is better by using discrete wavelet transform.

5   Conclusion and Future Work

Discrete cosine transform has the advantage of more energy compaction and thus useful in real-time applications. It gives good spatial resolution as well. Various Kekre's wavelet transform can be generated by changing the size of Kekre's transform. It provides good quality of fused image but increases in the mean square error. Discrete sine transform gives good energy compression with mean square error and spatial information. It also provides good signal-to-noise ratio.

Discrete wavelet transform provides good quality fused image and better signal-to-noise ratio. It also minimizes spectral distortion. But the fused image has less spatial resolution. Walsh transform has variation in energy from the origin. It gives better mean square error than Kekre's wavelet transform. But it minimizes the spatial resolution. By comparing all the mentioned methods, only one method not provides the good quality in all aspects. Thus, optimization of these methods is necessary which can be done by using evolutionary approach. In future work, image fusion using evolutionary algorithms by taking the base of these methods can be done to find the optimized fused image.

References

1. Arpita Das, Mahua Bhattacharya, "Evolutionary algorithm based automated medical image fusion technique comparative study with fuzzy fusion approach", Nature & Biologically Inspired Computing, pp: 269–274, 9–11 Dec 2009.
2. Tao Wu, Xiao-Jun Wu, Xiao-Qing Luo, "A study on fusion of different resolution images", Procedia Engineering, Vol. 29, pp: 3980–3985, 2012.
3. Peijun Du, Sicong Liu, Paolo Gamba, Kun Tan and Junshi Xia, "Fusion of difference images for change detection over urban areas", IEEE journal of selected topics in applied earth observations and remote sensing, Vol. 5, No. 4, Aug 2012.
4. Peijun Du, Sicong Liu, Junshi Xia, Yindi Zhao, "Information fusion techniques for change detection from multi temporal remote sensing images", Information Fusion 14, pp: 19–27, 2013.
5. Ren C. Luo, Chun Chi Lai, "Multisensor fusion based concurrent environment mapping and moving object detection for intelligent service robotics", IEEE transactions on industrial electronics, Vol. 61, No. 8, pp: 4043–4051, Aug 2014.
6. Y. Asnath Victy Phamila, R. Amutha, "Discrete cosine transform based fusion of multi focus images for visual sensor networks", Signal Processing 95, pp: 161–170, 2014.
7. Mohammd Bagher Akbari haghighat, Ali Aghagolzadeh, Hadi Seyedarabi, "Real time fusion of multi focus images for visual sensor networks", 6th Iranian conference of machine vision and image processing, pp: 1–6, 27–28, Oct 2010.
8. Liu Cao, Longxu Jin, Hongjiang Tao, Guoning Li, Zhuang Zhuang,Yanfu Zhang, "Multifocus image fusion based on spatial frequency in discrete cosine transform domain", IEEE signal processing letters, Vol. 22, No. 2, pp: 220–224, Feb 2015.
9. B. Roopa, Sunilkumar S. Manvi, "Image fusion techniques for wireless sensor networks: Survey", ITSI Transactions on Electrical and Electronics Engineering, 2320–8945, Vol. 2, pp: 13–19, 2014.
10. Shivdeep Kaur, Rajiv Mahajan, "Evaluating the short comings of digital image fusion techniques", International Journal on recent and innovation trends in computing and communication, Vol. 2, No. 5, pp: 1162–1167, May 2014.
11. Qian Zhang, Zhiguo Cao, Zhongwen Hu, Yonghong Jia, Xiaoliang Wu, "Joint image registration and fusion for panchromatic and multispectral images", IEEE geoscience and remote sensing letters, Vol. 12, No. 3, pp: 467–471, Mar 2015.

IoT-Based Smart Garbage Management System

Prajyot Argulwar, Suvarna Borse, Kartik N. Argulwar
and Udge Sharnappa Gurunathappa

Abstract In the present day scenario, it is seen that the garbage bins or dustbins are placed at different public places in the cities which are spilling over due to raise in waste every day. Due to which it creates unhygienic condition for the people and creates the bad smell around the surroundings. This makes the spreading of diseases and human illness. To avoid such a situation, we designed the IoT-based garbage management system. To keep up cities clean, this paper "IoT-based Smart Garbage Management System" gives a very new and useful system. With this system, the monitoring on garbage bins and the level of garbage available in the garbage bins can be acknowledged through a Web page. The page shows a notation level of waste bins and shows level of garbage collected in the bins in order to take an action on it. The Web page view can be seen with LCD which displays actual level of waste available in bins. If the garbage level crosses beyond the predefined limit, the system buzzes. This system is usefull to make an awareness to particular authority, so that an action can be perform to collect the garbage from bins. Hence, this smart system gives information about garbage levels of garbage bins which indirectly helps to keep the city neat and clean by only if the output at LCD with graphical representation via a Web page is available.

Keywords IoT · LCD · PIC microcontroller · RS232 module · Ultrasonic sensors

P. Argulwar · S. Borse · K.N. Argulwar (✉) · U.S. Gurunathappa
Pune, India
e-mail: kartik.argulwar08@gmail.com

P. Argulwar
e-mail: prajyotargulwar@gmail.com

S. Borse
e-mail: suvarnaborse088@gmail.com

U.S. Gurunathappa
e-mail: sharnappaudge@yahoo.in

© Springer Nature Singapore Pte Ltd. 2018
S.S. Dash et al. (eds.), *International Conference on Intelligent Computing and Applications*, Advances in Intelligent Systems and Computing 632,
https://doi.org/10.1007/978-981-10-5520-1_23

237

1 Introduction

The Internet of Things is a theory of surrounding things, matter, or objects which are managed via wired and wireless structures without any user invention. In the concept of IoT, the things or objects connect with each other and interchange the information to serve the advanced intelligent services to the users. With communication topologies such as Wi-fi and VoLTE, various sensors, communication methods, and advanced intelligent equipment such as mobile devices, the IoT has achieved proper interest in academics.

In our system, the smart garbage bins are connected to Internet to collect real-time information of the smart dustbins. In the recent years, there is a rapid growth in population which leads to more waste disposal. So a proper waste management system is necessary to avoid diseases and keep the city clean. Managing smart bins is taking care by monitoring the status of it. There are number of dustbins which located throughout the city or the campus (educational institutions, companies hospitals, etc.). These dustbins are interfaced with microcontroller-based system with ultrasonic sensors and RF modules.

2 Working Principle

The garbage monitoring system consists of ultrasonic sensors, PIC microcontroller, LCD display, power supply, Wi-fi modem. The power supply unit provides fixed 9 V supply to the PIC microcontroller and Wi-fi modem. The garbage monitoring system comprises of no. of ultrasonic sensors that capture the data from the dustbins or garbage collector. The ultrasonic sensors are fitted into the dustbin or garbage collector so that the level of the container can be monitored. These ultrasonic sensors are connected to one of the port of the PIC microcontroller in order to take the input signal.

3 Related Work

In [1], to manage the waste bins properly the integrated system is formed with the use of ZigBee, GSM (global system for mobile communication), and ARM7. The sensors are placed to different garbage bins at a common predefined place at different civic places. If the waste (garbage) reaches to the sensor, then that information will be provided to ARM7 controller. With this, ARM7 controller gives the warning level to truck driver to which bin is completely filled with garbage and wants critical awareness. If trash reaches a particular sensor level, then that information will provide to ARM7 controller. The controller will provide information to driver which indicating that garbage bin is completely filled and wants critical awareness. ARM7 will send the SMS using GSM module.

In [2], they gave the importance to increasing levels of waste production with societal concerns that how it could be reduced in proper manner. So that policy makers have positive toward recycling and reprocess strategies to cut the order for raw supplies and to reduce the amount of garbage available to landfill.

In [3], in this paper, it is proposed that the combination of integrated systems such as radio frequency identification, global position system, geographic information system, Web camera, general packet radio services will give the solution for the problem of waste management, and with this the performance, analysis is done through actual implementation. The process will be same as the sensors will provide the data indication level report to the controller and the controller will send the message to the user.

In [4], the objective in this paper is to study the categorization of garbage and existing system of waste supervision actions. This paper indicates an overview of existing management system of municipal solid waste, in which it shows with few suggestions, which will be helpful to the improvement of existing waste management system. The scenario of existing waste management system with needs of improvement is provided. This paper shows the working system of waste management of Thoubal Municipality.

In [5], the proposed system consists of three model sensors, microcontroller, and GUI. The sensors describe the level of garbage in the bins through the microcontroller. The sensors provide the information like the level of garbage in the bin to the microcontroller, and then the microcontroller sends this information on graphical user interface (GUI). From this GUI, the user can get details about different garbage bins and according to that the user will take the necessary action on it. This proposes the smooth traffic monitoring process of waste management.

In [6], here the description of smart bin is explained with its applications. This smart bin is designed with sensor, microcontroller, and GUI. The network sensors activate smart bins connected through the cellular network like GSM, GPRS, LTE which provides information, and this information will be analyzed and visualized at real time to get the situation of waste around the city. This paper gives the proper information for the research in the area of waste management system.

4 Problem Definition

As we have seen number of times, the dustbins are getting over flown and concern person does not get the information within a time and due to which unsanitary condition formed in the surroundings, at the same time bad smell spread out which looks city with deadly diseases with illness around the locality which is easily spreadable.

4.1 Disadvantages of Existing System

- Time consuming and less effective: Trucks go and empty containers whether they are full or not.
- High costs.
- Unhygienic environment and look of the city.
- Bad smell spreads which causes diseases.
- More traffic and noise.

4.2 Advantages of Proposed System

- Real-time information on the fill level of the dustbin.
- Deployment of dustbin based on the actual needs.
- Intelligent management of the services in the city.
- Effective usage of dustbins.

5 Material

Here, the PIC 18F4520 microcontroller used to fetch the information via sensor and process information received through sensor and same data transmitted to the PC using RS232.

5.1 PIC 18F4520 Microcontroller

- Instruction sets with C compiler.
- Interrupts and exceptions priority levels.
- Variable programmable range.
- In-circuit debug (ICD) via DIP.
- Variable voltage range—2 to 6 V.

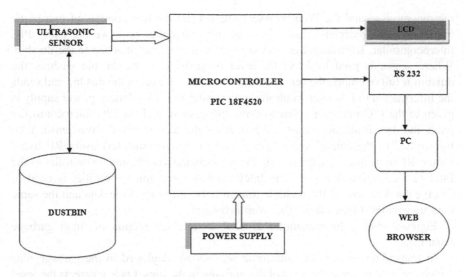

Fig. 1 Block diagram of smart garbage management system

5.2 Ultrasonic Sensors

The ultrasonic sensor is used to detect level in the dustbin whether the dustbin is full or not. The ultrasonic sensor consists of an emitter, detector, and associated circuitry. The circuit required to make an ultrasonic sensor consists of two parts: the transmitter circuit and receiver circuit.

5.3 Liquid Crystal Display

The use of LCD in a project is to show the output. We have used 16×2 LCD which indicates 16 columns and two rows. So, each line consists of 16 characters. Thus, LCD is important to display the results as well as errors occurred at different modules of management system if in case of system failure happened, the LCD shows the detected module so it can be easy to rectify the problem (Fig. 1).

6 Actual Working

The block diagram shows the different components used to make smart garbage management system, in which the different components are used like ultrasonic sensor, PIC microcontroller, power supply, transmitter as well as receiver, microcontroller, and the Web browser. The project module consists of two parts: transmitter and receiver section. Here in the transmitter part, we are using PIC microcontroller, RF transmitter, and sensors, and these are attached to the dustbin. Where sensor is used to detect the level of garbage in the dustbin whether the dustbin is full or empty, the sensor senses the garbage level of the dustbin and sends the information to PIC microcontroller power supply +9 V. Battery power supply is given to the PIC microcontroller to drive the system, and the PIC microcontroller reads the data from the sensor and processes the data received from sensor then transmitted to the central system (Intel Galileo microcontroller) using RF transmitter. RF transmitter is used to send the information via PIC microcontroller to the Intel Galileo microcontroller. The Intel Galileo Gen2 microcontroller is used to receive the data sent by the multiple transmitters and process the data and the same data transmitted to the client, i.e., Web browser.

Figure 2 shows the transmitter as well as receiver section of smart garbage dustbins.

In transmitter section, the ultrasonic sensors are deployed in the garbage bins which are used to sense the level of the garbage in the bins. Once it detects the level of garbage, it will send the detected information to the PIC controller. Then, the PIC controller checks the status of garbage bins and sends to the central server system.

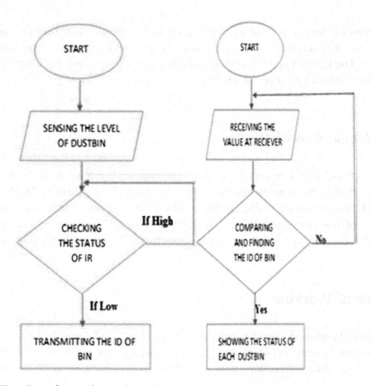

Fig. 2 Flowchart of transmitter and receiver action

The receiver always connects to the central system once the server updates through PIC controller the receiver always check the Bin identity if it match found with at server lever the message will be display at LCD screen with level of indication garbage in the bins, then the proper message will be sent to the vehicle driver so that the collection of garbage will takes place. In other way, the receiver sections receive the values sent by the sender through RF receiver to the central system, check all the dustbin status, and display on the browser.

7   Result and Discussion

This IoT-based waste management is very useful for smart cities in different aspects. it is seen that there are number of garbage is are allocated in the different area's and dustbins get over flown many times and the concerned people do not get information about this. Our system is designed to solve this issue and will give total information of which is situated at different areas throughout the city. The concerned authority can operate the data from different locations at any time with details. Accordingly, they can take the decision on this immediately (Figs. 3 and 4).

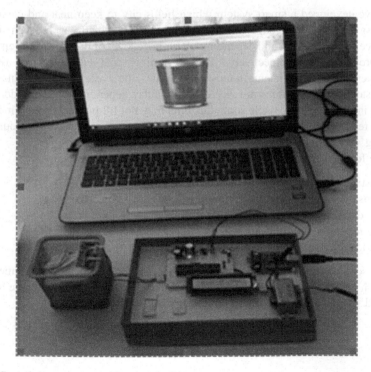

Fig. 3 Central server system with smart dustbin

(a)   (b)   (c)

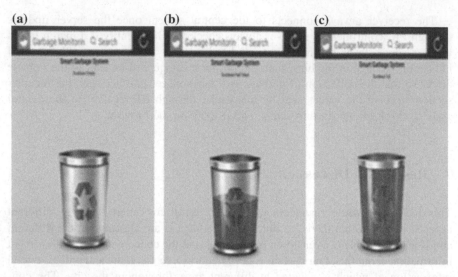

Fig. 4 Garbage indication from different bins: **a** an empty bin, **b** half-filled bin, and **c** full bin

8 Conclusion and Future Scope

We have developed a smart system which helps city to keep neat and clean. This smart system shows garbage level status of all bins which can be seen through anyplace by any person, and depending on garbage level of dustbins, the person can take proper action, respectively. This system is cost effective, less resource equipment, and very efficient. In most of the cities, the waste collection vans collect the waste from the dustbins in twice or thrice a day depending on living density of people in different area, while some dustbins may not fill properly which indirectly affects resources. This smart system helps to achieve a proper use of resources by sending vans to a particular dustbin only when it gets full properly.

The following are the results which are obtained from this work:

- Waste level detection inside the dustbin is possible.
- Transmit the information wirelessly to concern.
- The data can be accessed anytime and from anywhere.
- The real-time data transmission and access.
- Avoids the overflows of dustbins.

The scope for the future work is this system can be implemented with time stamp in which real-time clock shown to the concern person at what time garbage bins are full and which bins garbage will collected first via shortest route.

References

1. Thakor N. V., Webster J. G., and Tompkins W. J.: Design, implementation, and evaluation of a microcomputer-based portable arrhythmia monitor. Med. Biol. Eng. Comput., vol. 22, pp. 151–159, (1984).
2. Kanchan Mahajan, "Waste Bin Monitoring System Using Integrated Technologies", International Journal of Innovative Research in Science, Engineering and Technology, Issue 3, Issue 7, July 2014.
3. Raghumani Singh, C. Dey, M. Solid waste management of Thoubal Municipality, Manipur- a case study Green Technology and Environmental Conservation (GTEC 2011), 2011 International Conference Chennai 21–24.
4. York, J., and Pendharkar, P.C., "Human–computer interaction issues for mobile computing in a variable work context," International Journal of Human-Computer Studies, Vol. 60, No. 5–6, 2004, pp. 771–797.
5. Khattak, A. M. Pervez, Z. Jehad Sarkar, A.M., and Lee Y., "Service Level Semantic Interoperability," 2010 10th IEEE/IPSJ International Symposium on Applications and The Internet, saint, pp. 387–390, 2010.
6. Vikrant Bhor, "Smart Garbage management System International Journal of Engineering Research & Technology (IJERT), Vol. 4 Issue 03, March-2015 2000.
7. Narayan Sharma,, "Smart Bin Implemented for Smart City", International Journal of Scientific & Engineering Research, Volume 6, Issue 9, September-2015.
8. Barnaghi, P., Wang, W., Henson, C., and Taylor, K., "Semantics for the Internet of Things: Early Progress and Back to the Future," International Journal on Semantic Web and Information Systems, vol. 8, No. 1, 2012.
9. www.engineersgarage.com/sites/dcfault/files/LCD%2016x2.pdf.
10. www.embeddedrelated.com/usenet/embedded/.../26531-1.php.

Event-Driven SOA-Based IoT Architecture

Poonam Gupta, Trupti P. Mokal, D.D. Shah
and K.V.V. Satyanarayana

Abstract In twenty-first century, the size and scope of the Internet has been increasing as smart devices have ability to communicate with one device to another and share data over the Internet. This is known as Internet of things (IoT). Scalability of the entire system increases in case of more devices in IoT system. So in IoT system, main challenges are to maintain scalability and throughput. To meet these challenges, various architectures are used for implementing IoT applications such as SOAP-based, RESTful, broker-based SOA architecture. While service-oriented architecture provides system integration, interoperability, business agility. Though service-oriented architecture is mainly used for implementing various IoT applications, it faces some challenges such as security, interoperability, and context awareness. Event-driven service-oriented architecture (EDSOA) is a new architectural style which is an extension of broker-based architecture and most suitable for IoT applications. In this paper, we have tried to cover all these architectures.

Keywords Internet of things · Service-oriented architecture · SOAP
REST · MQTT · EDSOA

P. Gupta (✉) · K.V.V. Satyanarayana
Department of Computer Engineering, Koneru Lakshmaiah
Educational Foundation (K. L. University), Vijayawada, India
e-mail: poonam77gupta@gmail.com

K.V.V. Satyanarayana
e-mail: kopparti@kluniversity.in

P. Gupta · T.P. Mokal (✉)
Department of Computer Engineering, G.H. Raisoni College
of Engineering & Management, Pune, India
e-mail: mokaltrupti33@gmail.com

D.D. Shah
Department of E & TC Engineering, Imperial College of Engineering
and Research, Pune, India
e-mail: dilip.d.shah@gmail.com

© Springer Nature Singapore Pte Ltd. 2018   247
S.S. Dash et al. (eds.), *International Conference on Intelligent Computing
and Applications*, Advances in Intelligent Systems and Computing 632,
https://doi.org/10.1007/978-981-10-5520-1_24

1 Introduction

In this modern century, the introduction of network technologies has enhanced the growth of heterogeneous devices due to which application no longer sends the data and functionality not only to humans but also to machines [1]. Internet of things (IoT) can be thought of as billions of devices connected together where the devices have to be intelligent and they need to interact with environment and interact with people. And when we take these three things together and have an intelligent system you can react and do things with benefit, we call this nothing but IoT. For instance, consider one scenario of a medical shop. If a patient buys a particular medicine from a medical shop, then the radio frequency identification (RFID) sensor in medical shop gives alert message (e.g., two days left to expire) on the mobile of that specific patient and even if that patient went to Africa or some other country still the alert messages are provided. This is what IoT is. The "things" include people, location (of objects), time information (of objects), and condition (of objects). So considering above scenario we can say these things enable anytime, anywhere connectivity.

There are various IoT applications which are listed as follows: in health care (e.g., drugs tracking, ambulance telemetry, hospital asset tracking), for smart cities (e.g., smart streetlight, pipeline leak detection, traffic control), for wearable (e.g., smart watch, entertainment), for building and home automation (e.g., light and temperature control, energy optimization).

With the rise of IoT, there were some critical issues observed that the IoT was facing—interoperability, business agility, etc. Moreover, another issue is tracking of devices. There are some privacy issues around tracking and monitoring. What about maintenance of these devices? Maintenance is a critical issue as many different devices are connected over the Internet.

2 Various IoT Application Architectures

As IoT devices need to interoperate the service-oriented architecture (SOA) came into existence as a solution for interoperability issues. Moreover, a large number of sensor networks and applications face an integration problem. To achieve business integration, researchers have incorporated SOA technology with IoT [2]. SOA has the ability to scale operations to meet different demand levels. SOA provides improved manageability and security. SOA provides business intelligence, performance measurement, and security attack detection. SOA also makes enterprises being able to quickly respond to business changes with agility and support new channels of interactions with customer, partners, and suppliers. Thus, IoT requirements are fulfilled by SOA technology. This is why we need SOA.

SOA has two approaches: (1) SOAP (2) RESTful. In SOAP, we have SOAP ENVELOPE, SOAP BODY, and SOAP HEADER which is an overhead along with actual message payload. And in case of REST, we only deal with actual message that has to be exchanged. Since SOAP has some limitations such as tightly coupled invocation, no hyperlink support, which can be overcome by RESTful services, we can say that for IoT applications it is good to use RESTful-based Web services than using SOAP-based Web services.

The main challenges faced in IoT are integration of different technologies, which can be solved with N-to-N approach [3], but such approach is suitable only in case if it is concerned with a limited heterogeneity of technologies. To overcome such limitation we use broker-based architecture which is more scalable approach. In home networks, an automatic discovery mechanism of various devices is created by broker. Broker forwards understandable information to the interested subscribers after receiving such understandable information. This is why broker is required.

As of today, SOA is a well-known approach and is used for integration purpose and in business processes. In case of larger and more complex IoT service application, it is required for an IoT service system to be a strong integration capable and adaptable to a different and dynamically changing environment. Event-driven architecture (EDA) provides concurrent responsive processing and support sensing. However, EDA and SOA are different from each other; merging the advantages of both SOA and EDA service coordination can be achieved.

2.1 RESTful

Laine et al. [4] described an end-to-end IP and RESTful Web services-based architecture to integrate constrained devices with the Web. He discussed the utilization of RESTful Web services with IoT. CoAP has been designed in different way for M2M applications over constrained environment on the IoT. Processing power required by RESTful Web services is low and also has advantages over SOAP:

- lightweight
- less overhead
- statelessness
- less parsing complexity

Thus, applications that supports RESTful Web services are easy to learn and implement in comparison with SOAP.

Rathod et al. [5] stated that SOAP-based services follow an operation-centric approach whereas RESTful HTTP-based services follow resource-centric approach. He proposed Web service resource bundle (WSRB) to show contribution toward dynamic composition of RESTful Web service. He concluded that static binding of Web service increases network traffic and consumes more memory. He also concluded that the RESTful Web service gives best performance as compared to SOAP-based Web service.

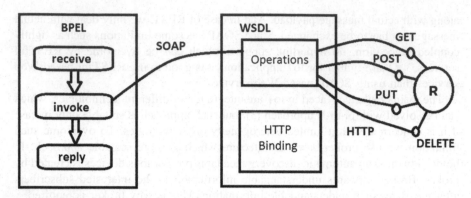

Fig. 1 Restful Web services architecture

Bohara et al. [6] explore the concept of RESTful service Mashup, which is related to farmer's queries, by integrating individual Web services which can fulfill the end user's requirements. He proposed a recursive algorithm which has complex queries processing capability using integration of different RESTful Web services. He described different RESTful Web services such as FarmerInformation, GeoCoding, CropRecommendation, DataWeave, and GoogleService. He also concluded a broad solution of RESTful Web service integration (Fig. 1).

2.2 SOAP

Lee et al. [7] described SOAP as a protocol specification to exchange structured information while implementing Web services. They also stated that to enable direct invocation and composition of SOAP, RESTful, and other services with JSON, java objects BPEL engine is used to extend and bundle with adaptors. Such a composition was made to execute complex tasks on various mobile devices. They also concluded that request and response messages of invocations of RESTful services can be directly transferred in between extended BPEL engine and the RESTful services. These messages need not to be transferred via server-side SOAP services (Fig. 2).

2.3 Broker-Based Architectures

Cheng and Chen [8] described a broker architecture called Grid quorum-based Pub/Sub system (GQPS). This architecture supports Pub/Sub systems in IoTs. This GQPS is compared with DHT-based approaches, after doing so it concluded that GQPS achieves better latency, good scalability, and high fault tolerance.

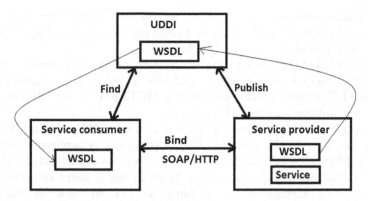

Fig. 2 SOAP-based web services architecture

However, for delay-sensitive IoT, it requires some space consumption which lowers the latency.

Govindan et al. [9] proposed MQTT-SN for Wireless Sensor Networks which provide end-to-end delay from the time the content was published to the time the content was delivered to the doctor and sufficient probability of content delivery in wireless healthcare IoT system. They also stated that MQTT-SN is used by sensors mounted on human body to PUBLISH the sensed data to gateway. PUBLISH data is sent by Gateway to server. Doctors' side gateway accesses the patient sensors published data after connecting to the server.

Gomes et al. [10] described a broker-based architecture called MQTT health broker architecture in which connection between Agents and Managers established and in home networks, an automatic discovery mechanism of new Personal health devices (PHDs) is created by broker. In this broker is used to find new devices. First, there is subscription to a particular topic they are interested in; this subscription is done by manager, and then a new agent connects and subscribes to the same data topic. After its subscription, an association request is sent by agent to all managers subscribed to, and by publishing the message, the interested managers reply to this association request. They also stated that MQTT is a lightweight protocol that holds short processing and memory capabilities.

Luzuriaga et al. [11] presented adoption of MQTT for IoT and M2M environments where efficient handling of mobility is of great importance for overall performance of IoT application. An MQTT publisher publishes the MQTT messages with a given periodicity to a predefined MQTT broker, which then forwards the incoming messages directly to the subscribers.

Collina and Corozza [12] proposed a new hybrid approach called QEST broker architecture by bridging REST and MQTT. They also stated that MQTT implements Pub/Sub while HTTP is a request/response protocol. QEST broker architecture was required to bridge the gap between the things and Web allowing existing developers to use their skills to interact with smart objects.

Rayes and Mohammadi [13] proposed an enhanced MQTT architecture where they realized a horizontal IoT integrated framework by improving MQTT protocol which in turn provided to hold up QoS features.

Chen and Lin [14] proposed a method to integrate MQTT protocol with ETSI M2M architecture via a new network called MQTT proxy. They compared MQTT proxy with HTTP proxy which concluded that MQTT proxy lowers the latency and saves power in better way than HTTP proxy.

Dhar and Gupta [15] proposed a novel framework for IoT vehicular information network based on MQTT protocol. They investigated authentication issues for vehicular information network architecture based on the communication principle of named data networking. This paper helps to make better content naming, addressing, data aggregation, and mobility for IVC in vehicular information network.

Avachat and Gupta [16] discussed ideal characteristics of a middleware, i.e., scalability, spontaneous interaction, context detection. They also described architecture of three middleware solutions, i.e., triple space-based semantic distributed middleware, UBIWARE, and finally SOA-based middleware. They also found which characteristics the three middleware possess.

2.4   EDSOA

As of today, SOA is a well-known approach; it provides greater speed and flexibility for IT organizations and is used for integration purpose and business processes.

Lodi et al. [17] discussed the level of dependability that can be currently obtained by adopting SOA- and EDA-based solutions, respectively. They also mentioned that ESB middleware is used to apply the SOA and EDA convergence which facilitates the internetworking among different financial organizations.

Adding event processing to an SOA through an event-driven architecture (EDA) gives your information systems the ability to sense and respond to events rapidly, either through an automated process or human interaction. Merging these concepts results in a new generation of middleware platforms that will inherit the best of both worlds. Its major aim is realized in the event-driven SOA (EDSOA) concept by SOA combining business functions and IT, and EDA focusing on data as well as business relevant event orientation [18].

Traditional SOA architecture is not enough in real-time response and parallel process of services execution. So this paper presents new system based on EDSOA architecture to support real-time, event-driven, and active service execution (Fig. 3).

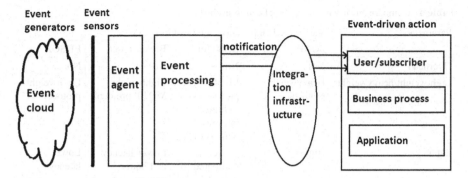

Fig. 3 Event-driven service-oriented architecture

3 Comparison and Generic Framework for IoT

In this framework, Fig. 4 represents devices such as CCTV, door access, DVR use broker for publishing the data to gateway. Gateway then sends PUBLISH data to the server. Users connect and SUBSCRIBE to the same data topic. Thus, broker establishes connection between these devices and user (Table 1).

Each of these devices may use distinct protocols to connect such as MQTT. The Internet of Things is all about communication and messaging. Devices connected to the IoT system have to connect to a kind of centralized hub that allows them to exchange their data with other devices and services. Messages are exchanged between protocol adapters and services using cloud.

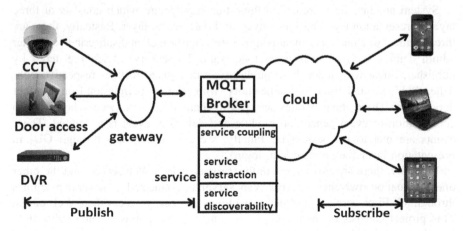

Fig. 4 Generic framework for IoT

Table 1 Comparison of various IoT application architectures

Parameters	Various IoT application architectures			
	SOAP	RESTful	Broker-based	EDSOA
Overhead	Yes	No	No	No
Lightweight/heavyweight	Heavyweight protocol	Lightweight protocol and tighter integration with HTTP	Lightweight MQTT protocol	Lightweight protocol
Latency	Higher	Lower as compared to SOAP	Lower latency and better power saving than REST	Lower latency
Important characteristics	Schema confirmation, xsd, wsdl, and WS-standards are key to use SOAP-based Web services	WSDL 2.0, REST principles are used	Multiple clients are permitted to subscribe to the same category, creates an automatic device discovery in home networks	Real-time event-driven response, parallel process of active service execution
Used in IoT	Less	Better than SOAP	Mostly used	Most suitable

4   Implementation

System architecture for weather forecasting system based on event-driven service-oriented architecture is shown in Fig. 5

System architecture represents a three-tier architecture which consists of three layers: presentation layer, business layer, and data access layer. Basically, there are three entities: weather department admin, publisher/broker, and subscriber. Weather admin defines services such as cyclone, rain. These services are registered by publisher. Subscriber under these publishers will register to their respective publisher. Publisher and broker are same entity, who accesses permission from weather department. These three entities are getting data from data access layer. From coordination of event generation module and database, event is generated. These events are met to business layer. Finally, we are getting data through GUI in presentation layer (i.e., registration, login).

Basically, there are two servers in this project: one is Web server, and the other one is situation awareness server. Web server is connected to several machines through Wi-Fi or wireless, and these machines can easily access their Web pages. This project consists of three entities: weather admin, publisher, and subscriber. Weather admin defines services such as cyclone, rain. These services are registered by publisher, and subscriber registers their respective publisher. Situation

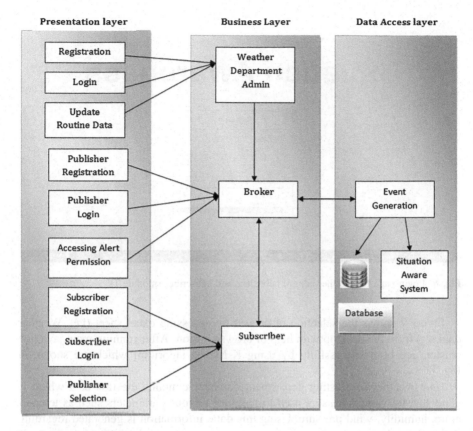

Fig. 5 Three-tier architecture

awareness server is purely a standalone system, and it takes all the weather data and performs following steps: preprocessing, K-Means clustering, HMM, fuzzy logic, pattern identification, Gaussian distribution. Hidden Markov model (HMM) extracts the hidden states by observing the observed states. For example, if observed states are heavy cloud, low wind, and high humidity, then hidden state will be rain. And if observed states are heavy cloud, high wind, and high humidity, then hidden state will be cyclone. Now, if we will make cluster of these then there will be two clusters based on K-Means clustering algorithm. Cluster 1 includes heavy cloud, low wind, and high humidity, whereas cluster 2 includes heavy cloud, high wind, and high humidity. Fuzzy logic is used for setting numerical values to every observed state. After pattern identification is done, publish/subscribe broker module broadcast event risks to all the subscribers using Gaussian distribution model via SMS or email.

Figure 6 shows login page for the three entities where we can login or register the details as weather admin or publisher or subscriber. If we login as a subscriber, then it allows us to update profile, select alert scenario, etc.

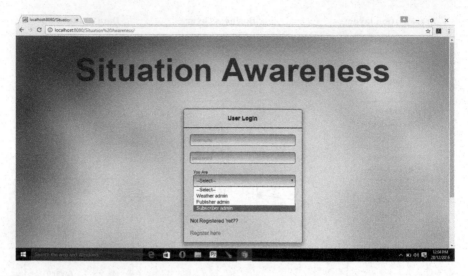

Fig. 6 Login page for weather admin, publisher, and subscriber (module 1)

Figure 7 shows the selection of dataset by accessing excel sheet (i.e., Weather data.xls) from the desktop with help of browse button. After running main.java five clusters are built successfully by using K-Means algorithm which are shown in Fig. 8.

Here weather forecasting department database contains the raw data which is generated by various sensors used to measure various parameters such as temperature, humidity, wind pressure. Using this data, information is generated regarding

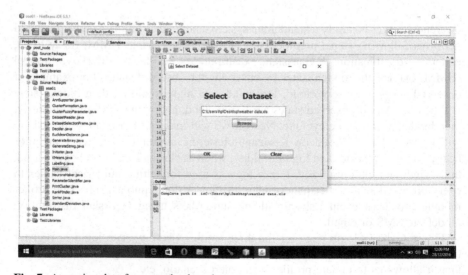

Fig. 7 Accessing data from weatherdata.xls

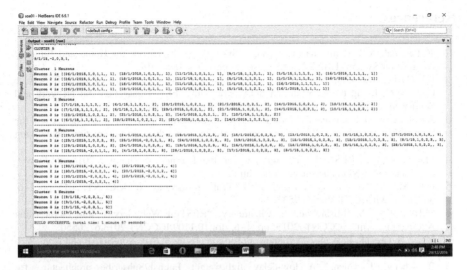

Fig. 8 Cluster formations by K-Means algorithm

forecasting of rain, cyclone and forwarded to the subscribers of the particular topics such as rain, cyclone whenever situation arises, proposed Event driven SOA to make this system efficient.

5 Conclusion

In this paper, detailed study of four IoT application architectures: SOAP, RESTful, broker-based architecture, and event-driven SOA architecture is given. While comparing SOAP and Restful architectures, it is observed that SOAP does the parsing of XML documents whereas in RESTful does not require. At the same time, SOAP requires significant amount of memory and processing compared to RESTful. So RESTful is more suitable than SOAP but still uses HTTP protocol which is heavier than MQTT. MQTT a lightweight broker-based protocol reduces the latency and saves power in better way and has many other support features as compared to HTTP protocol which is the basis of RESTful. So SOA approach using broker-based architecture is better than REST approach. While EDSOA is an extension to broker-based SOA approach, by making system event driven, each "service" can be completely decoupled from each other. Thus, even when a business process changes, there is no complex altering of Routing and system is both flexible and manageable. Thus, event-driven SOA architecture is best for IoT in comparison with SOAP and RESTful. In this paper, we have given generic framework for IoT application using event-driven SOA approach. Various module details about the weather forecasting system are also given where we have used proposed architecture, i.e., EDSOA. There are multiple MQTT brokers available, and which broker works well with EDSOA architecture can be studied in future.

References

1. T. Perumal, A.R. Ramli, and C. Leong, "SOA based framework for Home and Building Automation System (HBAS)", vol. 8, No. 5, pp. 197–206, 2014.
2. Bo Cheng, Da Zhu, Shuai Zhao and Junliang Chen, "Situation-Aware IoT Service Cordination Using the Event-driven SOA paradigm", IEEE Transactions on Network and Service Management", vol. 13, no. 2, pp. 349–361, June 2016.
3. A.C. Olivieri, G. Rizzo and F. Morard, "A Publish-Subscribe Approach to IoT integration: the Smart office use case", in 29th international conference on Advanced Information Networking and Applications workshops, 2015, pp. 644–651.
4. Markku Laine, "RESTful Web Services for the Internet of Things", 2014,http://www.sensinode.com/EN/products/nanoservice.html.
5. D.M. Rathod, "Towards composition of RESTful web services", 6th ICCCNT, 2015, pp. 1–6, doi:10.1109/ICCCNT.2015.7395237.
6. M.H. Bohara, M. Mishra and S. Chaudhary, "RESTful Web Service Integration using Android platform", ICCCNT, 2013, pp. 1–6, doi:10.1109/ICCCNT.2013.6726550.
7. J. Lee, S. Lee and P. Wang, "A framework for composing SOAP, Non-SOAP and Non- Web Services", IEEE transactions on Services Computing, Issue No. 01, 2013.
8. B. Cheng, J. Chen, "A low-delay lightweight Publish/Subscribe architecture for delay-sensitive IoT services", 20th IEEE International conference on Web Service, vol. 10, Issue 3, pages 60–81, July 2013.
9. K. Govindan, A.P. Azad, "End-to-end service assurance in IoT MQTT-SN", 2015 12th annual IEEE Consumer Communications and Networking Conference (CCNC), Las Vegas, NV, 2015, pp. 290–296, doi: 10.1109/CCNC.2015.7157991.
10. Y. Gomes, D. Santos and H. Almeida, "Integrating MQTT and ISO/IEEE 11073 for health information sharing in the IoT", 2015 IEEE International Conference on consumer electronics (ICCE), Las Vegas, NV, 2015, pp. 200–201, doi:10.1109/ICCE.2015.7066380.
11. J.E. Luzuriaga, J.C, Cano and C. Calafate, "Handling Mobility in IoT applications using the MQTT protocol", Internet Technologies and Applications (ITA), 2015, Wrexham, 2015, pp. 245–250, doi:10.1109/ITechA.2015.7317403.
12. M. Collina and G. Corazza, "Introducing the QEST broker: Scaling the IoT by bridging MQTT and REST", 23rd IEEE International Symposium on Personal, Indoor and Mobile radio communications (PIMRC), Sydney, NSW, 2012, pp. 36–41, doi:10.1109/PIMRC.2012.6362813.
13. A. Rayes and M. Mohammadi, "Toward better horizontal integration among IoT services", IEEE Communications Magazine, vol. 53, no. 9, pp. 72–79, September 2015.
14. H.W. Chen, F.J. Lin, "Converging MQTT resources in ETSI standards based on M2M platform", 2014 IEEE international conference on Internet of Things, Taipei, 2014, pp. 292–295, doi:10.1109/iThings.2014.52.
15. P. Dhar, P. Gupta, "Survey: IoT vehicular Information Network", International journal of Science and Research (IJSR), vol. 4, Issue 11, Nov. 15.
16. Avachat, P. Gupta, " A study of semantic middleware for IoT", International Journal of Advanced Research in Computer Science and Software Engineering, vol 4, Issue 12, Dec 14.
17. Giorgia Lodi, Leonardo Querzoni, Roberto Beraldi, "Combining Service-oriented and Event-driven architectures for designing dependable systems", March 2006.
18. Olga Levina, "A model and an implementation approach for Event-driven Service Orientation", International journal on Advances in Software, vol 2, 2009.

FPGA Implementation of AES Algorithm for Image, Audio, and Video Signal

C. Sapna Kumari and K.V. Prasad

Abstract The video, audio and image security, bandwidth, speed of transmitter, and reception of data are the main concepts of every communication system. The data that is being communicated can be safeguarded in many ways. The proposed technique creates an effective modified advance encryption standard (AES) algorithm for data security for encryption and decryption. In this Research work, proposes a method based on fuzzy logic principle to control the read and write operation of the overall S-box and also operations of memory related, that are used as inputs to AES and the lifting scheme wavelet is utilized to diminish the data transfer capacity. It is implemented using Vertex-2 Pro FPGA. The performance factors such as power, area, and speed are calculated and compared with existing techniques. In this work, we achieved 35% of power reduction and 46% increase in speed.

Keywords AES · Lifting scheme DWT · Fuzzy logic · Verilog
Vertex-2 Pro FPGA

1 Introduction

In today's world, transmission of data over the Internet and security of data from attacks is one of the major issues. In order to overcome these attacks, encryption/decryption on data is necessary. Thus, AES is best suitable algorithm for encryption/decryption. As the technology advances, data transfer rate increases. It is necessary to build algorithm that matches that data rate. Cryptography helps in providing security to the data, and it also enables us to store and transmit dedicated

---

C. Sapna Kumari (✉)
Jain University, Bangalore, India
e-mail: sapnakumaricc@gmail.com

K.V. Prasad
Department of ECE, Bangalore Institute of Technology, Bangalore, India
e-mail: drsvt@yahoo.com

© Springer Nature Singapore Pte Ltd. 2018   259
S.S. Dash et al. (eds.), *International Conference on Intelligent Computing and Applications*, Advances in Intelligent Systems and Computing 632,
https://doi.org/10.1007/978-981-10-5520-1_25

information across apprehensive networks so that the data cannot be accessed by unauthorized people. The technique where the data is translated into an undercover code is known as encryption. For secret communication, the technic of encryption is being used in military service areas and also this technique is used by government for secret transmission of data. In many kinds of civilian systems such as, Mobile networks, e-commerce, involuntary teller machine transactions, Copy protection i.e. software piracy the technique of encryption is being used. Only when one has the key, the encrypted data can be decrypted. The technic is used to translate data from a readable format to an encoded format using the key can be defined as encryption [1]. Based on the kind of key used, encryption procedures are approximately characterized into two types, specifically symmetric and asymmetric algorithms [2].

The private key encryption and public key encryption are the encryptions accomplished using above kinds of ciphers. Only one key is being used in symmetric ciphers. To preserve privacy, this key needs to be secret. The same key is required by the one who encodes the data and also by the receiver who decrypts the data. The public key and the private key are the two types of key used in asymmetric algorithm. The data is being encrypted by the public key and decrypted by private key. Compared to the symmetric algorithms, asymmetric algorithms are more intensive. To increase speed, symmetric ciphers are used. There are many cryptography algorithms available like DES (data encryption standard), 3DES (Triple DES), and AES (advanced encryption standards) [3].

The principle of AES algorithm is substitution and permutation. The AES algorithm is a symmetric block ciphers it process data chunks of 128 bits using a cipher key of length 128, 192, 256 bits [4]. Each data block consists of 4×4 array of bytes called state, in which the basic operation of the AES algorithm is performed. The AES algorithm is used for encryption and decryption of data and images and to protect them from an unauthorized access [5]. The operation like SubBytes (S-box)/Inv SubBytes (Inv S-box), Mix Columns/Inv Mix Columns and Key Scheduling operations are used to provided higher security and to increase speed of operation. The applications of AES is to provides simplicity flexibility easiness of implementation and provides high throughput [6].

Mandal and prakash [3] focused on implementation of DES and AES algorithms in MATLAB software. After implementation, he says that the utilization of memory and simulation time required for implementation of AES is less than DES. In AES, avalanche effect is very high. Chirag Parikh [6] has compared the results obtained to implement AES algorithm on different platforms like field-programmable gate array (FPGA), desktop, handheld devices. After comparison, the 32-bit hardware implementation of AES on FPGA consumes less power and chains the required throughput suitable for handheld devices. Banu et al. [7] in this paper hardware and software technic is used to increase throughput and provide security of AES algorithm for the applications like smartcard and Internet. In hardware technique like architectural optimization pipeline, loop unrolling and iterative design. In this, author focused on pipelining technique and parallelization technique with open MP standard which is used to increase throughput. Yoo [8] discussed about hardware implementation of an AES algorithm in order to increase throughput. Fully pipelined architecture achieves a throughput of 29.77 Gbps in encryption compared to previous work. In this, author focused on ROM Macro for S-Box implementation.

Karsannbhai [9] in this work, authors focused on implementation of AES algorithm for wireless transmission which works at 433 MHz frequency. Experiments are conducted to transmit the data up to range of 100 m with a speed of 4.6 Kbps. This architecture perfectly suits for a wireless communication, and a maximum speed of 4.6 Kbps is provided for wireless communication. Anitha Christy [10] focused on composite field arithmetic (CFA) used in 128 bit advanced encryption standard (AES) bit algorithm to reduce area. CFA concept of AES algorithm used in byte substitution block and inverse byte substitution block, key expansion block. This concept is used to reduce area compare to Look up Table (LUT) technic in S-Box/Inverse S-Box. Author focused on a FPGA implementation of AES algorithm using multistage sub-pipelined architecture is used to increase throughput and author compared this technic with the previous FPGA implementation. Hammed [11] presents implementation of AES algorithm on FPGA to increase throughput/area, efficiency compared with the previous loop unrolled pipelined technic. This design composite field S-Box implementation is used to relocate the mix column step and merging between the inverse isomorphic mapping, the affine transformation multiplication and the isomorphic mapping of the next encryption stage [12]. Here, author combines three operational blocks into one block to achieve higher efficiency by reducing the no of slices and less number of sub-pipelining stages required to achieve certain throughput. Hodjat [13] discusses about high-speed AES encryption; in this, author focused on the area/throughput of a fully pipelined architecture. Here, loop unrolling technic, inner and outer round pipelining technic are used in AES algorithm; this technique is used to get the throughput of 30–70 Gbits/s using a 0.18-m CMOS technology. Lin [14] in this work the implementation of different modes of operations like ECB, CBC, CTR and CCM modes of AES algorithm are used and this design includes two-stage pipelining for the CCM mode by a single data path to achieve the greatest throughput of 4.27 Gbps using a 0.13 µm CMOS technology with a 333 MHz clock rate. The cost of the hardware is 86.2 K gates with a power of 40.9 mW.

Good and Benaissa [15] author focused on implementation of AES algorithm to reduce area/power. This architecture uses resource sharing technic with split 8-bit data path between key and state processing of the SubBytes operation. This work shows the best power–latency–area compared with other design. Olteanu [16] focused on implementation of advanced encryption standard (AES); here for the security purpose, encryption cipher is used to provide the security of the transmission. The technic is used to transmit AES chipper in terms of frames. This method analyzes the trade-off between throughput and payload size, channel error when AES is used to encrypt the frames. From the earlier literature survey, it is inferred that the fully pipeline technology is used to increase speed and different techniques are used to reduce area (LUT, CFA, ROM MACRO). To create S-Box, the more number of computations is required in terms of multipliers, additions, and exponential. So to reduce the number of computations in the existing AES algorithm and also to overcome above discussed problems in this paper uses modified AES algorithm with fuzzy logic concepts to control S-Box.

2 Problem Statement

For every communication system, security is very important to transmitting and receiving of video, audio, and image. The proposed work is to develop an efficient AES algorithm with fuzzy logic principal which is used to control S-box in order to optimize the performance factors like speed and area for security concern and also lifting scheme wavelet transformation technique is used to decrease bandwidth.

3 Methodology

Figure 1 demonstrates an overall block diagram of proposed work. It consists of the following blocks namely memory module, lifting scheme which is used for compression, decompression, encryption, and decryption to decode original data. The video or audio or image is converted into pixels coefficients and stored in the random access memory (RAM). In each memory location stores 8-bit each pixel by providing address of memory location using counter. In one clock pulse, one pixel value is stored, and with same clock pulse, the data will be read from memory and applied to compression technique.

3.1 Advance Encryption Standard AES

AES key generation module as shown in Fig. 2 [4] is used to generate 128 bit key, S-Box and XOR gates are to create round keys on each positive edge of the clock, during enabled period and XOR operation is performing. In modified AES, shift operation does not require any logical gates and the key register done by port mapping according to the required shift and S-Box is created and controlled using fuzzy logic membership between logics 0 and 1, fuzzy logic will optimized area, power and delay. Also in this work data path is used for rearrange of bits from register to S-Box, dynamic memory stores each round constant and retrieved data on each clock as shown in Fig. 3 [4]. Reset, str_mix, stop are control signal. The indicate encryption "done" signal is used and it is possible to reduce hardware

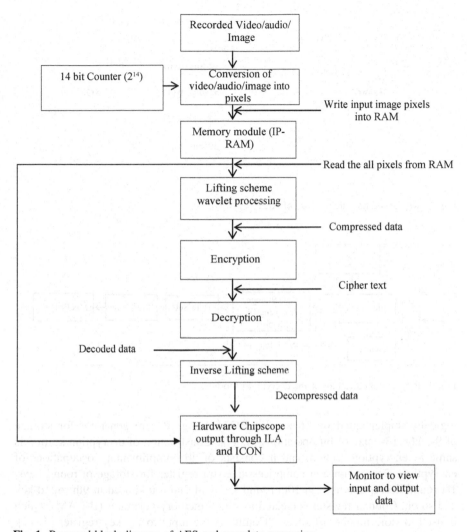

Fig. 1 Proposed block diagram of AES and wavelets processing

complexity of this architecture. The control unit is not designed separately. The four-bit counter is used to design control unit, key generation module and also to control entire module. The control unit is shared by both round key generation and encryption that give unique benefit to decrease hardware as compared to other applications [4].

Inverse fuzzy logic controls memory access and counter, inverse GF creates inverse S-box values. The created values are stored in LUT, and these values are applied to inverse sub-byte transformation operation. Mix column. Operations are enhanced by incorporating 256 × 8 read only memory (ROM). Galois's multiplication requires more memory to store results; in this work, XOR operation is

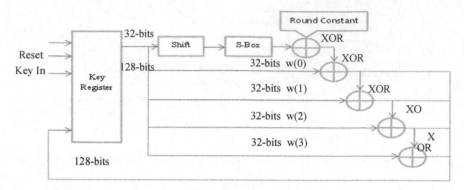

Fig. 2 Key generation architecture of AES

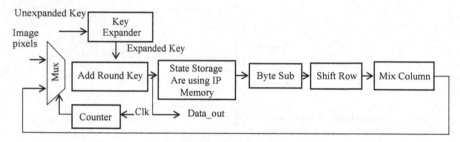

Fig. 3 Proposed architecture of the decryption module

logically implemented by 16×1 bit ROM using IP core generator for storage of 9-, 11-, 13-, and 14-bit operations [4, 17]. Construction of decryption section is same as encryption section, and it consists of all complimentary occupations of encryption. Decryption unit comprises an extra register for storage of round keys. 1st round decryption used in 10th round key and 2nd round used in 9th round key and so on. Counter register is created as block memory generator (B-RAM) which is used to store number of slices. "Count" is the input to the key register [4, 17]. CCM-AES mode is used for processing the image pixels and create key of 128-bits to produce an authentication in communications. To control the encryption mode operation CTR controller is used to count number of round. When counter value is 9, then AES-CCM mode algorithm will produce the cipher text as shown in Fig. 3 [4, 17], and its flowchart is shown in Fig. 4. The proposed AES-CCM mode is to get data confidentiality, integrity, and replay-attack protection, on the operation of plaintext [4] accessed [17]. The construction of decryption module is shown in Fig. 5 [4, 17].

This is because the core uses the expanded key backwards during decryption [13]. In some cases, a key expander is not required. This might be the case when the key does not need to be changed (and so, it can be stored in its expanded form) or

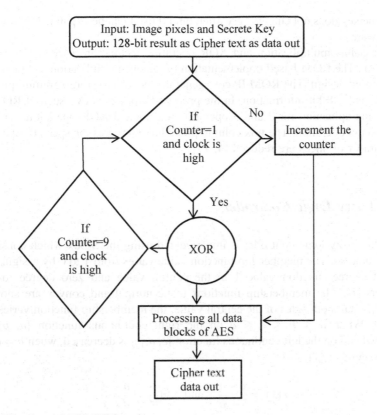

Fig. 4 Flowchart of AES-CCM mode

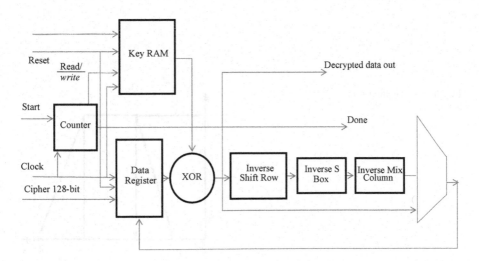

Fig. 5 Proposed architecture of decryption module

when the key does not change very often (and thus, it can be expanded more slowly in software).

The Galois multiplication and XOR operation technic are used in Mixcolumn operation. The ROM-based implementation of Galois multiplication technic is used in proposed design. The ROM IP core generator is used to store 256 multiplication conditions for 8-bit information. In the proposed work uses two such of ROMS of Galois multiplication for 2 and 3 operation and logical XOR operation is used in mix column operation. The mix column procedure increases the speed of operation, and number of slices are reduced [4, 17].

3.2 Fuzzy Logic Controller

Here, the fuzzy logic controller is implemented using the S-box which consists of fuzzy numbers. The membership function value varies from 1 to 0 by assigning the highest degree function value 1 to the central value and zero degree to other numbers [1]. The membership functions to be normal and convex are shown in Fig. 6 [1]. On both sides of the central value, the membership function varies from 1 to 0. When $a \leq b \leq c \leq d$ on the right continuous function $[a, b]$, f is increased and on the left continuous function $[c, d.]$ g is decreased, when $b = c, A$ is a fuzzy number [1].

$$f(x) = \frac{x - a}{b - a} \quad \text{and} \quad g(x) = \frac{d - x}{d - c}. \tag{1}$$

In Eq. (1), f and g are functions, when $b = c$ then A is fuzzy number [1].

Fig. 6 Fuzzification for S-box [1]

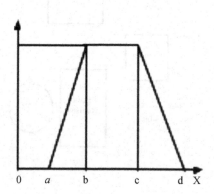

4 Results Analysis

The AES and lifting scheme DWT are implemented on FPGA devices and simulated, and the results are shown in Fig. 8. The arrow shown in simulation results that the pixels values of input and decrypted data are matched. The delay between input pixel and output pixel is 5 clock pulses, each pixel passes through AES and wavelets processing, and to get output, it will take minimum of 5 clock pulses; therefore, the delay is 5 ns. The design summary of AES algorithm is shown in the Fig. 7.

The modified AES with fuzzy logic and lifting scheme DWT are very effective to provide high security for image, audio, and video data through wireless or wired transmission. The proposed work was implemented using Verilog on Vertex-2 Pro field-programmable gate array (FPGA), and input pixels and output pixels are displayed on monitor using Chip scope tool which contains integrated logic analyzer (ILA) and integrated controller (ICON) to control ILA. It is also simulated using Modelsim 10.1 PE version, analysis by IP core and Chip Scope Pro tool in Xilinx 10.1 ISE tool for better performance comparison in terms of power, slices, memory usage, look-up table (LUT) and throughput for the various AES methods.

Slice Logic Utilization	Used	Available	Utilization	Note(s)
Number of Slice Registers	266	28,800	1%	
Number used as Flip Flops	265			
Number used as Latches	1			
Number of Slice LUTs	221	28,800	1%	
Number used as logic	166	28,800	1%	
Number using O6 output only	96			
Number using O5 output only	69			
Number using O5 and O6	1			
Number used as Memory	47	7,680	1%	
Number used as Shift Register	47			
Number using O6 output only	46			
Number using O5 output only	1			
Number used as exclusive route-thru	8			
Number of route-thrus	77			
Number using O6 output only	77			
Number of occupied Slices	154	7,200	2%	
Number of LUT Flip Flop pairs used	350			
Number with an unused Flip Flop	84	350	24%	
Number with an unused LUT	129	350	36%	
Number of fully used LUT-FF pairs	137	350	39%	
Number of unique control sets	54			

Fig. 7 Design utilization summary of AES

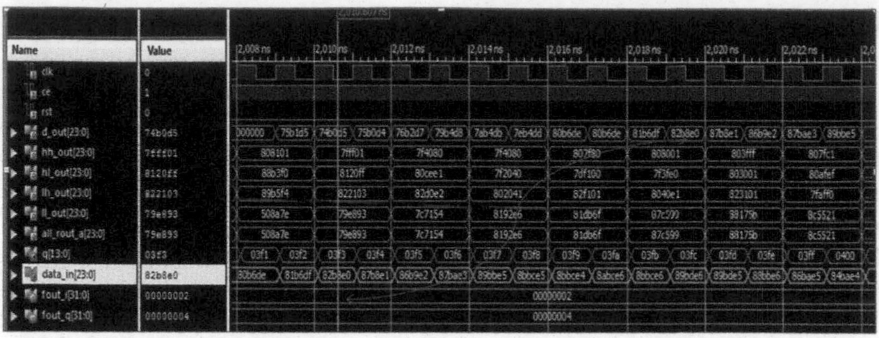

Fig. 8 Simulation results of AES

Table 1 Comparison results with existing and proposed methods

Author	Comparison between proposed and previous works					
	Power (MW)	No. of slices	No. of slice LUTs	Memory usage	No. of LUT flip-flop's pairs	Area ratio
Good and Benaissa [15]	0.692	5500	–	–	–	–
Samiee et al. [18]		7865	–	–	–	–
Yoo et al. [8]	2.083	5408	4884	–	–	–
Jyrwa [12]		6211		–	–	
Present work	0.460	266	221	3131	350	6

The proposed work consumption 460 mw of dynamic power, area utilization actual ratio is 6, number slices occupied 206 and LUT used are 221 are shown in Table 1.

Figure 9 shows top level of lifting scheme to extract four level decomposed sub-bands. High-frequency components are the compressed pixels and applied to AES encryption. The top-level RTL diagram, place and route of all modules are shown in Fig. 10.

Figure 11 shows input and output pixels of selected image from database. For every one clock pulse, one pixel values will be read and displayed. The AES and lifting scheme DWT are implemented on FPGA devices and simulated; the results are shown in Fig. 11. The arrow shown in simulation results that the pixels values of input and decrypted data are matched. The delay between input pixel and output pixel is 5 clock pulses, each pixel passes through AES and wavelets processing, and to get output, it will take minimum of 5 clock pulses; therefore, the delay is 5 ns. Results obtained from the simulation of this modified AES algorithm are compared with the results available with the literature survey and are available in Table 1.

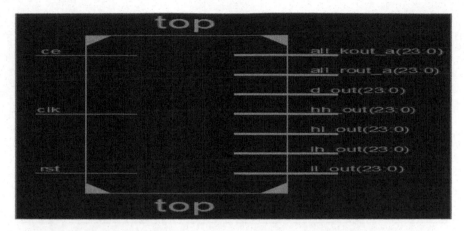

Fig. 9 Top-level diagram of RTL

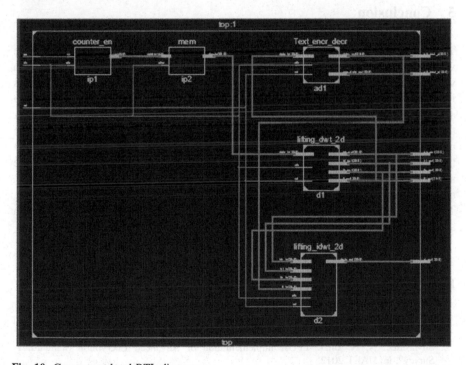

Fig. 10 Component-level RTL diagram

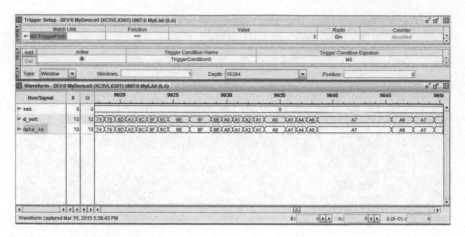

Fig. 11 Hardware Chip scope results

5 Conclusion

The modified AES with fuzzy logic and two-level subband lifting scheme DWT are very effective to provide high security for image, audio, and video data through wireless or wired transmission in communication system. The proposed work was implemented in real time using Verilog on Vertex-2 Pro field-programmable gate array (FPGA). The validation of proposed work is done in Chip Scope Pro analyzer to analyze the timing waveforms of cipher and decrypted results. The utilization parameters are found to be better performance as compared with existing works in terms of power, slices, no. of slice LUTs by comparing 4 papers and tabulated the same results but our work enhanced by adding another 3 parameters by memory, area ratio, no. of LUT Flip flop's pairs used and the results are shown in Table 1. The proposed work consumption 460 MW of dynamic power, area utilization actual ratio is 6, number slices 266 and slice LUT used are 221.

References

1. T. Akther, S. U. Ahmad "A computational method for fuzzy arithmetic operations", Daffodil International University Journal of Science and Technology, vol 4,issue 1 Jan 2009.
2. Shylashree. N, N. Bhat "FPGA implementation of Advanced Encryption Standard: A Survey", in: IJAET 2012.
3. A. K. Mandal, C. Prakash " Performance Evaluation of Cryptographic Algorithms: DES and AES" In IEEE Conference 2012.
4. Abhijith. P. S, et.al, "High Performance Hardware Implementation of AES Using Minimal Resources", 2013 International Conference on Intelligent Systems and Signal Processing (ISSP), 978-1-4799-0317-7/13/2013 IEEE.

5. A. K. Mandal, C. Parakash, Mrs. A. Tiwari Performance Evaluation of Cryptographic Algorithms: DES and AES.
6. Chirag Parikh, M.S. and Parimal Patel, Ph.D. Performance Evaluation of AES Algorithm on Various Development Platforms.
7. J. Saira Banu et.al. "Loop Paralllelization and Pipelining Implementation of AES algorithm using Open MP and FPGA" In IEEE International Conference 2013.
8. S. M. Yoo,*, D. Kotturi, D. W. Pan, J. Blizzard "An AES crypto chip using a high-speed parallel pipelined architecture" 2005 Elsevier.
9. G. R. Karsannbhai, M. G. Shajan "128 bit AES implementation for secured wireless Communication" In IEEE 2011.
10. N. Anitha Christy and P Karthigaikumar "FPGA Implementation of AES Algorithm using Composite Field Arithmetic", In ICDCS 2012.
11. I. Hammad, K. EI-Sankary "High speed AES Encryptor with efficient Merging Techniques", In IEEE Embedded Systems Letters, VOL. 2, NO. 3, September 2010.
12. B. Jyrwa, R. Paily, "An area Throughput Efficient FPGA Implementation of Block cipher AES algorithm", in international Conference on Advances in computing, control, and Telecommunication Technologies 2009.
13. A. Hodjat "Area Throughput Tradeoffs for Fully Pipedlined 30 to 70 Gbits/s AES Processors" in IEEE Transactions On Computers, Vol. 55, No. 4, April 2006.
14. S.Y. Lin, C. T. Huang "A High Throughput Low Power AES Copher for Network Applications" in: IEEE 2007.
15. T. Good and M. Benaissa "692-nW Advanced Encryption Standard (AES) on a 0.13- m CMOS" in IEEE Vol. 18, No. 12, December 2010.
16. A. Olteanu, Y. Xiao, Senior Member, IEEE, and Y. Zhang, Member, IEEE "Optimization Between AES Security and Performance for IEEE 802.15.3 WPAN" In IEEE Vol. 8, No. 12, December 2009.
17. K. Viswanath "Design and Implementation of Area-optimized 256-bit Advanced Encryption Standard for real time images on FPGA", *International Journal of Advances in Electrical and Electronics Engineering,* ISSN: 2319-1112/V1N2: 134-140 ©IJAEEE.
18. H. Samiee et al. "A Novel Area-Throughput Optimized Architecture for the AES Algorithm", 2011 International Conference on Electronic devices, system and applications.

Power Extraction from Small Hydropower Plant

Prawin Angel Michael, Marcin Czubala, Matthew Palmer
and Ajay Kumar

Abstract Acquisition of renewable energy is beneficial to the environment in which we live. We all know that in future run out of non-renewable energy sources like natural gas, coal, or petroleum is possible. Therefore, we are looking for new ways that allow in sufficient quantities to produce electricity and heat to the public. This is mainly due to nature that people have invented a new way of obtaining energy called as 'renewable energy'. In our work, we have focused on calculation of power potential of the Valara waterfall based on rainfall data for ten years (2001–2010). Having knowledge of the catchment area of the stream and also on the ground, we are able to calculate how much water will flow though the stream. These data are necessary to calculate power potential of flowing water. Energy, thus, generated can be stored and can be used in time of need. The results of energy potential gave us the idea of how much energy can be produced from the hydroplant and, therefore, how to make simulation using MATLAB software. Proper adjustment is crucial for later results which will be shown later in the paper.

Keywords Discharge · Hydro · Power · Simulink · Turbine

1 Introduction

Hydroelectric power plants are the most extensively used source of renewable energy. Hydroelectric power plants are comparably economical source of energy but the key point lies in the availability of rich river heads so as to construct the hydroplant. In addition to it, the development of dams for hydropower involves

P.A. Michael (✉) · M. Czubala · M. Palmer · A. Kumar
Karunya University, Coimbatore, India
e-mail: prawin@karunya.edu

M. Palmer
e-mail: gmpalmer5628@gmail.com

A. Kumar
e-mail: ajaykumarb123456@gmail.com

© Springer Nature Singapore Pte Ltd. 2018   273
S.S. Dash et al. (eds.), *International Conference on Intelligent Computing and Applications*, Advances in Intelligent Systems and Computing 632,
https://doi.org/10.1007/978-981-10-5520-1_26

huge constructions and the creation of reservoirs. However, in the case of small hydropower, impact on the environment is negligible. Based on the installed generation, we can coin the term as small hydropower projects. It also may depend on the size of bleed water. Therefore, state or organization determines the size of ingested power. This depends primarily on the degree of industrialization. Small hydropower plant (SHPP) often uses the potential of small rivers and small reservoirs, irrigation systems, water supply systems. SHPP is very useful, as the energy of them acquired can be used by local consumers with minimal transmission loss. Small hydropower plants are mostly runoff river schemes. The water here is not stored in the additional tanks, and turbine and generator are located in the river bed.

2 Discharge of the Stream

Power potential of the Valara waterfall was based on rainfall data for ten years from 2001 to 2010. To calculate discharge of the stream, catchment area and runoff coefficient have to be known. The catchment area is where the water comes together at a single point and the runoff coefficient (C) is a dimensionless coefficient. From the runoff coefficient and the total area of catchment, the discharge of water is calculated. Table 1 shows clearly that maximum discharge occurs during month of July and minimum during months of January, February, and December. From these data, it is clear that the river is completely depending on monsoon and in off monsoon that is in other months on an average around 10 cubic meter per second of discharge of water is observed which is more than sufficient for a reliable production of electricity.

To calculate discharge Q, the following equation has been used:

$$Q = C i A \tag{1}$$

where

Q Discharge of the stream in m^3/s
C Runoff coefficient
i Rainfall intensity m/s
A Catchment area in m^2

Recent research of Valara waterfall shows that catchment area is equal to 61 km^2. Designation of runoff coefficient has to base on geographical data of catchment area. Studies show that about 70% of the catchment area is under dense mixed forest and rest of the area is utilized for human settlements and cultivation of rubber, cardamom, tapioca, arecanut, and coconut. The rock type along the stream is found to be pyroxene granulite. About 80% of the area has an exposed rock presence. According to this data, runoff coefficient can be approximated as per Table 2. The runoff coefficient was approximate to 0.3. The average discharge for each month is shown in Table 3, and the discharge for every month for different years is depicted in Table 4.

Table 1 Rainfall data between 2001 and 2010 (mm)

Month	2001	2002	2003	2004	2005	2006	2007	2008	2009	2010	Average
Jan	14.4	0	0	4	53.6	12.4	2.2	0	33.3	0	11.99
Feb	4	3.8	46.2	6	27.6	0	3.4	14	0	24.2	12.92
Mar	11.2	120.6	93.2	20.9	79.8	130.8	1.4	155.4	50.2	77.7	74.12
Apr	627.6	199.2	179.6	218.4	232.8	127.6	139.4	68.6	126.33	123.4	204.293
May	349.8	405.8	169.6	728.8	211.8	605.2	224.8	76.9	333.2	779.6	388.55
Jun	821.1	773.6	577.1	1413.7	785.2	415.9	1018.8	338.2	548.8	1243.2	793.56
Jul	1365.8	600	1019.6	585.8	2474.8	945.9	1204.4	807.9	1025.2	1012.1	1104.15
Aug	668	1032.8	988	1246.1	752.4	543.4	755.8	658.8	798.2	654.2	808.87
Sep	457.2	320.4	178.4	393.8	553.8	546.6	171.1	371.2	615.7	457.2	461.14
Oct	396.2	611.6	554.4	4545.4	357.8	385.7	416.9	204.6	458.9	465.2	430.57
Nov	260.4	69.2	14149,89,8	111.1	213.8	34.5	29.6	156.1	156.1	179.2	143.35
Dec	26.2	25.2	10.8	0	97.2	0	63.4	6.4	70.3	66.8	36.63

Table 2 Runoff coefficient

Ground cover	C
Business	
Downtown areas	0.70–0.95
Neighborhood areas	0.50–0.70
Residential	
Single-family areas	0.30–0.50
Multiunits, detached	0.40–0.60
Suburban	0.25–0.40
Industrial	
Light areas	0.50–0.80
Heavy areas	0.60–0.90
Parks, cemeteries	0.10–0.25
Playgrounds	0.20–0.35
Railroad yard areas	0.20–0.40
Lawns	
Sandy soil, flat, 2%	0.05–0.10
Sandy soil, avg., 2–7%	0.10–0.15
Sandy soil, steep, 7%	0.15–0.20
Heavy soil, flat, 2%	0.13–0.17
Heavy soil, avg., 2–7%	0.18–0.22
Heavy soil, steep, 7%	0.25–0.35
Unimproved areas	0.10–0.30
Forest	0.05–0.25

Table 3 Average discharge

Month	Average rainfall	Q (m^3/s)
Jan	11.99	0.081921
Feb	12.92	0.097733
Mar	74.12	0.50642
Apr	204.293	1.442346
May	388.55	2.654744
Jun	793.56	5.602681
Jul	1104.15	7.544036
Aug	808.87	5.526554
Sep	461.14	3.255734
Oct	430.57	2.941843
Nov	143.35	1.012078
Dec	36.63	0.250272

Table 4 Discharge for each month of the year

Month	Discharge Q (m³/s)									
January	0.0984	0.0000	0.0000	0.0273	0.3662	0.0847	0.0150	0.0000	0.2275	0.0000
February	0.0273	0.0260	0.3157	0.0410	0.1886	0.0000	0.0232	0.0957	0.0000	0.1653
March	0.0765	0.8240	0.6368	0.1428	0.5452	0.8937	0.0096	1.0618	0.3430	0.5309
April	4.2880	1.3610	1.2271	1.4922	1.5906	0.8718	0.9524	0.4687	0.8631	0.8431
May	2.3900	2.7726	1.1588	4.9795	1.4471	4.1350	1.5359	0.5254	2.2766	5.3266
June	5.6101	5.2856	3.9430	9.6590	5.3648	2.8416	6.9609	2.3107	3.7496	8.4941
July	9.3317	4.0995	6.9664	4.0024	16.9089	6.4628	8.2290	5.5199	7.0046	6.9151
August	4.5647	7.0565	6.7504	8.5139	5.1407	3.7127	5.1640	4.5012	5.4537	4.4083
September	3.1238	2.1891	1.2189	2.6906	3.7838	3.7346	4.8995	2.5362	4.2067	3.1238
October	2.7070	4.1787	3.7879	3.1047	2.4446	2.6353	2.8484	1.3979	3.1354	3.1784
November	1.7792	0.4728	1.0235	0.7591	1.4608	1.5701	0.2357	0.2022	1.0665	1.2244
December	0.1790	0.1722	0.0738	0.0000	0.6641	0.0000	0.4332	0.0437	0.4803	0.4564
Total	2001	2002	2003	2004	2005	2006	2007	2008	2009	2010

Table 5 Average power potential for each month

Month	Energy potential (kW)
January	64,29154047
February	76,70096446
March	397,4386138
April	1131,953467
May	2083,442706
June	4396,98371
Julu	5920,559166
August	4337,239227
September	2555,099889
October	2308,758013
November	794,2784601
December	196,4136053

3 Power Potential

Power potential for stream such as Valara with a head of 100 m is calculated, the calculated values are presented in Table 5, and the same is shown graphically in Fig. 1.

Table 6 shows the calculated value of power potential for every month of every year from 2001 to 2010. Calculation of power potential of SHPP was done promptly. It is to be observed that every year we can see possibility of similar amount of average energy potential generated. Power potential is biggest in month July during highest rainfalls. During months January, February, and December, power potential is lowest. According to Table 4, energy potential may not be similar in the same month for different year.

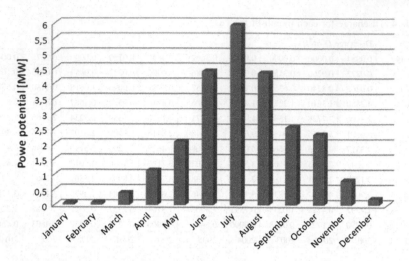

Fig. 1 Average power potential in MW for each month

Table 6 Power potential for each month of the each year

Month	Power potential (kW)									
Jan	77,21	0,00	0,00	21,45	287,41	66,49	11,80	0,00	178,56	0,00
Feb	21,45	20,38	247,73	32,17	147,99	0,00	18,23	75,07	0,00	129,76
Mar	60,06	646,67	499,75	112,07	427,90	701,36	7,51	833,27	269,18	416,63
Apr	3365,25	1068,13	963,03	1171,08	1248,30	684,20	747,48	367,84	677,39	661,68
May	1875,66	2175,94	909,41	3907,90	1135,69	3245,14	1205,40	412,35	1786,65	4180,29
Jun	4402,82	4148,12	3094,47	7580,40	4210,32	2230,10	5462,90	1813,46	2942,72	6666,16
Jul	7323,55	3217,26	5467,19	3141,12	4210,32	5072,01	6458,11	4332,04	5497,22	5426,98
Aug	3581,88	5537,97	5297,75	6681,71	4034,44	2913,76	4052,67	3532,55	4280,03	3459,62
Sep	2451,55	1718,02	956,60	2111,59	2969,53	2930,92	3845,16	1990,41	3301,44	2451,55
Oct	2124,46	3279,46	2972,75	2436,54	1918,56	2068,16	2235,46	1097,09	2460,67	2494,45
Nov	1396,29	371,06	803,24	595,73	1146,42	1232,21	184,99	158,72	837,02	960,89
Dec	140,49	135,12	57,91	0,00	521,20	0,00	339,96	34,32	376,96	358,19
Ave	2235,06	1859,84	1772,49	2315,98	2609,82	1762,03	2047,47	1220,59	1883,99	2267,18
Year	2001	2002	2003	2004	2005	2006	2007	2008	2009	2010

4 Simulation of a Hydropower Plant

The simulation for small hydroplant was prepared in MATLAB [1, 2]. MATLAB is a computer program which was created to perform scientific and engineering calculations, and to create computer simulations. Whole simulation was prepared in one of the MATLAB base tool—Simulink. Simulink is a part of the numerical MATLAB and is used to perform computer simulations. Simulink allows user to build simulation models using the graphical interface and blocks. With the help of Simulink, simulations can be carried out both discrete time and continuous [3]. The results from our previous calculations give us an idea of how to adjust each

Simulink block according to amount of potential energy provided by waterfall. Simulation was based on predefined Simulink block like turbine and governor, synchronous machine, and exciter. Simulation was based on per unit system which describes traditional SI unit modified by Eqs. 3 and 4.

$$\text{Base value in p.u.} = (\text{Quality expressed in SI units})/(\text{Base value}). \quad (2)$$

The base current and the base impedance are calculated by the formulas shown below by choosing the base power as the nominal power of the equipment and the base voltage as the nominal voltage of the equipment.

$$\text{Base current} = (\text{Base current})/(\text{Base power}) \quad (3)$$

$$\text{Base impedance} = (\text{Base voltage})/(\text{Base current})$$
$$= [(\text{Base voltage}] \wedge 2/(\text{Base power}) \quad (4)$$

The simulation was based on predefined block such as hydraulic turbine and governor block [4] as shown in Fig. 2.

Figure 3 shows the subsystem of the hydraulic turbine, and Fig. 4 is for the servomotor [5–7].

The gate servomotor is modeled by a second-order system [8–12] as shown in Figs. 4 and 5 shows the exciter model [9, 10].

The synchronous machine block can operate either as a generator mode or if dictated it can work as in motor mode [13]. The main factor deciding the operating mode is the polarity of the mechanical power. If the synchronous machine operated in the generator mode it means that the mechanical power is positive and viceversa if it is operated in the motor mode. The sixth-order state-space model is used to

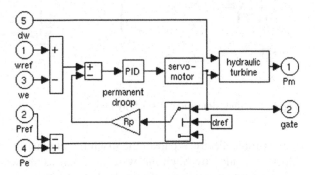

Fig. 2 Turbine and governor model

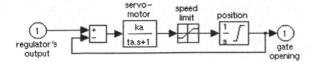

Fig. 3 Hydraulic turbine model

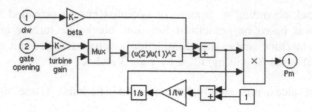

Fig. 4 Gate servomotor

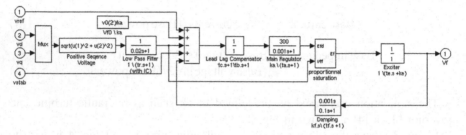

Fig. 5 Exciter model

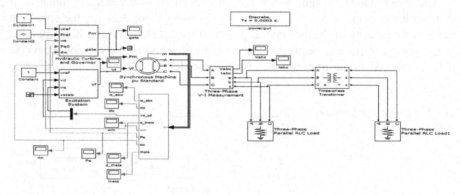

Fig. 6 Simulink simulation of small hydroplant

represent the electrical part of the machine and the mechanical system [14–18]. All blocks provided by Simulink were assumed as shown in Fig. 6. Specific way of connecting each block was forced by MATLAB structure. All blocks were adjusted according to previous calculations of power potential. In this simulation, mechanical power Pm is an input of synchronous machine which in this case is a generator. Field voltage (Vf) is coming directly from excitation block to generator. Right side of synchronous machine is the simulation of the grid.

5   Results

The simulation model shown in Fig. 6 was simulated, and by adopting scope, various electrical outputs such as electrical power versus time (Fig. 7), output active power versus time (Fig. 8), rotor speed versus time (Fig. 9), rotor deviation versus time (Fig. 10), field current versus time (Fig. 11), stator current versus time (Fig. 12), phase-to-phase voltage versus time (Fig. 13), line currents versus time (Fig. 14), rotor mechanical angle versus time (deg) (Fig. 15), rotor angle deviation versus time (rad) (Fig. 16) were obtained which validate the efficiency of the system.

Fig. 7 Electrical power versus time

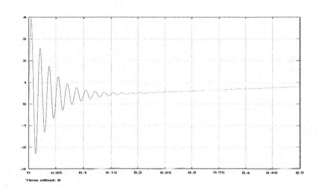

Fig. 8 Output active power versus time

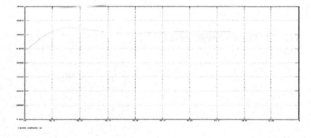

Fig. 9 Rotor speed versus time

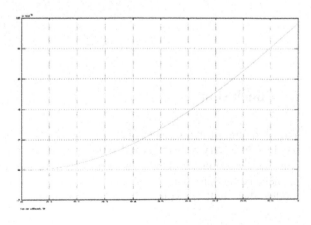

Fig. 10 Rotor deviation versus time

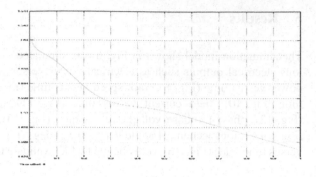

Fig. 11 Field current versus time

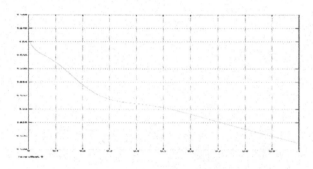

Fig. 12 Stator current versus time

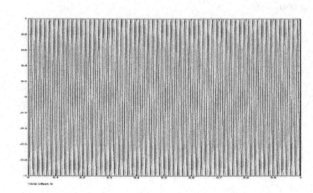

6 Conclusion

The modeling of hydropower plant is extensively reviewed in this paper. Initially, the literature survey was carried out on all aspects and equipments required to construct hydropower plants. The exploration of the model proposed started with elementary models that used system modeling. The paper presents a generalized model that will be utilized in the simulation of SHPP using Simulink.

Fig. 13 Phase-to-phase
voltage versus time

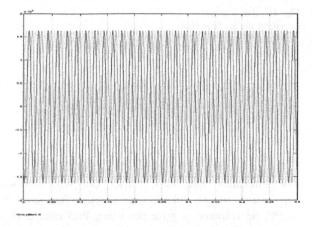

Fig. 14 Line currents versus
time

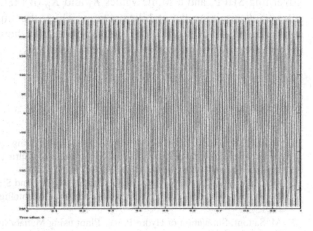

Fig. 15 Rotor mechanical
angle versus time (deg)

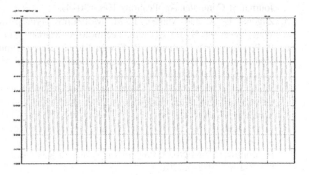

Fig. 16 Rotor angle
deviation versus time (rad)

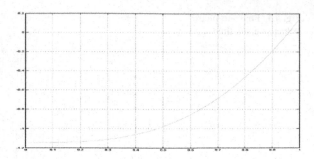

The simulation was, thus, carried out with various simulation tools. Proper adjustment of all blocks is crucial for correct working simulation of SHPP. Servomotor as governor using PID controller will be the best choice for governing SHPP, and also the values K_i and K_p of PID controller have significant roles in determining the stabilizing time. Thus, the small hydropower plant was simulated successfully and in all simulation results showed a perfect generation of energy from hydropower plants with high efficiency.

References

1. Auwal Abubakar Usman, R. A. Abdul Kadir, Modelling and Simulation of micro-hydro power plant using MatLab Simulink, International Journal of Advanced Technology in Engineering and sciences, vol. 3, pp. 260–272, (2015).
2. J. Tiwari, A. K. Singh, A. Yadav, R. K. Jha, Modelling and Simulation of Hydro Power using Matlab and WaterPro 3.0, PL International Journal of Intelligent Systems and Applications, (2015).
3. M. Sattouf, Simulation of Hydro Power Plant using Matlab/Simulink, Journal of Engineering Research and Applications, vol. 4, pp. 295–301, (2014).
4. A. Acakpovi, E. B. Hagan, F. X. Fifatin, Review of Hydro Power Plant Models, International Journal of Computer Applications 108, (2014).
5. Zagona, E. A. and Magee, T. M, Modelling Hydropower in River Ware. In Proc. of Waterpower 99: Hydro's Future: Technology, Markets and Policy pp. 1–10, (2013).
6. P. P. Sharma, S. Chatterjee, B. Singh, Matlab Based Simulation of Components of Small Hydro-Power Plants, VSRD International Journal of Electrical Electronics and Communication Engineering 3 (2013).
7. W. li, L. Vanfretti, and Y. Chompoobutrgool, Development and implementation of hydro turbine and governor models in a free and open source software package, Elsevier—Simulation Modeling Practice and Theory vol. 24, pp. 84–102, (2012).
8. Naghizadeh, R. A., Jazebi, S. and Vahidi, International Review on Modelling and Simulations 5, Cost Optimization of an Electrical Energy Supply from a Hybrid Solar, Wind and Hydropower Plant, No. 4. Power Conversion of Renewable Energy Systems, Springer, ISBN 978-1-4419-7978-0, (2012).
9. Hernandez, G. A. M., Mansoor, S. P. and Jones, Review of Hydro Power Plants, Springer, DOI 10.1007/978-1-4471-2291-312, 2012.

10. F. D. Surianu, "Mathematical Modelling and Numerical. Simulation of the Dynamic Behavior of Thermal and Hydro Power Plants", Politechnica University of Timisoara, Romania, INTECH Open Access Publisher, (2012).
11. A. A. Ansari, D. M. Deshpande, MATLAB Based Model for Analysis of the Effect of Equivalent Circuit Parameters of an Asynchronous Motor on its Dynamic Characteristics, International Journal of Engineering Science and Technology, vol. 2, pp. 1260–1267, (2010).
12. L. A. Lucero Tenorio, Hydro Turbine and Governor Modelling Electric-Hydraulic Interaction, Master of Science in Electric Power Engineering, Norwegian University of Science and Technology Department of Electric Power Engineering, (2010).
13. Aguero, J. L., et al. Hydraulic transients in hydropower plant impact on power system dynamic stability. In Power and Energy Society General Meeting—Conversion and Delivery of Electrical Energy in the 21st Century, IEEE. (2008).
14. Pérez, J. I., J. R. Wilhelmi, and L. Maroto, Adjustable speed operation of a hydropower plant associated to an irrigation reservoir, Energy Conversion and Management, vol. 49, pp. 2973–2978, (2008).
15. Hongqing, F. et al., Basic Modeling and Simulation Tool for Analysis of Hydraulic Transients in Hydroelectric Power Plants, IEEE Transactions on Energy Conversion, vol. 23, pp. 834–841, (2008).
16. Nicolet C., et al., High-Order Modeling of Hydraulic Power Plant in Islanded Power Network, IEEE Transactions on Power Systems, vol. 22, pp. 1870–1880, (2007).
17. Andy Ward, Jessica L. D'Ambrosio, and Jonathan Witter, Determining Discharge in a Stream, The Ohio State University Department of Food, Agricultural, and Biological Engineering and the Ohio NEMO Program, Environmental hydrology (2nd ed.), sections of Chapters 5 and 6 in Ward, A. D. And Trimble, S. T. Boca Raton: CRC Press, (2004).
18. Munoz-Hernandez, G. A. and Jones, Modelling and Controlling Hydro Power Plants, D. I. Control University, (2004).

Enhancement of Digital Distance Protection Scheme with Novel Transducer

A.N. Sarwade⊙, P.K. Katti and J.G. Ghodekar

Abstract Performance of distance protection scheme (DPS) depends on selectivity, sensitivity, and speed of distance relay (DR). Correct estimation of the impedance seen by DR depends on accuracy of current received via current transformer (CT). In practice, saturation of CT disturbs the reach and operating time of DR. A novel current transducer, Rogowski coil (RC) is attaining increased acceptance and use in power industry due to its inbuilt linearity, greater accuracy, and extensive operating current range. This paper presents use of RC for enhancing the performance of DPS affected by saturation of CT. The simulation results of DPS used for protection of 220 kV AC system show outstanding performance of RC during normal as well as abnormal conditions.

Keywords Distance protection · Relay reach · Rogowski coil · CT saturation PSCAD/EMTDC

1 Introduction

Distance protection scheme (DPS) is found to be more popular for protection of high-voltage transmission line worldwide, as it provides more secure, prompt, and reliable protection [1]. Frequency of fault occurrence is more in case of transmission line as compared to other parts of power system. Current transformer (CT) is used to reduce the magnitude of fault currents to a more suitable level on its secondary side [2]. CT saturation causes distance relays (DRs) to notice lower effective current than they would see and causes them to reach a shorter distance than they would, if

A.N. Sarwade (✉) · P.K. Katti
Dr. Babasaheb Ambedkar Technological University, Lonere, Raigarh, India
e-mail: asarwade@yahoo.com

P.K. Katti
e-mail: pk_katti2003@yahoo.com

J.G. Ghodekar
Government College of Engineering, Karad, India

© Springer Nature Singapore Pte Ltd. 2018
S.S. Dash et al. (eds.), *International Conference on Intelligent Computing and Applications*, Advances in Intelligent Systems and Computing 632,
https://doi.org/10.1007/978-981-10-5520-1_27

287

there were no CT saturation. This also causes the DR to issue its trip decision with certain time delay [3].

Modern digital/numerical relays make high-power output unnecessary and open the doors of novel measurement transducers such as Rogowski coils, which have many advantages over conventional CTs [4]. Rogowski coil (RC) can meet the requirements of protective relaying due to its greater performance, inbuilt linearity, exceptional dynamic response, extensive bandwidth without saturation [5]. RC is designed such that position of the conductor inside it and electromagnetic interference caused by adjacent live conductor will not deviate its output [6, 7]. Hence, performance of DPS can be enhanced by using RC as current transducer. This paper presents use of RC in 220 kV DPS to overcome under reach and time-delayed operation issue of DR. The results given by use of actual CT, ideal CT, and RC are compared.

2 Theory of Operation

2.1 Distance Protection Scheme and CT Saturation

DPS is used to detect line faults and isolate, as quickly as possible, the faulted line. DR relay collects voltage and current signals with the help of CT and VT and calculates impedance which can be used to acquire information regarding operating condition of the system and type and location of fault [8]. During abnormal conditions, line current exhibits more transients than line voltages. These transients saturate CT and distort the nature of secondary current signals. Distortion introduced by CT modifies calculated impedance and disturbs the reach of DR and causes it to under reach or over reach [9]. Hence, it is very important to search an alternative to CT, which will help to improve performance DPS. In this paper, RC is used as current transducer to overcome the issues related with DPS.

2.2 Rogowski Coil

In general, RC is toroidal coil wounded on insulated/air core. The flexible insulating material core increases their use in awkward environment. Air core makes them lighter in weight as compared to iron core CT. The insulated/air core in the coil structure makes the relative permeability (μ_r) of RC as unity. RC is always placed around the conductor whose current is to be measured. RC arrangement is shown in Fig. 1 [4].

Fig. 1 Rogowski coil

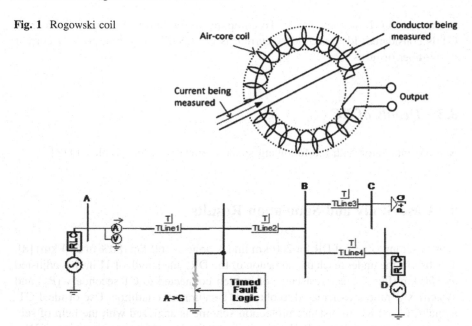

Fig. 2 PSCAD model of 220 kV AC system

The voltage induced in the coil ($V_{rc(t)}$) by current $i_p(t)$ is given by Eq. 1 [10].

$$V_{rc(t)} = -M \frac{di_p(t)}{dt} \qquad (1)$$

3 Modeling of 220 kV AC System

A part of 220 kV AC system is considered to find performance of DPS with RC which consists of two Thevenin's equivalent sources of 100 MVA, 220 kV, 50 Hz (Fig. 2). Source impedance is 32.15 ∠ 85° Ω. The transmission line of 200 km between bus A and B is divided in two parts to create in-zone faults. Positive and zero-sequence impedances of transmission line are 0.2928 ∠ 86.57° Ω and 1.11 ∠ 74.09° Ω, respectively. Load of (75 + j25) MVA is connected to bus C. Ground compensation factor for the line is 2.82 [11].

3.1 Details of CT

The actual CT is considered with CT ratio of 270. Its secondary winding resistance and inductance are 0.5 Ω and 0.8 mH, respectively. Burden of (0.5 + j0.251) Ω is connected to CT secondary [12]. To compare performance of actual CT, an ideal CT is considered. Ideal secondary current is obtained by dividing primary current by number of turns.

3.2 Details of RC

RC and integrator with the following specifications are used (Table 1) [13].

4 Case Study and Simulation Results

Zone 1 setting (Z_{1set}) of DR for 200 km line is done at 160 km (80% of 200 km) [8]. To observe the under reach phenomenon of the DR, line length of TLine1 is adjusted as 150 km (Fig. 2). The resistance of burden connected to CT secondary (R_b) and system X/R ratio are increased to obtain CT saturation condition. Use of ideal CT, actual CT, and RC in distance protection scheme is analyzed with the help of secondary current waveforms, B-H curves, Z_{ap} trajectories, and operating time of DR.

4.1 Impact of CT Burden and X/R Ratio

To observe the effect of unsaturated and saturated CT, R_b is varied from 0.5 to 10 Ω. System X/R ratio is increased from 14.097 to 28.09 (Table 2) [14, 15].

Table 1 Rogowski coil details

S. No.	Parameter	Spec	S. No.	Parameter	Spec
1	Mutual inductance	2 μH	6	R of integrator	100 Ω
2	L of Rogowski coil	7.8 mH	7	C of integrator	1 μF
3	R of Rogowski coil	186 Ω	8	No. of turns	270
4	C of Rogowski coil	235 pF	9	Output RMS	0.1 V/1 kA
5	Z of Rogowski coil	2 k Ω	10	Rated current	100 kA

Table 2 Details of source, line, and X/R ratio

S. No.	R_L (Ω)	X_L (Ω)	R_S (Ω)	X_S (Ω)	L_S (H)	Z_T (Ω)	X/R ratio
1	2.79	46.76	2.8	32.044	0.102	79	14.09
2	2.79	46.76	2.07	32.09	0.1022	79	16.22
3	2.79	46.76	1.18	32.14	0.1023	79	19.85
4	2.79	46.76	0.02	32.19	0.1025	79	28.09

4.1.1 Transient Response of CT and RC

Figure 3a–j shows the secondary current waveforms generated by use of ideal CT (blue), actual CT (red), and RC (green). When the fault is created at maximum value of voltage (Vmax), with R_b as 0.5 Ω and X/R ratio as 14.097, it is observed that ideal CT, actual CT, and RC produce symmetrical secondary current waveforms which are overlaying on each other (Fig. 3a, f). With same R_b and the same X/R ratio, when the fault is created at zero voltage, the current waveforms found to be shifted upwards from the reference due to DC offset and some distortions are observed in secondary current waveforms produced by actual CT (Fig. 3b, g).

When R_b is increased to 2, 5, and 10 Ω and X/R ratio to 16.22, 19.85, and 28.09, it is observed that the actual CT secondary waveform obtains more and more clipped and distorted shape (Fig. 3c–e and h–j), whereas it is found that RC transforms primary current faithfully on secondary side as its secondary current waveform overlaying on secondary current waveform produced by ideal CT.

Comparison of the secondary current effective values at different R_b and X/R ratio is given by Table 3. It is observed that effective value of the secondary current produced by ideal CT and RC is approximately equal, but in case of actual CT it goes on reducing with increase in R_b and X/R ratio.

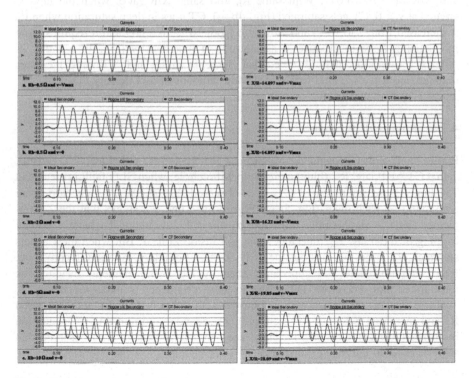

Fig. 3 Secondary current waveforms

Table 3 Secondary currents at different R_b and X/R ratio

Fault instant	Secondary currents (Amp)				
	v = Vmax	v = 0			
Relay burden (R_b)	0.5 Ω	0.5 Ω	2 Ω	5 Ω	10 Ω
Without CT	4.21	4.25	4.25	4.25	4.25
With CT	4.21	3.72	3.52	3.35	3.205
With Rogowski coil	4.22	4.26	4.26	4.26	4.26
Fault instant	Secondary currents (Amp)				
	v = Vmax	v = 0			
X/R ratio	14.097	14.097	16.22	19.85	28.09
Without CT	4.21	4.25	4.31	4.43	4.69
With CT	4.21	3.72	3.56	3.36	3.06
With Rogowski coil	4.22	4.26	4.31	4.436	4.71

4.1.2 B-H Curve of CT

Figure 4a–j shows B-H curves generated by magnetization of actual CT. CT gives linear B-H curve (Fig. 4a–f), when the fault is created at Vmax with R_b as 0.5 Ω and X/R ratio as 14.097. With same R_b and same X/R ratio, when the fault is created at v = 0, CT gets saturated (Fig. 4b–g). CT goes in deep saturation when R_b is increased from 2 to 10 Ω and X/R ratio from 16.22 to 28.09 and it requires more magnetizing force to produce same amount of flux density (Fig. 4c–e and h–j). After CT saturation, it is observed that, increase in R_b and X/R ratio increases magnetizing force required to produce same amount of flux density (Table 4).

4.1.3 Apparent Impedance

Figure 5a–j shows apparent impedance, Z_{ap} trajectories with ideal CT (green), actual CT (red), and RC (blue) along with Mho circle, when SLG fault is created at 150 km. Before saturation of CT, it is observed that all the Z_{ap} trajectories are overlaying on each other (Fig. 5a,g). Figure 5b–e and g–j shows that the Z_{ap} trajectory (red) is significantly impacted by the CT saturation. To have a correct tripping of the relay, Z_{ap} trajectory must fall inside Zone 1. But when the CT gets saturated, Z_{ap} trajectory lies outside of its Zone 1 boundary. As the CT comes out from saturation state, the impedance seen by DR (Mho element) matches the unsaturated plot. Therefore, DR shows to have a tendency to under reach.

Table 5 gives the values of Z_{ap} obtained with different fault instants, increased R_b and increased X/R ratio. The clipping of secondary current due CT saturation increases the magnitude of impedance seen by Mho element. It is observed that with increase in burden, Z_{ap} increases.

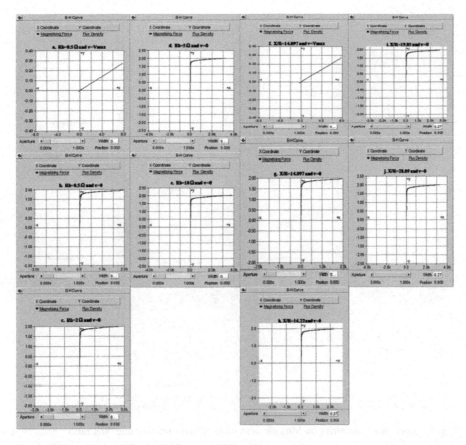

Fig. 4 B-H curves at different burdens and X/R ratio

Table 4 B-H parameters at last point of saturation with different burdens and X/R ratio

Fault instant	v = Vmax	v = 0			
Relay burden (R_b)	0.5 Ω	0.5 Ω	2 Ω	5 Ω	10 Ω
B (Wb/m²)	0.27	2	2	2	2
H (AT/m)	7.75	1955	2597	3296	3840

Fault instant	v = Vmax	v = 0			
X/R ratio	14.097	14.097	16.22	19.85	28.09
B (Wb/m²)	0.27	2	2	2	2
H (AT/m)	7.75	1955	2597	3296	3840

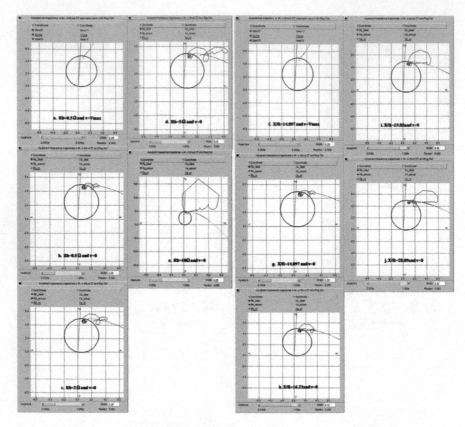

Fig. 5 Impedance trajectories on Mho element with different burdens and X/R ratio

Table 5 Apparent impedance at different burdens

	Apparent impedance (Z_{ap} (Ω))				
Fault instant	v = Vmax	v = 0			
R_b	0.5 Ω	0.5 Ω	2 Ω	5 Ω	10 Ω
Ideal CT	4.63 ∠ 79.21°	4.69 ∠ 78.16°	4.69 ∠ 78.16°	4.69 ∠ 78.16°	4.69 ∠ 78.16°
With CT	4.64 ∠ 79.01°	5.06 ∠ 65.61°	5.14 ∠ 65.30°	5.24 ∠ 63.84°	5.32 ∠ 63.10°
With RC	4.62 ∠ 79.34°	4.68 ∠ 78.30°	4.68 ∠ 78.30°	4.68 ∠ 78.30°	4.68 ∠ 78.30°
	Apparent impedance (Z_{ap} (Ω))				
Fault instant	v = Vmax	v = 0			
X/R ratio	14.097	14.097	16.22	19.85	28.09
Ideal CT	4.63 ∠ 79.21°	4.69 ∠ 78.16°	4.69 ∠ 78.16°	4.69 ∠ 78.16°	4.69 ∠ 78.16°
With CT	4.64 ∠ 79.01°	5.612 ∠ 56.61°	5.87 ∠ 53.20°	6.25 ∠ 48.97°	6.92 ∠ 42.94°
With RC	4.62 ∠ 79.34°	4.68 ∠ 78.30°	4.68 ∠ 78.30°	4.68 ∠ 78.30°	4.68 ∠ 78.30°

4.1.4 Operating Time

The operating time of a DR is considerable to make sure of high-speed tripping. Before CT saturation, all DR elements (Mho elements) issue their tripping signals at same instant (Fig. 6a, g). When R_b is increased from 2 to 10 Ω and X/R ratio from 16.22 to 28.09, CT goes in deep saturation. This CT saturation process causes the Zap to lie outside of Zone 1 for some time and to return back when CT comes out of saturation. It delays DR element operation connected to actual CT and results in slower than expected tripping times (Fig. 6b–e and g–j.

Table 6 gives the time required for the DR to operate, when the R_b is increased from 0.5 to 10 Ω and X/R ratio from 14.097 to 28.09. It is observed that increase in R_b increases the magnitude of the Zap, causing delay at the time of operation.

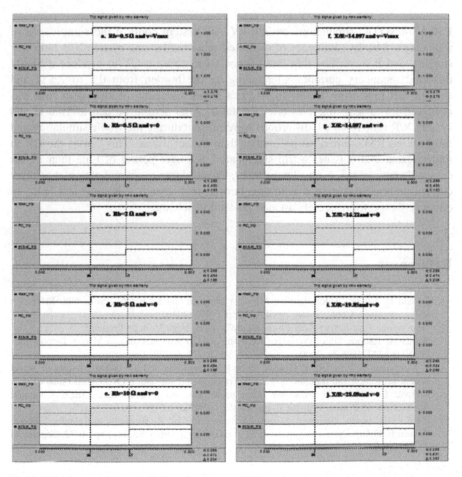

Fig. 6 Tripping signals with different burdens and X/R ratio

Table 6 Apparent impedance at different burdens

Fault instant	Tripping time				
	v = Vmax	v = 0			
R_b	0.5Ω	0.5 Ω	2 Ω	5 Ω	10 Ω
Without CT	Instantaneous	Instantaneous	Instantaneous	Instantaneous	Instantaneous
With CT	Instantaneous	After 0.183 S	After 0.186 S	After 0.196 S	After 0.204 S
With RC	Instantaneous	Instantaneous	Instantaneous	Instantaneous	Instantaneous
X/R ratio	14.097	14.097	16.22	19.85	28.09
Without CT	Instantaneous	Instantaneous	Instantaneous	Instantaneous	Instantaneous
With CT	Instantaneous	After 0.183 S	After 0.206 S	After 0.256 S	After 0.363 S
With RC	Instantaneous	Instantaneous	Instantaneous	Instantaneous	Instantaneous

5 Conclusion

PSCAD-EMTDC simulations on 220 kV AC system confirm the significance of replacing CT by RC. Influence of increased relay burden and system X/R ratio was studied, and it is verified that these aspects can cause a CT to produce a highly distorted secondary current. After changing the burden from 0.5 to 2.5 Ω and system X/R ratio from 14.097 to 16.22, a small indication of core saturation was observed for at least 6 cycles after the fault. After setting burden to 10.0 Ω and system X/R ratio to 28.09, distortions were present during the whole simulation and they caused RMS current to be smaller than in fact it was. This causes the DR to under reach, and it issues a trip signal after a longer period of time than it was originally anticipated. A novel transducer, RC produces exact replication of primary current without distorting it with any burden and system X/R ratio and helps to enhance the performance of distance protection scheme.

References

1. Donald D. Fentie, "Understanding Mho for distance characteristics", SEL, 69th Protective Relay Engineers Conf., Texas, April 4–7, pp 1–17, (2016).
2. "Instrument Transformer Application Guide", ABB, High Voltage Products.
3. Walter A. Elmore, "Pilot Protective Relaying", ABB Automation and Marcer Dekker Inc., ISBN: 0-8247-8195-3, (2000).
4. P. Mahonen, V. Virtanen, T. Hakola, "The Rogowski coil and the voltage divider in power system protection and monitoring", ABB Transmit and substation Oy, Vaasa, Finland.
5. Kojovic L. "Rogowski coils suit relay protection and measurement". IEEE Computer Applications in Power, Vol. 10(3), pp: 47–52, (1997).
6. IEEE PSRC report, "Practical Aspects of Rogowski Coil Applications to Relaying", Power System Relaying Committee of the IEEE Power Engineering Society, pp. 1–72, September 2010.

7. Veselin Skendzic and Bob Hughes, SEL, Inc. "Using Rogowski Coils Inside Protective Relays", 66th Annual Conference for Protective Relay Engineers College Station, Texas, 8–11 April 2013.
8. A. N. Sarwade, P. K. Katti and J. G. Ghodekar, "Optimum Setting of Distance Protection Scheme for HV Transmission Line", Journal of Power Electronics and Power Systems, STM, vol. 3(2), pp. 23–30, (2013).
9. S. R. Mohanty, V. Ravikumar Pandi, B. K. Panigrahi, Nand Kishor, Prakash K. Ray, "Performance Evaluation of Distance Relay with CT saturation", Journal of Applied Soft Computing, Vol. 11(8), pp: 4789–4797. (2011).
10. Kojovic L., Bishop M. T., "Field experience with differential protection of power transformers based on Rogowski coil current sensors", Actual trends in development of power system protection and automation, Moscow, Russia, 7–10 September 2009.
11. Power System simulation software, "PSCAD/EMTDC 4.2.1", Manitoba HVDC Research Centre Inc., Canada, 2008.
12. Dharshana Muthumuni, Lisa Ruchkall, and Dr. Rohitha Jayasinghe, "Modelling Current Transformer saturation for detailed Protection studies", Pulse Newletter, Manitoba HVDC Research Centre, pp. 1–4, (2011).
13. G. Sudha, K. R. Valluvan, T. Basavraju, "Fault Diagnosis of Transmission Lines with Rogowski Coils as Current Sensors", International Journal of Computer Applications, vol. 70 (25), pp. 19–25, (2013).
14. Piotr Sawko, "Impact of Secondary Burden and X/R Ratio on CT Saturation" Wroclaw University of Technology, Faculty of Electrical Engg., 1–3, (2008).
15. A. N. Sarwade, P. K. Katti, "Use of Rogowski Coil to overcome X/R ratio effect on Distance Relay Reach", Computation and Communication Technologies, ISBN: 9783110450101, De Gruyter, pp. 360–373, (2016).

A Hierarchical Underwater Wireless Sensor Network Design for Tracking Ships Approaching Harbors Using an Aerial Mobile Sink (AMS) Node

Madhuri Rao, Narendra Kumar Kamila and Kulamala Vinod Kumar

Abstract Underwater wireless sensor networks (UWSN) is an area that is gaining a lot of attention due to its immense possibilities of monitoring and tracking objects on open waters. Harbor monitoring can be achieved with the help of UWSN that can be laid under the shore up to a certain distance for tracking trespassing and approaching ships. Sensors using spatial and temporal correlation can classify objects as ships. Tracking mechanism in UWSN is characterized by a number of unique constraints such as the properties of the acoustic communication channel, longer network delays, node mobility due to water current, and very limited energy of nodes. Here, a hierarchical network design of the UWSN topology is proposed and verified to explore the possibilities of clustering for ensuring load sharing and to reduce delay and jitter, thereby saving energy of the nodes to prolong the network lifetime.

Keywords Underwater wireless sensor network (UWSN) · Hierarchical topology Relay nodes (RN) · Cluster head (CH) · Aerial mobile sink (AMS) node Harbor

1 Introduction

The researchers of [1] have proposed a technique for harbor surveillance using sensors equipped with three-axis accelerometer deployed on sea surface. Using cooperative signal processing techniques, [1] could detect and classify trespassing

M. Rao (✉) · K.V. Kumar
Siksha 'O' Anusandhan University, Bhubaneswar, Odisha, India
e-mail: madhurirao@soauniversity.ac.in

K.V. Kumar
e-mail: kulamalakumar@soauniversity.ac.in

N.K. Kamila
C.V. Raman College of Engineering Bhubaneswar, Bhubaneswar, India
e-mail: nkamila@yahoo.com

© Springer Nature Singapore Pte Ltd. 2018
S.S. Dash et al. (eds.), *International Conference on Intelligent Computing and Applications*, Advances in Intelligent Systems and Computing 632,
https://doi.org/10.1007/978-981-10-5520-1_28

Fig. 1 Model of network
topology design for tracking
ships in harbor

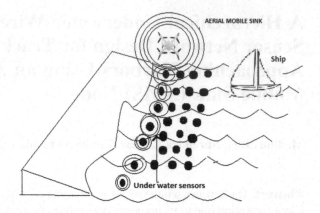

ships by exploiting spatial and temporal correlation of sensing data; however, they
had not considered the other challenges that are inherent due to ocean waves such
as node mobility due to water current. We extended the work of [1] and proposed
two algorithms in [2] and presented the impact of the proposed algorithms in
underwater sensors networks. However in [2], we considered a liner topology with
sensors deployed in a chain-like structure only on sea surface. Here, we propose a
harbor surveillance model as depicted in Fig. 1, which could address issues due to
water current and acoustic communications.

UWSN is characterized with a very dynamic network topology that changes
more frequently than WSN making the use of WSN routing protocols inefficient for
use in UWSN. The medium in UWSN is acoustic channel that results in more
failures, with very limited bandwidth and far more latency, as acoustic waves are
200,000 times slower than wireless waves [3]. The researches in [4–8] have suggested various clustering techniques for routing in UWSN; however, these are
meant to function in homogenous network. The very inherent differences between
UWSN and WSN were researched in [9, 10] present a survey of routing protocols in
UWSN which vary from distributed routing to centralized routing with autonomous
sink node. Protocols addressing issues such as mobility and delay of acoustic
channel have not been well addressed.

2 Literature Survey

The problem of tracking objects such as ships on harbors cannot be well addressed
by just deploying sensors on sea surface. We propose a hierarchical network design
by employing a set of heterogeneous nodes in three levels as illustrated in Fig. 2.
The researches in [4–8] have suggested various clustering techniques for routing in
UWSN; however, there are meant to function in homogenous network. Underwater
wireless sensor networks are quite different in approach when compared to traditional wireless sensor network. They are also larger in size due to packaging of

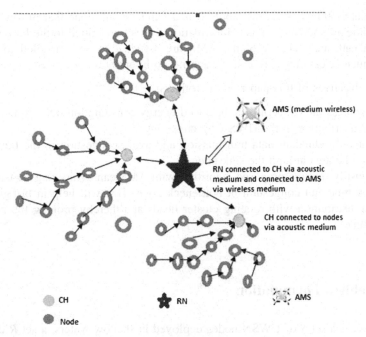

Fig. 2 Hierarchical network design of UWSN deployed in harbors

additional components such as the acoustic modems and triggerable air bladder. The very inherent differences between UWSN and WSN were researched in [9, 10] present a survey of routing protocols in UWSN which vary from distributed routing to centralized routing. We propose a hierarchical heterogeneous network set up here, with three types of nodes as represented in Fig. 2:

(a) Underwater wireless sensor node (UWSN) nodes: They are deployed mostly in shallow waters. These nodes comprise of acoustic modem and a triggerable air bladder in additional to power unit, processor, sensing, and memory. They are deployed with a greater density and require forming clusters and forwarding their data packets to cluster heads. Here cluster heads are proposed to be elected in successive rounds based on residual energy of the nodes as well as their distance from the relay node (RN). Clustering ensures that these acoustic nodes transmit only up to one hop while also rotating their role as cluster heads in order to balance communication and processing load that arise due to nodes drifting due to water current.

(b) Relay node (RN) deployed on sea surface: It is assumed to comprise of an acoustic modem and a transceiver for wireless communication. The acoustic modem enables communication with cluster heads (CH) using acoustic signals. RN is a device with greater abilities to store, process, and convert data from acoustic signals to wireless signals. The device is aware of path, velocity, and time of AMS arrival.

(c) Aerial mobile sink (AMS) node: It is a wireless sink node that moves toward each relay node and collects information that is transmitted to the base station. The path and velocity of the AMS are assumed to be scheduled to ensure frequent access to relay nodes ensuring no information is missed.

The objectives of this paper are as follows:

(a) To address the concern of unbalanced energy consumption due to imbalanced load distribution in the UWSN by clustering,
(b) To avoid redundant data transmission and avoid congestion in the bottleneck zone: the area around the sink node,
(c) To justify that the proposed algorithm with AMS can avoid long transmission delay time and congestion in bottleneck zone. This will help in bringing fair load distribution with rotating cluster heads and thereby prolong the network lifetime.

3   Problem Formulation

We consider a set S of UWSN nodes deployed in shallow waters, a set R of relay nodes (RN) deployed on sea surface, and an aerial mobile sink node. The harbor surveillance and monitoring area A is divided equally into R number of regions. Each region is further logically divided into clusters.

$$S = \{S_i | 1 \leq i \leq n, n = |S|\} \tag{1}$$

Hence, consider a set of UWSN sensors deployed in shallow waters with radius of acoustic communication as $c1$. Since these nodes are considered in a 3D space, the volume of acoustic transmission of signals is given by expression (2).

$V = c1^2 \times \pi \times h$, where h is the height of the three-dimensional cylinder the node coverage area creates. The height here is equivalent to the acoustic radius $(r2)$ of the relay node; hence,

$$V = C1^2 \times \pi \times r2 \tag{2}$$

A is the area of the harbor that needs to be monitored and is divided into l regions, where l is a set of relay nodes with sensing range of $r2$.

$$A = l_1 \cup l_2 \cup l_3 \cup l_4 \cup l_5 \ldots \cup l_i \tag{3}$$

The wireless communication range of each relay node (RN) is $r1$ which enables it to communicate with the aerial mobile sink (AMS) that has a communication radius of $a1$. Figure 3 illustrates the hierarchical network coverage proposed considering heterogeneous sensor nodes. The problem is to ensure that at a given point of time, one and only one node of the n nodes of the set is elected as a cluster head.

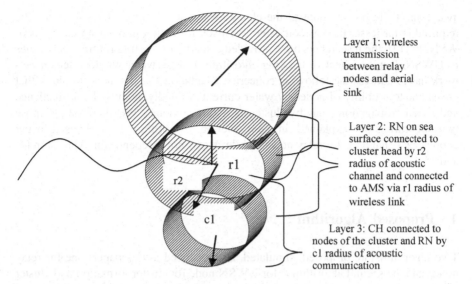

Layer 1: wireless
transmission
between relay
nodes and aerial
sink

Layer 2: RN on sea
surface connected to
cluster head by r2
radius of acoustic
channel and connected
to AMS via r1 radius of
wireless link

Layer 3: CH connected to
nodes of the cluster and RN by
c1 radius of acoustic
communication

Fig. 3 Hierarchical coverage for harbor monitoring

The cluster head selection and cluster formation are similar to LEACH-TLCH protocol [11] which considers average energy and distance of the cluster head from the base station. Let's assume that E_{curr} represents the current energy of the cluster head node and E_{avg} represents the average energy of all the sensor nodes in set S.

$$S.E_{\text{avg}} = \sum_{1}^{N} E(i)_{\text{curr}}. \qquad (4)$$

The researches of [12] have presented an energy consumption model for communication by considering various factors that result in transmission loss. We would incorporate these aspects in our future work. The second consideration for clustering and cluster head formation requires that the distance from the current cluster head or a given node to the relay node (RN) is less than the average distance, i.e.,

$$d_{\text{avg}} = \sum_{1}^{N} d_i \qquad (5)$$

If the CH energy $E_{\text{curr}} \geq E_{\text{avg}}$ and $d \leq d_{\text{avg}}$, then selecting a new cluster head is unnecessary, otherwise the algorithm for cluster formation and cluster head selection is needed to be performed with n rounds. In the second phase of the clustering algorithm, the selected cluster head broadcasts its information to all the neighboring nodes. Nodes join a given cluster while ensuring that their cluster head is only one hop away. Due to water current UWSN nodes are likely to be drifted away, and therefore to adapt to the changing topology, cluster formation will be required in far lesser time as compared to cluster formation in other applications of WSN. In [13], the researchers have provided a survey of existing routing techniques in UWSN. But, an effective tracking algorithm in underwater wireless sensor network has to address the following concerns: adaptation to a dynamic topology that arises due to mobility of nodes by water current; secondly, address load imbalance and signal propagation delay by clustering and relay nodes. We, therefore, propose two algorithms as explained in next section. Algorithm 1 is devised using expression (2) and (3) for relay nodes, while Algorithm 2 depends on expression (4) and (5).

4 Proposed Algorithm

Two algorithms are proposed, simulated, and analyzed in this paper, one for relay nodes and the second algorithm is for UWSN node for cluster formation and cluster head selection as presented below.

Algorithm 1 for relay nodes: for sensor $R_i \in R($ deployed on sea surface with acoustic and wireless transmission abilities)
Input: update_ams(ams,data_ch)// function that uploads data collected from cluster head to aerial sink node; Collect_ch(data_ch,c)// relay node collects data from i number of cluster heads.
Initialize: n= number of relay nodes, c= number of cluster heads, mode=0, x=0;
Begin
1. repeat the following for i=1 to n;
2. for every [i] if mode=0 goto step 3 else goto step 5
3. update_ams(ams,data_ch);
4. mode=1;
5. for x<=c goto step 6
6.   collect_ch(data_ch, c);
7.   x++;

Algorithm 2 for UWSN: for sensor $S_i \in S($ deployed in shallow water with acoustic modem)
Input: update_rn(data_c,rn)// function that uploads data collected from cluster nodes to relay node; forward(sensed_data,ch)//node forwards its sensing data to cluster head; Cluster_formation(d_{avg}, E_{avg},i,r[])//function for cluster head selection using LEACH[13];
Initialize: i= number of nodes in a cluster, mode=0, S= number of UWSN nodes, r[]= distance from relay node for each node i, d_{avg}= average distance of cluster from relay node, E_{avg}= average energy of all nodes
Begin
1. repeat the following for i=1 to S;
2. for every [i] if mode=0 goto step 3 else goto step 4
3. update_rn(data_c,rn);
4. forward(sensed_data,ch
5. if [i] (($E_{curr} \geq E_{avg}$) && ($d_i \leq d_{avg}$))
6.   mode=0;
7. else mode=1;
8. Cluster_formation(d_{avg}, E_{avg},i,r[])

5 Simulation Results and Analysis

The proposed algorithms were simulated in NS2 where a rectangular grid of dimension 6500 m × 3000 m was considered. Forty-five nodes were deployed as the UWSN nodes (layer 1) with 5 nodes as cluster head, 2 nodes as relay nodes, and one mobile sink node was considered, and in the second case a grid of 50 nodes was deployed with one mobile sink node. File transfer protocol (FTP) traffic source was employed for the traffic model. The simulation was executed for 200 ms, and the average delay, jitter, and throughput of data packets reaching the sink node for both the scenarios were obtained and plotted in Fig. 4. Our simulation result shows that with clustering and relay nodes, the packets arriving at the sink node reach with less delay than without clustering and relay nodes. The packet delivery fraction of the transferred packets with our approach is 84.1786%, whereas without clustering it was 88.609%. However, our approach yields less delay and jitter. The average delay and jitter were found to be 0.02656 and 0.039306959 s in our technique, while in the other scenario (without clustering and relay nodes) 0.17043 s of delay and 0.24851635 s of jitter. Hence, our technique is efficient in terms of reducing delay by 15.57% and jitter by 15.83%. We can see and confer from Fig. 4a, b that

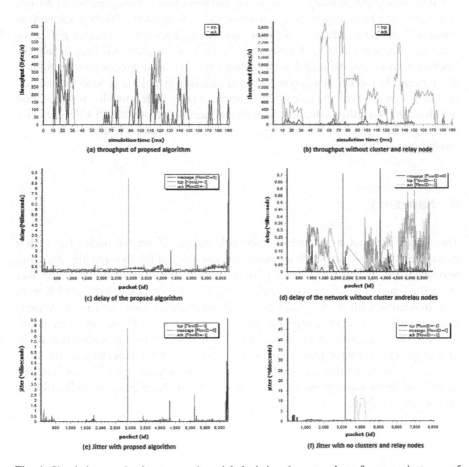

Fig. 4 Simulation results for proposed model depicting the network performance in terms of jitter, throughput, and packet delay

the throughput with our approach has yielded higher success. We like to draw the attention of our readers to the number of TCP packets generated in both algorithms. Due to clustering phenomena, there is an overhead of acknowledgment packets as can be seen in Fig. 4b. However, the number of data packets reaching the sink nodes successfully is higher as delay and jitter are reduced as can be interpreted from Fig. 4c–f. Delay and jitter are very important factors in underwater sensor network scenarios. These parameters define how long the nodes need to be performing. The reason the proposed algorithm yields lesser delay and jitter is due to the reduction in resource contention by the introduction of relay nodes and aerial mobile sink node. These factors also emphasize on the duty cycle of nodes and their energy consumption. Thus by reducing delay and jitter in transmission of packets, the overall energy consumption and network lifetime can be significantly improved. Packet delivery fraction though is a term that reflects on the number of packets successfully received to number of packets actually sent, it is not directly proportional to throughput in many scenarios. On the other hand, throughput of a network is the rate in which information is transmitted in a network. When a network is congested, the packet delivery fraction may not be a good performance indicative parameter; throughput in such cases may be far more resourceful. Packet delivery fraction however can be useful in analyzing concerns causing poor throughput in a given network. Further by applying our deterministic technique of load sharing and packet forwarding with the help of clustering and relay nodes, bandwidth utilization is enhanced. Acoustic communication has a narrow bandwidth 20 kHz only. By reducing delay, jitter and by enhancing throughput, maximum utilization of this limited bandwidth is possible.

6 Summary

This paper presents a hierarchical network design of sensor nodes for harbor monitoring and surveillance. It addresses the concerns of the dynamically changing network topology caused by mobility of the underwater sensor network nodes that are drifted due to water current. A clustering-based approach is presented here to help in dealing with mobility and essentially in reducing delay and jitter of transmission packets. Our simulations results suggest that with clustering relay nodes and aerial mobile sink, a network performance of roughly 15% is achieved in terms of lesser delay and jitter. Our future work lies in finding the trajectory of the aerial mobile sink node, finding the optimal number of relay nodes, and in exploiting the spatial and temporal correlation of sensing data for tracking and classifying objects approaching harbor.

References

1. Luo, H., Wu, K., Guo, Z., Gu, L., Ni, M.L.: Ship Detection with Wireless Sensor Network. IEEE Transaction on Parallel and Distributed Systems. Vol. 23, Iss. 7, pp. 1336–1343, (2012).
2. Rao, M., Kamila, N.K., Kumar, V.K.: Underwater Wireless Sensor Networks for tracking ships approaching Harbor. In Proceedings of International Conference on Signal Processing, Communication, Power and Embedded System. pp. 467–471, (2016).
3. Bahrami, N., Kamis, N.H.H, Baharom, A.B.: Study of underwater channel estimation based on different node placement in shallow water. IEEE Sensors Journal. Vol. 16, Iss. 4, pp. 1095–1102, (2016).
4. Liu L., Du J., Guo D.: Error beacon filtering algorithm based on K-means clustering for underwater Wireless Sensor Networks. 8th IEEE International conference on Communication Software and Networks. pp. 435–438, (4–6th June 2016). DOI:10.1109/ICCSN.2016. 7587196.
5. Ansari Z., Ghazizadeh R., Shokhmzan Z.: Gradient descent approach to secure localization for underwater wireless sensor networks. 24th IEEE Iranian Conference on Electrical Engineering. pp. 103–107, (2016).
6. Iqbal Z., Lee H-N.: Spatially Concatenated Channel-Network Code for Underwater Wireless Sensor Network. IEEE Transaction on Communications, vol. 64, Iss. 9, pp. 3901–3914, (2016).
7. Majid A., Azam I., Waheed A., Zain-Ul-Abidin M., Hafeez T., Khan Z.A., Qasim U., Javaid N.: An Energy Efficient and Balanced Energy Consumption Cluster Based Routing Protocol for Underwater Wireless Sensor Networks. 30th IEEE International Conference on Advanced Information Networking and Applications. pp. 324–333, (2016).
8. Cheng C-F., Li L-H.: Data Gathering Problem with the Data Importance Consideration in Underwater Wireless Sensor Networks. Journal of Network and Computer Applications http:// dx.doi.org/10.1016/j.jnca.2016.10.010.
9. Heidemann, J., Ye, W., Wills, J., Syed, A., Li, Y.,: Research Challenges and applications for underwater sensor networking. Proceedings of IEEE Wireless Communication and Networking conference, Nevada, USA, pp. 228–235. (2006).
10. Ayaz, M., Baig, I., Abdullah,A., Faye, I.: A survey of routing techniques in underwater wireless sensor networks. Journal of Network and Computer Applications. Vol. 35, Iss. 2, pp. 1908–1927. (2011).
11. Fu, C., Jiang, Z., Wei, W., Wei,A.: An Energy Balanced Algorithm of LEACH Protocol in WSN. International Journal of Computer Science Issues. Vol. 10, Iss. 1, pp. 354–359. (2013).
12. Pati. B, Sarkar. J.L, Panigrahi, C.R.: ECS: An Energy-Efficient Approach to Select Cluster-Head in Wireless Sensor Networks in Arabian Journal for Science and Engineering, Volume 42, Issue 2, February 2017.
13. Mukhtiar Ahmed, Mazleena Salleh, M. Ibrahim Channa,: Routing protocols based on node mobility for Underwater Wireless Sensor Network (UWSN): A survey in Journal of Network and Computer Applications, Vol. 78, pp. 242–252, 2017.

Development of Single and Multi-jet Conical Nozzle Based Open Jet Facility for Cold Jet Simulation

Kalakanda Alfred Sunny, Nallapaneni Manoj Kumar, Aldin Justin and M. Harithra

Abstract A significant negative impact is possible on practical high-speed propulsion applications due to shock wave and boundary layer interactions (SWBLI) when a supersonic jet is discharged out from a nozzle. So it is important to study the impacts associated with SWBLI. To study further, it is essential to analyze the physics of supersonic jet flow field by developing an open jet facility (OJF) in the laboratories. Supersonic jet can be produced in laboratories by allowing compressed air to escape through a nozzle into the atmosphere. Modeling, fabrication, and CFD simulation of nozzle-based open jet facility will help in understanding the supersonic jet flow. In this paper, an open jet facility is developed with single- and multi-jet conical nozzles in the Wind Tunnels Laboratory of Karunya University. The performance of this facility is evaluated theoretically and experimentally based on the runtime at different Mach numbers. Z-type schlieren technique is also applied to analyze the cold jet flow at a Mach number 2. CFD simulation is also carried out to verify the flow pattern that is visualized in experimental process.

Keywords Shock wave · Boundary layer · Shock wave and boundary layer interactions (SWBLI) · Open jet facility (OJF) · Runtime · Mach number Z-type schlieren technique · CFD

K.A. Sunny (✉) · A. Justin · M. Harithra
Department of Aerospace Engineering, Karunya University,
Coimbatore 641114, India
e-mail: alfredsunny@karunya.edu

A. Justin
e-mail: aldinjustin@karunya.edu

M. Harithra
e-mail: harithram@karunya.edu

N.M. Kumar
Department of Electrical & Electronics Engineering, Bharat Institute
of Engineering and Technology, Mangalpally, Ranga Reddy 501510,
Telangana, India
e-mail: nallapanenichow@gmail.com; nmanoj@biet.ac.in

© Springer Nature Singapore Pte Ltd. 2018   309
S.S. Dash et al. (eds.), *International Conference on Intelligent Computing
and Applications*, Advances in Intelligent Systems and Computing 632,
https://doi.org/10.1007/978-981-10-5520-1_29

List of Symbols

t	Runtime, s
V_t	Storage tank volume, m^3
A	Nozzle throat area, m^2
T_i	Initial temperature in the tank, K
P_i	Initial pressure in the tank, bar
P_f	Final pressure in the tank, bar
T_o	Stagnation temperature, K
P_o	Stagnation pressure, bar
n	Polytrophic index ($n = 1$)
γ	Specific heat ratio ($\gamma = 1.4$)
$\dot{m}$	Mass flow rate, kg/s
R	Universal gas constant, ($R = 287$ J/kg-K)
OJF	Open jet facility
CFD	Computational fluid dynamics
RPM	Revolution per minute
SWBL	Shock wave and boundary layer
SWBLI	Shock wave and boundary later interaction

1 Introduction

Jet is a stream of fluid that is projected into the atmosphere or surrounding medium from a nozzle. Jet flow with a Mach number greater than one is termed as supersonic jet. In a supersonic jet flow, shockwaves and boundary layers are obvious [1]. Also, the interactions between the shock waves and boundary layer commonly lead to negative impacts or significant problems [2, 3]. This happens when an externally generated shock wave impinges on the surface where there is a boundary layer. It may also arise if the slope of the body surface changes in such a way to produce sharp compression of the flow near to the surface, developing a shock wave that has its origin within the boundary layer. In any SWBL interactions, there is the possibility of thickening and separation of boundary layer. And this is because of the extreme unfavorable pressure gradient that is forced to act on a boundary layer by the shock wave resulting in viscous dissipation and flow separation [4].

Some of the consequences of SWBLI occurrence are as follows [5].

On transonic wings, SWBLI will increase the drag which is having a potential to cause the flow unsteadiness and buffet. They increase blade losses in gas turbine engines. Complicated boundary layer control systems must be installed in supersonic inlets to minimize the losses that they cause either directly by reducing the intake efficiency or by indirectly. This is due to the disruption they cause to the flow that is entering into the compressor. These control systems add weight to aircraft and absorb energy.

In hypersonic flights, SWBLI can be disastrous because at high Mach numbers, they have the potential to cause intense localized heating. This intense localized heating is sufficient to destroy a vehicle.

In the design of scramjet engines, SWBLI that occur in the intake and in the internal flow pose such critical issues that they significantly can limit the range of which vehicles using this form of propulsion can be deployed successfully.

Over decades, this subject has been an extensive research area. Studies [6–12] show the importance to know the physics of jet flows, causes for the consequences of SWBL, impacts of SWBLI on the other parts of the flow fields, influence on the parameters in the drag surface flux distributions, reattachment of the separated flows, and the possible ways in which it can be avoided so that there is no much effect. The interest and emphasis on this field have provoked the necessity in studying the behavior of jet flows. For this purpose, an open jet facility with supersonic nozzle is developed in the Wind Tunnels Laboratory of Aerospace Department, Karunya University.

An open jet facility represents a useful tool for aerodynamic research. This facility is designed especially for experimentation that usually involves scaled-down models. It can be used to stimulate flow phenomenon that is otherwise pertinent to a full-scale application under controlled environments. Mostly, the open jet facility is intended to study the nozzle flow characteristics.

This motivated for the development of an open jet facility with single- and multi-jet conical nozzles for cold jet simulation at different Mach numbers. Conical nozzles were designed and fabricated. Further to analyze the performance of this facility, both experimentation and CFD simulations were carried out and also for the verification of visualized patterns.

2 Configuration of an Open Jet Facility

Configuration of an open jet facility (OJF) is shown in Fig. 1. It consists of many components out of which the major are reciprocating air supply system with heatless dryer, storage tank, gate valve and high-pressure regulating valve, settling chamber, and nozzle.

All the components of an open jet facility are clearly explained in Fig. 1.

2.1 Reciprocating Air Supply System

Reciprocating air supply system consists of high-pressure 2-stage compressors and air dryer which provides highly compressed air for the tests. High-pressure air supply comes from two 20 HP electric-powered motors. Air is pumped up to 20 bar pressure, and this is the maximum available value for air storage facility that is established as a part of this open jet facility. Motor is totally enclosed with fan cooled provision. Three-phase 415 V and 50 Hz frequency motor is chosen for this

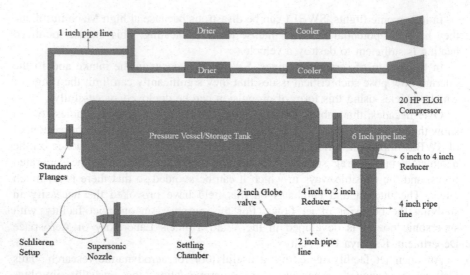

Fig. 1 Configuration of an open jet facility (OJF)

purpose, and it is a squirrel cage induction type having a speed of 1440 RPM. The overall dimension of the motor is 68 * 32 * 33 inch. In order to remove the moisture from the air, dryer and coolers were used. Once the moisture is removed, this air is stored in a storage facility having 14.14 m^3 volume. This moisture-free air is obtained from the dryers.

2.2 Heatless Dryer System

The dryer consists of 2 desiccant towers, pre-filter, and oil-removing filter. The wet-compressed air enters pre-filter, where moisture and dust contents are removed to 5 μm. Then, oil contents are removed from air upon entering into oil-removing filter by opening double-seat automatic drain valve which is fitted to the pre-filter.

2.3 Air Storage Tank

The runtime of the wind tunnel facility can be limited by parameters of a storage facility, i.e., peak pressure and size or volume of the tank. A storage tank was already available for supersonic wind tunnel.

This tank is 1.5 m in diameter, 8 m in length, having a volume of 14.14 m^3, and it is rated for a pressure of 150-psi. The atmospheric air is compressed using the 20 HP ELGI compressor. This compressed air at high pressures is pumped into the storage facility using the valves.

2.4 High-Pressure Regulating and Gate Valves

High-pressure regulating valve of 2 inch and gate valve of 6 inch are installed, respectively, for regulating the flow from the storage facility to the settling chamber through a 2-inch 300 class pipe.

2.5 Settling Chamber

It plays vital role in wind tunnel, and it helps in reducing the turbulence of airflow by increasing its strength. The direction of flow is axial. It is designed using number of screens such as structures and the honeycombs. A considerably less drop in the value of pressure in honeycomb structure helps in removing the fluctuations in flow across the axial length causing a comparatively less effect on the shear velocity. Apart from honeycombs, screen also available in settling chamber made of woven wires, whose refractive index is very small, and it is responsible for reducing the fluctuations in shear velocity, causing a negligible effect on the direction of flow. 0.2 cm diameter apart between each screen is maintained. This spacing allows the equal distribution of flow. This allows the first screen to settle before the flow reaches to the second screen. In order to maintain the flow speed constantly in the total cross section of a settling chamber, the last screen is placed at greater apart from the source screen.

2.6 Nozzle

Nozzle is a device that enables the user to achieve the desired flow velocities in both increasing and decreasing levels. This can be achieved when a fluid passes through the boundaries of the designed nozzles. These are regularly used in controlling many parameters of a fluid passing through a duct or pipeline-like structure. These parameters would be of mass of fluid, direction of the fluid, rate of flow, speed, pressure, and the shape. There are different varieties of nozzles in trend, and they are being used based on the applications. For this test facility, conical nozzles have been designed, simulated, and developed [13]. Here, a nozzle which is having simple design configuration is chosen, i.e., a conical nozzle. The design and fabrication process of conical nozzle are quite easy compared to other kinds of nozzles and are still having its importance in the present-day applications, especially in the case of small nozzle applications.

3 Design of an Open Jet Facility Components

One of the prerequisites of an open jet facility is the availability of compressed air to generate required speeds.

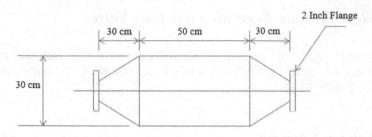

Fig. 2 Design of settling chamber

The compressed air for this setup is obtained from a dynamic facility associated with Karunya University Supersonic Wind Tunnel. A reducer from 4 to 2 inch is used to tap high-pressure air from the existing facility to the settling chamber. All the pipelines for the open jet facility are maintained at 2 inch. Also, a gate valve of 2 inch enables the pressure regulation at the settling chamber. Therefore, settling chamber and nozzle are designed in accordance with settling chamber pressures.

3.1 Design of Settling Chamber

The settling chamber that is designed for this purpose has a cross section of 110 * 30 cm^2 with a honeycomb structure and two meshes made up of stainless steel, which is shown in Fig. 2. Based on this design, it is fabricated and integrated with supersonic wind tunnel facility. A reducer from 6 to 4 inch forms a storage tank, and a reducer from 4 to 2 inch is used for tapping high-pressure air from the existing facility to the settling chamber, which is clearly shown in Fig. 1.

3.2 Design of Single- and Multi-jet Conical Nozzles

Two supersonic conical nozzles, single jet and multi-jet for Mach number 2 were designed and developed based on the following nozzle design equations [13].

The nozzle throat diameter is fixed to 0.4 inch and the chamber diameter as 2 inch for single-jet nozzle.

In the case of the multi-jet nozzle, the throat diameter was fixed to 0.2 inch and the chamber diameter to 1 inch.

The nozzle exit area for any given Mach number (but here, $M = 2$) is calculated using Eq. 1. [13]:

$$\left(\frac{A_e}{A_t}\right)^2 = \frac{1}{M^2}\left[\frac{2}{\gamma+1}\left(1+\frac{\gamma-1}{2}M^2\right)\right]^{\frac{\gamma+1}{\gamma-1}} \qquad (1)$$

The mass flow rate of the nozzle is calculated using the mathematical expression mentioned in Eq. 2. [13]:

$$\dot{m} = \frac{A^* P_o}{\sqrt{T_o}} \sqrt{\frac{\gamma}{R}} \left(\frac{\gamma+1}{2}\right)^{-\left(\frac{\gamma+1}{2(\gamma-1)}\right)}$$

(2)

The thickness required for withstanding the pressure is obtained by Eq. 3 [13]:

$$\sigma = \left(\frac{P * D_e}{2t}\right)$$

(3)

Based on the input conditions, the nozzle design parameters were calculated. These parameters are used for designing the single- and multi-jet nozzles in solid work software tool, and the final designed models are shown in Figs. 3 and 4.

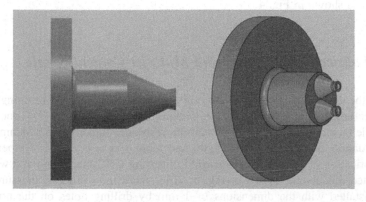

Fig. 3 3D models of single- and multi-jet nozzles modeled in solid works tool

Fig. 4 Fabricated setup of settling chamber

Fig. 5 Fabricated models of single- and multi-jet conical nozzles

3.3 Fabrication of Settling Chamber

The designed settling chamber shown in Fig. 2 is fabricated as per the parameters and is installed on the existing gas dynamic facility. The fabricated setup of settling chamber is shown in Fig. 4.

3.4 Fabrication of Single- and Multi-jet Conical Nozzle

As per the design equations [13], two supersonic conical nozzles, i.e., single- and multi-jet nozzles, are fabricated using stainless steel alloy SS-304 due to its favorable factors such as ease of fabrication, resistance to corrosion, contamination and oxidation, possible higher strength, and toughness at cryogenic temperatures. The fabrication of the nozzle was carried out using CNC machining and welding. Fabricated models are shown in Fig. 5. After fabrication process, pressure ports were installed with the dimensions of 1 mm by drilling holes on the nozzle at different distances. Pressure ports are inserted in the drilled holes. The *K*-type thermocouples are welded on the surface of nozzle at different distances.

4 Runtime-Based Performance Test

Runtime is referred as operational duration of the setup. It is one of the important factors to be considered for conducting any experiment in this facility.

4.1 Theoretical Variation of Runtime

$$t = \frac{V_t P_i}{\dot{m} R T_i} \left[1 - \left(\frac{P_f}{P_i} \right)^{\frac{1}{n}} \right]$$

(4)

The runtime is represented mathematically in Eq. 4 and it is a function of mass flow rate $\dot{m}$, which is represented in Eq. 5, the initial pressure P_i, the final pressure P_f, the initial temperature T_i, the tank volume V_t, polytrophic coefficient n, and the gas constant R [14]:

$$\dot{m} = \frac{A P_o}{\sqrt{T_o}} \sqrt{\frac{\gamma}{R}} \left(\frac{\gamma + 1}{2} \right)^{-\left(\frac{\gamma + 1}{2(\gamma - 1)} \right)}$$

(5)

4.2 Experimental Variation of Runtime

With the single-jet conical nozzle installed, the tunnel is made to run at different stagnation pressures. The runtime at each stagnation pressure was recorded using a stop clock. The variation in the runtime with respect to the stagnation pressure is plotted.

When compared to the theoretical runtime, it is observed that the practical runtime is less due to the account of losses in real time. The pressure drop across the valve, frictional loss, and heat loss accounts for the reduction in runtime. Reduction in runtime results in less time for the tests to be carried out. Hence, the possible ways in improvising the runtime have to be concentrated in future.

Assuming constant temperature, the runtime for different pressures is calculated and plotted. When the results are compared with experimental runtime, it is found that the experimental runtime is very short due to the account of loss in real-time conditions. Any increase in the runtime during future tests would be beneficial. Theoretical and experimental variation of the runtime in seconds with respect to the stagnation pressure in bar is shown in Figs. 6 and 7.

Fig. 6 Theoretical variation of runtime with stagnation pressure

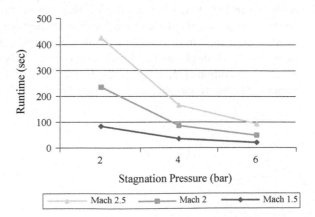

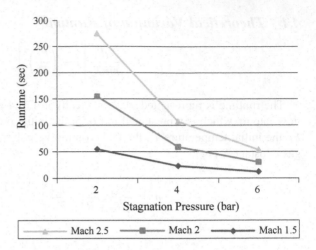

Fig. 7 Experimental variation of runtime with stagnation pressure

5 Analysis of Single- and Multi-jet Flow Patterns Using Schlieren and CFD

To analyze the performance of this facility, runtime-based experimental performance test is carried out in and the results of that test are discussed in Sect. 4.

In this paper, apart from the performance test, CFD simulation and schlieren technique were also used for analyzing the flow patterns whether they are visualized in similar manner both in simulation bed and in practical case.

5.1 CFD Analysis of Single- and Multi-jet Conical Nozzle

CFD analysis is carried out to analyze the flow patterns of supersonic jet discharged from the two nozzles which designed and developed.

The two nozzles, i.e., single and multi-jet nozzles, were modeled in solid works as shown in Fig. 3. Later, these model files were exported to GAMBIT 2.4.6 for creating the unstructured triangular mesh, and they are shown in Fig. 8.

And later, this unstructured mesh files are exported to ANSYS R 15.0. CFD analysis of both single- and multi-jet nozzles was analyzed by applying the desired boundary conditions [15, 16]. Figures 9 and 10 represent the simulation of density contours for the single- and multi-jet nozzles.

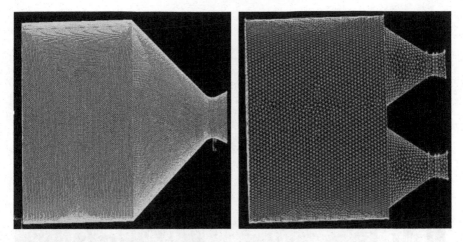

Fig. 8 Finely meshed single- and multi-jet nozzles in GAMBIT

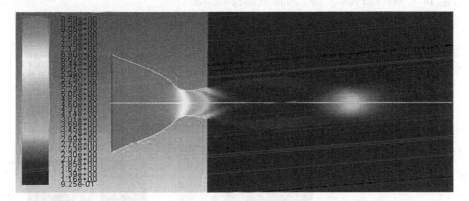

Fig. 9 Contours of density for a single-jet conical nozzle

5.2 Schlieren Analysis of Single- and Multi-jet Conical Nozzles

Jet flow visualization using schlieren technique is also carried out using the experimental setup shown in Fig. 11 helps in validating the CFD results. The images captured in schlieren [17] using camera which was kept at a distance of 1 m. Colored image was obtained by setting the knife edge to the colored filter lens, and shocks are clearly identified.

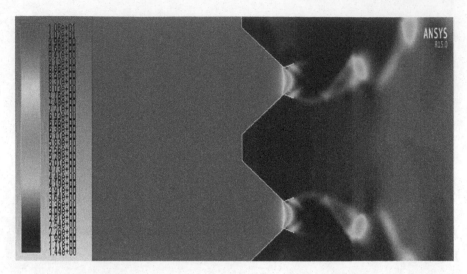

Fig. 10 Contours of density for a multi-jet conical nozzle

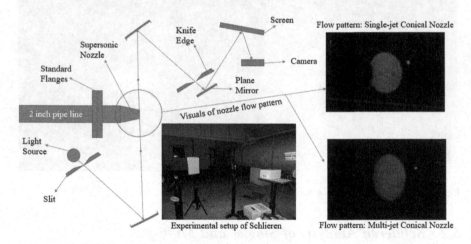

Fig. 11 Experimental setup for visualizing the jet flow patterns in schlieren system

The analyses of the results infer that the flow out through the single-jet nozzle in Figs. 9 and 12 show that the flow is under expanded. The flow pattern of multi-jet conical nozzle is shown in Figs. 10 and 12, and the flow after the nozzle appears to be pinched initially as the pressure is adjusted through the diamond shock waves, which is followed by series of expansion waves that reflect off the free jet boundary as the series of compression waves. This clearly validates the results obtained from schlieren with the CFD results.

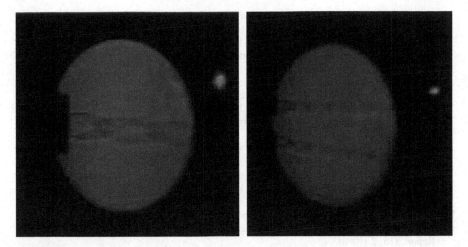

Fig. 12 Flow visualization of single- and multi-jet conical nozzles

6 Conclusion

An open jet facility for the study of SWBL interactions and jet flow simulations are engineered and built in this paper. Conventional design principles were used to develop two conical nozzles with single and multi-jet for the tests. Based on the performance test carried out, the runtime of the system is short and measures to minimize the pressure losses can be implemented in order to achieve better runtimes. Flow analyses of cold jet for Mach number 2 show the formation of the diamond shocks, which is the subject for the future SWBL studies. Further, flow analysis for different Mach numbers was carried out and the system discharged satisfactory performance.

Acknowledgements We thank the in-charge as well the technicians of the Wind Tunnels Laboratory, Karunya University for their excellent support during this work.

References

1. Jayahar Sivasubramanian and Hermann F. Fasel.: Numerical Investigation of Shockwave Boundary Layer Interactions in Supersonic Flows. 54th AIAA Aerospace Sciences Meeting, AIAA SciTech Forum, (AIAA 2016-0613), pp. 1–42. <http://dx.doi.org/10.2514/6.2016-0613>.
2. S. Pirozzoli, J. Larsson, J. W. Nichols, M. Bernardini, B. E. Morgan, S. K. Lele.: Analysis of unsteady effects in shock/boundary layer interactions. Proceedings of the Summer Program 2010, Center for Turbulence Research, pp. 153–164, (2010).
3. Green, J. E.: Interactions between shock waves and turbulent boundary layers. Progress in Aerospace Sciences, Vol. 11, pp. 235–340, December (1970).

4. Rose W. C., M. E. Childs.: Reynolds-shear-stress measurements in a compressible boundary layer within a shock-wave-induced adverse pressure gradient. Journal of Fluid Mechanics, Vol. 65, Issue. 01, pp. 177–188, August (1974).
5. Babinsky Holger, John K. Harvey.: Shock wave boundary layer interactions. Vol. 32. Cambridge University Press, (2011).
6. T. Pot, B. Chantz, M. Lefebvre, and P. Bouchardy.: Fundamental study of shock-shock interference in low-density flow: Flow field measurements by DLCARS. In Rarefied Gas Dynamics (Toulouse: Cepadues-Editions, 1998), Part II-545, (1998).
7. J. Delery and J. Marvin. Shock wave boundary layer interactions. AGAR Dograph 280, (1986).
8. J. E. Greene.: Interactions between shock waves and boundary layers. Progress in Aerospace Science, Vol. 11, pp. 235–340, (1970).
9. M. Honda.: A Theoretical Investigation of the interaction between shock waves and boundary layers. Journal of the Aerospace Sciences, Vol. 25, Issue. 11, pp. 667–678, (1958). <http://dx.doi.org/10.2514/8.7843>.
10. D. D. Knight, G. Degrez.: Shock wave boundary layer interactions in high Mach number flows-A Critical survey of current numerical prediction capabilities. AGARD Advisory Report 315, 2 (1998).
11. Holden Michael S.: A study of flow separation in regions of shock wave-boundary layer interaction in hypersonic flow. NASA STI/Recon Technical Report N 76 (1975): 17354. (1975).
12. L. Lees and B.L. Reeves.: Supersonic separated and reattaching laminar flows: I. General theory and application to adiabatic boundary layer- shock wave interactions. AIAA Journal, 2 (1964).
13. George P. Sutton, Oscar Biblarz.: Rocket propulsion elements. 8th ed., John Willey & Sons Inc. (2010).
14. Pope A., Goin K.: High-Speed Wind Tunnel Testing, New York-John Wiley & Sons, (1965).
15. Chris Bahr, Jian Li and Louis Cattafesta.: Aeroacoustic Measurements in Open-jet Wind Tunnels-An Evaluation of Methods Applied to Trailing Edge Noise. 17th AIAA/CEAS Aeroacoustic Conference (32nd AIAA Aeroacoustic Conference), 05 - 08 June 2011, Portland, Oregon, pp. 1–19, (2011).
16. Arun Kumar P., E. Radhakrishnan.: Triangular Tabs for Supersonic Jet Mixing Enhancement. The Aeronautical Journal, Vol. 118, Issue. 1209, pp. 1245–1278, November (2014).
17. G. S. Settles.: Schlieren and Shadowgraph Techniques: Visualizing Phenomena in Transparent Media. 2nd ed., Springer-Verlag Berlin Heidelberg, (2001).

Background Noise Identification System Based on Random Forest for Speech

Shambhu Shankar Bharti, Manish Gupta and Suneeta Agarwal

Abstract Background noise is acoustically added with human speech while communicating with others. Nowadays, many researchers are working on voice/speech activity detection (VAD) in noisy environment. VAD system segregates the frames containing human speech/only noise. Background noise identification has number of applications like speech enhancement, crime investigation. Using background noise identification system, one can identify possible location (street, train, airport, restaurant, babble, car, etc.) during communication. It is useful for security and intelligence personnel for responding quickly by identifying the location of crime. In this paper, using VAD G.729, a new algorithm is proposed for selecting an appropriate set of noisy frames. Mel-frequency cepstral coefficient (MFCC) and linear predictive coding (LPC) are used as feature vectors. These features of selected frames are calculated and passed to the classifier. Using proposed classifier, seven types of noises are classified. Experimentally, it is observed that MFCC is a more suitable feature vector for noise identification through random forest classifier. Here, by selecting appropriate noisy frames through proposed approach accuracy of random forest and SVM classifier increases up to 5 and 3%, respectively. The performance of the random forest classifier is found to be 11% higher than SVM classifier.

Keywords VAD · VAD G.729 · Random forest · SVM · Noise identification
Noise extraction

S.S. Bharti (✉) · M. Gupta · S. Agarwal
MNNIT Allahabad, Allahabad, India
e-mail: shambhu4u08@gmail.com

M. Gupta
e-mail: manishymca2007@gmail.com

S. Agarwal
e-mail: suneeta@mnnit.ac.in

© Springer Nature Singapore Pte Ltd. 2018
S.S. Dash et al. (eds.), *International Conference on Intelligent Computing and Applications*, Advances in Intelligent Systems and Computing 632,
https://doi.org/10.1007/978-981-10-5520-1_30

323

1 Introduction

Identification of background noise is a basic but tedious problem in audio signal processing. Till now, it has got less attention of the researchers working on audio signal processing. During communication through speech, background noise gets acoustically added with the speech signal. So, this noise signal is also communicated with the clean speech. These signals carry information about the background location (street, train, airport, restaurant, babble, car, etc.) of the person during communication. By identifying the type of background noise, one can easily identify the possible location of a person at the time of communication. For example, if a person is communicating using mobile phone, then the region of that person can be traced through mobile signal tower. Using background noise identification system, possible location within that region can be identified which will reduce the search space.

2 Previous Work

Chu et al. [1] used "composite of deep belief networks (composite-DBNs)" for recognizing 12 different types of common environmental sounds. Mel-frequency cepstral coefficient (MFCC) and matching pursuit (MP) are used as feature vectors. Maximum accuracy in environmental sound classification using composite-DBNs has been claimed as 79.6% by taking MFCC and MP features.

Frequency component of maximum harmonic weight (FCOMHW), local search tree, effective segment length of audio data (ESLOAD) and first-order difference mel-frequency cepstral coefficients matrix (D-MFCCM) features have been used by Li [2] to classify environmental sound.

Toyoda et al. [3] used combination of instantaneous spectrum at power peak and the power pattern in the time domain as features. Multi-layered perception neural system is used for the environmental sound classification. This classifier is used to classify 45 environmental sounds. Classification accuracy was about 92% claimed in the paper.

Pradeep et al. [4] used audio features like zero-crossing rate (ZCR), LPC, linear predictive cepstral coefficient (LPCC) and (log frequency cepstral coefficients) LFCC. Gaussian mixture model was used in the paper for modelling an event. Training audio data frame was taken of 50 ms. Using single Gaussian classifier, 76 and 80% accuracy for walking and running events, respectively, have been claimed.

Lozano et al. [5] used audio features like MFCCs, ZCR, centroid and roll-off point for audio classification. Maximum accuracy ratio by type of sound was about 81.42% claimed in this paper.

Han and Hwang [6] used traditional features (TFs) like ZCR, MFCC; change detection features (CDFs) like chirp rate spectrum; and acoustic texture features (ATFs) like discrete wavelet transform (DWT), discrete curvelet transform for sound classification. Nonnegative matrix factorization [7] has been used for dimensionality reduction, and SVM has been taken as classifier. Maximum accuracy was claimed as 86.09%.

Kraetzer et al. [8] used data mining tool WEKA with K-means as a clustering and naive Bayes as a classification technique for the classification process. AAST (AMSL Audio Steganalysis Toolset, version 1.03) [9] and 56 mel-cepstral domain-based features are used.

Following are the main objectives of this paper:

(i) To develop noise identification system using random forest classifier.
(ii) To develop an algorithm for noise extraction using VAD G.729.
(iii) To find suitable feature/features for noise identification system.
(iv) To find suitable classifier for noise identification system.

The rest of this paper is organized as follows. In Sect. 3, random forest is discussed. Section 4 explains the proposed approach. In Sect. 5, parameters used for performance evaluation of the system are discussed. Section 6 explains the experimental set-up followed by summarization of the results. Section 7 concludes the paper.

3　Random Forest

Random forest or random decision forest [10, 11] uses an "ensemble learning method" for classification. It works by constructing a multitude of decision trees at training time. It identifies the class using mode of the classes or mean prediction (regression) of the individual trees. The general method of random decision forests was first proposed by Ho in 1995 [10]. Two well-known methods are boosting [12] and bagging [13] used in classification trees. In boosting, successive trees give extra weight to points incorrectly predicted by earlier predictors. In the end, a weighted vote is taken for prediction. For example, if there are n classes for classification, then samples of all the classes are selected randomly for the training purpose. If there are K input variables in each sample, then k $(k < K)$ is specified at each node where k variables are selected at random. Each tree grows at its maximum extent without any pruning. New data are classified considering maximum vote gained from different trees.

4　Proposed Approach

Noise extraction plays an important role in noise identification system. Noise can be better classified if the noisy frames are extracted correctly. In speech communication system, a frame either contains both human speech and noise or noise only.

In this paper, a new algorithm has been proposed for better noise identification by selecting the subset of noisy frames obtained from VAD G.729. Here, a framework is also proposed for noise identification. This framework may also be useful for language identification. Classifiers are trained using the feature/features (LPC, MFCC, LPC and MFCC together) extracted from Aurora2 noise.

Proposed Algorithm for Noise Extraction:
INPUT: Speech signal S of size $N1 \times 1$.
OUTPUT: Noise signal of size $N2 \times 1$.
Assumptions and Notations:

(i) S is a vector of size $N1 \times 1$. It contains $N1$ samples of speech signal.
(ii) N is the total number of frames.
(iii) $N2$ ($N2 <= N1$) is the number of samples in noise signal.
(iv) out is a matrix of size $N \times 2$. Matrix out will have two columns. First column represents frame number and second column represents *VAD* technique output.
(v) ||, represents the logical *"OR"* operation.
(vi) flag, flag1 and flag2 are binary variables.

Procedure noise_extract(S):
Repeat step 1 and step 2 for i = 1 to N.

Step 1: *Apply VAD G.729 technique on ith frame and store its output with corresponding frame number.*
// VAD G.729 output will be either 1 or 0. 1 for those frames that contain human speech and 0 for others.

Step 2: Save the result in matrix out.

Repeat step 3 and step 4 for i = 3 to N − 1.

Step 3: *If out(i, 2) equals to 0, then*

flag1 = *out(i-1,2)* || *out(i-2,2)*.
flag2 = *out(i-1,2)* || *out(i + 1,2)*.
flag1 = *flag1* || *flag2*.
end If

Step 4: *If flag equals to 0, then*

Add the samples of that frame into noise signal.
end If

Step 5: *Return noise_signal.*
End Procedure.

Feature vectors are calculated for noise_signal. Noise is identified by trained classifier using these features.

5 Parameters Used for Measuring the Performance

Reliability ratio (RRN), accuracy ratio (ARN) [5] and confusion matrix for the noise are used for measuring the performance of the proposed system. Suppose in a given signal S, noise of K different classes is present. TF_i is the total numbers of frames with noise of class i in the given signal. FC_i is the total number of frames having noise of type i as classified by classifier while T_i is the number of frames truly classified as of class i by the classifier.

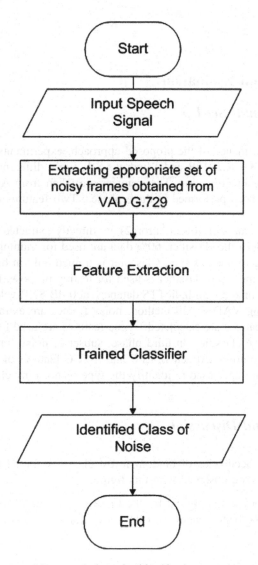

Block diagram of proposed framework for noise identification system

Reliability ratio by type of noise (RRN)
It is formulated as:

$$\text{RRN} = \frac{\sum_{i=1}^{k} \frac{T_i}{\text{FC}_i}}{K} \tag{1}$$

Accuracy ratio by type of noise (ARN)
It is formulated as:

$$\text{ARN} = \frac{\sum_{i=1}^{k} \frac{T_i}{\text{TF}_i}}{K} \tag{2}$$

6 Experimental Evaluation

6.1 Experimental Set-Up

To verify the performance of the proposed approach, experiments have been performed on MATLAB R.2015B and Windows 8.1. Eight different types of noises have been used for identification. These noises are taken from Aurora2 database. Experiments have been performed in three phases. Two features LPC and MFCC are used.

In first phase, features of noise (Aurora2) are directly extracted and are used for training and testing of the classifier. 60% data are used for training, and remaining are used for testing of the classifier. Classifier is trained only in first phase, and in remaining phases, the same classifier is used for testing purpose. In second phase, selected noises are mixed with IndicTTS database at 0 dB SNR value and named as noisy speech. Using VAD G.729 method, noisy frames are extracted from noisy speech and their features are calculated. These features are used for identifying the type of noise by the classifier. In third phase, subset of noisy frames is extracted using the procedure noise_extract(S). LPC and MFCC features of these frames are calculated which are later used to identify the type of noise by classifier.

6.2 Results and Discussion

In Fig. 1, average accuracies of random forest classifier for all three phases are shown. Two points are observed with this figure:

1. Average accuracy is the highest for the first phase of the experiment. It means that noise can be better classified if the noisy frames are extracted correctly.

Fig. 1 Average accuracy of
random forest classifier

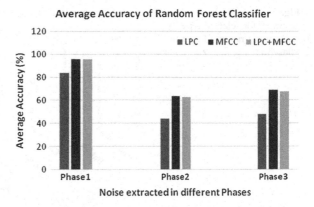

2. Random forest classifier gives more accurate result using MFCC feature rather than using LPC alone or LPC and MFCC both. It shows that accuracy of the classifier may not be enhanced by fusing LPC and MFCC.

In Fig. 1, it is found that the performance of the same classifier is higher for the Phase3 compared to the Phase2 of the experiment. It confirms that in Phase3, noisy frames are extracted more appropriately than Phase2.

Figure 2 shows that for selected types of noises, random forest classifier performs better using MFCC feature rather than LPC feature. It is also observed that this classifier performs better for noises like airport, babble, car and station noise with MFCC feature while for others with LPC and MFCC together.

By comparing Figs. 2 and 3, it is observed that the classification accuracy of random forest classifier is better for noises extracted using procedure noise_extract (S) rather than using VAD G.729.

In Fig. 4, average accuracies of SVM classifier for all eight types of noises are shown. Average accuracy is the highest for first phase of the experiment. It means that noise can be better classified if the noisy frames are extracted correctly.

Fig. 2 Average classification
accuracy for selected types of
noise using random forest in
Phase2

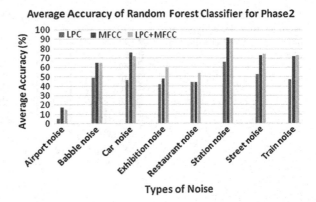

Fig. 3 Average classification
accuracy for selected types of
noise using random forest in
Phase3

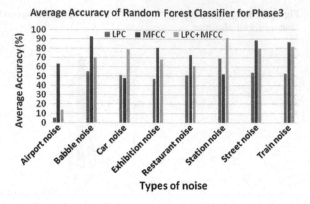

Fig. 4 Average accuracy of
SVM classifier

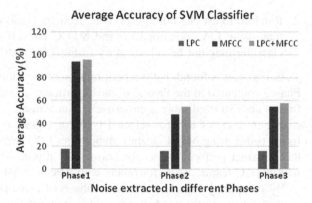

By comparing Figs. 2 and 5, it is observed that the performance of random forest
classifier is better than the SVM classifier for Phase2 of the experiment.

By comparing Figs. 3 and 6, it is observed that the performance of random forest
classifier is better than SVM classifier for Phase3 of the experiment.

Fig. 5 Average classification
accuracy for selected types of
noise using SVM classifier in
Phase2

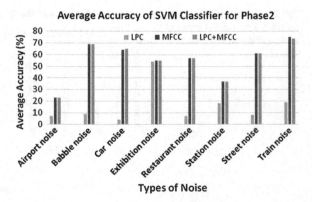

Fig. 6 Average classification accuracy for selected types of noise using SVM classifier in Phase3

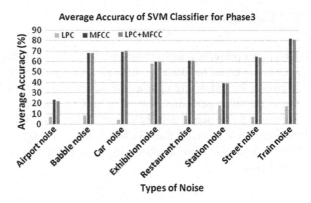

Average accuracy for random forest classifier in Phase1 of the experiment is 96%. It is obtained using MFCC feature alone/LPC and MFCC features together. Average accuracies for random forest classifier are 64 and 69% for Phase2 and Phase3, respectively. Maximum accuracy is obtained using MFCC feature. Thus, average performance of the random forest classifier increases 5% from Phase2 to Phase3, respectively. It confirms that noise extracted using procedure noise_extract (S) is more accurate than VAD G.729 procedure.

Average accuracy for SVM classifier in Phase1 of the experiment is 96% which is obtained using LPC and MFCC features together. Average accuracies for SVM classifier are 55 and 58% for Phase2 and Phase3, respectively. Maximum accuracy is obtained using LPC and MFCC feature together.

Thus, average performance of the SVM classifier increased 3% from Phase2 to Phase3. It again confirms that noise extracted using procedure noise_extract(S) is more accurate than VAD G.729 procedure.

The maximum average accuracy of random forest classifier is 9 and 11% higher than SVM classifier for Phase2 and Phase3, respectively. Thus, one can conclude from the experiment that as noise in speech can better be identified through selection of appropriate noisy frames as shown in Table 1.

Table 1 Classification results of random forest and SVM classifiers for different phases of noise

Classifier	Random forest						SVM					
Features	LPC		MFCC		LPC + MFCC		LPC		MFCC		LPC + MFCC	
Noise	Parameters											
	RRN	ARN	RRN	ARN	RRN	ARN	RRN	ARN	RRN	ARN	RRN	ARN
Phase1	0.94	0.82	0.99	0.96	0.99	0.95	0.18	0.17	0.99	0.94	0.99	0.96
Phase2	0.54	0.43	0.75	0.63	0.75	0.63	0.15	0.16	0.60	0.55	0.61	0.55
Phase3	0.57	0.48	0.78	0.69	0.78	0.68	0.15	0.16	0.61	0.58	0.62	0.58

7 Conclusion

This paper presents a new framework for noise identification in speech. As noise in speech can better be identified through appropriate noisy; therefore, in this paper, an approach for extracting appropriate set of noisy frame is also proposed. Random forest classifier performs best with MFCC feature alone while to give best performance SVM needs both LPC and MFCC together. The overall performance of random forest is better than SVM for noise identification. Thus, random forest is better choice for noise identification with respect to accuracy as well as computational cost.

References

1. Selina Chu, Shrikanth Narayanany and C.-C. Jay Kuo, "Composite-DBN for Recognition of Environmental Contexts," Proc APSIPA Hollywood CA, 2012.
2. Y. Li, "A classification method for environmental audio data," 2nd International Conference on Advanced Computer Control (ICACC), pp. 355–361, 2010.
3. Y. Toyoda, J. Huang, S. Ding, Y. Liu, "Environmental sound recognition by multilayered neural networks," Proceedings of the Fourth International Conference on Computer and Information Technology, CIT'04, IEEE Computer Society, Washington, DC, USA, pp. 123–127, 2004.
4. Pradeep K. Atrey, Namunu C. Maddage and Mohan S. Kankanhalli, "AUDIO BASED EVENT DETECTION FOR MULTIMEDIA SURVEILLANCE," Acoustics, Speech and Signal Processing (ICASSP), pp. 813–816, 2006.
5. Hector Lozano, Inmaculada Hernaez, Artzai Picon, Javier Camarena and Eva Navas," Audio Classification Techniques in Home Environments for Elderly/Dependant People", International Conference on Computers for Handicapped Persons (ICCHP), pp. 320–323, 2010.
6. Byeong-jun Han and Eenjun Hwang, "IEEE International Conference on Multimedia and Expo (ICME)," pp. 542–545, 2009.
7. D.D. Lee and S. Seung, "Learning the parts of objects by non-negative matrix factorization," Nature, vol. 401 no. 6755, pp. 788–791, 1999.
8. Christian Kraetzer, Andrea Oermann, Jana Dittmann and Andreas Lang, "Digital Audio Forensics: A First Practical Evaluation on Microphone and Environment Classification," Proceedings of the 9th workshop on Multimedia & security, pp. 63–74, 2007.
9. C. Kraetzer and J. Dittmann., "Mel-cepstrum based steganalysis for voip-steganography," E. J. Delp and P. W. Wong, editors, Security, Steganography, and Watermarking of Multimedia Contents IX, Electronic Imaging Science and Technology, SPIE Vol. 6505, San Jose, CA, USA, SPIE and IS&T, SPIE, 2007.
10. Ho, Tin Kam, "Random Decision Forests (PDF)," Proceedings of the 3rd International Conference on Document Analysis and Recognition, pp. 278–282, 1995.
11. Ho, Tin Kam, "The Random Subspace Method for Constructing Decision Forests", IEEE Transactions on Pattern Analysis and Machine Intelligence, pp. 832–844, 1998.
12. R. Shapire, Y. Freund, P. Bartlett, and W. Lee. "Boosting the margin: A new explanation for the effectiveness of voting methods," Annals of Statistics, pp. 1651–1686, 1998.
13. Leo Breiman, "Bagging predictors", Machine Learning, pp. 123–140, 1996.

Spectral Analysis of Speech Signal Characteristics: A Comparison Between Healthy Controls and Laryngeal Disorder

S. Shamila Rachel, U. Snekhalatha, K. Vedhasorubini
and D. Balakrishnan

Abstract The vocal cords pathologies such as hyper-functional voice disorder, spasmodic dysphonia, and vocal nodules affect the speech production completely or partially. The speech signal analysis is performed to detect the pathological conditions and used as a preliminary diagnostic method to identify the vocal cords pathologies or pathologies of the larynx. This study implements the identification of larynx condition by certain speech signal features. The main aim of the study is to assess the lesions of the larynx using the voice signal of the patient in order to extract the acoustic features from the recorded voice signal and to perform the comparison between semi-automated and automated methods. The voice signal features are calculated from the recorded voice signal and analyzed using three different software such as PRAAT, Dr. Speech, and automated analysis using MATLAB. The voice signal features are obtained from sustained phonation /a/ and continuous speech. The difference between normal and abnormal voice samples was predicted in both semi-automated and automated analyses. The acoustic features such as shimmer, pitch, F0 tremor, jitter were extracted and obtained the significance ($p < 0.05$) difference between normal and abnormal speech signal. This study reveals that significant variations in voice parameters such as spectrogram, jitter, and shimmer of the patients with laryngeal disorders in comparison with the healthy control. The voice samples of laryngeal disorder patients show clear deviation from the normal people, and the result shows that the automated analysis and speech spectral analysis methods can be used to detect lesions in larynx non-invasively.

S. Shamila Rachel (✉) · U. Snekhalatha (✉)
Department of Biomedical Engineering, SRM University, Kattankulathur,
Tamil Nadu, India
e-mail: rachelhelen03@gmail.com

U. Snekhalatha
e-mail: sneha_samuma@yahoo.co.in

K. Vedhasorubini · D. Balakrishnan
Department of Audiology and Speech Language Pathology,
SRM Medical College Hospital and Research Centre, SRM University,
Kattankulathur, Tamil Nadu, India

© Springer Nature Singapore Pte Ltd. 2018
S.S. Dash et al. (eds.), *International Conference on Intelligent Computing
and Applications*, Advances in Intelligent Systems and Computing 632,
https://doi.org/10.1007/978-981-10-5520-1_31

333

Keywords Autocorrelation · Formant · Short-time Fourier transform
Speech signal · Spectrogram

1 Introduction

The speech signal is produced by alteration of air pressure in the lungs and vocal cords. In normal speech, the vocal cords in the larynx touch each other smoothly. Any disturbance or interference with the movement of vocal cords causes voice disorder or abnormal voice. The voice disorders can be classified as follows: (i) structural, (ii) neurogenic, (iii) functional, and (iv) psychogenic disorders. The most common occurrence of voice disorder in population is due to structural and functional disorders [1]. Approximately 17.9 million adults have trouble in their voice in the USA [2]. In India, around 7.5% people were affected by speech disorders [3]. The main causes for the voice disorders are abnormal growth of tissue in the vocal cords, inflammation and swelling, nerve problems, hormones, and voice abuse. The voice disorders are diagnosed using laryngoscope, stroboscope, laryngeal EMG, and other imaging modalities such as X-ray and MRI.

The voice signal and its characteristics such as pitch, frequency, and jitter vary for each pathology of the larynx (vocal cords). The laryngoscope cannot be taken often, i.e., at a regular significant interval of time to assess the response to speech therapy, as it was time-consuming, cost-effective, and causes discomfort to the patient. Hence the spectral and formant analyses are used to assess the patient response using the voice-recording process. The features of voice pathology from the recorded voice can be assessed using speech processing algorithms such as PRAAT, Dr. Speech, and MATLAB. This study implements the speech processing algorithms to assess and monitor the progress of the larynx of the patient at regular intervals of time which are a boon to the speech therapies.

Several researchers [4–8] have studied and diagnosed various speech signal pathologies using the PRAAT and Dr. Speech software and obtained the quantitative results in differentiating the normal and abnormal conditions. Gelzinis et al. [9] performed automated analysis of a voice signal to diagnose the laryngeal diseases. Vikram et al. [10] presented a novel method for the adaptation of Mel-frequency cepstral coefficients (MFCC) and support vector machine (SVM) for the diagnosis of neurological disorders using voice signal. Karunaimathi et al. [11] implemented a method to analyze the fundamental frequency of the voice F0, which is the subjective pitch of the voice, and hence it plays a major role in clinical voice research. The main aim and objectives of this proposed method are given as follows: (i) to assess the lesions of the larynx and extract the acoustic features from the recorded voice signal using semi-automated method using PRAAT and Dr. Speech software; (ii) to extract the sound wave features using automated feature extraction method for the comparison of healthy controls and laryngeal disorders.

2 Methodology

2.1 Subjects

The study was carried out on ten healthy controls and ten laryngeal disorder patients who were attending as the outpatients in the audiology and speech-language pathology department, Faculty of Medical and Health Sciences, SRM Hospital and Research Center (Chennai, Tamil Nadu, India). The patients were diagnosed by performing laryngoscope and audiometry test for the confirmation of speech pathology. People who were healthy with a perceived good voice were included in the study. Patients with lung diseases, asthma, and heart diseases were excluded from the study. The Ethics Committee of SRM hospital approved the study. Informed consent was obtained from each patient.

2.2 Speech Signal Acquisition

The speech signal is recorded using a voice-recorder model proton BOOM 815 live sound wired microphone (NX Audio, Mumbai). The voice recording is done in a sound-attenuated (sound proof) room with the subject's persistent tone level. The condenser mike is placed 5 cm away from the subject so that the vocal sound production is not affected. The speech signals were sampled at 44.1 kHz frequency with 16-bit resolution. The subjects were solicited to phonate the sustained vowel // a// for the duration of 10–15 s to record the continuous voice assessment. The speech signals were analyzed using PRAAT, Dr. Speech, and MATLAB software.

2.3 Semi-automated Methods for Speech Signal Analysis

In PRAAT software version 6.022, the speech signal is recorded in two different modes such as mono sound recording mode and stereo sound recording mode. The operating frequency of voice signal recording ranges from 8 to 192 kHz, from which 44.1 kHz is selected as sampling frequency [12]. The voice signal is recorded and saved as wave file and can be visualized as the pitch curve, an intensity curve, a spectrogram, and formant tracks. From the entire signal, a segment is selected and the features such as pitch, jitter, shimmer, and HNR (harmonic-to-noise ratio) were extracted.

The voice signal is recorded continuously, and a segment is selected for analysis using the Dr. Speech software version 4 (Tiger DRS, Inc, Seattle, USA). From the selected segment, the voice features such as habitual F0, jitter, shimmer, F0 tremor, HNR, and SNR (signal-to-noise ratio) were extracted. The voice quality was estimated using the parameters such as hoarseness, harshness, and breathiness.

2.4 Automated Methods for Voice Feature Extraction

The recorded voice signal is selected and plotted in time domain using MATLAB software. The spectrogram is obtained for both healthy control and laryngeal disorder. Then, the spectral analysis is computed by short-time Fourier transform using Hanning window to overcome the signal discontinuities. If the obtained FFT is symmetric, then the first half of the signal is considered by eliminating the second-half signal. The parameters such as minimum, maximum and mean signal peaks, dynamic range (difference between the soft sound and loud sound), autocorrelation, were extracted. The algorithm was given below in flow chart as follows:

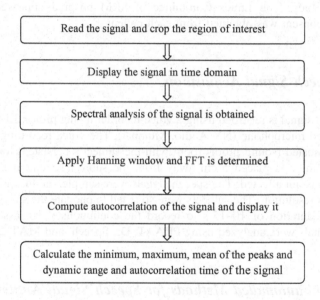

3 Results and Discussions

The acoustic features such as pitch, formant frequency, jitter, shimmer, HNR, and voice breaks are analyzed and the difference between the normal and abnormal values is studied using PRAAT software. Table 1 indicates the voice features extracted from PRAAT software for the total population ($N = 20$). Among all the features extracted, shimmer exhibits significant ($p < 0.01$) difference between normal and abnormal voice samples. The acoustic features such as pitch, formant frequency, jitter, shimmer, HNR, F0 tremor, SNR, harshness, hoarseness, and breathiness of the voice signal are analyzed using Dr. Speech software. Table 2 represents the features extracted from Dr. Speech software for the total population

$(N = 20)$. Among all the voice features extracted from the Dr. Speech software, formant F0 and jitter produced 79 and 65.6% difference between the normal and abnormal voice samples. Table 3 depicts the feature extracted from the automated method for the total population $(N = 20)$. From the automated analysis using MATLAB software, the parameters such as RMS value and autocorrelation provide the significant difference between the normal and abnormal voice samples.

The spectrogram of normal voice samples and abnormal voice samples is indicated in Figs. 1 and 2. Significant variation in the time period was observed in abnormal voice sample compared to normal. The autocorrelation of normal and abnormal voice sample was observed in Figs. 3 and 4.

Shilaskar and Ghatol [13] in their study extracted the statistical features such as jitter, HNR, formant frequencies $f1$, $f2$, and $f3$. They obtained the mean of jitter value using PRAAT software for normal and pathological cases as 0.05 and 0.21, respectively. The parameter HNR was found to be decreased in the pathological case compared to normal. Mishra et al. [14] in their study obtained the jitter mean % for normal and abnormal CVD as 1.35 and 0.86, respectively. They also found the shimmer mean % for normal and abnormal to be 8.058 and 4.98, respectively. Hence it was observed that the jitter and shimmer parameters exhibit decreased

Table 1 Voice features extracted from PRAAT software for the total population $(N = 20)$

S. No.	Features	Normal mean ± SD	Abnormal mean ± SD	% difference	P-value
1	Pitch (Hz)	180.25 ± 62.99	201.23 ± 67.9	10.42	0.478#
2	Jitter (%)	0.463 ± 0.377	0.64 ± 0.574	28.12	0.461#
3	Shimmer (dB)	0.384 ± 0.443	4.225 ± 3.24	90.91	0.0005**
4	HNR (dB)	20.34 ± 5.57	20.51 ± 6.29	0.82	0.886#

$*p < 0.05$ and $**p < 0.01$
$^{#}P > 0.05$

Table 2 Features extracted from Dr. Speech software for the total population $(N = 20)$

S. No.	Features	Normal mean ± SD	Abnormal mean ± SD	% difference	P-value
1	Pitch (Hz)	190.84 ± 51.54	201.23 ± 67.9	5.16	0.703#
2	Jitter (%)	0.229 ± 0.05	0.64 ± 0.574	65.62	0.037*
3	Shimmer (dB)	2.202 ± 0.5	4.225 ± 3.24	47.88	0.036*
4	HNR (dB)	23.85 ± 2.62	20.51 ± 6.29	−16.28	0.13#
5	F0 Tremor (Hz)	1.281 ± 0.30	6.192 ± 4.44	79.31	0.002**
6	SNR (dB)	22.323 ± 2.65	19.564 ± 6.05	−14.11	0.203#

$*p < 0.05$ and $**p < 0.01$
$^{#}P > 0.05$

Table 3 Features extracted from automated method for the total population ($N = 20$)

S. No.	Parameters	Normal mean ± SD	Abnormal mean ± SD	% difference	P-value
1	Maximum value	0.288 ± 0.49	0.253 ± 0.28	−13.83	0.081[#]
2	Minimum value	−0.65 ± 0.33	−0.52 ± 0.26	−25	0.025*
3	Mean	−0.024 ± 0.007	−0.46 ± 0.035	−94.7	0.3[#]
4	RMS	0.186 ± 0.11	0.46 ± 0.035	60.86	0.001**
5	Dynamic range (dB)	82.98 ± 5.25	74.47 ± 8.65	−11.4	0.16[#]
6	Auto correlation time (s)	5.172 ± 1.55	8.925 ± 11.74	42.05	0.159[#]

*$p < 0.05$ and **$p < 0.01$
[#]$P > 0.05$

Fig. 1 Spectrogram of a normal voice sample

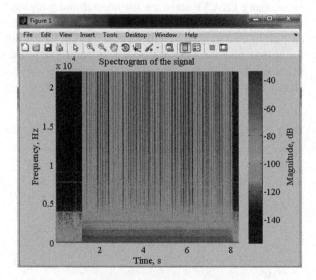

Fig. 2 Spectrogram of an abnormal voice sample

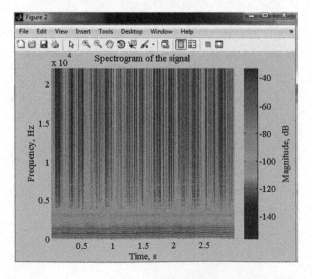

Fig. 3 Autocorrelation of a normal voice sample

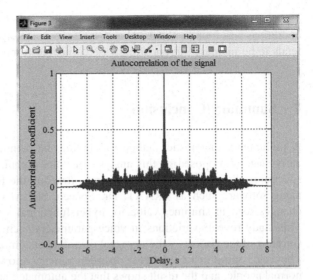

Fig. 4 Autocorrelation of an abnormal voice sample

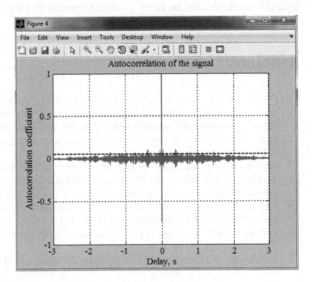

value in abnormal cases compared to normal. Rusz et al. [15] predicted the mean value of jitter for normal and abnormal pathological case as 0.33 and 0.91, respectively, and obtained the HNR (db) for a normal and abnormal conditions as 22.55 and 14.05, respectively. In our study, the jitter mean % value for normal and abnormal voice pathologies was found to be 0.46 and 0.64, respectively. Lopes et al. [16] compared the speech signal for normal and dysphonia patients and obtained the jitter value for normal and dysphonia patients as 0.193 and 2.44. Also, they obtained the shimmer value for normal and dysphonia patients as 4.03 and

12.95, respectively. In our study, the mean shimmer % for the normal and abnormal conditions was found to be 0.384 and 4.225, respectively. Also, the parameter HNR tends to be increased in abnormal case compared to normal in our study.

4 Summary/Conclusion

A preliminary diagnostic method for finding the pathology of vocal cords in the early stage is proposed by this analysis of voice signal. This model is based on a spectral analysis of the voice signal and extracting the features of the voice signal using software such as PRAAT, Dr. Speech, and MATLAB. There are distinct changes seen in shimmer value for hyper-functional voice disorders particularly. This study reveals variations in voice parameters such as spectrogram, jitter, and shimmer of the patients with laryngeal disorders in comparison with normal people. The voice samples of laryngeal disorder patients show clear deviation from the normal people, and the result shows that the automated analysis and speech spectral analysis methods can be used to detect lesions in larynx non-invasively.

References

1. Carvalho RT S, Cavalcante C, and Cortez PC (2011) Wavelet transform and artificial neural networks applied to voice disorders identification. Proceedings of 3rd world conference on nature and biologically inspired and computing, pp 371–376.
2. http://www.nidcd.nih.gov/health/what is voice-speech language, accessed on 28th October 2016.
3. http://www.languageindia.com/disability in India, accessed on 28th October 2016.
4. Teixeira P, and Fernandez PO (2015) Acoustic analysis of vocal dysphonia, procedia computer science, vol 64, pp 466–473.
5. Teixeira JP, Oliveira C, and Lopes C (2013) Voice acoustic analysis-jitter, shimmer and HNR parameters. Procedia Technology, vol 9, pp 1112–1122.
6. Teixeira JP, and Gonclaves A (2014) Accuracy of jitter and shimmer measurements, procedia Technology, vol 16, pp 1190–1199.
7. Teixeira JP, and Fernandez PO (2014) Jitter, Shimmer and HNR classification within gender, tones, and vowels in healthy voices, Procedia technology, vol 16, pp 1228–1237.
8. Dixit, Mittal V, and Sharma. Y (2014) Voice parameter analysis for disease detection, IOSR Journal of Electronics and communication Engineering, vol 9, pp 48–55.
9. Gelzinis A, Verikas A, and Bacauskiene M (2008)Automated speech analysis applied to laryngeal disease categorization, computer methods programs biomed vol 91, pp. 36–47.
10. Vikram CM, and Umarani (2013) Pathological Voice Analysis to Detect Neurological Disorders Using MFCC and SVM. International Journal of Advanced Electrical and Electronics Engineering, (IJAEEE), vol 2, pp 87–91.
11. Karunaimathi V, Gladis D, and Balakrishnan D (2015) An analogy of F0 estimation algorithms using sustained vowels, proceedings of third International symposium on women in computing and Informatics, pp 217–221.
12. Boersm P and Weenink Z (2003) PRAAT: Doing Phonetics by Computer. http://www.fon.hum.uva.nl/praat.

13. Shilaskar S, and Ghatol A (2014) Feature Enhancement for Classifier Optimization and Dimensionality Reduction Speech Pathology Detection – A case study, India conference (INDICON), IEEE, 2014 Annual IEEE, pp 1–6.
14. Mishra S., Balakrishnan S., and Babitha M (2016) Coronary Heart Disease Detection from Variation of Speech and Voice, Proceedings of 2nd International Conference on Intelligent Computing and Applications, Advances in Intelligent Systems and Computing pp 467, DOI 10.1007/978-981-10-1645-5_49.
15. Rusz J, Cmejla R, Ruzickova H, Klempir J, Majerova V, Picmausova J, Roth J, and Ruzicka E (2011) Acoustic analysis of voice and speech characteristics in early untreated Parkinson disease, Proceedings of 7th International Workshop: Models and Analysis of Vocal Emissions for Biomedical Applications, pp 181–184.
16. Lopes LW, Cavalcante D P, and Costa PO (2014) Severity of voice disorders: integration of perceptual and acoustic data in dysphonic patients, Codas, Vol 26, pp 382–388.

Analysis of Job Scheduling Algorithms and Studying Dynamic Job Ordering to Optimize MapReduce

Ahmed Qasim Mohammed and Rajesh Bharati

Abstract As there was a big rise in the Big Data field, Hadoop became one of the most used platforms in many applications like clinical data analysis, Facebook, Amazon, in which Big Data processing and utilization is required. One of the most important features that made Hadoop, one of the most popular platform, is adopting MapReduce, which made large changes in the market by processing huge amount of data in parallel technique by distributing data across multiple TaskTracker nodes and each node splits data by Map slots and shrinks the output of Map slots (key/value) by Reduce slots. However, different algorithms are used for job scheduling in MapReduce which is used to improve resource utilization, job allocation, and minimizing processing time. However, research is still underway to propose optimal method to improve MapReduce model, and there are still several major drawbacks that are still not well studied. In this paper, we studied and analyzed different job scheduling, job ordering algorithms, and dynamic slot configuration. The study discussed the most popular and efficient systems design to find the most efficient improvements for MapReduce and review the corresponding solutions. In our proposed method, we are applying classification algorithm which is going to classify job into highest utilization and poor utilization. After that, the highest utilization job will be forwarded to PRISM algorithm, which will schedule into phase level as there are variations in processing time and resource requirement in each phase, for different kinds of jobs along with the application of dynamic slot configuration, which helps to improve resources utilization and reduce time.

Keywords MapReduce · Hadoop · Scheduling algorithm · Job ordering PRISM · Resource utilization · Dynamic slot

A.Q. Mohammed (✉) · R. Bharati
Dr. D.Y. Patil Institute of Engineering & Technology, Pimpri, Pune 411018, India
e-mail: ahmed.altaj@yahoo.com

R. Bharati
e-mail: rdbharati@gmail.com

© Springer Nature Singapore Pte Ltd. 2018
S.S. Dash et al. (eds.), *International Conference on Intelligent Computing and Applications*, Advances in Intelligent Systems and Computing 632,
https://doi.org/10.1007/978-981-10-5520-1_32

1 Introduction and Related Work

This era has witnessed a huge amount of data all around, where research survey says that 90% of the data has been generated in the last five years. This data returns to cloud as it is the most efficient storage for data and computation.

Big Data refers to different types of huge datasets that require high-speed computing to process massive amount of data with minimum possible time. With the enormous evaluation of sources that generate Big Data, a lot of work is required for analyzing this huge data and storing it. Huge data means more challenges to operate or perform such operations. And here arise the needs for an efficient framework to process this data. One of the most popular frameworks is Hadoop [1] which employs MapReduce [2] model, and nowadays, it has become the most popular in processing Big Data because it is an open source model and free, and, there is Dryad, [1] an another, computing project proposed by Microsoft researchers.

As there is a need for large distributed processors to enhance the performance of processing and storing this huge amount of data, Hadoop came to life in 2006 with the help of Yahoo after few publications written by smart developers in 2004, one of the reason that Hadoop became a model for intensive processing is MapReduce [2]—a technique to process any Big Data, as this has been an open source platform made by AFS, i.e., HDFS [3], an application which stores data in a distrbuted system.

Hadoop has two main programing models in which one works as a memory, where data is stored, and is known as HDFS (Hadoop distributed file system). MapReduce is the heart of Hadoop ecosystem which contains two main phases: Mapper in which Map Task gets processed and the other is Reducer which shrinks intermediate output (key/value) of the mapper phase. Figure 1 shows Hadoop processing diagram.

Both Mapper and Reducer have sub-phases [4] (Map, Merge, Shuffle, Sort, Reduce), for achieving efficient result, and programmers have to take care of tasks

Fig. 1 Hadoop processing diagram

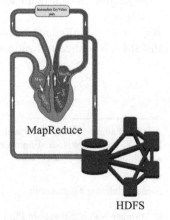

assigning jobs in each phase with understanding the relation between phases and the data dependency between phases.

There are two constraints [5–7] in MapReduce: First is Map Task should be processed before Reduce task and output of Mapper phase is the intermediate key/value pairs that will be the input for Reducer; second Map tasks can be processed only in Map slots and the same for Reduce tasks which must be processed only in Phase slots.

Sometimes Mapper output becomes larger than the original input so to overcome this, there is combiner who is known as "Mini-Reducer" [8] which will combine the key value pairs and forward to the reducer phase and the output will be stored in HDFS.

Now, we go back to our main focus, MapReduce, which is providing all the features of parallel [9] programming which is user-friendly.

Scheduling [10–12] is a technique of allocating system resources for coming tasks by assigning task to the free slots. The main scheduling algorithms used by Hadoop are FIFO which is a default scheduler for Hadoop v1, and Fair scheduler, which is a default for Hadoop v2. In Job Scheduling Algorithms section, we explain the workings of both the schedulers. Also, many adaptive schedulers are proposed as in the literature [11, 12].

JobTracker takes responsibility for scheduling and allocating free slots in TaskTracker for the coming jobs [10], and communication between JobTracker and TaskTracker is periodically through heartbeat message which contains all details of processing task status in TaskTracker and free slots.

Each TaskTracker node consists a number of slots where there are Mapper slots and Reducer slots which are of a fixed number by default in Hadoop. Each slot typically contains a CPU core and memory management.

2 Overview of MapReduce

Many classical and traditional data processing and data management softwares had an issue with Big Data analytics for the complexity and large volume of the datasets that it included. So the need for parallel process became a necessity and important to processing a huge volume of data in an exceedingly timely manner. MapReduce was one of the best solutions especially after it became part of Hadoop, now MapReduce is a highly known model used for processing huge quantity of data on a cluster of computers.

MapReduce model is originally created for distributed system, and it is an open source implementation. The evolution of Hadoop as a complete infrastructure was made so that MapReduce could provide more simplicity to the user for processing SQL and NoSQL dataset, and the examples of these applications are Pig languages and Hive. On the other hand, there is Zookeeper, which is used for coordination service and HBase for distributed table store.

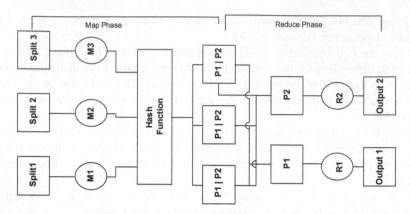

Fig. 2 Working of MapReduce

MapReduce considers the following features: minimal abstraction, automatic parallel computing programming, and fault tolerance via duplication to improve performance and takes care about 3Vs.

Typically, MapReduce model has two main phases, Mapper and Reducer, where Mapper works to split and map input data and, in the end, it will merge and generate intermediate output which is input for Reducer.

Figure 2 shows the basic working of MapReduce.

Phases of MapReduce

MapReduce mainly performs two tasks, Map and Reduce, which represent phases of MapReduce. Map phase contains two sub-phases: Map and Merge.

A. **Map**:

First phase of Map where data chunks are first processed; chunks or data blocks are coming from HDFS; size of these data blocks is 64 or 128 MB; in this phase, mapping functions start and key/value pairs are generated, which it gets collected into Buffer until this buffer gets full, and then, it starts writing on the local disk.

B. **Merge**:

Second Phase of Map gets input from data already recorded into local disk and starts grouping these key/value pairs and sends it to another buffer to be ready for Reduce processing (Fig. 3).

Reduce phase contains three sub-phases:

A. **Shuffle**:

It is nothing but only a phase that fetches output of Map phase and makes it ready for the second phase of Reduce.

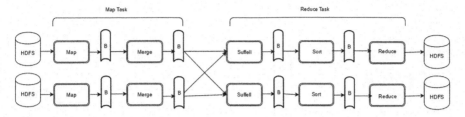

Fig. 3 Phases of MapReduce

B. **Sort**:
This phase is just a step which uses to save Reduce phase time, where each key/value pair gets sorted in an orderly way. It is like a secretary that sorts files for the manager for signing.

C. **Reduce**:
Last step of MapReduce where data gets reduced and output will store in HDFS as it shows in Fig. 1.

3 Job Scheduling Algorithms

Multiple users issue the jobs on Hadoop distributed network, and to control uses of recourses, there are many scheduling algorithms would be employed at JobTracker, in case of MapReduce v1, and ResourceManager and JobTracker in case of MapReduce v2.

Many algorithms proposed to improve performance of job scheduling in MapReduce model, after which we study default scheduler for MapReduce v1 and MapReduce v2. Next, we took a look at new scheduling algorithms that deal with phase level instead of task level.

Taking care about data dependency is an issue that causes time delay. To study each of the phases and understanding its requirement, it is really important because each phase requires different resources and consumes different time depending on type of task assigned to it.

3.1 FIFO (First-In First-Out) Scheduler

The early version of Hadoop employed this algorithm in the simplest form. Fig. 4 shows the technique used in this algorithm for assigning jobs, where the first job that comes will have priority to use all resources and get serviced first.

This algorithm suffers from issues of small jobs that will wait for an unreasonable time if resources are already processing larger jobs.

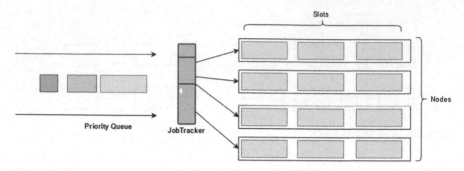

Fig. 4 FIFO scheduler

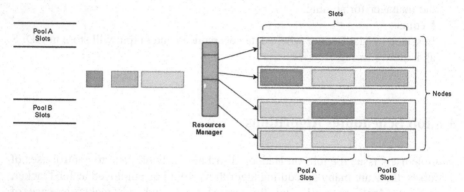

Fig. 5 Fair scheduler

3.2 Fair Scheduler

Fair scheduler is a default scheduling algorithm for MapReduce v2, and this algorithm is conceptually very similar to capacity scheduler with a minor difference.

In this scheduler, ResourceManager will pick up jobs from pool and assign these to respective slots which are already assigned to process jobs for a specific pool.

In case any job comes while processing other jobs, ResourceManager will pick up this job and assign it to a free slot (Fig. 5).

3.3 PRISM Scheduling Algorithm [4]

PRISM is an algorithm proposed to use phase-level scheduler instead of using task-level scheduler to maximize resource utilization and decrease processing time by assigning each job to one of the five phases that we have explained before.

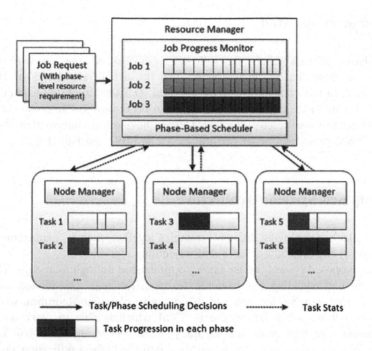

Fig. 6 Architecture of PRISM [4]

However, PRISM still has an issue with classifying tasks and understanding the exact requirement to process tasks in minimal time.

For that, we have to merge one of the classification techniques such as Starfish [13] for profiling jobs. In our proposed system, we are going to use SVM algorithm for classification.

In this scheduler, scheduling will be done into phase level. This will improve the average job running time, as compared to task-level resource-aware schedulers. Figure 6 shows architecture of PRISM algorithm for MapReduce workload.

PRISM consists of the following components:

1. Phase-based scheduler: It is used at the master node; phase-level scheduling algorithm is used by phase-based scheduler. Phase-level scheduling algorithm is designed to achieve the highest resource utilization and job performance.
2. Local node managers: This module is responsible for managing phase transitions with scheduler by sending permission request with typical heartbeat message, each which goes in 3 s from TaskTracker to JobTracker and waits for scheduler decision. This component is located inside the TaskTracker.
3. Job progress monitor: This component is responsible for capturing phase-level progress information [3].

4   Comparative Study

By studying different papers, the following graph shows the response time of PRIMS algorithm that is always better than the default schedulers of Hadoop MapReduce. In this comparative study we used 90 Tasks; the FIFO response time was 152 s, while Fair scheduler was lesser than FIFO with 3 s. In comparison, the PRISM response was exactly 27 s less. Despite having little information about the tasks, PRISM provides a better performance than FIFO and Fair (Fig. 7).

5   Proposed System Architecture

There is a lack in deploying different strategies together, such as, management [14] of data flow, classification of tasks and scheduling algorithms.

In our proposed system, we are taking optimization through four steps. The first step is by understanding our resource features, job requirement, and overload details. In this step, we are going to use adaptive classifier algorithm, where the algorithm that we are going to use is SVM which is able to work on online environment with high prediction accuracy; where, SVM algorithm will work to classify jobs according to the collected data into the highest utilization tasks and poor utilization tasks.

The second step is that we are going to arrange jobs in an orderly manner with the help of Johnson's rule [15].

The third step involves scheduling jobs to exact phase and control progress of job processing by using Prism algorithm.

Finally, slot configuration will be done dynamically and will provide most accurate slot configuration for job under process. Using second and fourth steps to minimize Makespan and total completion time where Makespan time is the time required to finish all jobs while time completion time is summation of time required to finish each job, there is a drawback between both, and utmost care has to be taken (Fig. 8).

Fig. 7 Response time graph

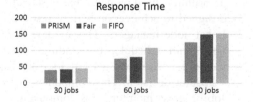

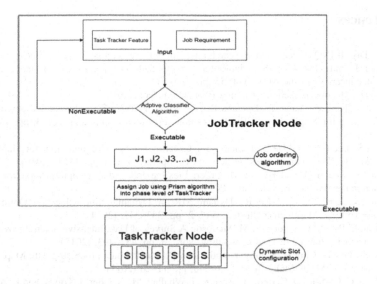

Fig. 8 Proposed system

6 Conclusion

After analyzing and studying different papers, we found that PRISM, always gives better results than Fair scheduler and FIFO in CPU utilization, except, near the end of execution. In the locality of tasks, PRISM completes lower task locality first, while it delivers better performance than FIFO and Fair schedulers. According to our analysis, PRISM reduces execution time by 24 s than Fair scheduler. However, we also found using dynamic job ordering, before scheduling jobs into phases By PRISM algorithm, will definitely minimize total completion time.

From the literature and the previous proposed work, we got an idea to use several optimizations together to improve performance of Hadoop MapReduce.

For optimizing Hadoop MapReduce performance, we have to take care of the following techniques, tuning system parameters in the best way. Slot configuration must be dynamic as it affects a lot on the performance of the system. Dynamic configuration of MapReduce slots improves resource utilization and follows management rules to reduce Makespan time.

In the end, we suggest to use dynamic slot configuration with efficient scheduling algorithms and follow management rules to sort jobs. All this will lead to a better performance.

References

1. The DryadLINQ project. http://research.microsoft.com/research/sv/DryadLINQ/.
2. Dean J. and Ghemawat S.: MapReduce: Simplified data processing on large clusters, Communications of the ACM, vol. 55, pp. 107–113, (2008).
3. C. Lam: Hadoop in Action, Manning Publications Co, (2010).
4. Zhang, Zhani M., Yang Y., Boutaba R. and Wong: PRISM: Fine-Grained Resource-Aware Scheduling for MapReduce. IEEE Transactions on Cloud Computing. vol. 3, pp. 182–194, (2015).
5. Tang S., Lee B.S., and He B.: Dynamic Job Ordering and Slot Configurations for MapReduce Workloads. IEEE Transactions on Services Computing. vol. 9, pp. 4–17, (2016).
6. Tang S., Lee B.S., and He B.: MROrder: Flexible Job Ordering Optimization for Online MapReduce Workloads, vol. 8097, PP. 291–304, (2013).
7. Tang S., Lee B.S., and He B.: DynamicMR: A Dynamic Slot Allocation Optimization Framework for MapReduce Clusters. Vol. 2, pp. 333–347, (2014).
8. Ji Liu, E. Pacitti, P. Valduriez, M. Mattoso.: A Survey of Data-Intensive Scientific Workflow Management, Journal of Grid Computing, vol. 13, pp. 457–493, (2015).
9. Lee K., Lee Y., Choi H., Chung Y. and Moon B.: Parallel data processing with MapReduce: A survey. ACM SIGMOD Record, vol. 40, pp: 11–20 (2011).
10. J. Polo, C. Castillo, D. Carrera, Y. Becerra, I. Whalley, M. Steinder, J. Torres, and E. Ayguad´ e.: Resource-Aware Adaptive Scheduling for MapReduce Clusters, ACM/IFIP/USENIX Middleware, pp. 187–207, (2011).
11. J. S. Manjaly and V. S. Chooralil.: Tasktracker aware scheduling for Hadoop MapReduce, Third International Conference on Advances in Computing and Communications, pp. 278–281, (2013).
12. Wolf J., Rajan D., HildrumK, Kumar V., Parekh S., Wu K., and Balmin A.: FLEX: a slot allocation scheduling optimizer for MapReduce workloads, Middleware '10 Proceedings of the ACM/IFIP/USENIX 11th International Conference on Middleware, pp. 1–20, (2010).
13. H. Herodotou, H. Lim, G. Luo, N. Borisov, L. Dong, F. Cetin,and S. Babu.: Starfish: A self-tuning system for big data analytics, Conference on Innovative Data Systems Research (CIDR11), (2011).
14. J. Polo, Y. Becerra, D. Carrera, M. Steinder.: Deadline-Based MapReduce Workload Management, IEEE Transactions on Network and Service Management., vol. 10, pp. 231–244, (2013).
15. Johnson S.: Optimal Two-and Three-Stage Production Schedules with Setup Times Included, Naval Res. Log. Quart, vol. 1, pp. 61–68, (1954).
16. Hadoop. http://hadoop.apache.org.

Performance Analysis of QoS Against Packets Size in WSN

**Dattatray Waghole, Vivek S. Deshpande, A.B. Bagwan,
Shahistha Pirjade, Sayali Kumbhar, Apurva Mali
and Gitanjalee Chinchole**

Abstract In wireless sensor network (WSN), a number of different sensors are put in environmental field. The sensors capture the climatic conditions and changes. All captured data are sent to the destination node. The parameters of WSN are energy efficiency, congestion, throughput, scalability, latency, and delay. Energy efficiency and congestion are the challenging issues of a network. For efficient performance of the network, Mac and routing protocols are used. This paper studies MDSA Mac protocol, ABSD algorithm, TRMAC, and on-the-fly (OTF) bandwidth reservation algorithm. There are also certain techniques introduced, such as cross-layer operational model and CSMA protocol with polling technique. It provides advantage to increase the energy efficiency, improve network performance, and schedule bandwidth according to the traffic load. This paper is useful for beginners for studying and achieving the various QoS factors using different Mac protocols and techniques to improve the network performance.

Keywords WSN · QoS · Throughput · MAC · Energy efficiency
Latency · Packet size · End-to-end delay · Energy efficiency

1 Introduction

In Sensor network, a numbers of sensors are put for observing the climatic changes and alerting the control system. The sensors put in sensor area capture the physical conditions like temperature, pressure, humidity from environment. It performs

---

D. Waghole (✉) · V.S. Deshpande · S. Pirjade · S. Kumbhar · A. Mali · G. Chinchole
VIT, Pune, India
e-mail: dattawaghole10@gmail.com

V.S. Deshpande
e-mail: vsd.deshpande@gmail.com

D. Waghole · A.B. Bagwan · S. Pirjade · S. Kumbhar · A. Mali · G. Chinchole
Kalinga University, Raipur, India
e-mail: aliakbar.bagwan@gmail.com

© Springer Nature Singapore Pte Ltd. 2018
S.S. Dash et al. (eds.), *International Conference on Intelligent Computing
and Applications*, Advances in Intelligent Systems and Computing 632,
https://doi.org/10.1007/978-981-10-5520-1_33

analog to digital processing on the data using ADC. In sensor node, it is difficult to replace the battery. Every sensor nodes have a limited energy resource. In wireless sensor networks, heavy traffic and congestion are the main problem of energy consumption in the network.

Mac protocol has an important role to carry out communication between the sensor nodes. It is used for packet transmission and receiving between the source nodes and destination nodes. Some MAC protocols are working with active and sleep mechanisms. WSN focuses on the scenario when a traffic load increases in the network. TRMAC is energy-efficient mac protocol; it reduces the wastage of energy during the communication. It is used in less data rate and low duty cycle [1]. To reduce energy consumption and increase throughput of the network, Cd-mac is used. Cd-mac is an energy-efficient and robust duty-cycled Mac for various WSN applications. This improves throughput and completely avoids data collision with low overhead [2]. In this paper, ABSD algorithm is also studied for minimizing network contention and for increasing throughput. The ABSD is Adaptive Beacon Order, consists of (order) super-frame as well as cycles (duty). The main contribution is to suggest an adaptive packet transmission protocol to hold up the EE for uncertain traffic load. Two categories of energy and QoS packet transmission algorithm are incoming traffic load estimation and node's queue state estimation [3]. In WSN collision occurs due to multiple sensors transmitting data packets to a node simultaneously. This issue can be resolved by certain Mac protocols. One of them is the MDSA Mac protocol. It is a protocol which takes responsibility of scheduling each node in a network. The MDSA Mac protocol uses three dimensions: time, code, and frequency. Increasing throughput and decreasing the collision rate are the main design objectives of MDSA protocol [4]. Over the network, traffic is increased due to transmission and receiving of packet. The priority Mac also manages the different types of congestion and network load. The priority Mac has been replaced by default CSMA/CA Mac protocol [5]. A new Mac protocol is also studied in the following, on the basis of perceived data and the spatial correlation between the connected nodes of the network. This protocol provides minimum distortion and maximum energy efficiency, when compared with Smac and Cc-mac [6]. Mac protocols play an important role in managing the entire network. Smac and IEEE 802.11 are Mac protocols studied in this paper. In comparison of Smac and IEEE 802.11, Smac gives better results for certain parameters [7]. In slotted Csma, collision of packet is occurred. So for reducing collision and increasing the packet transmission Sift protocol is designed. Slotted Csma protocol is combined with polling technique and is used for analysis of different QoS like energy consumption, throughput, and fairness [8]. For a good performance of the network, maximum energy efficiency is required. For which various techniques and models are introduced. One of them is network operation cross model for avoiding extra usages of energy, which is studied in the following. The model consists of position of sensor nodes, packet routing, MAC as well as (physical) layers. To reduce more energy consumption in the network, control packets are broadcasted over the network. It is also helpful to reduce channel occupancy in the network [9]. To make a network less power consumable, harvested energy resources can also be used. For which a power manager (PM) is embedded into the sensor. The power management will provide buffering and usage of the energy according to need hence, improving the life performance of the network and creating the most energy-efficient nodes for a network [10]. One of the protocol named as X-mac protocol is short preamble Mac protocol. Receiver Initiates X-mac (Rix-mac) protocol gives better result for QoS like delay, throughput and energy consumption than X-mac protocol [11]. For increasing the speed of transmission of the data through the network, maximum throughput is required. Throughput is a rate of packet sending through single communication channel. Once throughput is improved, parameters like latency and delay are also improved. So as to provide high throughput for the network, multicasting over Mac layer is done. This provides various limitations. To overcome the limitations of multicasting over Mac layer, HiMac protocol was introduced. It consists of the Ucf and Unf mechanism, which provides better efficient and scalable performance of the network [12]. Maximum throughput gives better network performance. The aim of WSN is that the network should give high throughput for an efficient network performance. High level throughput is shown in this paper because author uses cross-layer technique in this paper. In multi-hop mesh-network, cross-layer technique is used for increase throughput of the network. Expected multicast transmission count (EMTX.) combines the things of link quality perception and wireless broadcast advantage. It also reduces the transmission overhead [13]. The combination of WPAN and WLAN is studied in the following. It is found that the WLAN scenario shows better results than WPAN. WLAN is further combined with ZigBee to give maximum throughput [14]. In smart industrial appliances, there is always a need to provide efficient and reliable communication. Primitively various protocols were used. Recently introduced is the on-the-fly (OTF) bandwidth reservation algorithm, which uses the 6TiSCH architecture. OTF is used for scheduling bandwidth over a distributed network and allocating the maximum bandwidth within the selected channel. This provides energy efficient and reliable communication over the network [15, 16].

In Fig. 1, the last level node that is source node will sense information, send the data to the second level node and from the level 2, it will go to the level one (1) nodes. The level one (1) sensors send data to the sink. This is a similar mechanism which takes place in a wireless sensor network with the help of protocols. Mac

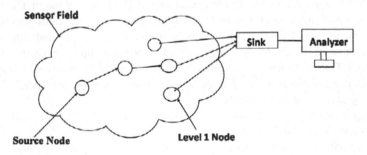

Fig. 1 Overview of sensor network

and routing protocol are important to carry out communication between two nodes in a wireless network. MAC and routing protocols are used for communication between the sensors in the area. The routing protocol is used for a searching small distance between sender to the receiver node. Before creating communication link, routing protocol decides through which path the source node must transmit information to receiver node and then only communication will take place [1, 4].

2 Literature Survey

Author proposes the traffic adaptive congestion control mechanism which effects on network throughput. Low data rate is main reason of low throughput and maximum latency in the sensor network. But in case of event-driven scenario, traffic may be rapidly increased in the network. So due to the low data rate, average network throughput will be decreased. So to solve this problem authors implement TRMAC. This mac is working with low duty cycle mechanism for communication. In this technique, nodes periodically change the states from sleep to active and vice versa.

TRMAC is suitable for less data rate and low duty cycle scenarios. The protocol has different operating states like synchronised and unsynchronised states. At the initial stage, transmitter node works in unsynchronized states and will send the combine data packets, then waits for response from receiver node. Transmitter sensor does same procedure again and again until receiving response from node. After getting packets from sender nodes, it sends acknowledgement to transmitter node. Traffic-based adaptive duty cycle as well as burst data transmission are the two techniques used in this paper. The protocol maximizes the throughput and energy efficiency. It is also used for scalability of a network. The duty cycle adaption mechanism helps to achieve energy efficiency. Techniques used in this paper helps to reduce network congestion—traffic and reduce loss of packet at the time of transmission [1]. Reliable data delivery as well as energy-efficient packet transmission are two important reasons for the use of low duty-cycle technique. In case of event-driven communication, packets are rapidly increased in the network. It suffers from packet collision specifically under busy traffic. We started contention detectable mac (Cd-Mac), robust duty-cycled Mac, and an energy efficient for normal WSN application. The implementations of Cd-Mac is done in TinyOS and classify the execution on an indoor tested with single-hop and multi-hop network. The outcome shows Cd-Mac can extremely improve throughput and can correlate with the state-of-the-art receiver-initiated Mac protocol under busy traffic load. Traditional TDMA reservation approaches holds certain limitations. Tdma-based reservation approaches are hard to conclude. With very low overhead, Cd-Mac effectively avoids data collision [2].

Energy efficiency is a big issue in wireless sensor networks. Mac and physical layer protocols are used for to improve the energy efficiency. Main purpose of ABSD algorithm is to minimize network contention, improve throughput, and provide energy efficiency. Mac controls the transmission of packet, and it also plays important role in increasing EE. Using Mac, the sensor nodes could be the following modes like receiving, transmitting, idle listening, sleeping. The life span of WSN could be extended by minimizing idle listening and increasing sleeping time. 802.15.4 i.e. ZigBee protocol is used for low power communication. ZigBee (802.15.4) protocol is operates with low power and low data rate in network. The ABSD is Adaptive Beacon order, super-frame order as well as duty cycle. The main contribution is to raise the EE to develop an adaptive packet transmission protocol for changing traffic load. It also keeps the QoS requirement. The ABSD algorithm tracks queue status of node, burst (busy) traffic load, and make IEEE 802.15.4 super-frame parameter of sensor nodes. IEEE 802.15.4 supports small area communication. It can operate in non-beacon enabled and beacon enabled modes. Low duty cycle and use of guaranteed time slot (GTS) are allowed in beacon enabled mode. For Low duty cycle & GTS are unable to use in Non - beacon mode. By adaptive DC mechanism, expansion of sleeping time can be managed. Three methods to obtain values are Fixing BO while varying SO, Varying BO while fixing SO, and Varying BO while varying SO. The studied energy and QoS-aware packet transmission algorithm gives incoming traffic load estimation. By counting the number of incoming packets, a PAN evaluates the traffic flow because of the variation of incoming traffic load. Also it gives node's queue state estimation. On the basis of the values of BI (Beacon Interval) and DC, the energy consumption is built. ABSD algorithm offers lower energy consumption, while keeping higher throughput, lower end-to-end delay, lower packet loss [3]. The study of multi-dimensional slotted Aloha (MDSA) states that it is a combining advantage of TDMA, FDMA, and CDMA. To overcome the drawback of Mac protocol that is to overcome collision rate and to increase the throughput of slotted Aloha protocol is the target of a paper. First protocol (MAC) is Pure Aloha. To overcome the disadvantage of Pure Aloha, slotted Aloha came into existence. The disadvantage of Pure Aloha is, it does not use 100% of channel capacity. Second, the nodes do not check whether the channel is idle or busy before sending the packets. Third, it uses 18% of total resources All the above disadvantage has overcome in slotted Aloha. Slotted Aloha is simple and low power consumption, there is no idle listening or overhearing in such protocols, Slotted Aloha Mac protocol can be adopted in such a way that the throughput can be increased and collision rates can be decreased efficiently. Slotted Aloha uses one dimension, i.e., time. In the proposed system, MDSA utilizes the slotted Aloha with 3 dimensions: CDMA, TDMA and FDMA. In MDSA, the base station takes responsibility to schedule each node in a network. First, the scheduling process can start with broadcasting the total number of slots. That is, nodes can randomly choose their data transmission. Second, every node in a network randomly chooses one reservation time slots, code set, and frequency band. Third, the data transmission process where every node sends its packet at the chosen reservation time slot to a central node. Nodes which have the same time, code, and frequency packet will collide and retransmit. Nodes which have a different time, code, and frequency have the opportunity for successful transmission. A packet with same time and code will also be transferred; the central node distinguishes between the packets. The MDSA protocol is the conspicuously effective protocol for short data packets and delays tolerant wireless communication system such as WSN [4].

3 Performance Analysis

Number of sensor nodes are put using uniform random topology in NS-2. Node densities (number of nodes) are varying from 15 to 90 nodes. Packet size in bytes also vary from 50 to 250 bytes. Two types of protocols are used for communication (Csma, i.e., 802.11 Mac) and finding shortest path, i.e., AODV (Ad hoc on-demand Distance Vector Routing Protocol). Sensor nodes are deployed in 1000 m * 1000 m area. Results are taken for varying different node density and packet size with various QoS of wireless sensor nodes.

Figure 2 indicates that average PDR for 15 nodes is constant for varying node density as well as packet size. Average PDR for 30 nodes is slightly less as compare to 15 nodes. Behavior of 60 nodes PDR shows decreasing and increasing order at one threshold and threshold value of 30 nodes. Average PDR for 90 nodes is very less and slightly decreasing when change node density 15–90 and packet size 50–250 bytes.

As node density from 15 to 90 nodes with packet size 50 to 250 nodes, average PDR also drastically decreases near about 40–45%. When number of nodes are more in network then, average PDR will be decreases due to the heavy traffic near the destination node.

For node density 15 and 30 nodes, average PDR is constant and drastically better as compare to other node density. Node density 50 and 60 are behaving like increasing and decreasing order in the graph. It gives average PDR for different packet size.

Average (PLR) for a packet size is shown in Fig. 3. Average PLR for 15 nodes is slightly constant when node density as well as packet size increases. For 30 nodes it is changing in increasing and decreasing order but it is slightly more than 15 nodes. In case of 15 node density average PLR drastically increasing for various packet size for 50–250 bytes. Average PLR for 60 nodes is initially less for 50 packet size. But drastically increases when it varies from 15 to 30 nodes. After threshold of 30

Fig. 2 Average PDR for of packet size

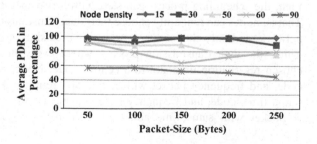

Fig. 3 Average PLR for
packet size

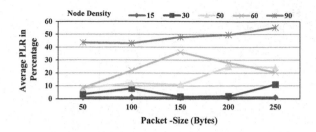

Fig. 4 Average end-to-end
delay for packet size

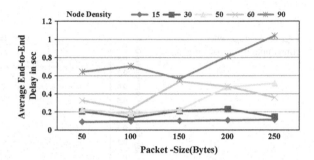

nodes average PLR for 30–90 nodes drastically decreases for varying size of
packets 50–250 bytes. In case of 90 nodes, average PLR is high as compare to other
node density and it increases 20% when packet size is from 50 to 250 bytes. Due to
heavy traffic network and congestion, average PLR will increase. Many times due
to simultaneously packet transmission in the same communication channel, collision will increase so average PLR also increases. So, graph shows that average PLR
for 15 nodes is very less and for 90 nodes is drastically high in the network.

Figure 4 indicates that average end-to-end delay for 90 nodes is more when
changing packet size from 50 to 250 packet size. Average end-to-end delay for
15–30 nodes near about same and less where packet size is varying from 50 to 250
packets per second.

For 60 nodes average delay (latency) is average as compare to 90 and 15 nodes.
But is behaving like increasing and decreasing order at threshold point 100 and 150
packet size. In case of 90 nodes due to the maximum nodes deployed in the sensor
area. Maximum data packets are generated and continuously transmitted to the
communication channel. So, maximum number of packet loss, due to the heavy
congestion in the network and less communication channel capacity the average
E-E delay increases.

Figure 5 indicate that the Avg throughout for 60 nodes drastically increases
when varying packet size 150–250 bytes. Graph shows that average throughput
drastically increases as increasing node density from 15 to 90 nodes. Maximum
number of packets transmitted through channels and due to full channel utilization
average throughput also increases. For less node density use of communication
channel is less due to the less packets are created by the sensor nodes. But when

Fig. 5 Average through put
for packet size

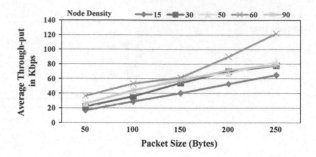

Fig. 6 Average energy
consumption for packet size

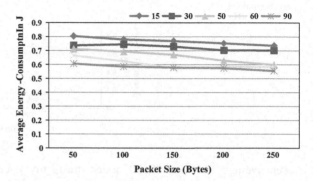

max nodes are put in network. So, channel utilization, i.e., rate of packet generation
and transmission also increases as per the graph. Performance of average
throughput is drastically increased for 60 nodes because of less traffic and proper
utilization of channel. So, Average throughput for 60 nodes is 10–50% increases
when packet increases from 50 to 250 packets.

Figure 6 indicates that average energy consumption for 60 and 90 nodes is
constant as well as less for changing packet size from 50 to 250 packets/s. For node
density 15–50, average energy consumption is more compare to 60, 90 node
densities. For varying packet size from 50 to 250 nodes. Packet generation and
transmission ratio is high for the 15 nodes. So, average energy consumption is also
high in the graph. Changing packet size 50–250 bytes and changing node density
from 15 to 90, average energy consumption is the constant for the same.

4 Conclusion

Average PDR graph when node density is 15 and packet size is changing from 50 to
250 packets. Then Average PDR is 35–40% better as compared to 90 node density
and 10% better as compared to 50 and 60 node densities. So, average PDR for 15
nodes is drastically better, i.e., near about 98% varying (packet size) 50–250

packets per second. Average PLR for 15 and 30 node density drastically decreases for varying packet size from 50 to 250. For 15 and 30 nodes, only 2–5% packets are lost in a network. So, Average PLR is about 40–45% less for 15 and 30 nodes as compared to 90 nodes. In case of node density 60, Average PLR drastically increases up to the threshold point 150 packet size and later it decreases when packet size changing from 150 to 250 packets per second. When packet size is varying from 50 to 250 bytes packet, Average end-to-end delay for 15 nodes is drastically less, i.e., 40–50% comparing 90 node density result. Average end-to-end delay in case of 90 nodes is very poor as shown in graph. It increases 20–40% as compare to 15–30 nodes. Average throughput for 60 nodes with 250 packet size is drastically increased as comparing to other nodes. Average throughput for 60 nodes increases near about 10% at initial stage for 50 packets and (50–60%) for 250 packet size. Node density 15, 30, 50, and 90 gives 10–20% less average throughput as compared to 60 nodes. Average energy consumption in case of 90 nodes is 10–20% less as compared to 15 nodes, 10% less as compared to 30, 50, 60 node density. Result shows parameters like Average PDR, average throughput, and average PLR are drastically better for 15 nodes, when it is varying from 50 to 250 packet size. Average throughput for 60 nodes is drastically increased when packet size increases from 50 to 250 bytes. Average energy consumption in case of 90 nodes is less as compared to other nodes. In future, we will implement a new novel Hybrid Mac protocol for achieving various QoS parameters of WSN.

References

1. S-Morshed, M-Baratchi, G Heijenk, "Traffic-Adaptive Duty Cycle adaption in TRMAC protocol for WSN", The Netherlands, Wireless -Days (-WD), 2016.
2. D, Liu, X., Wu, Z., Cao, M., Liu, Y-Li, M. Hou, "CD-MAC: A Contention-Detectable (MAC) for Low- Duty-Cycled WSN's", 12th-Annual IEEE International-Conf on Sensing, Comm, &, Networking (SE-CON)-2015.
3. Thien, (D), Nguyen, Jamil-Khan, & Duy-Ngo, "An Energy & (QoS) Aware Packet Transmission Algorithm for IEEE (802_15_4) Networks", IEEE-, 26th Annual International Symposium on Personal, Indoor &Mobile-Radio Communications - (PIMRC): MAC and Cross, Layer Design, 2015.
4. F. Alassery, Walid K. M. Ahmed, and Victor Lawrence, "Multi-Dimensional Slotted Aloha MAC Protocol for Collision High Throughput Wireless Communication System", 36th IEEE Sarnoff Symposium 2015.
5. M-Zeeshan, A-Ali, A, Naveed, AX. Liu, A. Wang and H.K. Qureshi, Zeeshanetal, "Modeling - packet losses probability, & busy time in (multi-hop) wireless, networks", EURASIP Journal on WC and Networking (2016).
6. D. Zhang, Chen-peng Zhao, Yan-pin Liang, Zhao-jing Liu, "A new medium access control protocol based on perceived data reliability and spatial correlation in wireless sensor network", Elsevier 2012, Computers and Electrical Engineering 38, P.P. 694–702 (2012).
7. A. P. Dave, S. Singh, "Comparative Analysis between IEEE 802_11 and SMAC", International Conf on Pervasive Computing (ICPC) 2015.

8. ZAEu, Hwee-Pink-Tan, Winston-KG. Seah, "Design & performance analysis of (MAC) schemes for WSN Powered by Ambient-, Energy -Harvesting", Elsevier -/- Ad Hoc Networks, 9, PP, 300–323, (2011).
9. M_AlJemeli, FA Hussin, "An Energy-Efficient Cross_layer Network; Operation Model for IEEE 802_15_4-based Mobile(WSN)", (IEEE-) Sensors- Journal (2013).
10. T-Nhan Le, A, Pegatoquet, O-Berder, O-Sentieys, "Energy Efficient Power- Manager & (MAC) Protocol for Multi-, Hop WSN's Powered-by Periodic Energy-Harvesting Sources", IEEE SJ- 2015.
11. H. Lee, "Modeling & Analysis of an Energy-efficient MAC-Protocol for WSNs", ICOIN - 2016.
12. A. Chen, D. Lee, G. Chandrasekaran, P. Sinha, "High throughput MAC layer multicasting over time-varying channels", Computer Communications 32, P.P 94–104, (2009).
13. X, Zhao, J, Guo, CT.-Chou, A, Misra, S, Jha, "High-Through-put Reliable Multi-cast in Multi-Hop WMN", IEEE Transactions on Mobile Computing 2015.
14. P Luong, TM. Nguyen, L. Bao Le, "Through-put analysis for co-existing IEEE 802_15_4 & 802_11 networks under un-saturated traffic", EURASIP J on WC & Networking, Issue -: 2016.
15. W-Shen, T, Zhang, F, Barac, M, Gidlund, "Priority-(MAC): A Priority-Enhanced MAC Protocol for Critical-Traffic in Industrial WS & Actuator Networks", IEEE,-Transaction On Industrial Information, VOL. (-10), Num. (1), Feb-(2014).
16. MR. Palattella, T Watteyne, Q Wang, "On the Fly Bandwidth-Reservation for 6-TiSCH WIN", IEEE SJ, 2015.

A Novel Method for the Conversion of Scanned Electrocardiogram (ECG) Image to Digital Signal

Macline Crecsilla Lewis, Manjunatha Maiya
and Niranjana Sampathila

Abstract Electrocardiogram (ECG) is the record of origin and propagation of electrical potential through cardiac muscles. It provides information about heart functioning. Generally, ECG is printed on thermal paper. The person having heart abnormalities will have to maintain all the records for the diagnosis purpose, which requires large storage space and is minimized by storing in the computer using scanner. The stored data is processed manually, which is time consuming. So an automatic algorithm that is developed does the conversion of the ECG image to digital signal. In order to convert the image, image processing methods like binarization, morphological techniques have been used. Usage of morphological skeletonization helps in converting the image to digital signal form by finding the skeleton of the ECG signal. The performance of the conversion algorithm is analyzed using root-mean-square error (RMSE), and it was found good. The average error found between the binarized image and the skeletonized image is nearly 7.5%.

Keywords Electrocardiogram · Binarization · Morphology · Skeletonization
RMSE

M.C. Lewis (✉) · N. Sampathila
Department of Biomedical Engineering, Manipal Institute of Technology, Manipal
University, Manipal, India
e-mail: lewismacline@gmail.com

N. Sampathila
e-mail: niranjana.s@manipal.edu

M. Maiya
Philips India Pvt. Ltd. (BOP), Manipal, India
e-mail: manjunatha.maiya@philips.com

© Springer Nature Singapore Pte Ltd. 2018
S.S. Dash et al. (eds.), *International Conference on Intelligent Computing
and Applications*, Advances in Intelligent Systems and Computing 632,
https://doi.org/10.1007/978-981-10-5520-1_34

1   Introduction

Heart-related diseases are one of the common causes of death. It kills millions of people worldwide each year. There are various kinds of heart-related problems. Arrhythmias are one of the problems related to heart. It occurs due to the abnormality in heart rhythm. They are accessible when the heart's electrical impulses run with the pack of heartbeats which do not function properly. Different types of arrhythmias include tachycardia, bradycardia, premature contraction and fibrillation. Generally, arrhythmia occurs when the electrical signal which controls the pulses is delayed or some time blocked. This can occur only if the nerve cells that deliver electrical signals won't work appropriately. Once the doctor has documented that the subject has an arrhythmia, subject will need to find out whether it is normal or abnormal or merely reflects the heart's normal processes [1].

Electrocardiogram (ECG) is the most important and generally used strategy to think about the heart diseases. The restorative condition of the heart is found by the state of the ECG waveform. This will help in separating the sorts of diseases. Over the previous decades, considerable work has been done to ease cardiologists' task of diagnosing the ECG recordings. The real test confronted today is the early identification and treatment of arrhythmias [2]. The representation of electrical activity of the heart is known as ECG. It demonstrates the standard contraction and relaxation of the heart muscle. The recorded data contains vital data about rhythmic attributes. It is the most effectively available bioelectrical signal that gives the sensibly precise information with respect to the state of the heart. The heart condition is analyzed by an equipment generally known as electrocardiography. The examination of ECG waveform will help in diagnosing the various abnormalities.

ECG comprises of five fundamental waves P, Q, R, S, and T and sometimes U waves. The typical ECG outline is appeared in Fig. 1. The P wave represents the atrial depolarization. The QRS complex represents the ventricular depolarization. The T wave represents the repolarization of the ventricles. It follows each of the QRS complexes. Normally, it is isolated from the QRS by a steady interval [3].

The reported work is to read the JPEG form of ECG, from the printed form of ECG and to convert that image to digital time series signal. The conversion procedure involves image processing approach for converting the scanned image into digital signal form by using morphological operations and skeletonization technique.

Fig. 1 Typical ECG waveform

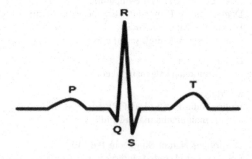

2 Literature Review

ECG is a noninvasive, transthoracic diagnostic technique. Usually, 12-lead ECG signal is used for the diagnosis of heart. There are 12 different segments. They are limb leads (1, 2, and 3), chest leads ($V1$, $V2$, $V3$, $V4$, $V5$, and $V6$), and peripheral leads (aVR, aVL, and aVF). The limb lead consists of four leads. They are located on the left and right wrist, followed by left and right ankle. The lead connected to the right ankle is a neutral lead. The six leads which are labeled as 'V' leads are positioned on the rib cage. The generated ECG signal using 12 leads is traced on a thermal paper using stylus. It is difficult to keep the paper-recorded signals as the number of patients is increasing day by day, and also it takes lot of storage space. In order to check the records as and when required, they are scanned and stored in the electronic devices such as computer in the form of PDF files or JPEG images. This form of ECG has to be digitized in order to get the extracted signal so that it can be used for further process such as QRS complex detection [1].

In the study of digitization of paper-recorded ECG, MATLAB is used to change the ECG data from paper printout into digital signals. A strategy is built that includes preparing of ECG paper records by an effective and iterative arrangement of digital image processing methods. The transformation of ECG image information to digital signal brings about less storage and less recovery of data. In this study, the methods like de-skewing, enhancement of image, color-based segmentation, and region-based segmentation, signal representation and filtration have been used. De-skewing uses Hough transform, and it is done to rotate the image. The color image segmentation techniques used here are to segment the set of ECG wave region. This can be used in ECG wave analysis and detection application. This technique also involves the binary image as preprocessing [4].

In general, there are three types of ECG paper charts which are divided upon their backgrounds: uniform background, background with colored grid and with black grid. The developed algorithm consists of morphological operations to retrieve ECG data present in the image. The results show that the method erases the background noise and acquires the digital ECG signal from ECG paper [5]. A binary image of the ECG record is found by applying thresholding technique. It is applied to remove the background grid present in the image. It is also applied in order to get the digitized signal by determining the pixel scale. Automatic methods used in the process will make the analysis easier by detecting the characteristic waves in simpler way. In addition, these files contain one-dimensional signals which are smaller in size compared with the image ones. This will help in simplifying the patients' record storage space [3].

From the past few years, several researchers have been working on developing the accuracy of the conversion of the ECG image. Mainly, all researchers are trying to reduce the time of execution for the conversion of image to digital signal. Some of these methods had a few drawbacks such as less accuracy, high computational time, or more manual interaction. In order to overcome these drawbacks, the new technique which involves image binarization, morphological operations, and some filtering techniques have been used.

3 Methodology

Considered printed thermal paper of ECG with 12-lead signal is recorded at 25 mm/s. In order to convert the ECG paper image to a digital signal, it is necessary to convert the paper to image form. In order to convert the paper-printed ECG to JPEG form, first need is to scan the paper. During digitization, ECG scan will resample the waveform. Also it rescales the digitized waveform to the required sampling rate. It also helps in calculating the amplitude resolution [6]. The flow diagram for digitization of the ECG image is shown in Fig. 2.

After scanning the ECG paper, first step is to binarize the scanned image of ECG in order to get the image in terms of 0's and 1's. This will help in finding the pixel values in the binarized form. The flow diagram of the binarization process of scanned ECG image is shown in Fig. 3.

Thresholding converts over an information set containing values that shift over some range into another information set containing values that differ a smaller range [6]. Thresholding is used to find binary images from a gray scale image.

Let the image intensity be $I_{i,j}$. If $I_{i,j}$ is less than some fixed constant, then the image is replaced as black pixel or else as white pixel. This can be done by known gray levels. Thresholding is defined as an operation that involves test against a numeric function T [7]

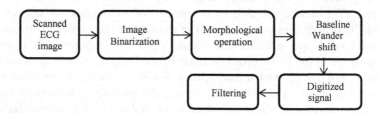

Fig. 2 Flow diagram of digitization algorithm

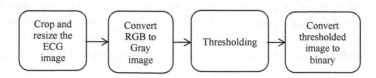

Fig. 3 Flow diagram of the process of binarization operation

$$T_0 = T\{X, Y, P(X, Y), f(X, Y)\}, \tag{1}$$

where $P(X, Y)$ represents local property of the point (X, Y). The resulting thresholding image $g(X, Y)$ is then defined as:

$$g(x, y) = \begin{cases} 1 & \text{if} \quad f(X, Y) > T_0 \\ 0 & \text{if} \quad f(X, Y) \leq T_0 \end{cases}. \tag{2}$$

Binary image is also a digital image. It has two possible values for each pixel. It is referred as an image, since it takes just binary digital to represent the every pixel. The binarized image from the scanned ECG image is then morphologically processed in order to find the skeleton of the binarized ECG image. The morphological operations that are carried out in this work are erosion, dilation, and skeletonization. Figure 4 shows the flow diagram of morphological operations carried out in this work.

Erosion is the key operation for all the other morphological operations involved in the technique of morphological image processing. In binary image morphology, the image is considered as a subset of a Euclidean space. The fundamental is to test the image with a simple and with the pre-characterized shape of the image. Let A be a binary image in the given Euclidean space E. Let B be the structuring element for A. The erosion of A by B is defined as:

$$A \ominus B = \{z \in E | B_z \subseteq A\}, \tag{3}$$

where B_z is the translation of B by vector z:

$$B_z = \{b + z | b \in B\}, \quad \forall z \in E. \tag{4}$$

Dilation is an essential operation in a mathematical morphology. It uses structuring element for examining and extending the shapes of the input image. In binary image morphology, it works as a shift-invariant operator. The dilation of A by B is defined as:

$$A \oplus B = \{z \in E | (B^s)_z \cap A \neq \emptyset\} \tag{5}$$

where B^s denotes the symmetric of B, that is

$$B^s = \{x \in E | -x \in B\}. \tag{6}$$

A morphological skeleton technique is used to find the skeletal version of the image. Morphological skeletons can be of two sorts, and the first one is characterized by method for morphological openings to form the original shape. Another

Fig. 4 Flow diagram for skeletal image of ECG

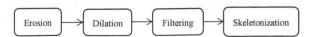

alternate method, suggested to use, is hit-or-miss transform. In this reported work, the method of morphological openings is used for the reconstruction of ECG from the printed form of the ECG. The idea of skeleton $S(A)$, which is a subset of A, is naturally simple. Consider a point z in $S(A)$. $(D)_z$ is the largest disk in A, and it is called as maximum disk. The $(D)_z$ touches the boundary of A at two or more different locations [7]. The skeleton of A is expressed in terms of erosions and openings.

$$S(A) = \bigcup_{k=0}^{k} S_k(A). \tag{7}$$

The skeletonized image of ECG is then used to find the end points and the branch points of the skeletal image. These two points are then cascaded, and again the image is regenerated. This regenerated image will have some information loss. This regenerated image is the de-masked using geodesic distance transform. Geodesic distance transform is mainly used for binary images. The skeletal image formed is also a binary image, so this transform can be used to compute the distance between the binary image that is skeletal image and seed locations specified by the mask. Also the distance between the branch point and end point is calculated in order to find de-mask of the skeletal image. The difference between the de-masked image and the skeletal image of ECG will help in getting the digitized signal.

In instances where a constant, linear, and curved offset is available, detrend technique is used to expel these impacts. Detrend fits a polynomial of an offered request to the whole digitized ECG signal and subtracts this polynomial. This calculation fits baseline points in the signal. This calculation also fits the polynomial to all points, baseline, and the signal. It tends to work just when the biggest source of signal in every sample is background obstruction. In estimations, the detrending tends to evacuate varieties which are valuable in demonstrating. They even make nonlinear reactions from generally direct ones.

Furthermore, the way that an individual polynomial is fit to every range expands the measure of meddling difference in an information set. Because of these reasons, usage of detrend is recommended just when the general signal is controlled by backgrounds which are for the most part similar shape. Normally, the baseline is approximated by lower order polynomial. A particular baseline reference is provided in order to shift the signal to that baseline. This baseline reference is referred as 'basis.' When the basis is given, the background will be evacuated by subtracting each of these bases to acquire a low background result. The result found by detrend technique is without negative peaks.

Smoothing is a low-pass filter. It is used for expelling the high-frequency noise from the digitized ECG signal. This is done independently on every line of the information grid. It accepts that the factors which are close to each other in the information grid are identified with each other and contain comparable data which can be arrived at the midpoint of together to remove noise without critical loss of the signal of interest. The implemented smoothing is the Savitzky-Golay (SavGol) algorithm. The algorithm basically fits singular polynomials to windows around every point in the range. These polynomials are then used to smooth the information. The calculation requires choice of both the extent of the window that is the channel width and the order of the polynomial. The bigger the window and lower the polynomial order, the additionally smoothing happens. The algorithm approximates and removes some abnormal components present in the ECG signal.

The performance of the algorithm is evaluated with the root-mean-square error (RMSE) measures. That is the error between the binarized and the skeletal form of ECG image is found. The error is found by subtracting these two images. The RMSE is computed for n different predictions, and here the predicted value is $\hat{x}_t$ for times t and a dependent variable x_t.

$$\text{RMSE} = \sqrt{\frac{\sum_{t=1}^{n}(\hat{x}_t - x_t)^2}{n}}. \tag{8}$$

4 Result Analysis

The proposed algorithm is applied on a scanned ECG paper of a patient of 46 years old of a male sex, by using all the 12 segments from the 12-lead ECG recording. Proposed method of skeletonization of signal contains high-frequency noises. In order to remove the noise present in the signal, filters are applied. The accuracy of the found result is done by calculating the root-mean-square error between the binarized image or original image and the skeletonized image. RMSE is applied only to check that the resultant signal found in this project is almost same as that of the original scanned document.

Usually, ECG is printed on a thermal paper. And this thermal paper is scanned and stored in the form of JPEG files, and these files are given as an input for the digitization process. The image of scanned ECG printout which is given as input is shown in Fig. 5. Among 12-lead image, only lead 2 has been selected and the cropped form of input image is shown in Fig. 5.

The scanned image is in the color form. To discretize the waveform present in the image, the image is converted to gray scale format. The conversion of color image to gray scale form is shown in Fig. 6a, b. This gray scaled image is then converted to binary image by applying a suitable threshold. The values of binary image will be in terms of 0's and 1's. The values above threshold will write as 0 s. Find the pixel values in terms of 1 and 0 s. The binarized image from the gray scale image is shown in Fig. 6c.

Morphological operations include erosion and dilation. The result of erosion operation on the binary image tends to loss of information present in the image. During this process, the upper and lower limit in the Cartesian coordinates are recorded. In order to make the image signal even, filtering is applied. Figure 7 shows the eroded, dilated and filtered, and the complement of the filtered image.

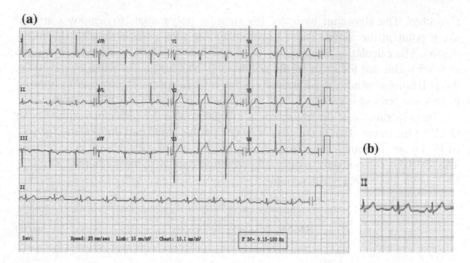

Fig. 5 **a** Scanned image of ECG with 12 segments as input and **b** cropped image (*Source* Adapted from Philips, CHC Hospital, Hebri)

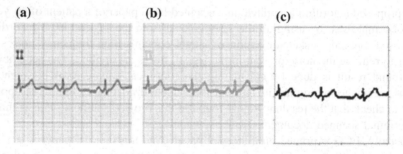

Fig. 6 **a**, **b** Gray scale and **c** binary form of image

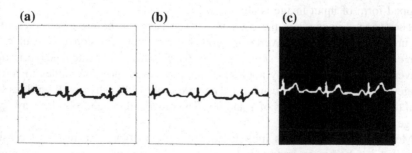

Fig. 7 **a** Eroded image, **b** dilated image, and **c** complement of filtered image

The image formed by the above method has two edges. In order to get the digitized signal, there must be only one edge. Thus, skeletonization method helps us in finding the mean of the two edges of the binarized signal image. The branch points and end points of the skeletonized ECG image are shown in Fig. 8. By cascading the branch point image and end point image, we get the thinned skeletonized image. Figure 8c shows thin skeletal of the binarized ECG image.

The waveform of ECG is categorized by black pixels. The locations of the pixels are denoted by x-y Cartesian coordinates. The image shown in Fig. 9a is the

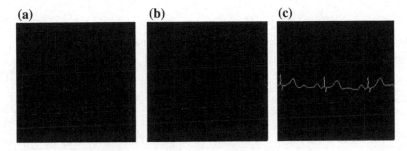

Fig. 8 a Branch points, b end points, and c thin skeletal image of ECG

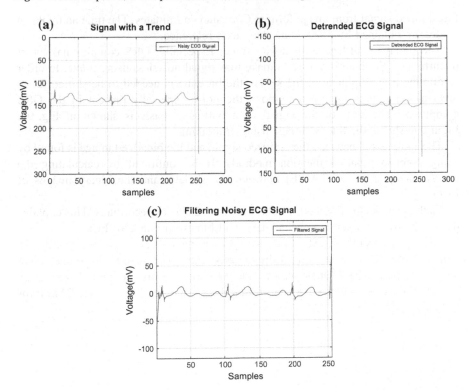

Fig. 9 a Digitized signal and b detrended signal, and c filtered signal

Fig. 10 Average
root-mean-square error of the
25 samples

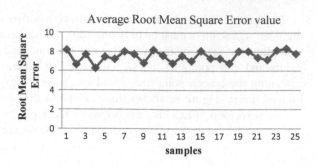

Average Root Mean Square Error value

Table 1 RMSE between the skeletal image and the binary images of different patients

Root-mean-square error (RMSE)												
Patient	Leads			aVR	aVL	aVF	V1	V2	V3	V4	V5	V6	Mean RMSE (%)
	1	2	3										
P1	6.84	6.77	10.2	5.96	10.7	7.99	9.19	8.29	8.21	7.51	8.39	8.12	8.17
P2	6.45	6.10	5.97	6.88	8.87	6.62	6.55	6.40	7.45	6.62	5.89	5.94	6.65
P3	6.24	7.52	7.17	8.83	6.09	9.18	9.84	9.40	7.62	5.92	8.04	6.42	7.68
P4	5.83	6.49	6.41	6.04	5.79	6.50	7.19	6.10	6.55	6.14	6.07	6.06	6.26
P5	7.95	8.30	7.69	7.37	8.70	8.79	7.19	6.33	8.26	6.96	6.10	5.79	7.45

transformed signal from image form to Cartesian coordinates. The time and voltage values are obtained by first setting the y-axis reference.

The signal found here is of noisy signal. Especially, QRS complex has lot of disturbances. So there is a need to make that signal noiseless using filters. In order to find the true amplitude of the signal, the baseline is needed to be shifted to '0' level and that it is filtered to remove the noise especially present in the QRS complexes. The shift of the signal to a particular basis is shown in Fig. 9b. Figure 9c shows the filtered image using smoothing.

The comparison between the digitized signal and the binarized image is found by using point-by-point verification methods. It is confirmed by calculating the root-mean-square error (RMSE) between the skeletal and the binary images of ECG.

Table 1 shows the RMS error values for the 12 chosen recordings. This explains the small error between the skeletonized and binarized image of ECG.

Table 1 shows that RMSE values for different patients are very less. Totally, 25 patients' report was taken and all the records taken here were all have age above 45 years. There were 11 male records and 14 female records. Figure 10 shows the average root-mean-square error chart for all the 25 samples. The average RMS error found is 7.5%.

5   Conclusion

The developed algorithm converts a 12-lead ECG image to digital signal. All the 12 segments in the image do not contain same shape of ECG waveform. In order to find the digital signal of whole image, a part by part or each single segment has to be taken one after another. That means at a time only one segment is executable. This is the main disadvantage of the work. But the advantage of this project is that by executing the each segment individually, the accuracy of getting the digitized signal is more. Other than this, the digitized signal gives the accurate signal or similar signal as that of the original signal in the image. Here, the skeletonization technique used will help in the conversion process. And also it makes the algorithm very simple. If an algorithm is simple, then the execution time will also be less. So the execution time taken for this algorithm is very less. The digitized signal derived can be further used for the detection of P, T waves and QRS complex of the ECG signal. All these detections will help in the classification of ECG signal.

Acknowledgements The authors would like to thank the supports of the Department of Biomedical Engineering, MIT, Manipal University, Manipal and also the Department of Cardiology (Philips), CHC Hospital, Hebri for providing the ECG data needed for the study. The authors would also like to thank Mr. Nandish S. from School of Information Science for his time and help.

References

1. Benjamin Wedro, and Charles Patrick Davis: Heart Disease (Cardiovascular Disease, CVD), www.medicinenet.com.
2. Jasminder Kaur and J.P.S. Raina: An intelligent diagnosis system for Electrocardiogram images using Artificial Neural Network(ANN), International Journal of Electrical, Electronics and Computer Engineering, vol. 1(1): 47–51, 2012 ISSN:2277-2626.
3. Skander Bensegueni, and Abdelhak Bennia: R-Waves Localization from an Electrocardiogram Scanned Paper, International Journal of Computing, Communications & Instrumentation Engg. (IJCCIE), Vol. 2, Issue 1 (2015) ISSN 2349-1469 EISSN 2349-1477.
4. Priyanka R.K. Shrivastava, Shraddha Panbude and Geeta Narayan (2014): Digitization of ECG paper records using MATLAB, International Journal of Innovative Technology and Exploring Engineering(IJITEE), Vol. 4, Issue 6, 2014, ISSN:2278-3075.
5. Tsair Kao, Len-Jon Hwang, Yui-Han Lin, Tzong-Huei Lin and Chia-Hung Hsiao: Computer analysis of Electrocardiograms from ECG paper recordings, Proceedings of the 23rd Annual EMBS International Conference, Vol. 5, 25–28, 2001, IEEE 0-7803-7211.
6. Fabio Badilini, PhD, Tanju Erdem, PhD, Wojciech Zareba, MD and Arthur J. Moss, MD.: ECG Scan: a method for conversion of paper electrocardiographic printouts to digital electrocardiographic files, Journal of Electrocardiography (ScienceDirect), Vol. 38, Issue 4, 2005, pages 310–318.
7. Gonzalez, R.C. and P. Wintz, "Digital image processing, second edition", Addison-Wesley, USA, 1987.

Fuzzy Logic Modeling for Strength Prediction of Reactive Powder Concrete

Akshay Nadiger, C. Harinath Reddy, Shankar Vasudevan
and K.M. Mini

Abstract Compressive strength forms the major property which ensures safety and stability in the design of any concrete structure. Addition of admixtures makes concrete of higher strength, which is based on trial-and-error combinations. In the present study, an attempt is made for developing a tool for compressive strength prediction of reactive powder concrete by Mamdani-based fuzzy logic interface system (FIS). The eight main parameters which influencing the strength of concrete were considered as input variables. Database set of 100 data was collected from different literature reviews and worked with trial permutation and combination with different order of material inputs, and 125 rules are set. Twenty-five test results are examined to check the efficiency of the proposed tool and compared with the FIS output by applying various membership functions using both centroid and bisector methods of defuzzification. The predicted results show the potential efficiency of FIS in prediction of compressive strength for reactive powder concrete. The results obtained were satisfactory with high accuracy ranging from 95 to 99%.

Keywords Reactive powder concrete · Compressive strength · Fuzzy logic
Membership functions · Prediction

A. Nadiger · C. Harinath Reddy · S. Vasudevan · K.M. Mini (✉)
Department of Civil Engineering, Amrita School of Engineering,
Amrita Vishwa Vidyapeetham, Amrita University, Coimbatore, India
e-mail: k_mini@cb.amrita.edu

A. Nadiger
e-mail: civilakshay17@gmail.com

C. Harinath Reddy
e-mail: harichinnavula@gmail.com

S. Vasudevan
e-mail: shankar271994vasudevan@gmail.com

© Springer Nature Singapore Pte Ltd. 2018
S.S. Dash et al. (eds.), *International Conference on Intelligent Computing
and Applications*, Advances in Intelligent Systems and Computing 632,
https://doi.org/10.1007/978-981-10-5520-1_35

1 Introduction

Technology has been accelerating the interactions of science and multidisciplinary engineering in different fields of problem solving, decision making, diagnostics, health monitoring, design, and rehabilitation of structures with enhanced knowledge. The complexity of most problems in any discipline can be solved by using mathematical modeling methods or by using intelligent systems for unpredictable problems involving lot of nonlinearity. Lately, fuzzy logic and artificial neural network are popularly used in civil engineering like structural health monitoring, prediction of material properties, design details, and so on.

Concrete is most abundant and globally used building construction material which consists of binder, fine aggregate, coarse aggregate, and water in certain proportions to form a designed mix. Properties and proportions of these materials majorly affect the mechanical properties of concrete. Concrete design is mainly dependent upon compressive strength, and hence, the compressive strength plays a vital role for structural safety and stability of structure. Lot of research works are going on in concrete to improve the compressive strength, which involves the addition of mineral and chemical admixtures. Reactive powder concrete (RPC), well known as ultra-high-performance concrete, is a newly developing fine powder-based material in concrete industry. It is produced on high percentage of cement, sand, fine crushed quartz powder, high dosage of super plasticizer, and very low water–binder ratio. Every small variation in material parameters changes the properties of RPC in major extent. Highly efficient RPC has compressive strength ranging from 200 to 800 MPa and flexural strength of 60–140 MPa with excellent durability and workability [1].

Comparative studies of RPC and OPC composed of ultrafine powders of sand, quartz powder, rice husk ash, and silica fume with low water–binder ratio less than 0.2 were studied by Sahani and Ray [2]. Incorporation of locally available industrial additives in concrete increases its durability and reduces carbon emission [3]. Improvement in mechanical property of RPC with different admixtures is reported by various researchers [4–7]. Preparation of green RPC of 200 MPa and its static dynamic behavior was carried out by Hang Yunsheng et al. [8]. Stress–strain relation of RPC in quasi-static loading on elevated temperature and mechanical properties of steel fiber-reinforced RPC exposed to elevated temperature was carried out by Tai et al. [9] and Bashandy [10]. Analytical study for predicting compressive strength of self-compacting concrete, containing various proportions of fly ash by FIS and ANN was reported by Belalia Douma [11], Gencelo et al. [12] and, shrinkage effect by Da Silva and Stemberk [13]. Prediction of heavyweight concrete, cement strength, rubberized concrete properties and ready mix concrete, lightweight concrete, effects of GGBS on strength modeling using ANN, and other soft computing methods are reported by many researchers [14–21]. Durability studies were predicted using fuzzy logic modeling by Nehdi and Bussuani [22], boundary conditions for structural tuning and monitoring by Muthukumaran et al. [23].

The present study aims in the production of RPC having higher strength with the replacement of cement by various waste materials with an emphasis on sustainability. As there is no specified mix design procedure for attaining higher strength, it becomes important to work on a trial-and-error method which leads to higher variations in strength. This also involves lot of experimentation with high amount of materials and manual labor. The scope of study focuses on preparing tool of prediction which ensures the specified proportion of input and expected output using Mamdani-based fuzzy logic interface system (FIS), thus proves the potential efficiency of FIS in prediction of compressive strength for RPC. The proposed model eliminates lot of experimental works and expenditure involved in the production of RPC.

2 Development of Fuzzy Logic Model

Fuzzy logic is a multivalued logic, which ranges from 0 to 1 and handles multivalued partial truth, and lies between completely true and completely false. It provides easy and modified way of dealing problems in which systematic mathematical formulations are set to deal with information investigation and solving uncertainty problems. It performs mathematical computations using crisp data simulated by membership functions. Identification and prediction of parameter as input to expected output is the major task in fuzzy logic model. Structure detection consists of issues such as selection of input variables, choosing fuzzy interface systems, rule set, type, and number of membership functions [24].

2.1 Selection of Database

Compressive strength of concrete depends on many parameters such as material properties, mix proportions, mixing methods, transporting and placing methods, compaction, curing conditions, and testing methods. The various factors affecting the compressive strength of reactive powder concrete are cement content, fine aggregate, admixtures, fibers, water–binder ratio, sand–binder ratio, super-plasticizers, etc. Concrete strength is controlled by concrete mix design, which involves proportioning of various constituents like cement, sand, water, various binder, super-plasticizers. Optimum water is needed for proper chemical action and hardens the cement paste; extra water increases the fluidity and reduces strength of concrete. Compressive strength is majorly dependent on quality control, placement, inspection and compaction, and testing methods should confirm to IS 516-1959.

In this study, the data collected are from available literatures and experimental investigation with uniform conditions except material mix proportions. Hence, study is reported to investigate the effect of various proportions of input materials to expected output. Output is compared with available experimental test results.

Table 1 Range of input and output parameters

Parameter	Range
Input parameters	
Cement (kg/cum)	600–1200
Silica fume (kg/cum)	100–400
Quartz powder (kg/cum)	200–400
Steel fibers (kg/cum and %)	0–350 and 0–10
W/B	0.12–0.24
S/B	0.6–1.8
Fly ash (kg/cum)	180–670
GGBS (kg/cum)	200–330
Output parameter	
Compressive strength (MPA)	70–230

Model accuracy in prediction depends on comprehensiveness of membership functions according to training data and fuzzy rules. A large number of experimental data are collected from literature surveys and experimental analysis to develop proper relationships between input and output, taking consideration of all proportions and properties of material. The input parameters considered in the present study are cement (kg/cum), silica fume (kg/cum), quartz powder (kg/cum), steel fibers (kg/cum and %), water–binder ratio, sand–binder ratio, fly ash (kg/cum), and GGBS (kg/cum). A database of 100 experimental data was collected from [25–28] and some experimental study to develop FL model. A total of 75 sets were applied for training the set, and 25 were taken to compare with output as testing data. Table 1 shows the range of input and output data used for the development of model, and Table 2 presents a representative sample from the training data set.

Table 2 Sample training set of input and output

S. No.	Cement	Silica fume	Fly ash	GGBS	Quartz powder	Steel fibers	W/B	S/B	Strength
1	1000	250	0	0	330	0	0.216	0.616	97
2	1000	140	300	330	330	280	0.169	1.255	106
3	1000	300	0	0	350	220	0.192	0.962	168.5
4	1000	300	0	0	350	340	0.192	0.931	156.5
5	1000	320	0	0	360	20	0.167	1.136	151
6	1000	0	0	0	0	60	0.18	1.761	72.5
7	1000	150	0	0	0	60	0.157	1.531	100
8	1000	0	0	0	0	50	0.18	1.639	75.5
9	1000	150	0	0	0	50	0.157	1.427	115
10	1000	300	0	0	0	50	0.138	1.185	128.5
11	1000	200	0	0	350	0	0.183	1.292	93
12	1000	250	0	0	350	0	0.176	1.241	97

(continued)

Table 2 (continued)

S. No.	Cement	Silica fume	Fly ash	GGBS	Quartz powder	Steel fibers	W/B	S/B	Strength
13	1000	300	0	0	350	0	0.169	1.192	116
14	1000	320	0	0	350	0	0.167	1.174	121
15	1000	300	0	0	350	0	0.154	1.153	126
16	1000	270	0	0	350	0	0.236	1.221	81
17	720	216	0	0	252	0	0.218	0.962	173.1
18	720	216	0	0	252	1%	0.218	0.934	198.3
19	720	216	0	0	252	2%	0.218	0.906	187.3
20	720	216	0	0	252	3%	0.218	0.877	181
21	745	132	0	219	0	0	0.148	1	153.2
22	884	221	0	0	0	0	0.125	1	217.4
23	737	184	184	0	0	1%	0.122	1	226.7
24	884	221	0	0	0	2%	0.125	1	218.8
25	714	216	0	0	252	6.9%	0.194	0.956	168.5

3 Results and Discussion

Data collected from various experimental results are analyzed using fuzzy logic tool in MATLAB R2013a, with different membership functions and defuzzification methods. Predicted testing data were compared to actual experimental data. Results obtained are highly accurate and satisfactory when Triangular, Gaussian, Pi, D Sigmoid, Generalized Bell membership functions are used (Figs. 1, 2, 3, 4 and 5). Accuracy of prediction was found to be above 99% by centroid and bisector methods of defuzzification. Centroid method gave precise results compared to

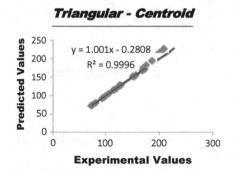
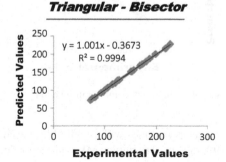

Fig. 1 Triangular membership function

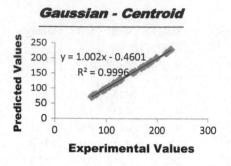

Fig. 2 Gaussian membership function

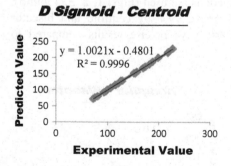

Fig. 3 Pi curve membership function

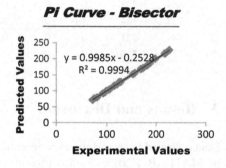

Fig. 4 *D* Sigmoid membership function

bisector method with negligible variations. Maximum accuracy obtained was 99.48% by Triangular MF with centroid method.

The sample representation of S curve is presented in Fig. 6. Similar curve is obtained for Sigmoid membership function. The prediction is based on the range of values within the S curve and Sigmoid curve, where lesser range is taken care at higher strength which results in higher accuracy, and vice versa. The corresponding

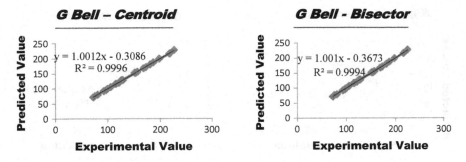

Fig. 5 *G* Bell membership function

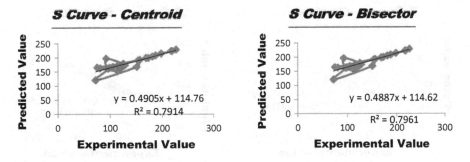

Fig. 6 Sample representation of *S* curve

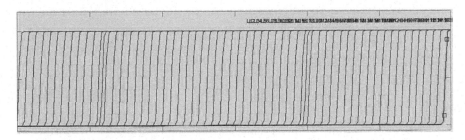

Fig. 7 *S* curve membership function

comparison of predicted and experimental results is shown in Figs. 7 and 8. A careful analysis of the predicted result shows nonlinearity at lesser strength values and linear variation at higher values resulting in unsatisfactory output.

Z Curve membership function is represented in Fig. 9 which could predict more accurate results at lower strength values due to its curve nature. The comparison between the predicted and experimental results is shown in Fig. 10 which shows a nonlinear variation at average range values.

382   A. Nadiger et al.

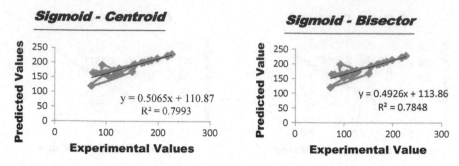

Fig. 8 Sigmoid membership function

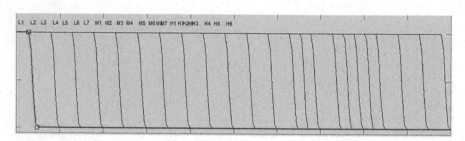

Fig. 9 Sample representation of Z curve

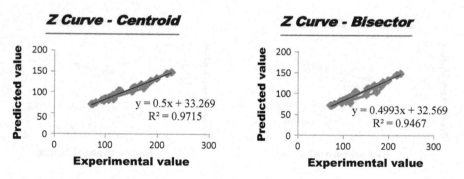

Fig. 10 Z curve membership function

The average accuracy of 25 testing data corresponding to different membership functions for both the defuzzification methods is reported in Table 3. A better performance is observed for Triangular, Gaussian, Pi curve, D Sigmoid, and G Bell membership functions. Z curve produces satisfactory results, whereas S Curve and Sigmoid curve almost failed in the prediction.

From the analysis of experimental data, it is observed that out of 8 input data which represents the behavior of RPC, the major parameters contributing to the compressive strength (output) are cement, sand–binder ratio, water–binder ratio,

Table 3 Average accuracy of 25 test data

Membership function	Centroid	Bisector
Triangular	99.48	99.24
Gaussian	99.46	99.24
Pi Curve	99.42	99.22
D Sigmoid	99.46	99.24
G Bell	99.48	99.24
Z Curve	76.43	75.85
S Curve	59.95	60.25
Sigmoid	38.64	37.41

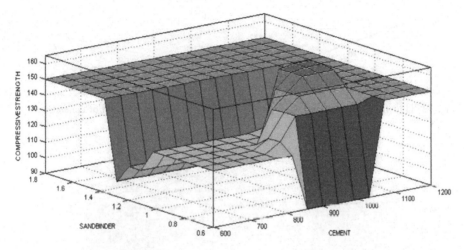

Fig. 11 Influence of cement and sand–binder ratio on compressive strength of concrete

and silica fume (admixture). Hence, a study is also carried out in fuzzy interface system to understand the various combination effects of these parameters on the compressive strength and is represented in Figs. 11, 12 and 13.

4 Conclusion

A novel approach in prediction of compressive strength of reactive powder concrete using fuzzy interface system is developed, which is economical and time saving, since lot of trial-and-error experimentation is needed to achieve high strength of concrete. The predicted results obtained from Triangular, Gaussian, D Sigmoid, G Bell, and Pi curve are very accurate to experimental values by more than 99% on an average basis where as an accuracy of 95–99% is obtained for individual test data. Also the prediction of the input parameters for a desired strength is possible by trial-and-error method within the specified range. Large variations in prediction are

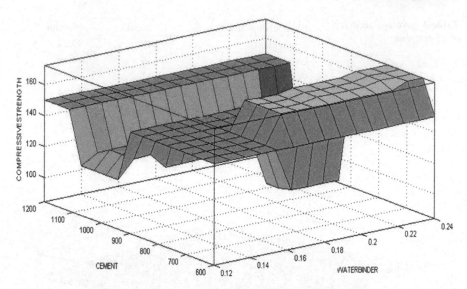

Fig. 12 Influence of cement and water–binder ratio on compressive strength of concrete

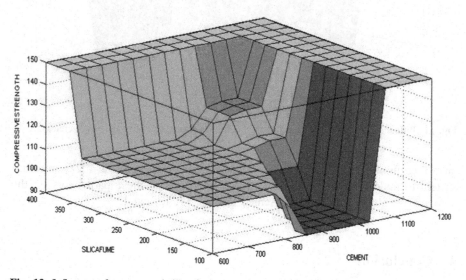

Fig. 13 Influence of cement and silica fume on compressive strength of concrete

observed using Sigmoid, *S* curve, and *Z* curve because of its membership function shape and range. Though *Z* Curve was satisfactory compared to other two, it failed at average compressive strength prediction. Thus, application of fuzzy logic can provide an optimal solution in prediction of compressive strength of reactive powder concrete. The developed fuzzy algorithm can adjust itself to any combination of input parameters within the specified range. From the results obtained, it can be concluded that fuzzy models are efficient and suitable in solving complex problems accurately.

References

1. P. Richard, Marcel Cheyrezy, "Composition of RPC", cement and concrete research vol 2, no 7, pp 1501–1511, 0608-8846(95) 00144-1 (1995).
2. B.S Sahani, N.H.S Ray "A comparative study of RPC and OPC by ultra high strength technology", IJR Vol 1, Issue-6, ISSN-2348-6848 (2014).
3. A. Zenati, K. Arroudj, M. Lanez, M.N Oudjit "Influence of cementatious additions on rheological and mechanical properties of RPC", ELSEVIER, Science direct (2009).
4. Halit Yazici, Mert.Y Yardimici, Huseyin Yigiter, Serdar Aydin, Selcuk Turkel, "Mechanical properties of RPC containing high volume of GGBS" ELSEVIER, Cement and concrete composites 32, 639–648 (2010).
5. Halit Yazici, Huseyin Yigiter, Anil K, Bulent Bardan "Utilization of fly ash and GGBS as an alternative silica source in RPC", ELSEVIER, Fuel 87, 2401–2407 (2008).
6. A. Cwirzen, V. Penttala, C. Varnanen "Reactive powder based concretes: Mechanical properties, durability and hybrid use with OPC", ELSEVIER, Cement and concrete research 38, 1217–1226 (2008).
7. Mr Arjun Kumar, Dr Asha Udaya Rao, Dr Narayan Subhahit "Reactive powder concrete properties with cement replacement using waste material", International Journal of Science and Engineering Research, Volume 4, Issue 5, ISSN2229-5518 (2013).
8. Z. Hang Yunsheng, Sun Wei, Lin Sifeng, Jiao Chujie, Lai Jianzhong "Preparation of C200green RPC and its static-dynamic behaviours" ELSEVIER, Cement and concrete composites 30, 831–838 (2008).
9. Yuh-Shiou Taai, Huang-Hsing Pan, Ying-Nienkung "Mechanical properties of steel fiber reinforced RPC exposed to high temperature reaching 800 degree centigrade", ELSEVIER, Nuclear engineering and design 241, 2416–2424 (2011).
10. Alaa.A. Bashandy "Influence of elevated temperature on behavior of economical RPC", Journal of civil engineering research, 3(3); 89–97, DOI:10.5923/J.JCE20130303.01 (2013).
11. O. Belalia Douma, B. Boukhalem, M. Ghrici "Prediction of strength of self compacting concrete containing fly ash using FIS", (2014).
12. Gencelo, Ozelc, Koksal F, Martinez Barrera. G, Brostow. W, Polat. H "Fuzzy logic model for prediction of properties of fiber reinforced self compacting concrete", Material science, Vol 19, no 2, pp-203–215, (2013).
13. Da Silva WRC, Stemberk P, "Predicting self compacting concrete shrinkage based on a modified fuzzy logic model", Engineering mechanics, vol 229, pp-1173–1183, (2012).
14. C. Bas.yigit, Iskender Akkurt, S. Kilincarslan, A. Beycioglu " Prediction of strength of heavy weight concrete by ANN and FL models, Neural Comput & Applic, 19:507–513, DOI 10.1007/s00521-009-0292-9 (2010).
15. Hang-Guang N, Ji-Zang W, "Prediction of Compressive strength of concrete by ANN", Cem Cancr Res; 30(8) 1245–50. (2000).
16. Gao F.L, "A new way of predicting cement strength fuzzy logic" Cement and concrete research, Vol 27, n 6 pp. 883–888, (1997).
17. Topcu I.B, Saridemir M "Prediction of rubberized concrete properties using ANN and FL", Construction built mater 22:532–540 (2008).

18. Deka P.C, Diwate S.N "Modeling compressive strength of RMC using soft computing techniques", International journal of earth sciences and engineering, Vol 4, no 6, pp 793–796, (2011).
19. Tanyildizi. H "Fuzzy logic model for prediction of mechanical properties of lightweight concrete exposed to high temperature", Master Des 30:2205–2210, (2009).
20. Saridemir M et.al "Prediction of long term effects of GGBS on compressive strength of concrete by ANN and FL", Construction Build mater 23:1279–1286.
21. E.H Mamdani "Fuzzy logic control of aggregate production planning", S Assilian International Journal of man machine studies, vol7, pp 1–13. (1975).
22. Nehdi M.L, Bussuani M.T "Fuzzy logic approach for estimating durability of concrete", Proceedings of ICE-Construction materials vol 62, n2, pp-81–92, (2009).
23. Muttukumaran P, Demirli K, Stiharu I, Bhat R.B "Boundary conditioning for structural tuning using fuzzy logic approach", Compact struct, vol 74, n5, pp-547–557, (2000).
24. Sugeno.M, Kang G.T, "Structure identification of fuzzy model", Fuzzy sets system cyber, vol 23, pp. 665–685, (1993).
25. LIU Shu-hua, LI Li-hua and FENG Jian-wen "Study on mechanical properties of RPC", Journal of Civil engineering and construction 1:1 (2012).
26. Yuh-Shiou Tai, Huang-Hsing Pan, Ying-Nien Kung "Mechanical properties of steel fibers reinforced RPC following exposure of high temperature", ELSEVIER, Nuclear engineering and Design 241, 2416–2424 (2011).
27. YS Tai "Uniaxial compression tests at various loadingrates for RPC", ELSEVIER, Theoritical and Applied fracture mechanics 52, 14–21 (2009).
28. Jagannathasan Saravanan, Saranya Poovazhagen "Analytical study on compressive strength of RPC", International journal of recent scientific research, Vol 6, Issue 5, pp-3872–3880, May (2015).

Real-Time Intelligent NI myRIO-Based Library Management Robotic System Using LabVIEW

Anita Gade⑩ and Yogesh Angal⑩

Abstract Library administration is a subcontrol of institutional administration that focal point of consideration on particular issues confronted by library administration. Human being has always tried to give life qualities to its artifacts in an effort to find choice for human to complete tasks by intimidating situation. The prominent thought of robot is work and appearance similar to person. Today's exceptionally forming human advancement, time and labors are impediment for culmination of assignment in expansive scales. The robotics is assuming vital part to spare human endeavors in an expansive bit of the standard and much of the time conveyed works. Frequently, we need labor to pick the book and give up it to the issuing counter. People take additional time and exertion for issuing and returning the book. To conquer this bother, we have developed automation in library for quick conveyance of books utilizing robot with a few degrees of freedom. The usage of robots portrays some of cutting-edge patterns in robotization of the present day process. This work presents automation in library using robot. To accomplish this work, task planning algorithm is used. RFID technology is used to identify the book. This system is NI myRIO-based mechatronic framework recognizes the book, picks a book from source area, and places at desired location using LabVIEW software.

Keywords Book · IR sensor · LabVIEW · myRIO · National instruments
RFID · Robotic arm

A. Gade (✉) · Y. Angal
JSPM's Bhivarabai Sawant Institute of Technology & Research, SPPU, Pune, India
e-mail: anita.gade@yahoo.co.in

Y. Angal
e-mail: yogeshangal@yahoo.co.in

© Springer Nature Singapore Pte Ltd. 2018   387
S.S. Dash et al. (eds.), *International Conference on Intelligent Computing and Applications*, Advances in Intelligent Systems and Computing 632,
https://doi.org/10.1007/978-981-10-5520-1_36

1 Introduction

The procreative displaying method allows evaluating the understanding endeavors given faulty questionable data, and furthermore dissents and handles assurance in task-organized way. Robot mechanism have transformed into the game plan, lacking limits as cost work wage and customer's demand. The utilization of robot has expanded seriously; mechanical arms are exceptionally adaptable with the more exact and productive sensors, and we can incite the robot for particular and exact needs. This framework is utilized keeping in mind the end goal to replace human to perform the tasks. Robotic arm is a kinematic chain of open or closed robust links interrelated by movable joints. Robotics autonomy is related with mechanics, gadgets, and programming. These days Robotics Research is focusing on creating frameworks that exhibits adaptability, adaptation to internal failure, measured quality, coordinated idea, programming condition, and perfect availability to different components. In this outline, connections are considered to compare with humanoid structure [1]. Final element of arm is a wrist joint which interfaces a gripper. All troubles required in library administration process have been strongly assessed. In this expect, we are working up a structure using sensors, as demonstrated by the sensor records, the advancement of the robot is controlled. Using mechanical arm, this system picks the book from source zone and spots at fancied range [2]. LabVIEW program empowers the robot to move from source point to destination point keeping away from undefined obstacle present in the path. Robot is working on principle of Sense, Think, and Action, avoiding the obstructions for achieving the goal.

2 Related Work

2.1 System Block Diagram

Figure 1 shows implementation of system consisting of ultrasonic sensor which is used for obstacle detection. This sensor transmits and receives the ultrasonic waves reflected from an object. Once an electrical pulse is applied to the sensor, it vibrates over particular range of frequencies and generates sound waves. As obstacle comes in front of the ultrasonic sensor, the sound waves will revert in the form of echo along with generation of electric pulse. It estimates the time taken between transmitting sound waves and receiving echo. The echo patterns and sound waves patterns will be compared to decide detected signal's condition. The myRIO (my reconfigurable input-/output-embedded controller) is a real-time processor. Benefit of this device is its capability to gain and processing information in actual world. The robot's integral sensors and motors are controlled through the FPGA. The robot

Fig. 1 System block diagram

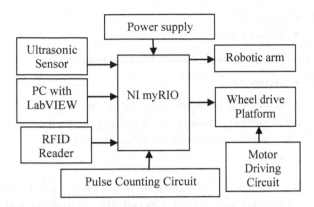

uses four wheels in drive [3]. Rotation of wheel is measured with optical quadrature encoders with pulse width modulation. The driving framework consists of a main frame, a dc motor, a pair of spur gears, and a flange. The spur gears convert rotary motion into linear motion by operating the dc motor. An end effector is the last part of the robotic arm, which is designed to interrelate with the situation. End effector consists of gripping tool which is utilized for holding the book. The robotic movement is a collaborative action of forward, reverse, left, and right direction to perform the desired task of book griping. We have systemized Wi-fi communication in between robotic system and PC (Fig. 2).

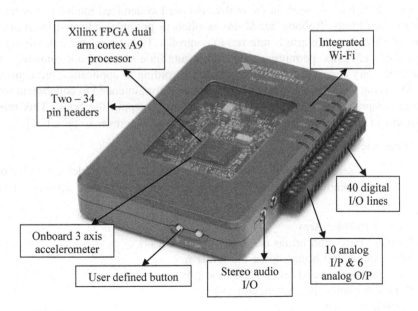

Fig. 2 NI myRIO device

- myRIO
- Power source

We are utilizing battery for power source.

- Actuation

Actuators are strengths of a robot, the elements which transform power into development. Actuators are electrically driven motors that turn a steering wheel and conventional motors control robot in variables.

- RFID

Radio-frequency recognizable proof makes use of electromagnetic radiation to interchange data which consequently distinguish and track labels appended to books. Our framework executes book induction and book-based handle arranging reasonable for a manufactured operator with a particular epitome, by utilizing RFID innovation. The labels hold electronically put away book information. The label data is put away in memory. The RFID tag fuses programmable justification for setting up the transmission and sensor data, independently. It can likewise go about as a security gadget. Truth be told, library spending plans are being lessened for staff, making it essential for libraries to add robotization to adjust for the diminished representatives size. If RFID sensor data matches with entered book then robot performs the action.

- Manipulation

Robots which have to work in the real world need systemized modular concept to control the objects. Robotic hands are as often as possible indicated as gripping tool, whereas the robot arm is referred to controller. Robot arms have flexibility of replaceable effectors, permitting them to execute little scope of assignments. The length of every connection has been planned according to application prerequisite. Gripper is the gadget toward the end of automated arm, intended to collaborate with nature comprising of holding device, utilized for grasping the book. This robot advances in reverse, left, and right, so we can travel wherever to pick book.

- Ultrasonic Sensor

The ultrasonic sensor empowers the robot to practically see and identify and avoid obstacles and compute distance. The working slope of ultrasonic sensor is 10 to 30 cm (Fig. 3).

- Axes of Robotic Arm
- Shoulder raises and brings down the upper arm [4].
- Elbow raises and brings down the forearm.
- Wrist pitch raises and brings down the gripper.
- Wrist roll rotates the end-effector gripper.
- Kinematic Chain.

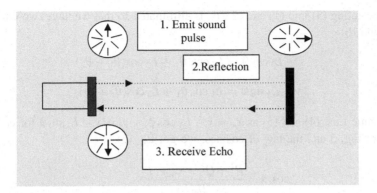

Fig. 3 Ultrasonic sensor

Robot kinematics controls the activity of the controller.

Figure 4 shows the axes of robotic arm. Basics of trigonometry give the joint coordinates of the robot arm for location and direction of the end effector as follows:

$$x = L_1 \cos\theta_1 + L_2 \cos(\theta_1 + \theta_2) + \theta_1 + L_3 \cos(\theta_1 + \theta_2 + \theta_3) \qquad (1)$$

$$y = L_1 \sin\theta_1 + L_2 \sin(\theta_1 + \theta_2) + L_3 \sin(\theta_1 + \theta_2 + \theta_3) \qquad (2)$$

$$\phi = \theta_1 + \theta_2 + \theta_3 \qquad (3)$$

Equations (1), (2), and (3) gives the correlation between the effector coordinates and combined coordinates. To find the joint coordinates to the position of end-effector coordinates (x, y, ϕ), we needs to evaluate the nonlinear equations for θ_1, θ_2, and θ_3.

Fig. 4 Axes of robotic arm

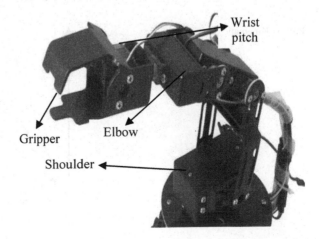

Substituting (3) into (1) and (2), θ_3 is eliminated so that we have two equations in θ_1 and θ_2:

$$x - L_3 \cos \emptyset = L_1 \cos \theta_1 + L_2 \cos(\theta_1 + \theta_2) \tag{4}$$

$$y - L_3 \sin \emptyset = L_1 \cos \theta_1 + L_2 \cos(\theta_1 + \theta_2) \tag{5}$$

Rename Eqs. (4) and (5) as $x_p = x - L_3 \cos \emptyset$, $y_p = y - L_3 \sin \emptyset$ for ease. From Fig. 5 and the law of cosines, we get Eq. (6).

$$\cos \alpha = \frac{P^2 + Q^2 - L_1^2 - L_2^2}{2L_1 L_2}$$

$$\propto = A \cos \left(\frac{P^2 + Q^2 - L_1^2 - L_2^2}{2L_1 L_2} \right)$$

$$\theta_2 = 180 - \alpha \tag{6}$$

$$\theta_2 = A \tan 2 (y_p, x_p) + A \sin \sqrt{\frac{L_2 \sin \theta_2}{x_p^2 + y_p^2}} \tag{7}$$

From Eq. (3)

$$\theta_3 = \emptyset - \theta_1 - \theta_2 \tag{8}$$

By executing the Eqs. (6), (7), and (8) using LabVIEW, we acquired the robot arm end-effector position. By executing Eqs. (7), (9), and (10), we got the correct joint angles [5].

$$\theta_2 = \theta_2 - 270 \tag{9}$$

$$\theta_2 = 180 - (\theta_3 + 270). \tag{10}$$

Fig. 5 Kinematic chain

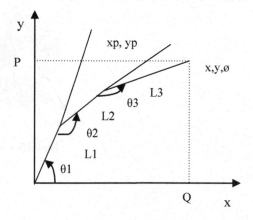

Fig. 6 System development

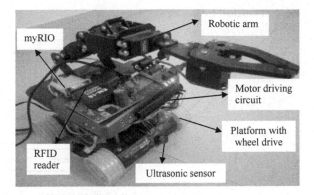

2.2 System Development

Figure 6 shows developed library management robotic system. This system contains NI myRIO device which is processing element. LabVIEW software is used to control and monitor the framework [6]. The arms are directed in X and Y directions to place the book. For multidirectional robotic movement, DC motors are fitted under the base of robotic chassis platform. Each book placed in rack is tagged by RFID encoder. The robot performs a brute force method search, and when the RFID tag information is matched with desired book, the robotic arm will close jaws to get a hold of the book. The arm is designed so that the book which it grips should not fall down. Suppose user wants to select particular book, then user has to give specific number which is tagged to the book. At that same time, controller starts the RFID module and starts book detection. RFID reader sends particular book tag information to myRIO and then robot starts to travel. If book is detected, then it will proceed for the verification. Arm will pick that book. After picking the book from the rack, robot will return to the book-issuing counter and place the book.

3 Flowcharts for Execution of System

3.1 Flowchart for Issuing Book

After entering the user login details, enter the required book to be searched. At that point, framework will show availability of book. Then, user needs to check out book. If searched book is accessible at that time, myRIO provides the flag to robot for fetching the book from specific rack to issuing kiosk (Fig. 7)

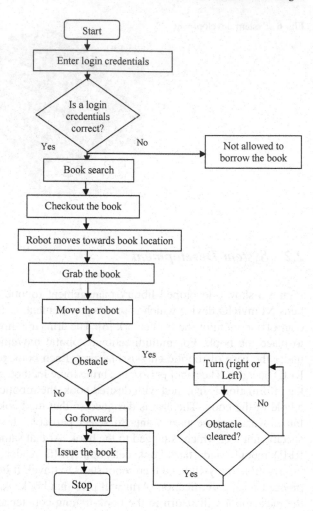

Fig. 7 Flowchart for issuing book

3.2 Flowchart for Returning Book

While returning the book, robot grips the book and detects location of book and moves toward this book location to place book (Fig. 8).

4 Experimental Results

While accessing library records, enter the login details for getting the record data. If login credentials are correct, then user can access the library. Following results show user interface VI, issuing book VI, returning book VI, searching book VI, admin access VI, etc.

Fig. 8 Flowchart for returning book

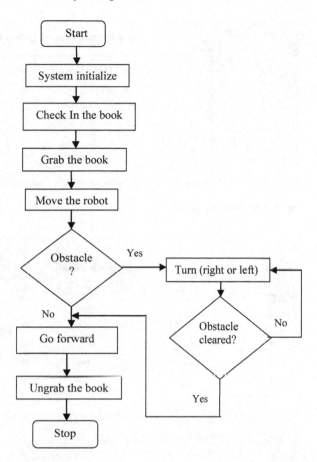

4.1 User Interface Display

See Fig. 9.

4.2 Issuing Book Display

See Fig. 10.

4.3 Returning Book Display

See Fig. 11.

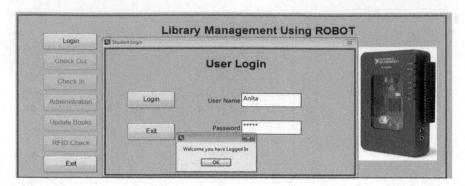

Fig. 9 User interface display

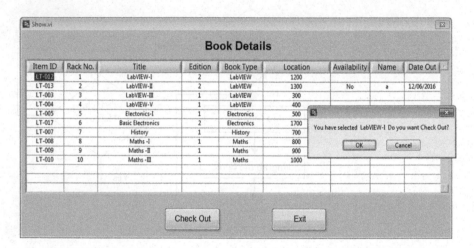

Fig. 10 Issuing book display

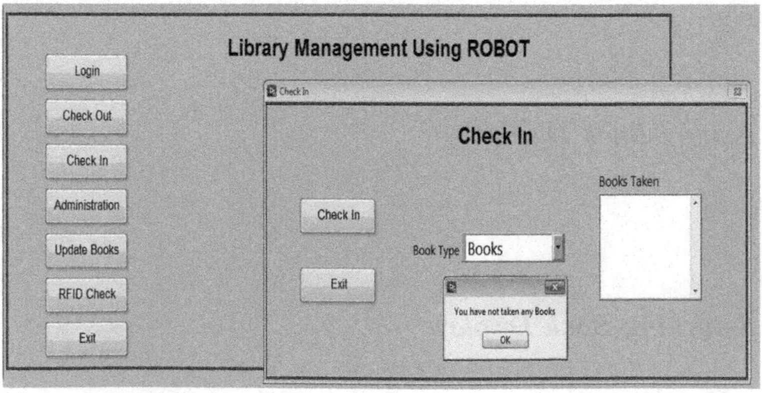

Fig. 11 Returning book display

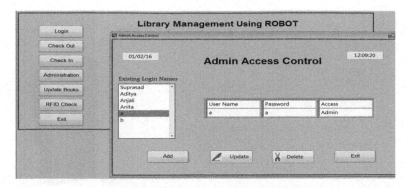

Fig. 12 Admin access display

4.4 Admin Access Display

See Fig. 12.

4.5 Book Update Display

See Fig. 13.

4.6 RFID Test Display

See Fig. 14.

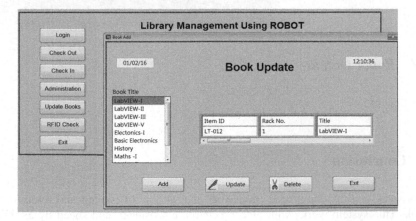

Fig. 13 Book update display

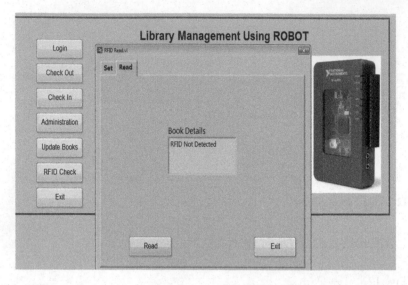

Fig. 14 RFID test display

Fig. 15 Robot output

4.7 Robot Output

See Fig. 15.

5 Conclusion

This work presents automated framework that is able of issuing and returning of book. This system works with high precision, consistency and speed by avoiding obstacle. This scheme eliminates the use of paper work by managing the book database electronically. Admin can update database of new books in library and their accessibility. This system has well systematized and analytically organized the books in various categories in scheme; thus, client can simply access the library.

Acknowledgements Authors would like to thank Savitribai Phule Pune University for partially financed BCUD research grant, Sanction Letter, OSD/BCUD/113/48. We express our gratitude to JSPM's BSIOTR, Wagholi, to support this work.

References

1. Charles M. Best, Phillip Hyatt, Levi Rupert, Marc D. Killpack, Vallan Sherrod: New Soft Robot Control Method: Using Model Predictive Control for Pneumatically Actuated Humanoid, IEEE Robotics and Automation Magazine, vol. 23, no. 3 (2016). doi:10.1109/MRA.2016.2580591
2. Jianing Chen, Melvin Gauci, Wei Li, Andreas Kolling and Roderich Grob: Occlusion based Co-operative Transport with a Swarm of Miniature Mobile Robots, IEEE Trans. on Robotics, vol. 31, no. 2 (2015). doi:10.1109/TRO.2015.2400731
3. Junmin Wu, Xiangyu Yue, and Wei Li: Integration of Hardware and Software Designs for Object Grasping and Transportation by a Mobile Robot with Navigation Guidance via a Unique Bearing-Alignment Mechanism, In: IEEE Trans. on Mechatronics, vol. 21, no. 1 (2016). doi:10.1109/TMECH.2015.2429681
4. M. Ramirez Neria, N. Lozada Castillo, M.A. Trujano Cabrera, J.P. Campos Lopez, A. Luviano Juarez: On the Robust Trajectory Tracking Task for Flexible Joint Robotic Arm with Unmodeled Dynamics, vol. 4. IEEE (2016). doi:10.1109/ACCESS.2016.2618373
5. Wang Zeyang, Zhao Ziang, Pang Zhifeng, and Zhang Chunlin: Kinematics Analysis and Simulation of a New 3 Degrees of Freedom Spatial Robot Mechanism Composed by Closed Chain, In: 2nd International Conference on Mechanic Automation and Control Engineering, (IEEE 2011). doi: 10.1109/MACE.2011.5987200
6. Anita Gade, Yogesh Angal: Automation in Library Management Using LabVIEW, In: International Conference on Computing Communication Control and automation, IEEE (2016). doi:10.1109/ICCUBEA.2016.7860133

A Novel Approach for Dynamic Decision Making by Reinforcement Learning-Based Cooperation Methods (RLCM)

Deepak A. Vidhate and Parag Kulkarni

Abstract A novel approach for dynamic decision making by reinforcement learning-based cooperation methods (RLCM) is proposed in this paper. Cooperation methods for reinforcement learning depending on multi-agent system are projected and executed. Various coordination techniques for cooperative RL are projected here, i.e., simpleGroup technique, simpleDynamic technique, simpleGoal technique. Performance outcome has established that the recommended coordination techniques are capable to speed up the learning of agents that bring out excellent exploitation plans. The methods are derived for changing items accessibility in a three seller stores in the marketplace. Dealers can help one another so as to obtain maximum revenue from coordination data by their individual strategies that exactly characterize their purpose and benefit. The dealers are the knowledgeable agents in the study with employed Q-learning to find out cooperation in the state. Significant assumptions are made on the dealer's supply plan, restock time, and entrance process of the consumers. The situation is converted into Markov decision process model to make it possible to use learning algorithms.

Keywords Cooperation techniques · Dynamic buyer behavior · Multi-agent learning · Reinforcement learning

D.A. Vidhate (✉)
Department of Computer Engineering, College of Engineering Pune,
Pune, Maharashtra, India
e-mail: dvidhate@yahoo.com

P. Kulkarni
iKnowlation Research Lab. Pvt. Ltd., Pune, Maharashtra, India
e-mail: parag.india@gmail.com

© Springer Nature Singapore Pte Ltd. 2018
S.S. Dash et al. (eds.), *International Conference on Intelligent Computing and Applications*, Advances in Intelligent Systems and Computing 632,
https://doi.org/10.1007/978-981-10-5520-1_37

1 Introduction

Many stores all over the county selling products to large number of consumers are an excellent prototype of marketplace. The retailing counters of dealer verify the data of every transaction, i.e., date, consumer number, items procured, and their spending amount. It physically capitulate enormous quantity of documentations each day. If collected documentation is examine and convert into data, then it develops into functional data. It will be used a design to construct predication [1]. It treats a process to make easy the exhibition of the experiential data. When it is unknown then emphasize on the process liable to establish the data—for instance, purchaser actions are identified as an absolutely unplanned. People do not go to marketplace and procure products at random. It might be incapable to identify the practice completely, but even practical and superior guesses might be constructed. Such provisional calculation may not provide actual of the whole thing, although could still capable to build some division of the information [2]. Dealers have for all time come across the complexity of trade and the accurate commodities that would generate the utmost returns for them. During cooperative occasion, dealers would propose extraordinary collection of items, only tailored for every consumer, only for the immediate on the entire items [2, 3]. Special factors are to be thought here, i.e., disparity in period, the dependence of things, unusual concessions, cut rates, and market circumstances. Dealers can coordinate with one another for highest revenue in various conditions [3, 4]. A marketplace prototype in the viewpoint of dynamic consumer actions is calculated in the paper. The following are the exact value addition of this paper.

Three dealer retail stores are considered which sell a selected item and provide quantity discounts for consumers purchasing many products. Seller's supply plan, restock time, and the entrance process of the consumers are calculated. A Markov decision process (MDP) model is suggested for this system. A new way for context-based dynamic decision making by cooperative multi-agent learning algorithms is proposed. A novel move toward dynamic decision making by reinforcement learning-based cooperation methods (RLCM) is proposed here. Coordination techniques for reinforcement learning based on multi-agent scenario are projected and executed [4, 5].

The article is arranged as: Sect. 2 illustrates novel move toward dynamic decision making by reinforcement learning-based cooperation methods (RLCM), coordination techniques are in Sect. 3, the organization of retail stores designed through Markov decision process is described in Sect. 4. Section 5 gives results of all four techniques having long-term price being the profit factor, and conclusion is in Sect. 6.

2 Reinforcement Learning-Based Cooperation Methods (RLCM)

Coordination in multi-agent system may create a sophisticated group of fulfillment achieved by the way of agents' procedures. A section of fulfillment set (a whole action policy) be allocated amongst all agents via a part action policy (Q_i). Usually, these part policies seize the limited information regarding the condition. Such strategies may be collected to improve the sum of the partial rewards obtained by means of sufficient coordination prototype. The action policies are derived using multi-agent simple Q-learning algorithm by integrating all rewards and constructing the agents to reach to the excellent policy Q^*. When policies $Q_1, \ldots, Q_x$ are collected, it is possible to construct new strategy so as to is whole action policy (WAP = \{WAP$_1, \ldots,$ WAP$_x$\}), in which WAP$_i$ show the best reinforcements received by agent i all over the learning techniques [5, 6].

sPlan algorithm given below is used to allocate agents' shared knowledge information. Policies are evaluated using a simple Q-learning for every technique. Best reinforcements are stored in the WAP that structures a group of the best-gathered rewards by each agent. Such rewards then disseminated by the added agents [4–6]. Coordination is accomplished through the revolution of division rewards as WAP is forecasted through the best reinforcements. A utility function is utilized to discover the excellent strategy amongst the beginning states and target state for the known policies to calculate the WAP with the excellent reinforcements. The utility function is found out by measuring the phases an agent needs to reach the final state and the total of the received cost in a strategy amongst all initial state and final state. Figure 1 gives coordination amongst all agents [6].

Figure 1 explains capability of agents to preserve through data of all agents in the coordination. Agent i utilizes simple Q-learning algorithm to construct and build up rewards in Q_i. The moment the agent i reaches through initial state to the final state with decreased rate, an agent probably competent to assign such rewards by fresh agents using communication techniques. Agents may restore their data after

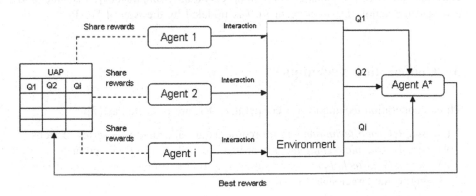

Fig. 1 Coordination scheme

receiving the rewards transmitted of each partial strategy Q_i and cooperate into the situation utilizing WAP [6, 7].

Multi-agent RL algorithm

```
algorithm sPlan (I, technique)
initTable: Qᵢ, Q*, WAP;
while agent i ∈ I do
    while state s ∈ S & a ∈ A do
            Qᵢ(s, a)← 0;
            WAPᵢ(s, a)← 0;
    done
done
episode ← 0;
while agent i ∈ I do
 if not simpleQ-Learning end_state is reached do
    event ← event +1;
    choice s ∈ S, a ∈ A
        restore Q(s, a) by Q(s, a)← Q(s, a) +α (r + γ
Q(s', a') - Q(s, a))
        fun_coop (event, tech, s, a, i);
    done
end if
while agent i ∈ I do
    Qᵢ←WAP;
done
return
```

Algorithm 1 is clarify as: Initialize $Q_i(s, a)$ and $WAP_i(s, a)$ and coordination through an agents $i \in I$; an Agent coordination until the final_state is achieved with renew policy found out the reward cost; *fun_coop* function decide a coordination techniques. *event, tech, s, a, I* are the factors, where *event* is current iteration, communication *tech* is {*simpleGroup, simpleDynamic, simpleGoal*}, *s* is state, and *a* is selected action; Q_i of an agent $i \in I$ is updated by the way of WAP_i.

3 Cooperation Techniques

Three cooperation techniques for cooperative RL are projected as [7]:

(i) *simpleGroup technique*—reinforcements are allocated in a series of steps.
(ii) *simpleDynamic technique*—rewards are distributed in each action.
(iii) *simpleGoal technique*—distributing the total reinforcement for agent reaching to the target state (Sgoal).

```
Algorithm 2 Cooperation model
fun_coop (event, tech,s,a,i)
q : count of sequence
switch technique
        Case "simpleGroup":
            if episode mod q = 0 then
                    get_Policy(Qᵢ, Q*,WAPᵢ);
            end if
    Case "simpleDynamic":
                    r ← Σˣⱼ₌₁ Qj(s,a);
                    Qᵢ(s,a)← r;
                    get_Policy(Qᵢ, Q*,WAPᵢ);
    Case "simpleGoal":
            if S =  S_goal then
                    r ← Σˣⱼ₌₁ Qj(s,a);
                    Qᵢ(s,a)← r;
                    get_Policy(Qᵢ,Q*,WAPᵢ);
            end if
end switch


Algorithm 3 get_Policy
Function get_Policy(Qᵢ, Q*,WAPᵢ)
while agent i ∈ I do
        while state s ∈ S do
            if utility(Qᵢ, s) ≤ utility(Q*,s) then
                    WAPᵢ(s,a) ← Qᵢ(s,a);
            end if
    done
done
```

simpleGroup technique: Agents gather rewards obtained from its actions throughout the learning sequence. Each agent put up the utility of Q_j to WAP during the last part of the progression (step q). If reward value is correct, that is, it improves the effectiveness of other agents in support of known state. The agents will subsequently supply to these rewards. Agent will continue to employ its rewards with the purpose intended for assemble newest cost [7, 8].

simpleDynamic technique: The coordination in the simpleDynamic technique is achieved as: each action performed by agent produces a reinforcement cost (+ or −), that is, sum of received rewards to all agents to an action a achieved in a state *s*. Each agent coordinates toward making the reward cost fulfilling its personal strategy [7, 8].

simpleGoal technique: The cooperation takes place as agent reaches to the destination at its target state. Agent coordinates during condition proposed to collect the highest quantity of reinforcements. It is necessary for the cause so as to in the simpleGoal technique the agent allocates its rewards by means of a variable number of happening. This cooperation techniques uses as a fast group of rewards taken together by an agent during the cooperation. As soon as agent reaches to a target state, it gives value of acquired rewards in condition to the WAP [8, 9].

4 Model Design

Consider the occasion of marriage season development. Starting with choosing the wedding location, finalizing the menu, beautification, shooting, makeup, purchasing of clothing, jewels, and supplementary items for bride and groom, so many activities are involved [10]. Such seasonable condition can be practically employed as given: consumer who would choose to purchase the cloths from cloth store surely purchase jewels, footgear, and supplementary items. Dealers of different items may get jointly and mutually fulfill consumer needs and would accomplish the advantage of an enhancement in the item selling [10, 11]. Figure 2 gives diagrammatic representation of the system.

Below are mathematical notations for above model.

- Consumers come into the marketplace by following a Poisson flow rate λ.
- Seller posts per unit item price p to the incoming consumers.
- Seller has limited store capacity I_{max}. It maintains a permanent restructure strategy for refilling.

States: Assume maximum stock level at each store = I_{max} = $i1$, $i2$, $i3$ = 20.

State for agent 1 becomes (x_1, i_1), e.g., (5, 0) that means 5 customer requests with 0 stock in store 1. State for agent 2 becomes (x_2, i_2), State for agent 3 becomes (x_3, i_3).

State of the system becomes **Input** as (x_i, i_i).

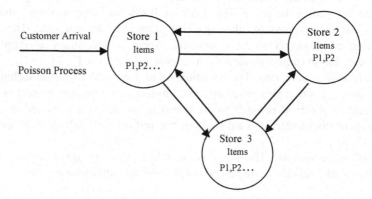

Fig. 2 Retail store model with three dealers

Actions: Assume set of possible actions, i.e., action set for agent 1 is (that means Price of products in store 1) $A1$ = Price p = {8–14} = {8.0; 9.0; 10.0; 10.5; 11.0; 11.5; 12.0; 12.5; 13.0; 13.5}. Set of possible actions, i.e., action set for agent 2 is $A2$ = Price p = {5–9} = {5.0; 6.0; 7.0; 7.5; 8.0; 8.5; 9.0}. Set of possible actions, i.e., action set for agent 3 is $A3$ = Price p = {10–13} = {10.0; 10.5; 11.0; 11.5; 12.0; 12.5; 13.0}.

Output is the possible action taken, i.e., price in this case. It is now the state-action pair system can be easily modeled using Q-learning, i.e., $Q(s, a)$. There is need to define the reward calculation [11, 12].

Rewards: Reward is calculated in the system as $R(s, a)$ during Q update function.

Whenever a consumer put up a request for an item, a decision needs to be completed regarding whether to allow or refuse the request. Another dealer studies the actions followed by first dealer and are ready to sell his items. In this way, as sale in one store increases automatically other stores get informed so they can sell their products [12].

5 Results

Algorithms are tested on one year's transaction dataset of three different retail stores and results are observed. Figure 3 shows that profit margin versus number of state transactions given by four methods. Profit obtained by cooperative methods, i.e., simpleGroup, simpleDynamic, and simpleGoal techniques is much more than that of without cooperation method, i.e., simple Q-learning for agent 1. Cooperation by simpleGroup and simpleDynamic technique is more suitable for agent 1 to obtain the maximum profit.

Figure 4 shows that profit margin versus number of state transactions given by four methods. Profit obtained by cooperative methods, i.e., simpleGroup,

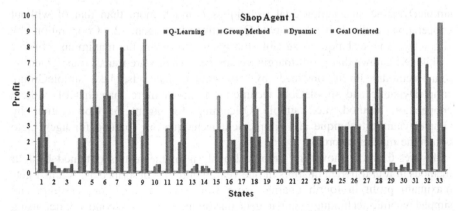

Fig. 3 Profit obtained by store agent 1 using four learning methods

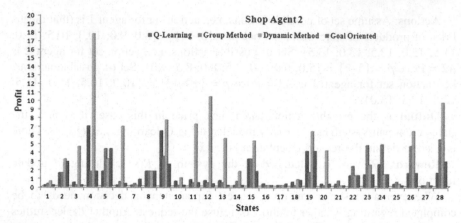

Fig. 4 Profit obtained by store agent 2 using four learning methods

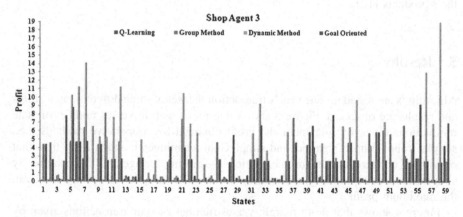

Fig. 5 Profit obtained by store agent 3 using four learning methods

simpleDynamic, and simpleGoal techniques is much more than that of without cooperation method, i.e., simple Q-learning for agent 2. Cooperation by simpleDynamic technique is suitable for agent 2 to obtain the maximum profit.

Figure 5 shows that profit margin versus number of state transactions given by four methods. Profit obtained by cooperative methods, i.e., simpleGroup, simpleDynamic and simpleGoal techniques is much more than that of without cooperation method, i.e., simple Q-learning for agent 3. Cooperation by simpleDynamic technique and simpleGoal technique are suitable for agent 3 to obtain the maximum profit.

Figures 6 and 7 shows the graphical analysis of the results obtained by four methods in four different quarters in one year. Figure 6 is described as: Agent 1 gets maximum profit in fourth quarter using simple Q-learning, simpleGroup, and simpleDynamic techniques, and it gets maximum profit in second quarter using

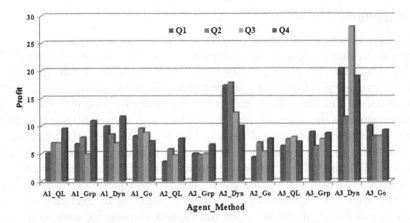

Fig. 6 Quarterly profit obtained by all store agents by four learning methods

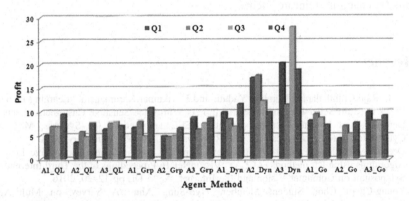

Fig. 7 Quarterly profit obtained by all store agents by four learning methods

simple Goal technique. Agent 2 gets maximum profit in first, second, and third quarter using simpleDynamic technique, whereas it gets average profit in fourth quarter using simple Q-learning, simpleGroup, and simpleGoal technique. Agent 3 gets maximum profit in first, third, and fourth quarter using simpleDynamic technique, whereas it gets average profit in second quarter using simple Q-learning, simpleGroup, and simpleGoal technique.

Figure 7 is described as: Agent 1 gets more profit as compared to agent 2 and agent 3 using simple Q-learning method and simpleGroup technique in fourth quarter. Agent 3 gets more profit as compared to agent 1 and agent 2 using dynamic in third quarter. Agent 3 gets more profit as compared to agent 1 and agent 2 using simpleGoal technique in first quarter.

6 Conclusion

Dynamic consumer activities are noticeably learned by this novel method. The results obtained by the projected cooperation methods show with the intention of these techniques get to a quick convergence of agents which interchange the reinforcements. The Paper shows cooperative methods give a strong results in more density, incompletely, and composite circumstances. It offers a help to interchange of best rewards to obtain a good whole action plan. All cooperation methods are able to guarantee best rewards which were acquired along learning process and change with a group of best rewards received in incomplete action strategies. The interchange of Q-function throughout three techniques, i.e., simpleGroup, simpleDynamic and simpleGoal technique, a store agent evaluates the most excellent possible items so as to get the highest revenue for each store agent. Dynamic decision making using reinforcement learning-based cooperation methods (RLCM) demonstrates the techniques may lead to a quick interaction between agents by changing reinforcements.

References

1. V. L. Raju Chinthalapati, Narahari Yadati, and Ravikumar Karumanchi, "Learning Dynamic Prices in Multi-Seller Electronic Retail Markets With Price Sensitive Customers, Stochastic Demands, and Inventory Replenishments", IEEE Transactions On Systems, Man, And Cybernetics—Part C: Applications And Reviews, Vol. 36, No. 1, January 2008.
2. Deepak A. Vidhate, Parag Kulkarni "New Approach for Advanced Cooperative Learning Algorithms using RL methods (ACLA)" VisionNet'16 Proceedings of the Third International Symposium on Computer Vision and the Internet, ACM DL pp 12–20, 2016.
3. Young-Cheol Choi, Student Member, Hyo-Sung Ahn "A Survey on Multi-Agent Reinforcement Learning: Coordination Problems", IEEE/ASME International Conference on Mechatronics and Embedded Systems and Applications, pp. 81–86, 2010.
4. Deepak A. Vidhate, Parag Kulkarni "Innovative Approach Towards Cooperation Models for Multi-agent Reinforcement Learning (CMMARL)" in Smart Trends in Information Technology and Computer Communications, Springer Nature, Vol. 628, pp 468–478, 2016.
5. Zahra Abbasi, Mohammad Ali Abbasi "Reinforcement Distribution in a Team of Cooperative Q-learning Agent", Proceedings of the 9th ACIS Int. Con. on Software Engineering, Artificial Intelligence, and Parallel/Distributed Computing 978-0-7695-3263-9/08 pp 154–160, IEEE 2008.
6. La-mei GAO, Jun ZENG, Jie WU, Min LI "Cooperative Reinforcement Learning Algorithm to Distributed Power System based on Multi-Agent" 3rd International Conference on Power Electronics Systems and Applications Digital Reference: K210509035, 2009.
7. Deepak A. Vidhate, Parag Kulkarni "Enhancement in Decision Making with Improved Performance by Multiagent Learning Algorithms" IOSR Journal of Computer Engineering, Volume 1, Issue 18, pp 18–25, 2016.
8. Adnan M. Al-Khatib "Cooperative Machine Learning Method" World of Computer Science and Information Technology Journal (WCSIT) ISSN:2221–0741 Vol.1, 380–383, 2011.
9. Liviu Panait, Sean Luke "Cooperative Multi-Agent Learning: The State of Art", Journal of Autonomous Agents and Multi-Agent, 11, 387–434, 2005.

10. Jun-Yuan Tao, De-Sheng Li "Cooperative Strategy Learning In Multi-Agent Environment With Continuous State Space", IEEE Int. Conf. on Machine Learning, 2006.
11. Deepak A. Vidhate, Parag Kulkarni "Multilevel Relationship Algorithm for Association Rule Mining used for Cooperative Learning" in International Journal of Computer Applications (IJCA), Vol. 86 No 4, pp. 20–27, 2014.
12. Deepak A. Vidhate, Parag Kulkarni "Improvement In Association Rule Mining By Multilevel Relationship algorithm" in International Journal of Research in Advent Technology (IJRAT), Vol 2 No 1, pp. 366–373, 2014.

Interaction of Cell Phone Radiations and Human Nervous System

Manish Kr. Gupta, R.K. Khanna, K.J. Rangra and Y.K. Vijay

Abstract Conduction of an external sensed signal in the human nervous system can be defined in the form of electrical RC circuit equivalent which is realized very well by the researchers. In this paper, experimental results due to the continuous exposure to external cell phone radiation on the nervous system are being reported. Experimental data are stored using DSO and analysed by ORIGIN software. Comparison of rise/fall time of electrochemical KCL in the presence and absence of cell phone radiation is shown for this purpose.

Keywords Nerve conduction · Cell phone radiations · EM signal
Electrochemical transmission

1 Introduction

The well-known behaviour of the nerve conduction is realized in the form of RC transmission line by the biomedical researchers. It is shown that conduction in this transmission line is dependent on excitation and de-excitation of Na and K ions [1, 2]. Change in behaviour in this conduction process and then on the overall performance of the brain due to the presence of any external signals is still unknown.

Mechanism for the penetration of EM signal inside the skin of humans is developed, but only thermal effects due to these signals are reported available worldwide. Specific absorption rate (SAR) is used to define and find the penetration of the EM field inside the skin. These effects are minimal in amount and can be further optimized by changing the way of use of cell phones.

M.Kr. Gupta (✉) · R.K. Khanna · Y.K. Vijay
Department of Physics and Electronics, Vivekananda Global University, Jaipur, India
e-mail: mkg.met@gmail.com

K.J. Rangra
Sensors & Nano-technology Group, SDA, CSIR-Central Electronics Engineering
Research Institute, Pilani, India

© Springer Nature Singapore Pte Ltd. 2018
S.S. Dash et al. (eds.), *International Conference on Intelligent Computing
and Applications*, Advances in Intelligent Systems and Computing 632,
https://doi.org/10.1007/978-981-10-5520-1_38

413

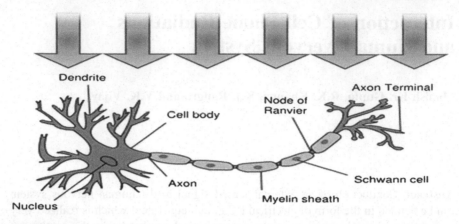

Fig. 1 A neuron with its electrochemicals under the exposure of external EM field [1]

Effects of EM radiations on human nervous system have already been simulated using MATLAB® [3]. Simulation results show significant variation in the propagation time of the signal through the nerve. In this paper, based on the concept of EM field penetration in the skin, an experimental study is reported using direct imposition of cell phone radiations on the signal propagation in a solution of an electrochemical (KCl) body. Affects of the presence of the radiations are measurable and show significant variation in the signal transmission through the solution. This work may lead to strengthen the non-thermal effects of the EM waves (Fig. 1).

2 Theory

Conduction in the nervous system is a well-defined process and can be understood by the ion-exchange mechanism in the inner and outer side of axons [4, 5]. Na^+, K^+, Cl^- and proteins are major responsible ions in conduction process (Fig. 2).

Lloyd Hodgkin and Andrew Huxley developed Hodgkin–Huxley model to explain nerve conduction and received the Nobel Prize for this in 1963 [6, 7]. This model characterizes the initiation and propagation of neural signals in giant axons of squids and is able to describe very well the dynamic behaviour of channel kinetics [6] (Fig. 3).

This mathematical model also explains the experimental results of the voltage clamp experiment [8, 9]. With this mathematical model, the prediction of stimulus response of nerve system is possible (Fig. 4).

Fig. 2 Movement of charges inside the axon in response to a stimulus [1]

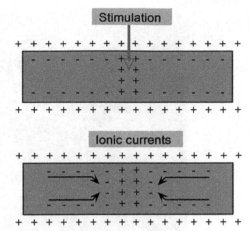

Fig. 3 An equivalent electrical circuit for the neuron [1]

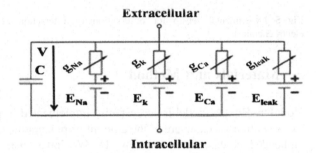

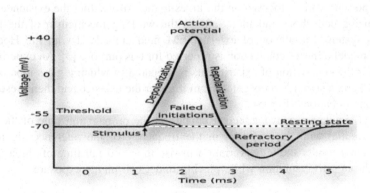

Fig. 4 Different phases of action potential [10]

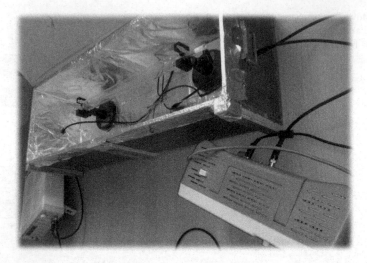

Fig. 5 Experimental set-up for observation of interaction of cell phone radiations with electrochemical

3 Materials and Method

Hodgkin–Huxley model is successfully implemented in past using MATLAB® [2, 7]. A continuous exposure of the amount of penetrating external EM field has been embedded in this simulation (Fig. 1). We have taken a 5 mV AC signal as a reference for the penetrated signal [3].

The present work is focused on the investigation of whether the existence of the non-ionizing cell phone radiations affects the working mechanism of the human nervous system. Simulation of external EM field embedded with the Hodgkin–Huxley model of nerve conduction is reported for this purpose [3]. An experimental set-up for the conduction of experiments is prepared in which a glass tube is filled with KCl and a signal is propagated from it under the absence and then presence of cell phone radiations (Fig. 5).

Model simulation shows effective change in the propagation time of the membrane potential with the variation in the frequency of the reference signal, and experimental results show the variation in rise time and fall time of signal propagation through KCl solution while placing under cell phone exposure.

4 Experimental Results and Conclusion

A significant variation in rise and fall times of KCL solution is recorded in the presence of cell phone radiations.

Fig. 6 Variation in fall time of pulse signal propagation through KCL solution in the absence/presence of cell phone radiations: **a** original real-time propagation and **b** zoomed view for clear vision

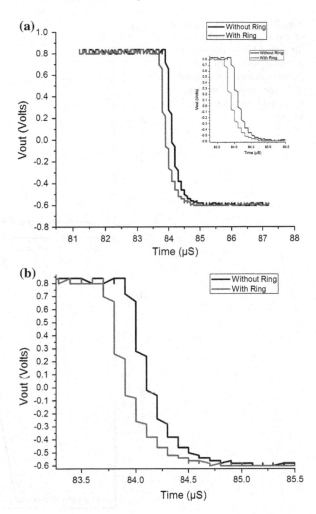

It is clear that both rise and fall time decreases when the observation is recorded in the presence of the radiations. This behaviour clarifies that the working of human nervous system is not totally independent of the high-frequency non-ionized radiations (Fig. 6).

This external signal works as catalyst in the case of the KCl electrochemical solution. Because of that, the response time of nerve to any stimulus is decreased.

Similar behaviour is found in the rise time of the KCL solution in the presence of the external cell phone radiations. Significant reduction in rise time is recorded while the signal processing is checked in the presence of the radiations (Fig. 7).

Action potential and its transmission in a human nerve are an electrochemical process and can be realized by equivalent electrical circuit which occurs due to the flow of variation in concentration of different chemicals inside and outside of the axon. Decrement in the rise and fall time of KCl solution affects the rate of flow of

Fig. 7 Variation in rise time of pulse signal propagation through KCL solution in the absence/presence of cell phone radiations: **a** original real-time propagation and **b** zoomed view for clear vision

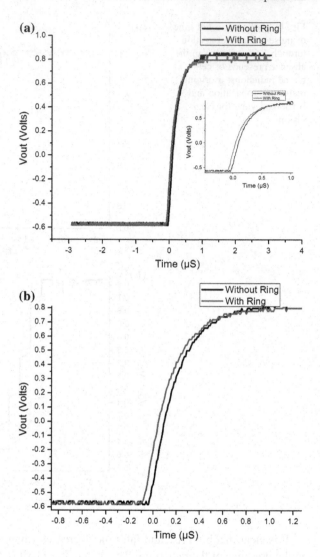

K^+ and Cl^- ions which results as variation in the propagation of stimulus inside neuron.

5 Work Under Process

Simulation based results [3] related to effects of non-ionizing radiation on human health and experimental results reported in this work motivated us to enhance the experimental set-up for the verification of the effects of cell radiation exposures on other electrochemical. We are currently working with different electrolyte solutions and checking the propagation of electric signal with and without the presence of cell radiation exposures. Individual study on each electrochemical may provide important answers of how cell phone radiations are affecting the nervous communication and brain response process.

6 Future Work

Effect of EM radiation on human health is a topic of great debate. This work evidences the effect of radiations on working of human nervous system. Further experimental verifications are under process.

Development of the artificial membrane and analysis of this effect on the membrane may be carried out for the detailed study.

Variation in peak action potential and propagation time may lead to some advancement in electronic neurology.

Further studies may lead to further development in control on the nerve signal processing.

References

1. Gupta MK, Khanna RK, Rangra KJ, "Study and Analysis of Human Nervous System and Effects of External EM Signals on Conduction in Human Nerves", International Journal on Recent and Innovation Trends in Computing and Communication", Volume: 4 Issue: 4, 448–458.
2. Forehand C J., Ph.D., "The Action Potential, Synaptic Transmission, and Maintenance of Nerve Function", Cellular Physiology, Chapter 3, 2009.
3. Gupta MK, Khanna RK, Rangra KJ, "Study of Effect of Non-ionizing External EM-Field on Nerve Conduction" Advanced Science, Engineering and Medicine (ASP), Accepted.
4. Zechari Ryan Tempesta, "Action Potential Simulation of the Hirudo Medicinalis's Retzius Cell in Matlab", M.Sc. Thesis, California Polytechnic State University, 2013.
5. Thompson Kathryn, Stewart Meredith, Rodriguez Juan, "Nerve Conduction", 2004.
6. Kenneth Leander Anderson Jr., Jackie Chism, Quarail Hale, Paul Klockenkemper, Chelsi Pinkett, Christopher Smith, and Dr. Dorjsuren Badamdorj "Mathematical Modeling Action Potential in Cell Processes" June, 2013.
7. Doi S, Inoue J, Pan Z, Tsumoto K, "The Hodgkin–Huxley Theory of Neuronal Excitation", Chapter 2, Springer, ISBN 978-4-431-53861-5.
8. Ingalls Brian, "Mathematical Modeling in Systems Biology: An Introduction", University of Waterloo, 2012.
9. C. Minos Niu, Sirish K. Nandyala, Won Joon Sohn Terence, D. Sanger, "Multi-scale Hyper-time Hardware Emulation of human Motor Nervous System Based on Spiking Neurons using FPGA", Advances in Neural Information Processing Systems 25, 2012.
10. BP Bean, " The action potential in mammalian central neurons", Nature Reviews, Neuroscience, June 2007.

Simultaneous Scheduling of Machines and Tools Considering Tool Transfer Times in Multimachine FMS Using CSA

N. Sivarami Reddy, D.V. Ramamurthy and K. Prahlada Rao

Abstract This article addresses simultaneous scheduling of machines and tools considering tool transfer times between machines, to produce best optimal sequences that minimize makespan in a multimachine flexible manufacturing system (FMS). The performance of FMS is expected to improve by effective usage of resources, proper amalgamation and synchronization of their scheduling. Crow search algorithm (CSA) is a potent tool which is a better choice to solve optimization problems like scheduling. The proposed heuristic is tested on various problems with makespan as objective, and the results are compared with results of existing methods. The results show that CSA has outperformed.

Keywords Flexible manufacturing systems · Crow search algorithm
Priority dispatching rules · Simultaneous scheduling of tools and machines
Tool transporter

1 Introduction

FMS is an integrated manufacturing system which includes various facilities such as computer numerically controlled (CNC) machines, automated guided vehicles (AGVs), automated storage/retrieval systems (AS/RSSs), central tool magazine (CTM), robots and automated inspection under the control of a central computer [1, 2]. Various subsystem flexibilities are integrated together to have an overall

N. Sivarami Reddy (✉)
JNTUA, Ananthapuram, AP, India
e-mail: siva.narapureddy@gmail.com

D.V. Ramamurthy
GIET, Rajahmundry, AP, India
e-mail: ramdwivedula@gmail.com

K. Prahlada Rao
JNTUACEA, Ananthapuram, AP, India
e-mail: drkprao1@yahoo.com

© Springer Nature Singapore Pte Ltd. 2018   421
S.S. Dash et al. (eds.), *International Conference on Intelligent Computing
and Applications*, Advances in Intelligent Systems and Computing 632,
https://doi.org/10.1007/978-981-10-5520-1_39

flexibility in FMS. One of the recent techniques in industrial automation is FMS, and several researchers have been attracted over the last 30 years towards FMS. FMS has various advantages such as greater productivity, low work-in-process inventory, high machine utilization, production with least supervision, increased product variety and high quality to satisfy customer requirements. The employing of pallets, fixtures, tool transporter (TT) and CTM nearly eliminated the job setting time [3].

The higher flexibility of FMS results in better utilization of resources and better scheduling, and routing enhances the productivity [4]. Broadly, FMS is categorized into four groups: single flexible machines (SFMs), flexible manufacturing cells (FMCs), multimachine FMS (MMFMS) and multicell FMS (MCFMS). FMS aims at combining the advantages of elevated efficiency in high-quantity mass production and better flexibility in job shop production.

In FMS, in order to attain the elevated efficiency and flexibility, different scheduling decisions such as allotment of machines to jobs and selection of tools are made. Appropriate scheduling plays a critical task in FMS.

2 Literature Review

For improvement of shop floor productivity, scheduling is recognized to be a crucial task. For "p" jobs and "q" machines "$(p!)^q$", different sequences are to be inspected with regard to a performance measure, to suggest the best sequence in the problem of scheduling. This implies that the search region is increased exponentially for problem of larger size that makes the problem of scheduling an NP-hard problem. In FMS, different jobs are to be assigned to machines to optimize the FMS performance. This is analogous to job shop scheduling. The main difference between them is that the job shop considers only jobs and machines, whereas FMS considers resources such as AGVs, CTM, robots, AS/RS, fixtures and pallets besides jobs and machines. Hence, problems of scheduling in FMS are also NP-hard.

Jerald and Asokan [3] presented various optimization algorithms for solving FMS scheduling problems. In scheduling area of FMS, for optimization, earlier the researchers had recognized scheduling of machines and scheduling of tools as two different problems, whereas of late much interest has been noticed for collective effect of scheduling of machines and scheduling of tools.

Several researchers have studied tool scheduling and allocation. Tsukada and shin [5] addressed scheduling of tool and problem of borrowing tool in FMS and also tool sharing approach to take into account the unforeseen coming of job in dynamic environment by employing a distributed artificial intelligence method. It is observed that sharing of tools among different cells of FMS cuts down the cost of tooling and improves effective utilization of tools.

Jun et al. [6], for provisioning problem and scheduling of tools in FMS, proposed a greedy search algorithm to find the tools with required number from each type for minimizing makespan objective. This method also gives information about extra tools to be brought when FMS configuration alters due to product-mix change.

Sureshkumar and Sridharan [7] dealt with the problem of sharing and scheduling of tools for minimizing the objectives such as mean tardiness, conditional mean tardiness and flow time by employing priority rules of scheduling and job scheduling. Sureshkumar and Sridharan [8] investigated the problem of tool scheduling in FMS by minimizing mean flow time, mean waiting time, mean tardiness for tool and percentage of tardy parts by using various priority dispatching rules. Agnetis et al. [1] probed a problem of concurrent part/tool scheduling in FMC. They proposed that all the tools are placed in a central tool magazine and are moved throughout the cell by an automatic tool transporter. When the same tool is required by two machines, Tabu Search was employed to address the conflict and production schedule preparation for makespan minimization and lateness maximization. The FMC overall performance may be enhanced by combined scheduling by taking tools and fixtures into account. Prabhaharan et al. [9] attempted on combined tool operation scheduling problem in FMC which consists of a CTM and "m" identical work cells. They proposed simulated annealing for makespan minimization for concurrent job scheduling and tool scheduling. Udhaykumar and Kumanan [10] proposed ant colony optimization for job and tool scheduling problems. Aldrin Raj et al. [11] addressed concurrent tool scheduling and machine scheduling in a FMS which has machines and a CTM. They proposed four different algorithms and AIS algorithm, to solve concurrent tool and machine problems with makespan minimization as objective. They found that AIS algorithm yielded superior results for concurrent scheduling of tools and machines.

Automated tool sharing system is a technological reply to high tool cost in FMS by allowing different machines to employ the same tool by shifting them automatically between machines as tooling needs evolve. In the previous studies, some assumptions have been made about concurrent tool and machine scheduling in FMS which consist of machines and CTM. One of those assumptions is that the transferring time of tools between different machines is negligible. However, it is highly not practical not to consider the tool handling activities in reality, especially when the movement of tools is completely dependent on tool transporter and travelling times are considerable. Omitting the tool transfer times will make the scheduling result impossible to be implemented.

In this work, a new metaheuristic search algorithm CSA is used to minimize makespan by simultaneous scheduling of jobs and tools considering transferring times of tools between machines and is explained in the following sections.

3   Problem Formulation

Generally, CTM is provided in FMS for storage of tools. The tool required by a machine is transported from the CTM or shared from other machines to this machine by a TT during the machining of job. CTM cuts down number of tools that are required in system and hence brings down the tool cost, whereas tool transfer time considerably influences the makespan. The problem definition, assumptions and constraints are given in the following sections.

3.1 Problem Definition

Consider "n" jobs $\{J_1, J_2, J_3 \ldots J_n\}$ to be processed through "m" machines $\{M_1, M_2, \ldots M_m\}$ requiring "t" tools $\{T_1, T_2, \ldots T_t\}$ from CTM. The best sequence by joint selection of machines, jobs and tools is to be found which minimizes the makespan. In the present work, CSA is employed to produce best schedules with makespan minimization as objective. The same problem set that was analysed with methods explained in [11] is considered, and the CSA results are compared with those results.

The procedure employed is explained with the example problem. In Table 1, the jobs, tools and machines given are for job set 1. The job set 1 consists of 5 jobs, the first three jobs have three operations and remaining two jobs have two operations. The system considered has four machines and four tools. An entity in the table gives information about the machine, tool and processing time needed for the operation of a job. For example, T3-M1 {8} shows that Operation I of Job I requires tool T3, machine M1 and 8 units of processing time. The objective is to determine an operations sequence of jobs for makespan minimization by taking machine and tool constraints into account. Making decision on selecting a machine and tool for every operation of a job is required during the process scheduling. Machine and CTM will have a number of requests from incomplete jobs in the form of queue. A right operation of a job with a request has to be picked up to minimize the makespan. Hence, a sequence of operations is obtained which minimizes total elapsed time.

3.2 FMS Environment

FMS considered in this work has 4 machines, 4 tools in a CTM, automatic tool changer (ATC), AGVs and TT. There is a load/unload (L/U) station on one end. Jobs are stored in the buffer storage provided at each machine before and after processing. The system is shown in Fig. 1 with the elements.

Table 1 Job set 1

Jobs	Operation I	Operation II	Operation III
I Job (J1)	T3-M1 {8}	T4-M2 {16}	T1-M4 {12}
II Job (J2)	T2-M1 {20}	T3-M3 {10}	T1-M2 {18}
III Job (J3)	T1-M3 {12}	T4-M4 {8}	T2-M1 {15}
IV Job (J4)	T3-M4 {14}	T4-M2 {18}	–
V Job (J5)	T2-M3 {10}	T1-M1 {15}	–

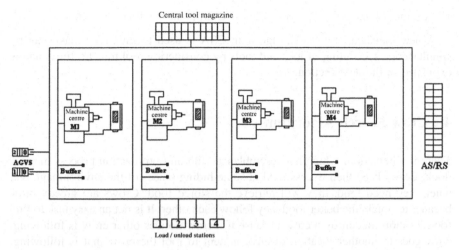

Central tool magazine

Load / unload stations

Fig. 1 Considered FMS environment

- **Assumptions and Constraints**.
- The following assumptions are made for the problem under study.
- Each job has J different operations.
- Required machines and tools are known prior to scheduling to process each operation.
- Operations in a job have processing order of its own.
- Each job has the pre-specified operations sequence and its corresponding processing times.
- Each machine can process one job at a time.
- Tools are stored in CTM.
- Tool transporter moves the tools throughout the system.
- Tools are shared between machines in the system.

The problem constraints are given below.

- Precedence constraints exist, that is a set of pre-specified operations sequence will be there for every job that cannot be changed.

Consider the operation 4143
4—Job number
1—First operation of J4
4—First operation of J4 is performed on machine 4.
3—First operation of J4 requires tool 3.
The second operation of J4 can be processed only after first operation completion, and hence, operation 42XX will not be processed before 41XX. This constraint is known as precedence constraints.

- A same job cannot be processed on two different machines at a time.

A new metaheuristic search algorithm CSA is used to minimize makespan by simultaneous scheduling of jobs and tools by considering tool transfer times and is explained in the next section.

4 Crow Search Algorithm

There is a behaviour which has resemblance with an optimization process in crow flock. Crows keep their excess food in some hiding places of the environment, and when they need food, they will retrieve the stored food. Crows are greedy birds because to obtain the better food they follow each other. It is not an easy task to find food location hidden by a crow because if a crow notices other crow is following, crow goes to another location of environment to fool the crow that is following. CSA tries to imitate the behaviour of crows to find solution for optimization problems [12].

The following are the principles underlying in this algorithm.

- Crows live in group.
- Crows remember the location of hiding places of their food.
- Crows go behind other crows to do pilfering.
- Crows guard their food by a probability from being pilfered.

A d-dimensional environment is assumed which includes number of crows. N is the number of crows (group size), and position of *crow i* at iteration iter in search space is represented by a vector.

$x^{i,\text{iter}}$ ($i = 1, 2\ldots N$, iter $= 1, 2\ldots$ iter$_{\text{max}}$) where $x^{i,\text{iter}} = [x_1^{i,\text{iter}}, x_2^{i,\text{iter}}, x_3^{i,\text{iter}} \ldots x_d^{i,\text{iter}}]$ and iter$_{\text{max}}$ is the maximum number of iterations. Each crow remembers its hiding place in its memory. Hiding place position of *crow i* is shown by $m^{i,\text{iter}}$ at iteration iter.$m^{i,\text{iter}}$ is the best position obtained so for by *crow i*. Crows will move in environment and search for superior food locations that are hiding places. Assume that *crow j* wishes to go to its hiding place $m^{i,\text{iter}}$ at iteration iter.*crow i* decides to follow *crow j* to reach the hiding place of *crow j* at this iteration. Two situations may occur.

Situation 1: *Crow j* does not notice that *crow i* is following it. So *crow i* approach to the hiding place of *crow j*. The new location of *crow i* is obtained as follows.

$$x^{i,\text{iter}+1} = x^{i,\text{iter}} + r_i \times \text{FL}^{i,\text{iter}} \times (m^{j,\text{iter}} - x^{i,\text{iter}}) \tag{1}$$

where r_i is the random number between 0 and 1, and FL$^{i,\text{iter}}$ is the flight length of *crow i* at iteration iter.

This situation and the effect of FL on search capability are shown in Fig. 2. Large values of flight length result in global search, whereas small values of FL lead to local search.

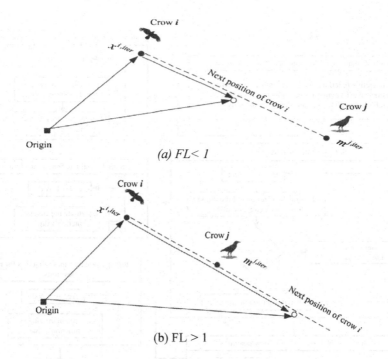

Fig. 2 Effect of flight length on search capability

Situation 2: Crow j notices that *crow i* is following it, as a consequence *crow j* goes to another search space position in order to protect its food being pilfered.

Entirely, situations 1 and 2 are expressed as follows:

$$x^{i,\text{iter}} = x^{i,\text{iter}} + r_i \times \text{FL}^{i,\text{iter}} \times \left(m^{j,\text{iter}} - x^{i,\text{iter}}\right) \quad if \ r_j \geq \text{AP}^{j,\text{iter}}$$
$$= \text{a random position otherwise} \tag{2}$$

where $\text{AP}^{j,\text{iter}}$ is the awareness probability of the crow j at iteration iter. Awareness probability controls the intensification and diversification. Using small values of AP increases intensification, whereas using large values of AP increases diversification.

The flow chart for CSA is given in Fig. 3.

5 Results and Discussion

Initially, makespan optimization in FMS by concurrent scheduling of jobs, tools and machines without considering tool transfer times has been executed by the proposed algorithm. Totally 22 job sets are considered in the work, and the data of these job sets are given in [11]. The job sets with different numbers of jobs, tools

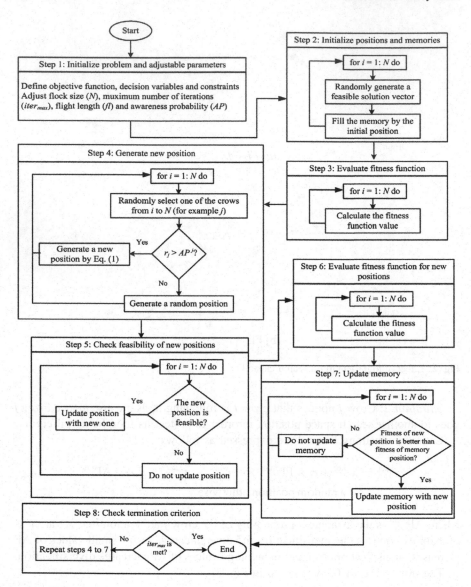

Fig. 3 Flow chart for CSA

and machines with different processing times have been taken into account to test the efficacy of CSA method. The results are compared with the results of existing methods [11] and are shown in Table 2.

From Table 2, it is obvious that proposed CSA method is yielding better results. The best makespan is indicated in bold, and it is observed that CSA outperforms all existing methods for all 22 job sets. For majority, job sets' improvement is noticed. For job set 22, the improvement is maximum and is 45.43%. For job set 21, the

Table 2 Makespan comparison of CSA and existing methods for 22 job sets

Job set	MWR	LWR	LPT	SPT	MOR	LOR	ND-MWR	ND-LWR	ND-LPT	ND-SPT	ND-MOR	ND-LOR	MGTA	AIS	CSA	% Improvement
1	104	116	77	100	125	116	90	101	86	83	88	73	77	69	69	0.00
2	112	133	111	125	153	133	96	107	98	90	98	87	90	82	80	2.44
3	90	151	148	139	121	141	87	115	108	105	115	105	87	80	80	0.00
4	80	152	119	132	77	154	74	83	89	73	73	78	84	72	61	15.3
5	72	105	78	66	60	87	66	66	75	66	72	75	66	48	48	0.00
6	100	109	100	100	108	140	95	95	95	104	115	101	98	95	88	7.37
7	120	150	107	139	89	127	84	74	101	74	74	84	87	74	70	5.41
8	215	213	211	204	215	213	160	153	153	151	154	165	204	145	131	9.66
9	182	146	184	156	158	158	139	126	160	134	144	130	145	122	113	7.38
10	239	238	244	183	224	217	164	152	182	158	165	164	158	149	136	8.72
11	128	218	153	171	150	153	105	137	109	124	142	101	104	96	93	3.13
12	134	134	134	134	134	134	71	83	77	76	75	82	72	71	65	8.45
13	209	226	195	204	211	236	137	152	139	161	166	136	161	126	113	10.3
14	100	160	121	127	105	132	84	92	82	71	83	84	132	70	70	0.00
15	140	177	162	178	177	184	104	130	119	129	139	106	127	104	100	3.85
16	123	137	96	114	108	123	89	83	90	88	90	80	89	75	75	0.00
17	109	89	75	121	94	99	74	82	76	82	81	74	83	72	61	15.28
18	82	146	98	116	130	151	71	68	88	81	86	70	79	64	64	0.00
19	113	187	148	142	157	187	91	109	109	127	121	104	127	89	89	0.00
20	115	183	163	139	172	160	107	125	115	114	127	112	98	107	92	14.0
21	862	876	755	655	709	623	652	661	661	623	623	644	626	582	325	44.2
22	779	300	768	757	1264	1255	784	742	781	748	780	804	787	733	400	45.4

Table 3 Makespan for job sets with and without considering tool transfer times

Job set	Makespan without tool transfer times		Makespan with tool transfer times								
			Case I (original processing time) by CSA				Case II (double processing time) by CSA				Case III (triple processing time) by CSA
	AIS	CSA	L1	L2	L3	L4	L1	L2	L3	L4	L4
1	**69**	**69**	116	**95**	95	123	170	**159**	158	175	–
2	82	**80**	120	**100**	104	137	185	**174**	180	192	270
3	**80**	**80**	118	**102**	106	131	188	178	**180**	189	277
4	72	**61**	116	**98**	99	131	160	148	**153**	180	225
5	**48**	**48**	96	**77**	82	112	133	**119**	120	147	188
6	95	**88**	107	**99**	100	117	188	**181**	182	195	–
7	74	**70**	113	**95**	99	126	183	170	**173**	193	259
8	145	**131**	160	**151**	149	172	272	**270**	269	281	–
9	122	**113**	135	**129**	130	141	245	**246**	245	248	–
10	149	**136**	171	**165**	166	182	303	293	**302**	310	–

improvement is 44.16%. The proposed method has given same result for 7 job sets out of 22 job sets.

The same algorithm is now employed for scheduling jobs, machines and tools by considering tool transfer times between machines. The transfer times of tools between machines are taken as 70% of AGV travelling times. It is tested on first 10 job sets of aforementioned 22 job sets with four different layouts and with different processing times. These are the benchmark instances in the literature [13]. These results are presented in Table 3 and the best makespan is indicated in bold.

Three cases are considered here to show the influence of tool transfer times on makespan with increasing processing times. In case I, original processing times are used; in case II, processing time is taken as double the original processing time; and in case III, processing time is taken as triple the original processing time. The Gantt chart for optimal sequence produced by CSA for job set 5 and layout 2 in case I is shown in Fig. 4. The operations that are assigned to each machine as well the start and finish times of each operation are shown in the Gantt chart. Utilization of tools for various operations of jobs is also shown in the Gantt chart. The Gantt chart also indicates loaded trip times, empty trip times and waiting times of TT for transferring tools in the system. The loaded trip of TT is labelled as "*L*", empty trip is labelled as "*E*", and waiting time of TT is labelled as "*W*" in Fig. 4. The Gantt chart shows the correctness of the solution provided by the proposed CSA method.

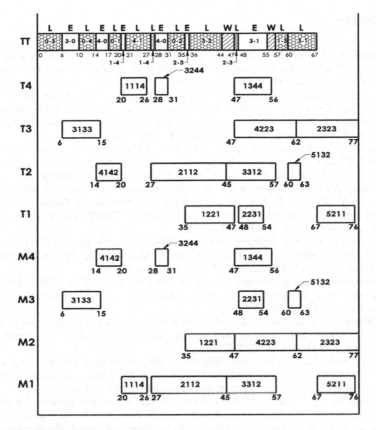

Fig. 4 Gantt chart for job set 5 and layout 2 in case I

6 Conclusion

Scheduling of jobs, tools and machines with and without considering tool transfer times is performed with the proposed CSA. It is noticed that CSA outperforms other algorithms in minimizing makespan without considering tool transfer times. The same CSA is tested on 22 job sets to show its constancy. It is observed that tool transfer times have a considerable impact on makespan in all three cases, and hence, any schedule without considering tool transfer times cannot be implemented in reality. The future scope of this work can be considering AGVs, as travelling time of jobs between L/U station and machines and between machines influences the makespan.

References

1. Agnetis, A., A. Alfieri, P. Brandimarte, and P. Prinsecchi. "Joint Job/Tool Scheduling in a Flexible Manufacturing Cell with No On-Board Tool Magazine." *Computer Integrated Manufacturing System* **10** (**1**), pp. 61–68(1997).
2. Baker, K. R Introduction to Sequencing and Scheduling. New York, Wiley (1974).
3. Jerald, J., and P. Asokan. "Simultaneous Scheduling of Parts and Automated Guided Vehicles in an FMS Environment using Adaptive Genetic Algorithm." *International Journal of Advanced Manufacturing Technology*, **29** (**5**), pp. 584–589 (2006).
4. Lee, D., and F. Dicesare. "Integrated Scheduling of FMSs Employing Automated Guided Vehicles." *IEEE Transactions on Industrial Electronics*, **41** (**6**), pp. 602–610, (1994).
5. Tsukada, T. K., and K. G. Shin. "Distributed Tool Sharing in Flexible Manufacturing Systems." *IEEE Transactions on Robotics and Automation*, 14 (3), pp-379–389 (1998).
6. Jun, H., Y. Kim, and H. Sub. "Heuristics for a Tool Provisioning Problem in a Flexible Manufacturing System with an Automatic Tool Transporter." *IEEE Transactions on Robotics and Automation*, 15 (3), pp-488–497 (1999).
7. Sureshkumar, N., and R. Sridharan. "Simulation Modeling and Analysis of Tool Flow Control Decisions in Single Stage Multimachine Flexible Manufacturing System." Robotics and Computer Integrated Manufacturing 23, pp-361–370 (2007).
8. Sureshkumar, N., and R. Sridharan "Simulation Modeling and Analysis of Tool Flow Control Decisions in a Flexible Manufacturing System." *Robotics and Computer Integrated Manufacturing* **25**, pp. 829–838 (2009).
9. Prabaharan, T., P. R. Nakkeeran, and N. Jawahar "Sequencing and Scheduling of Job and Tool in Flexible Manufacturing Cell." *International Journal of Advanced Manufacturing Technology*, **29** (**3**), pp. 729–745 (2006).
10. Udhayakumar, P., and S. Kumanan. "Sequencing and Scheduling of Job and Tool in Flexible Manufacturing System Using Ant Colony Optimization Algorithm." *International Journal of Advanced Manufacturing Technology*, **50** (**9**), pp. 1075–1084 (2010).
11. J. Aldrin Raj, D. Ravindran, M. Saravanan and T. Prabaharan "Simultaneous scheduling of machines and tools in multimachine flexible manufacturing system using artificial immune system algorithm" *International Journal of Computer Integrated Manufacturing*,**27**(**5**), pp. 401–414 (2014).
12. Alireza Askarzadeh, "A novel metaheuristic method for solving constrained engineering optimization problems: Crow search algorithm" Computers and Structures, 169, pp. 1–12, 2016.
13. Ulusoy G, Sivrikaya-Serifoglu F, Bilge U. "A genetic algorithm approach to the simultaneous scheduling of machines and automated guided vehicles", Journal of Computers & Industrial Engineering, 24(4): 335–351 (1997).

Energy-Aware Multi-objective Differential Evolution in Cloud Computing

Archana Kollu and V. Sucharita

Abstract Cloud computing (CC) could be a massive distributed computing driven by business, during which the services and resources are area unit delivered on request to external consumer via the Web. The distributed computing environment comprises of physical servers, virtual machines, data centers, and load balancers which are appended in an efficient way. With the increasing size of a number of physical servers and utilization of cloud services in data centers (DC), the power consumption is a critical and challenging research problem. Minimizing the operational cost and power in a DC becomes essential for cloud service provider (CSP). To resolve this problem, we introduced a novel approach that leads to nominal operational cost and power consumption in DCs. We propose a multi-objective modified differential evolution algorithm for first placement of virtual machine (VM) in the physical hosts and optimize the power consumption during resource allocation using live migration. The experimental results reveal that our proposed method is significantly better against state-of-the-art techniques in terms of limited power consumption and SLA for any given workload.

Keywords Cloud computing · Energy efficiency · Differential evolution
Virtual machine · Physical server

1 Introduction

Cloud computing (CC) could be a massive distributed computing driven by business, during which the services and resources are area unit delivered on request to outside customers via the Web [1]. The distributed computing environment

A. Kollu (✉) · V. Sucharita
Department of Computer Science & Engineering, KL University,
Guntur, AP, India
e-mail: sai.archna@gmail.com

V. Sucharita
e-mail: drvsucharita@kluniversity.in

© Springer Nature Singapore Pte Ltd. 2018
S.S. Dash et al. (eds.), *International Conference on Intelligent Computing
and Applications*, Advances in Intelligent Systems and Computing 632,
https://doi.org/10.1007/978-981-10-5520-1_40

433

comprises of servers, virtual machines, data centers, and load balancers which are appended in an efficient way. The customers and operators are generating a huge volume of data on different services/minute in a CC environment, which gradually shows all premises of big data. The main benefits of CC, such as scalability, flexibility, reliability, and availability, have made them a well favoured alternative to restore individual in-house IT infrastructures [2]. It has proceeded as a favorable for the organization, industry, and government to resolve the growing computing and storage issues. Presently, there are various commercial cloud providers (CSPs) such as Google, Amazon EC2, Azure services platform, and provided ICT resources are "virtualized" as DC facilities.

There are unlimited measures of resources that can be accessed with regard to CC environments. However, commercial CSP demands the consumer by an hourly based pricing method. Accordingly, the price is not analyzed in view of the genuine measure of assets utilized, however, as indicated by the time unit, implying that the consumer need to reimburse for the entire rented hour although they tenure the instances/second [3]. To the extent we know, the vast majority of the CSP offers different kinds of VM at various costs. Consequently, it is tough for cloud scheduler to develop an ideal strategy to perform technological work process applications within time constraint.

Reducing carbon emission (CE) and power consumption by the cloud computing DC requires energy-aware policies and developing of software, as well as hardware, which arise as one of the most influential research topics both in academia and in industry. Moreover, the power required by the DCs for its working, energy supply, illumination, and cooling subscribes fundamentally to the aggregate operational expenses. Hence, minimizing the energy consumption and power dispersal had become pivot study for making CC ecologically sustainable. The power utilization in worldwide DC represented around 1.3% of aggregate power use around the world by 2010. As per a McKinsey report, the aggregate-assessed power bill for a DC in 2010 was $11.5 billion and power costs in a run of the mill DC double every 5 years.

Recently, it is accounted for that 50% administration spending plan of Amazon DC is utilized for cooling and powering the hosts or servers, and the power consumption in cloud DCs is now consider for almost 0.5% of the world total energy usage. In addition to the power cost, DCs also generate the substantial amount of CO_2 emissions, which will specially grant to the increasing environmental concern of global warming [4]. Reducing the power consumption is essential due to the growing factor from the perspective of cost and environmental factors. Briefly, it is essential to enlarge a power-efficient workflow scheduling approach to decrease power consumption as much as possible.

- The power costs for powering ICT assets and cooling systems are taking more expenditure to the actual expenditure of buying the ICT resources.
- DCs are causing environmental concerns due to increasing power usage and CO_2 emissions.

- The increasing power usage and high-temperature dissipation have harmful impacts on reliability, scalability, and density of DC assets.
- The power-efficient asset allocation problem defined as selecting, deploying, and time management of DC assets in a way that hosted applications accomplish their QoS constraints.

In order to optimize physical resources and load balancing, increase performance, and improve visibility in large-scale DC, efficient resource scheduling algorithms are needed. Some traditional scheduling algorithms are first-fit, best-fit, linear programming. [5–7] are limited in use and may not optimize the system because of the dynamic way of the DC circumstances. The main objective of our research is to improve resource utilization with reducing financial and environmental costs and minimize power consumption in a DC by reducing the server idle conditions and enhancing the host utilization.

The rest of the paper is organized as follows. Section 2 discussed related work. Theoretical background related to our work is presented in Sect. 3. The modeling of resource allocation model and proposed method is presented in Sect. 4. Section 5 presents performance evaluation and followed by conclusions in Sect. 6.

2 Related Work

Many different techniques have been proposed for scheduling resources in the cloud, but energy efficiency is an important factor.

In order to reduce energy consumption, many researchers investigated the concept of VM consolidation to observed overload, underload, and VM selection. Generally, the VM placement problem is considered as an NP-hard problem [8], and an optimal arrangement of VM cannot be obtained within polynomial time. Shi et al. [5] proposed a first-fit heuristic algorithm for maximizing the profit under the SLA and minimizing the power consumption. Beloglazov et al. [7] developed a Modified Best-Fit Decreasing algorithm to place VMs in most energy-efficient and maximizing the profit under the SLA and minimize the power consumption. Younge et al. [9] presented a new greedy-based algorithm which maximized resource utilization and minimized total power consumption of the servers. Henceforth, the said algorithms have mentioned minimization of migration policy which accomplished well concerning VM migrations, power consumption, and service level agreements. There are various studies that have proved that meta-heuristic algorithms are giving the results very faster and increasing the convergence rate within the minimal time. Fortes [10] developed multi-objective GA (genetic algorithm) VM placement algorithm to reduce power consumption, thermal dissipation costs, and total resource wastage. However, GA consumes more time in order to converge, and there is no proof that GA will converge to the global optimum. Some other researchers have proposed improved PSO [11, 12] in order to reduce power consumption. However, all said approaches do not reduce physical machine/host over utilization, and SLA is violated as well PSO gets trapped at local optima and not often good at diversification.

However, in approaches as mentioned earlier, some have considered load balancing, resource wastage, and load migrations. But they have not taken into account high power consumption in VM placements and high convergence rate. In contradiction, we proposed a multi-objective differential evolution (MODE) model which considers fast convergence rate, high power consumption, and utilization of resources.

3 Theoretical Background

The differential evolution (DE) method is a simple and well evolution proposed by Stron and Price [13]. The main idea behind DE is generating trail parameter vector. This algorithm first produces a population consisting of NP D-dimensional initial vectors, also called as individual which is encoded as individual solutions.

$$A_{r_i} = A_{r_1}, A_{r_2}, \ldots\ldots\ldots A_{r_m}, m = 1, 2, \ldots\ldots\ldots, \text{NP}.$$

The population involves over generations through operations such as selection, crossover, and mutation up satisfying the condition. The trail vector P_i corresponding to an A_i individual (target vector) is produced through mutation and crossover operations. The following mutant vector y_i is most frequently used mutation operator listed as follows:

$$"best/1" [15] : y_i = A_{\text{best}} + F(A_{r_2} - A_{r_3})$$
$$"rand/1" [15] : y_i = A_{r_1} + F(A_{r_2} - A_{r_3})$$
$$"best/2" [15] : y_i = A_{\text{best}} + F(A_{r_1} - A_{r_3}) + F(A_{r_3} - A_{r_4})$$
$$"rand/2" [16] : y_i = A_{r_1} + F(A_{r_2} - A_{r_3}) + F(A_{r_4} - A_{r_5})$$

where r_1, r_2, r_3, r_4, and r_5 are random mutually exclusive integers between [1, NP]. A_{best} represents the best individual of current population, F represents scaling factor that controls the difference vector amplification, and k represents scaling factor or parameter that is randomly chosen between [0, 1].

$$"current - to - rand/1" [16] : y_i = A_i + k(A_{r_1} - A_i) + F(A_{r_2} - A_{r_3})$$

Based on A_i (target vector) and its corresponding mutant vector (MV) y_i, the offspring/trail vector $P_i = P_{i_1}, P_{i_2}, \ldots P_{i_D}$ is produced by crossover operation, whose element P_{ij} can be defined as follows:

$$P_{ij} = \begin{cases} y_{ij} & \text{if rand } (0,1) \geq \text{CR or } j = j_{\text{rand}}, \ j = 1, 2, \ldots D \\ A_{ij} & \text{otherwise} \end{cases} \tag{1}$$

where y_{ij} is the jth element of MV y_i, and j_{rand} is randomly chosen integer value [0, D]. D is dimension of solution space. CR is crossover rate [0; 1] fraction of parameter values mimic from MV.

4 Modeling of Resource Allocation Model

The main goal of our proposed approach is to decrease the energy consumption by considering VM allocation and workload scheduling at DCs. The notations that are used throughout the paper are mentioned in Table 1.

Let VM be the set of VMs and PS be the set of physical machines/hosts.

$$VM = Vm_1, Vm_2, \ldots, Vm_n, PS = Ps_1, Ps_2, \ldots, Ps_m \qquad (2)$$

Each physical machine/host consists of id, CPU, power storage, bandwidth. Similarly, each VM consists of as follows:

$$VM_i = (Vm_i^{id}, CPU_i, Mem_i, bw_i), PS_j = (Ps_j^{id}, CPU_j, Mem_j, bw_j) \qquad (3)$$

The main goal of our proposed approach is to maximize the physical resource utilization and minimize the power consumption. The overall utility function describes as follows:

Table 1 Notations

Symbol	Meaning
Vm_i	The ith virtual machine in data center, $i = 1, 2, 3\ldots$
VM	Set of virtual machine
PS	Set of physical machine/host
Ps_j	The jth physical machine/host in data center, $j = 1, 2, 3\ldots$
r_i^{CPU}	The maximum CPU utilization of ith VM
r_i^{Mem}	The maximum Mem utilization of ith VM
C_j^{CPU}	The CPU utilization of jth physical host/machine
C_j^{Mem}	The Mem utilization of jth physical host/machine
C_i^{bw}	The maximum bandwidth utilization of ith VM
Cj^{bw}	The maximum bandwidth utilization of jth physical host/machine
PS_j^{id}	Physical machine jth id number
CPU_j	Physical machine processing power
Mem_j	Physical machine amount of memory
bw_j	Physical machine amount of bandwidth

$$\text{Utility function } (x) = \frac{\text{Max}\left(U_j^d\right)}{\text{Min}\left(\sum_{j=1}^{n}\sum_{i=1}^{m} x * U_P^{j(i)}\right)} \quad (4)$$

subject to:

$$U_j^d = \frac{\sum_{i=1}^{n} x_{ij} * r_i^d}{c_j^d}, \forall d \in \{\text{CPU}, \text{Mem}, \text{bw}\}$$

$$\forall j \in 1, 2, \ldots, m \quad (5)$$

$$\sum_{i=1}^{n} x_{ij} = 1, \forall j \in 1, 2, 3 \ldots, m$$

where x_{ij} is allocated only one host for each VM. Equation 5 represents that the load on each physical machine is not greater than its total capacity. $x * U_P^{j(i)}$ represent the difference between the current utilization of physical machine placed with v_i placed in U_P^j. In order to find out the total utilization of physical machine/host, the calculation is done based on the following equation:

$$U_P^j = \sum_{i=1}^{m}\left(U_P^{j(i)}\right) \quad (6)$$

where $U_P^{j(i)} = U_j^d * 100$ if $x_{ij} = 1$ else 0. It means that ith VM is allocated to jth physical machine, otherwise not allocated to physical machine. The overall power consumption (OPC) of a DC can be determined as follows:

$$\text{OPC} = \sum_{j=1}^{n} U_P^j \quad (7)$$

where U_P^j is an estimated power consumption of physical machine with utilization.

4.1 VM Placement by Using Modified DE

In this subsection, we propose a multi-objective modified DE algorithm to find the optimal energy-aware VM placement. Each candidate solution or individual map of VM is to be placed in available physical machine in DC.

Initially, in conventional DE, the initial populations of NP individuals are initialized randomly and then directly generated mutant vector. In our proposed method, after initializing the initial population of NP individuals, each population number is randomly allotted with one mutation strategy (MS) and its corresponding values are picked randomly form pool. The modified DE algorithm is illustrated in Algorithm 1.

In our proposed method, each individual is mapped to VM-host that represents the placement of a VM on physical machine. Initially, all VMs are initialized in various physical machines. We calculated fitness of each individual in the population of NP individuals by using Eq. 4. Afterward, we pool the MS and its corresponding parameter values. The each VM is randomly set to MS from the pool, and corresponding parameter values are picked randomly. For each iteration, mutant step, crossover, and selection steps are used to find the optimal solution in the population, which provides the minimal power consumption. Thereafter, update each VM to host by using update step until either algorithm terminates or achieves the improvements. The VM placement using modified DE is described in Algorithm 1.

Algorithm 1: Modified Multi-objective DE

Begin

Step 1: Set generation G = 0;

Initialize a population of NP individuals randomly;

Calculate initial population using equation 4;

Step 2: Select a pool MS and its associated values assigned;

While stopping criteria is not satisfied **do**

Muation Step: //Generate a MV i for each target vector

For i=1 to NP **do**

Generate a MV y_i corresponding to the trail vector y_i by using equation 8;

end For

Crossover Step: // Generate a trail vector i for each target vector

For i=1 to NP **do** j_{rand} = [rand[0; 1], D]

    For j=1 to D **do**

$$P_{ij} = \begin{cases} y_{ij} & \text{if rand }(0,1) \geq \text{ CR or } j = j_{rand}, j = 1,2,3,....D \\ A_{ij} & \text{Otherwise} \end{cases}$$

    end For

end For

Selection Step: Selecting competition between target and offspring vector using equation 4;

Updating Step:

For i=1 to NP **do**

Randomly select a new MS and store the combinations;

end For

increment the generation g=g+1;

end While

end

5 Performance Evaluation

In this section, we compare our proposed approach with some state-of-the-art approaches, energy-efficient resource scheduling approaches, i.e. [7, 10, 11], in term of power consumption. With increasing number of workloads, increasing number of VMs, increasing number of migrations in terms of power consumption. We evaluated our experiments in a virtualized DC using CloudSim [14]. In our method, we simulated a DC consisting 50 heterogeneous physical hosts. Each physical host is distinguished by the speed of processing, the number of virtual cores, storage, and a scheduling policy. The physical host and their description are illustrated in Table 2. The characteristics for each VM are described in Table 3. In this model, the VM provisionary plays a significant role in allocating VMs to physical machines. The VMs are varied from microinstances (100 MIPS, 400 MB RAM) to large instances (400 MIPS, 1840 MB RAM). Each VM processes time-varying workload.

In order to perform our proposed method against the other approaches, simulation platform is set to the initial parameters. We adopted [15] to use the mutation strategy DE/current - to - rand/1/bin, and initial parameters setting NP vary from 20 to 50, $k = 0.5$, F-0.8, CR = 0.9. The initial population size of DE is set to 40, which gives optimal solution and linear convergence rate with increasing number of VMs. The number of VMs placed in each physical machine is represented in Fig. 1.

Table 2 Physical machine/host configurations

Resource Id	Configuration	Operating system	RAM (GB)	Band width (GBPS)	MIPS	Storage (GB)
R1	Dell Vostro 460, Intel Core i5	Windows	4	40	3500	320
R2	Dell Vostro 460, Intel Core i5	Windows	8	40	3500	320
R3	Intel Core 2 Duo-2.4 GHz	Linux	4	40	3500	320
R4	Intel G4 Xeon 52407-2.2 GHz	Linux	4	40	3500	320

Table 3 VM configurations

Resource Id	RAM (MB)	Band width (GBPS)	MIPS	Storage (GB)
R1	520	100	100	50
R2	670	100	200	50
R3	1560	100	300	50
R4	1780	100	400	50

Fig. 1 Active physical machines for different algorithms

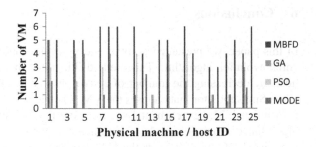

Fig. 2 Energy consumption performance

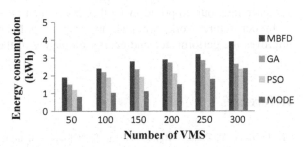

Fig. 3 Convergence rate with respect to other algorithms

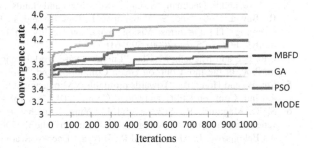

From Fig. 1, we observed that our proposed approach required a minimum number of physical machines that satisfy the VM requirements compared to other approaches. To compare the power-consuming performance, we used 400 VMs and performed our proposed approach against other approaches. The performance results are shown in Fig. 2. From Fig. 2, we observed that our proposed approach required 3.42 kWh and other approaches require 5.89 and 6.21 kWh. This result shows that our proposed approach reduces the power consumption up to 40% over PSO and 45% over MBFD.

The convergence rate of our proposed method against other approaches is shown in Fig. 3. From Fig. 3, we observed that our proposed method shows fast convergence than other approaches. We also performed the statistical analysis, based on the results from our proposed method which is most significant than the other methods.

6   Conclusions

In this paper, we presented a novel energy-aware multi-objective modified differential evolution algorithm. This method illustrated that high utilization of physical hosts by decreasing the number of active physical hosts. In this method, we used CloudSim simulator that chooses the VM when a physical machine is over or under over utilized. The experimental results show that the VM placement and selection algorithm combination leads to significant reduction in power consumption in a virtualized DC by 40% over the GA and 45% over the MBFD. In term of convergence rate, our proposed method is faster than other methods.

In our future work, we plan to propose a hybrid heuristic algorithm that improves the performance and energy efficiency of the huge amount of servers.

References

1. I, Foster, Y, Zhao, I, Raicu, and Lu, S. "Cloud computing and grid computing 360-degree compared" in: Proc. of the *Grid Computing Environments Workshop*, pp. 1–10. IEEE, 2008.
2. G, Juve, and E, Deelman, "Scientific workflows and clouds", Crossroads 16 (3) (2010) 14–18.
3. R., Buyya, C. S., Yeo, S., Venugopal, J., Broberg and, I., Brandic, "Cloud computing and emerging {IT} platforms: Vision, hype, and reality for delivering computing as the 5th utility", FGCS 25 (6) (2009) 599–616.
4. Cao, Fei, and Michelle M. Zhu. "Energy-aware workflow job scheduling for green clouds", in: Proc. of the Intl. Conf. on Green Computing and Communications, 2013, pp. 232–239.
5. Shi, W. and Hong, B., "Towards profitable VM placement in the data center", in: Proc. of the 4th Intl. Conf. on Utility and Cloud Comp., 2011, pp. 138–145.
6. S. T., Maguluri, R., Srikant, and L., Ying, "Stochastic models of load balancing and scheduling in cloud computing clusters", in:, Proc the INFOCOM. IEEE, 2012, pp. 702–710.
7. A., Beloglazov, J., Abawajy and R., Buyya, "Energy-aware resource allocation heuristics for efficient management of data centers for cloud computing", FGCS 28 (5) (2012) 755–768.
8. Gary, M. R., and David S. Johnson, "computers and intractability: A guide to the theory of np-completeness" (1979).
9. A. J., Younge, G., Von Laszewski, L., Wang, S., Lopez-Alarcon and W., Carithers, "efficient resource management for cloud computing environments", in: Proc. of the Intl. Conf. on Green Comp.", 2010, pp. 357–364.
10. J., Xu, and J. A., Fortes, "Multi-objective VM placement in virtualized data center env.", in: Proc. of the Intl. Conf. on Green Comp. and Comm. & Intl. Conf. on Cyber, Physical and Social Comp., IEEE, 2010, pp. 179–188.
11. A. P., Xiong and C. X., Xu, "Energy efficient multi resource allocation of VM based on PSO in cloud data center", Math. Prob. in Engg. 2014.
12. S. E., Dashti and A. M., Rahmani, "Dynamic VM placement for energy efficiency by pso in cloud computing", JETAI 28 (1–2) (2016) 97–112.
13. R., Storn and Kenneth P., "DE—a simple and efficient heuristic for global optimization over continuous spaces", JGO 11 (4) (1997) 341–359.
14. R. N., Calheiros, R., Ranjan, A., Beloglazov, De Rose, C. A. and R., Buyya, "cloudsim: a toolkit for modeling and simulation of cloud computing environments and evaluation of resource provisioning algorithms", Softw Pract Exp. 41 (1) (2011) 23–50.

15. Qin, A. Kai, Huang, Vicky L., and Suganthan, P. N., "DE algorithm with strategy adaptation for global numerical optimization", IEEE TEC.13 (2) (2009) 398–417.
16. Sawant, S, "A GA scheduling model for VM resources in a cloud comp. environment".

GRAPHON Tamil to English Transliteration for Tamil Biomedicine

J. Betina Antony and G.S. Mahalakshmi

Abstract Cross-Language Information Retrieval is a fast-growing field that attracts many researches. In a field with humongous application, basic understanding and accessibility of words is a crucial task. Transliteration is one such vital task that paves way for a wide range of improvements. In our work, we focus on deploying transliteration to retrieve essential information from concealed English words in a spool of unstructured Tamil text. These English words written in Tamil are identified, and their correct form is retrieved by performing statistical search in a collection of built-in database. This GRAPHON (Grapheme + Phoneme)-based Tamil to English transliteration gave an accuracy of 68% being the first of its kind.

Keywords Transliteration · Tamil biomedicine · Cross-lingual Information Extraction · Phonetic algorithm · Soundex code

1 Introduction

Tamil, one of world's oldest languages, is famous not only for its rich morphology and vocabulary but also for the opulent knowledge its literature bears. Of the many information that are passed on from generation to generation, the knowledge about indigenous medicines and their uses has always been preserved and practised for centuries. This knowledge, also known as Siddha System of Medicine (SSM), is put to practice even today. Siddha is predominantly a collection of alchemy texts that is believed to be invented by 18 *Siddhars*. It uses different minerals, metals and chemical products of nature to heal human ailments, both physical and mental.

J.B. Antony (✉) · G.S. Mahalakshmi
Department of Computer Science and Engineering, College of Engineering Guindy,
Anna University, Chennai 600025, Tamil Nadu, India
e-mail: betinaantony@gmail.com

G.S. Mahalakshmi
e-mail: gsmaha@annauniv.edu

© Springer Nature Singapore Pte Ltd. 2018
S.S. Dash et al. (eds.), *International Conference on Intelligent Computing and Applications*, Advances in Intelligent Systems and Computing 632,
https://doi.org/10.1007/978-981-10-5520-1_41

445

The main achievement of this ancient system is its application even in the present medicine system. The original Siddha texts are in the form of poems. Hence, a number of works have been carried out to convert these poems to prose for better understanding. These translations have been done for decades. As a result of cultural advancement, a number of contemporary words and languages have also mingled with the native language in the process of translation. Our work focuses on identifying these non-Tamil words that are otherwise lost as noisy data in our original information extraction system. Therefore, we seek the help of transliteration to locate these non-Tamil yet significant named entities.

Transliteration is the process of converting characters from one script to another without changing its phonetic structure. Transliteration is widely used in various cross-lingual applications such as Cross-Language Information Retrieval, WebSearch, Machine Translation. Transliteration is many a times confused with translation which also involves transmuting from one language to another. But translation strictly adheres to keeping the meaning of the word or phrase intact. Transliteration stresses on semantic correctness but not on meaning matching. In our work, we suggest a method that can identify and convert English terms that are concealed inside normal text. A statistical search approach is carried out to determine the corresponding English words for the identified Tamil words and to eliminate unrelated words. For words with colliding search results, a phoneme matching-based tie breaker is applied.

2 Background

This work is a part of Information Retrieval for Tamil Biomedicine [1]. The objective of the system is to retrieve medicinal information from the unstructured text. That is, for a query (name of a disease or an ingredient), a collection of data is obtained that contains information about the medicinal ingredients used, the diseases or disorders it can cure (both these fall into the field-named entities), and the preparatory procedure for the medicine along with information about the ingredients and directions to be followed to consume the medicine is provided. Many challenges were encountered on the way [2], and our research has been carried out to eradicate as many limitations in the system as possible.

Here, for the retrieval process, the identification of named entities is involved [3]. Named entities in the context of Tamil Biomedicine refers to terms that denote an ingredient element or its related items, its by-products, name of a disease or disorder, the symptoms, parts of a body, etc. Our data set is a collection of Siddha-related information obtained from published books, magazines and also to some extent information obtained from blogs and Web portals. The latter contemporary information gathered contains many words from other languages predominantly from English. However, these words did to large portion share of valuable information as they served as queries in the retrieval process. They also indicated words that could be understood by the current generation. For example, most people may not recognize பற்றுயிரி (*paRRuyiri*) which is the proper Tamil word for பாக்டீரியா (*paaktiriyaa*), (Bacteria). These words were lost in processing for Information Extraction. Hence, we proceeded with this work to locate non-Tamil words from text and identify their English alternates accurately.

3 Related Works

The works related to Tamil transliteration started in the late 90s or early 2000. Viswanadha [4] suggested a universally accepted ICU (International Components for Unicode) that involves Romanization of Tamil alphabets. This laid the foundation for transliteration in Indic languages. Since then it has been put to test in many multilingual querying and information processing [5–7].

When transliteration in Tamil is taken into picture, two types of analysis have been carried out. One includes conversion of English terms to Tamil which includes statistical as well as machine learning approaches. One particular group of researchers have worked on the different methods of transliterating from English to Tamil where they started by considered it as a sequence labelling problem, and multiclassification was done based on memory learning [8] with 84.16% accuracy. They further modelled a C4.5 decision tree classifier, in WEKA Environment [9] with 84.82% accuracy of English names. Finally, a One Class Support Vector Machine Algorithm was developed in 2010 [10] that outperformed both the previous methods. Finch et al. [11] suggested a bidirectional neural model to transliterate with 62.9% accurate English to Tamil transliterate.

The other set of works includes transliteration between Indian Languages. They were comparatively trivial as most of the Indian languages are morphologically rich and have similar phoneme structure. Keerthana et al. [12] proposed a sequence labelling-based method to transliterate from Tamil to Hindi. Their system produced an accuracy of about 84% in spite of not addressing the variation in the pronunciation and sound of different letters in both the languages. In [13], transliteration of Hindi to 7 different Indic languages was carried out based on word alignment and Soundex matching. The system, however, performed comparatively badly for Tamil when compared to the other languages with an accuracy of 68%. A number of these transliteration works considered Soundex code matching some of which altered the codes to favour Indian Languages. After a thorough study of all these works, it was evident that none of them was carried out for Tamil to English transliteration which paved way for our research.

4 System Description

The steps involved in our transliteration work and their operations are listed in the following sections (Fig. 1).

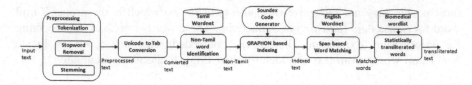

Fig. 1 Biotransliteration procedure

Table 1 Tamil romanization scheme

Vowels		Vowels		Consonants		Consonants		Special Letters	
Tamil	English	Tamil	English	Tamil	English	Tamil	English	Tamil	English
அ	a	எ	e	க்	k	ம்	m	ஷ	sh
ஆ	aa	ஏ	ee	ங்	ng	ய்	y	ஸ்	s
இ	i	ஐ	ai	ச்	c	ர்	r	ஜ	j
ஈ	ii	ஒ	o	ஞ்	nj	ல்	l	க்ஷ	ksh
உ	u	ஓ	oo	ட்	T	வ்	v	ஸ்ரீ	sri
ஊ	uu	ஒள	au	ண்	N	ழ்	zh		
		ஂ	q	த்	t	ள்	L		
				ந்	nd	ற்	R		
				ப்	p	ன்	n		

4.1 Preprocessing

The input to the system is given in the form of unstructured Tamil sentences containing Tamil biomedical instructions. Since the transliteration is done at word level, the initial step is splitting the words to tokens. After tokenization, the common stopwords are removed. Finally, the individual words are stemmed to remove variation.

4.2 Unicode to Tab Conversion

The first step towards transliteration is changing the text format of source language to the target language. It is also necessary in our case as the external DB to be used in the later stages involves words represented in target language. Hence, we convert our unicode Tamil text to their corresponding English representation. This representation is universally accepted standard representation (Table 1).

4.3 Non-Tamil Word Identification

Identifying non-Tamil words from the given lot is the strenuous part in the entire process. We have used a Tamil WordNet, assuming they contain most of the words in Tamil dictionary. Words that are not present in the DB are considered non-Tamil.

4.4 GRAPHON-Based Indexing

GRAPHON is a combination of grapheme- and phoneme-based phonetic algorithm that we suggest. Generally in any language, alphabetic letters are involved. However, when Tamil alphabets are represented in English, they involve more than one letter as they are the agglutinated form of a vowel and a consonant in most cases. Example க (ka) is an amalgamation of the consonant க் (k) + vowel அ (a).

Phoneme indexing, however, involves grouping words based on their phonetic representation. A phonetic algorithm generally involves giving a common code for each phoneme based on their sound and matching words with similar codes for indexing, comparison or other purposes. The very common phonetic algorithm used for various applications is Soundex Algorithm [14]. Soundex is a hashing system for English that uses a code to represent a word based on their sound. Each value of the code is given in such a way that all similar sounding words starting with the same alphabet are given the same value, thus identifying words with similar phonetics. In our system, we have first grouped our words into graphemes based on their byte code and then assigned Soundex codes to the phonemes.

4.5 Span-Based Word Matching

Now, the Soundex code for a given Tamil word is mapped onto their corresponding English words based on the sounding similarity. For this mapping, we have created a table with Soundex codes for all the words from English WordNet. Initially, when all the words with same code were considered, the system produced highly noisy results. Hence, the first level of filtration was done by confining the search to words with acceptable length. After few trials, the length was fixed to $\leq$ that Length + 5. All the words are now saved as an Array List.

4.6 Statistical Transliteration Based on Domain Knowledge

The final step in the transliteration process is to filter out the unrelated terms from the retrieved list. The first level was scrutiny may have been the word length. But that only filters noisy data. The ability to identify the correct words from the lot is the most challenging step. For our work, since almost 90% of the words are related to English medicines and biomedicine, we employ a biomedicine wordlist[1] that has more than 98,100 words. Hence words that are present in the dictionary are given a score to indicate its candidacy. Note that the values are numeric and are not Boolean, as more than one term in the list can be present in the dictionary and this leads to clash in assigning the terms.

To overcome the problem of multiple word assignment, each word is given a score based on its occurrence and weight-age. Weight-age here means phoneme match score or longest common subsequence score. Every term is assigned a score which is the length of the longest matched subsequence with the original transliterated word. The word with the maximum score is selected.

Let t be the term to be iterated and B be the list of terms $\{b_1, b_2,..., b_n\}$ that are present in the biomedical dictionary. Then, the final term t' is assigned to the term b_i which has the maximum normalized LCS value with t (Eq. 1):

$$t' = \{b_i \in B | \text{LCS}_{\text{norm}}(t, b_i) \geq \text{LCS}_{\text{norm}}(t, b_j) \forall b_j \in B\} \qquad (1)$$

5 Results and Discussion

5.1 Data Set

The experiment was initially started with 668 files containing Siddha Medicinal information collected from various sources. The sources include genuine publications dating from the 1980s to details obtain from recent Web pages using Web crawler. As suspected, the files from earlier decades did not contain any English texts though they had certain non-Tamil (mostly Sanskrit) words. These diluted the accuracy of the system. Hence, the data set was narrowed down to 85 files containing a total of 20,344 words. Two different WordNets (Tamil and English) and one biomedical name list were also involved in processing.

5.2 Discussion

Since the identification of the concealed English terms is the predominant task, our discussion revolves around the steps involved in refining the search for these terms. The identification task involves trying to locate a term in the Tamil WordNet.

---

[1]https://github.com/Glutanimate/wordlist-medicalterms-en.

```
+-------------------------+--------+-------+---------+----------+---------+---------+-------------+--------+
| nodeindex               | label  | gloss | example | relation | feature | english | indexlength | pos    |
+-------------------------+--------+-------+---------+----------+---------+---------+-------------+--------+
| 0,0,0,1,2,4,9,0         | paTi   | NULL  | NULL    | 4        | 1       | NULL    | 8           | Noun   |
| 2,0,1,5,4,0,4           | paTi   | NULL  | NULL    | 3        | 1       | NULL    | 7           | Noun   |
| 2,0,28,0,11             | paTi   | NULL  | NULL    | 3        | 1       | NULL    | 5           | Noun   |
| 2,0,37,3,2              | paTi   | NULL  | NULL    | 3        | 1       | NULL    | 5           | Noun   |
| 0,0,0,1,0,2,32,1,0,18   | paTi   | NULL  | NULL    | 1        | 1       | NULL    | 10          | Noun   |
| 8,1,1,2,3,0,2,10,1      | paTi   | NULL  | NULL    | 4        | 1       | NULL    | 9           | Noun   |
| 5,1,0,3,11,0,0,2        | paTi   | NULL  | NULL    | 3        | 1       | NULL    | 8           | Noun   |
| 5,1,0,9,0,8,7           | paTi   | NULL  | NULL    | 3        | 1       | NULL    | 7           | Noun   |
| 5,1,0,9,1,1,4,2,0,1     | paTi   | NULL  | NULL    | 4        | 1       | NULL    | 10          | Noun   |
| 5,1,0,12,19,0,10        | paTi   | NULL  | NULL    | 4        | 1       | NULL    | 7           | Noun   |
| 1,4,2,0,1               | paTi   | NULL  | NULL    | 4        | 1       | NULL    | 5           | Verb   |
| 1,8,1,2,0,10            | paTi   | NULL  | NULL    | 3        | 1       | NULL    | 6           | Verb   |
+-------------------------+--------+-------+---------+----------+---------+---------+-------------+--------+
```

Fig. 2 Twn table entries for word 'paTi'

Fig. 3 Morphtable entries for word 'paTi'

```
| Inflated_word          |Root_word |
+-----------------------------------------+
| paTikkakuuTiya         | paTi     |
| paTital                | paTi     |
| paTittirundta          | paTi     |
| paTittaanatumaana      | paTi     |
| paTittee               | paTi     |
| paTindtavaNNam         | paTi     |
```

The Tamil WordNet includes 4 different tables out of which only 2 are taken into consideration:

(i) twn—this table contains 50,497 labels or words along with their sense details and indexes to identify their synonyms, antonyms, hypernyms, hyponyms and troponyms. A sample tuple for entry படி (*paTi*) (to study) is shown in Fig. 2.

(ii) morphtable—this table has a collection of 434,849 words and their corresponding root word. For the example படி (*paTi*), there are about 574 different variations of the word. Some of them are shown in Fig. 3.

It is to be noted that the number of words in morphtable is more than the number of tables in twn. This is because the morphtable has most (but not all) of the variations of a given word. In our system, the word was initially mapped with any of the inflated words in the morphtable. If present, its corresponding root word was located in the twn table which connotes the fact that the word is Tamil. Note that the identification is purely based on words present in dictionary. This procedure had 2 major difficulties. Firstly, certain commonly used English words were already present in the dictionary. Example, the word ஆக்ஸிஜன் (*aakSijan*) (oxygen) was present in both the tables, hence was ignored in processing and hence adding to false negatives (Fig. 4). Secondly, the morphtable may not contain the particular agglutinated form of the word. This lead to the inclusion of actual Tamil words in the retrieved non-Tamil terms, thereby increasing the false positives and hence greatly affecting the precision of the system.

After tokenization and stemming, the first run of operation produced 5783 words as non-Tamil terms. This was a noisy result as the words had hidden junk bytes in them. After elimination of these bytes, the number of junks was largely reduced to 2982.

```
morphtable:
+----------------+------------+
| inflated_word  | root_word  |
+----------------+------------+
| aaksijan       | aaksijan   |
+----------------+------------+
twn:
+---------------------+-----------+-------+---------+----------+---------+---------+-------------+----------+------+
| nodeindex           | label     | gloss | example | relation | feature | english | indexlength | pos      |
+---------------------+-----------+-------+---------+----------+---------+---------+-------------+----------+------+
| 0,0,0,1,1,1,1,1,3,1 | aaksijan  | NULL  | NULL    | 4        | 1       | NULL    | 11          | Noun     |
+---------------------+-----------+-------+---------+----------+---------+---------+-------------+----------+------+
```

Fig. 4 Table entries for ஆக்ஸிஜன் (*oxygen*)

The non-Tamil words are changed to their corresponding English representation by Romanization. In this case, the letters are initially split based on their graphemes denoted by their byte codes. The units are then treated as phonemes for assigning Soundex codes. For ease of operation, the Soundex codes for all the words (203,147 words) in the English WordNet were stored priorly. Now, the Soundex for the English and Tamil words are mapped and all the words are retrieved:

e.g. விட்டமின்டி (*viTTaminTi*): [Vedanta, vitamin_D].

Here, the Soundex codes for all the words are V353. The related words are saved in the form of ArrayList to avoid duplication. Few challenges were encountered when mapping the Soundex codes.

(i)   Some letters in Tamil have different sounding patterns in different places. For example, the letter பி(*pi*) can be applied for பிரெட் (*pired*) (bread) as well as பிரஷர் (*pirashar*) (pressure). Their difference cannot be given individually. One solution we came up with was to replace few letters in the beginning of the word and find Soundex for them as well. Some of the words that were replaced are *b* for *p*, *c* for *k*, *o* for *aa*, *s* for *c*, etc.

(ii) The agglutination of non-Tamil words is difficult to uncouple as the root words are not present in the original dictionary. For example, கான்சரால் (*kaansaraal*) (due to cancer): [conger_eel, common_sorrel, censorial] did not identify the correct word as the Soundex for கான்சர் (*kaansar*) (cancer) is C526 and for கான்சரால் (due to cancer) is C5264.

(iii) Since Tamil does not support acronyms, the English acronyms are identified wrongly. One such example is *Urinary Tract Infection* (*UTI*) யூடிஐ (*yutiai*) (UTI): [youth, yodh, youth, Yeddo, Yedo, yeti, youth, yautia, youth, youth, youth, yet, yet, yet, yet, yet, yet].

The final step in the process is to determine the final English word to the term in question. The words in the list are checked with the medical wordlist, if present weighted score is given based on the total number of related words retrieved. This is to give a normalized value for all the words irrespective of the number of candidate terms identified. Consider the following:

ஆஸ்டியோபொரோசிஸ் (*aastiyooporoosiS*): [osteoporosis 1].

Table 2 Statistics of the transliteration process

Total number of words (including duplicates)	20,344
Number of non-Tamil words retrieved (including duplicates)	2,755
Number of unique non-Tamil words	1,832
Number of unique English words	185
Number of correctly transliterated words	126
Precision (TP/TP + FP)	0.681

In this case, the retrieved term is an exact match. Hence, the terms are assigned. Consider the following case:

மீசோதேலியம் (*miisootheeliyam*): [musteline 0, mesothelium 0.5, Mazatlan 0, Magdalena 0, magdalen 0, mesothelioma 0.5].
மீசோதேலியோமா (*miisootheeliyoomaa*):[musteline 0, mesothelium 0.5, Mazatlan 0, Magdalena 0, magdalen 0, mesothelioma 0.5].

Here, both mesothelioma and mesothelium match the term in question when considering the Soundex code. Here, the tie is broken by matching phonemes and assigning ranks. Hence, after matching, the results are:

மீசோதேலியம் (*miisootheeliyam*): [mesothelium—0.81, mesothelioma—0.75].
மீசோதேலியோமா (*miisootheeliyoomaa*): [mesothelium—0.81, mesothelioma—0.916].

Note that the values are normalized to balance the varying length of the words. The statistical results are listed in Table 2. Of the 2695 words that were retrieved, only 185 words were found to be English words. The major dip in the accuracy of retrieval is due to the lack of agglutinated form in the morphtable. This can be rectified by using a different stemmer, but the cost of false negative should also be taken into consideration. The precision value was calculated to determine the correct assigning of English words to Tamil-written entities. The recall and hence f-measure for system are almost impossible to determine as the total number of words to do manual checking is comparatively high.

6 Conclusion

The Tamil to English transliteration system, the first of its kind, was successfully built and performed effectively for a small yet productive data set. Certain difficulties such as agglutinated Tamil words can be resolved by applying different stemming algorithms on tagged non-Tamil words by non-dictionary-based methods. Also to overcome the ambiguity in pronunciation and sounding pattern, modification to the Soundex Algorithm can be made to adjust to the language in consideration. Even without the domain wordlist, the system will perform

J.B. Antony and G.S. Mahalakshmi

efficiently for data from any corpus as word length matching is also considered. The system obtained an agreeable result for the statistical method that was used, but the fact that it might perform better if a machine learning strategy is used is still debatable. Experiments to extend this work to machine translation for enhancing Information Extraction are in progress.

Acknowledgements This work is part of our research supported by **Department of Science and Technology's INSPIRE fellowship** Programme, Ministry of Science and Technology, India.

References

1. Betina Antony, J., & Mahalakshmi, G. S.: Content-based Information Retrieval by Named Entity Recognition and Verb Semantic Role Labelling. j jucs, 21, pp. 1830–1848, (2015).
2. Antony, J. B., & Mahalakshmi, G. S.: Challenges in Morphological Analysis of Tamil Biomedical Texts. Indian Journal of Science and Technology, 8(23), pp. 1–4, (2015).
3. Antony, J. B., & Mahalakshmi, G. S.: Named entity recognition for Tamil biomedical documents. In Circuit, Power and Computing Technologies (ICCPCT), 2014 International Conference, pp. 1571–1577, (2014, March).
4. Viswanadha, R. Transliteration of Tamil and Other Indic Scripts. INFITT, *Tamil Internet 2002* 277–285, (2002).
5. Ganesan, K., & Siva, G.: Multilingual Querying and Information Processing. Information Technology Journal, 6(5), pp. 751–755, (2007).
6. Kumaran, A.: MIRA: Multilingual information processing on relational architecture. In International Conference on Extending Database Technology, pp. 12–23, (2004, March).
7. Saravanan, K., Udupa R, and A. Kumaran: Crosslingual information retrieval system enhanced with transliteration generation and mining. Forum for Information Retrieval Evaluation (FIRE-2010) Workshop. (2010).
8. Vijaya, M. S., Shivapratap, G., Dhanakshmi, V., Ajith, V. P., & Soman, K. P.: Sequence labeling approach for English to Tamil Transliteration using Memory based Learning. In Proceedings of Sixth International Conference on Natural Language processing, (2008).
9. Vijaya, M. S., Ajith, V. P., Shivapratap, G., & Soman, K. P.: English to tamil transliteration using weka. International Journal of Recent Trends in Engineering, 1(1), (2009).
10. Vijaya, M. S., Shivapratap, G., & Soman, K. P.: English to Tamil Transliteration using One Class Support Vector Machine. International Journal of Applied Engineering Research, 5(4), pp. 641–652, (2010).
11. Finch, A., Liu, L., Wang, X., & Sumita, E.: Target-Bidirectional Neural Models for Machine Transliteration. ACL 2016, pp. 78–82, (2016).
12. Keerthana, S., Dhanalakshmi, V., Kumar, M. A., Ajith, V. P., & Soman, K. P.: Tamil to Hindi Machine Transliteration Using Support Vector Machines. In International Joint Conference on Advances in Signal Processing and Information Technology, pp. 262–264, (2011, December).
13. Srivastava, R., & Bhat, R. A.: Transliteration Systems across Indian Languages Using Parallel Corpora. Sponsors: National Science Council, Executive Yuan, ROC Institute of Linguistics, Academia Sinica NCCU Office of Research and Development, pp. 390–398, (2013).
14. Jacobs, J. R.: Finding Words That Sound Alike-The Soundex Algorithm. Byte, 7(3), pp. 473–474, (1982).

Prediction of Stock Market Indices by Artificial Neural Networks Using Forecasting Algorithms

Snehal Jadhav, Bhagyashree Dange and Sajeeda Shikalgar

Abstract Application of artificially intelligent methods for predictions is a fairly old area, although it is also the one in which there is always room for improvement in performance and in consistency, given the escalating nature of information and the varying efficacy of prediction logics. A hybrid of simple statistical methods coupled with intelligent computing (here artificial neural networks) is most likely to yield the closest prediction values with modest error rates. We propose to build an analytical and predictive model for estimating the stock market indices. This model can guide any kind of a user with or without experience in the stock market to make profitable investments. The forecasting done is by way of three statistical algorithms and an adaptive, intelligent algorithm, thus making the process fairly robust. Training and testing the neural network will be done with two-month stock market index values for some of the companies listed with the Bombay Stock Exchange. A comparative result of the four algorithms is calculated, and the one with best precision is suggested to the user with a sale/buy/hold answer.

1 Introduction

To explain very simply, a stock market is a place where companies that wish to generate capital sell their ownerships in the form of shares and equities. People who wish to invest in a particular company buy its stock, sell it when its price rises, thus earning money without any labour. Brokers or dealers guide consumers to gain maximum profits. The reason it makes share investments risky is the innate nature

S. Jadhav (✉) · B. Dange · S. Shikalgar
Department of Information Technology, Maharashtra Institute of Technology,
Pune, India
e-mail: snehaljadhav417@gmail.com

B. Dange
e-mail: dange.bhagyashree8@gmail.com

S. Shikalgar
e-mail: sajeeda.dsr@gmail.com

© Springer Nature Singapore Pte Ltd. 2018   455
S.S. Dash et al. (eds.), *International Conference on Intelligent Computing and Applications*, Advances in Intelligent Systems and Computing 632,
https://doi.org/10.1007/978-981-10-5520-1_42

of stock prices. They are highly volatile and dynamic making it impossible for even an experienced broker to predict its succeeding value.

Stock prices data is in a time series pattern and predicting real-time series values is an elaborate and complex task to even artificially intelligent systems because of its 'random walk nature'. A lot of economic factors like demand–supply, earnings, investors' sentiments, expected growth cannot be quantized into a single theory or a model that predicts flawlessly.

Human brain is the best prediction machine, and artificial neural networks are modelled on the human neural system. ANNs can be algorithms or softwares that study patterns and approximate functions, eventually predicting unknown variables.

Artificial neural networks can help in these predictions by recognizing patterns in the training data. The choice of model, learning algorithms, hyperparameters and cost functions contributes greatly towards making most accurate predictions and robust systems [2]. Experimenting with just enough amount and variety of data also decides the precision of the predicting network.

2 Literature Survey

Empirical studies carried out and as published by Sabaithip Boonpeng and Piyasak Jeatrakul in their paper 'Decision Support System for Investing in Stock Market by using OAA-Neural Network' have studied the use of several multi-class classification techniques using neural networks [3]. The multi-binary classification experiments using one-against-one (OAO) and one-against-all (OAA) are conducted on the historical stock data of Thailand stock exchange and further compared with traditional NN systems. The authors have calculated the accuracy of each of these models in different classes. Conclusively, the results show that OAA-NN outperforms OAO-NN and the multi-class classification using a single NN.

Meryem Ouahilal, Mohammed El Mohajir, Mohamed Chahhou, Badr Eddine El Mohajir Murugan in their paper 'A Comparative Study of Predictive Algorithms for Business Analytics and Decision Support systems: Finance as a Case Study' carried out extensive predictive experiments on L'Oréal financial data set [4]. They have analysed the performance of three different algorithms: multiple linear regression, support vector regression and decision tree regression. The results obtained showed the superiority of SVR model over MLR and DTR models.

3 Artificial Neural Network Architecture

Computational systems that have a linear path of execution also called as procedural computational systems follow a sequence of steps to eventually reach an end state or output. A real neural network being a manifestation of such a computational system is also additionally tethered to what are called the nodes (neurons). This non-sequential and linked architecture results in collateral and collaborative processing of information. A network of artificially intelligent neurons may contain a few hundred to millions of these units. Hierarchically, the network may contain an input, an output and a variable number of hidden layers. The input layer is an interface to the outside world information and transfers this received information to the inner layers. The inner layers decide the behaviour of the network from what it has learned. A wholly linked neural network will have each of the units connected to every other unit in every layer. Every unit has two natural behaviours to excite or to suppress another linked neuron. The link between these units called a 'weight' is a number that depicts one of these behaviours: positive for exciting and negative for suppressing. Weight is directly proportional to a unit's influence. This working is a simulation of how actual brain cells activate each other over synapses.

3.1 Algorithm

The neural networks are broadly classified into feed-forward and feed-backward networks. A non-recurrent network, which contains inputs, outputs and hidden layers, is called the feed-forward network. The signals are allowed to travel in only one direction. Input data is passed onto a layer of processing elements where the calculations are performed. Each processing element makes a computation based upon the weighted sum of the inputs. The newly calculated values thus become the new input values, which feed the next layer. This process continues until it has passed through all the layers and the final output has been decided. The output of a neuron in the output layer can be quantified by a threshold function. Perception (linear and nonlinear) and radial basis function networks are also integral part of the feed-forward networks.

Feed-backward networks have only feed-backward paths. This means that they can have signals travelling in both directions using loops. Neurons can have all possible connections. Due to the presence of loops, a feed-back network becomes a nonlinear dynamic system, which changes until a state of equilibrium is reached. Associative memories and optimization problems can be best solved by the feed-back network, where the networks search for the best possible arrangement of interconnected factors (Fig. 1).

4 Methodology

A hybrid approach to financial time series prediction was observed to have better prediction, and hence, we have adopted an ensemble of statistical and artificially intelligent methods. The statistical algorithms capture the fundamental analysis of the historic data, while the ANN analyses the data technically. Both the methods

Fig. 1 ANN architecture

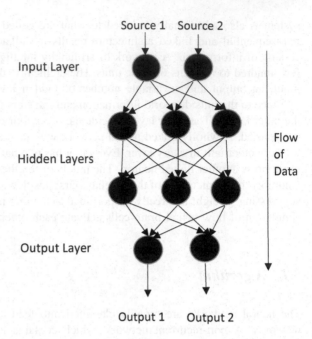

have been known to yield almost accurate results under different circumstances. An integration of the two distinct techniques into a single technological model encapsulates most of the approaches known for prediction of stock markets.

Indian stock market unlike other international markets is affected by factors other than economic, for example, festivals. The market makes exception on days like Diwali in terms of working hours. We have tried to capture this effect by taking two data sets for a user-selected date prediction: (1) stock market data for the preceding 30 days from the selected date of the current year and, (2) stock market data for the same period in the preceding year. This means a cumulative data of two months. Data for the same period from two different years essentially guarantees more varied input to the neural network that embodies all possible economic fluctuations while avoiding overfitting.

5 The Workflow

The Yahoo API is a part of Yahoo's finance Web services that provide us with real-time stock market data from BSE and NSE. Every time the system is logged into and has an active Internet connection, and this data is stored into a database and further used to train the neural network for every new prediction request. To keep the delay here in loading data fairly minimum, we store data for two months only.

Month-wise prediction

The user is given a date choice to check predictions; it is possible to view the prediction results and analysis for all the available working days of the month selected.

Day-wise prediction

Every time the user selects a date and an algorithm, and the closing and adjusted closing values for the day are predicted along with a graphical trend of that stock (Fig. 2).

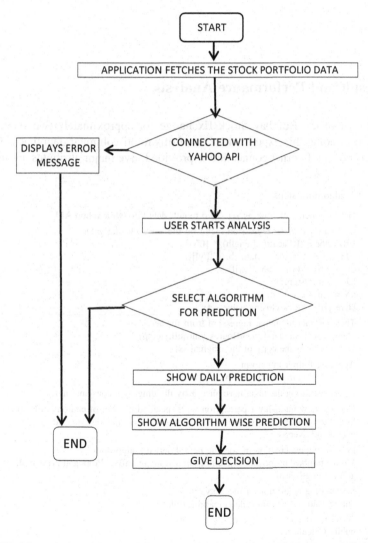

Fig. 2 Workflow chart explains the prediction phases from data acquisition to analysis and result display

6 Training and Calculating

The four algorithms used: moving averages algorithm, forecasting algorithm, regression algorithm and a neural network algorithm. The first three being static algorithms normalize and mathematically process the two-month data fed to the network and arrive on an output value.

6.1 How the Neural Network Algorithm Works

See Table 1.

7 Result and Performance Analysis

The data set of the Bombay Stock Exchange for approximately two months has been used in doing this experiment. Out of this multi-columnar data that is fetched online every time Internet connection is provided, five major attributes are used for

Table 1 NN algorithm steps

Step 1:	Get the stock data training set. Two-month data through a yahoo API.
Step 2:	Calculate the linear regression factor on the stock closing value LRFactor = (RFactor/y.Length) * 100.0 RFactor = RFactor + Math.abs(diff/y[j]) diff = y[j] − (aX + bX * x[j]) bX = sum1/sum2 aX = ym − bX * xm Here y[j] = stock closing value set.
Step 3:	Then subtract the linear regression from the set ytemp = LR.SubtractLinearRegression(null,ytemp) Here ytemp is the copy of the original set
Step 4:	Applying fourier transform interval1 = fa.transform(ytemp)
Step 5:	After processing the data, normalize it by dividing by a constant value.
Step 6:	Create a new multi-layer perceptron with specified number of neurons in layers. MultiLayerPerceptron(TransferFunctionType transferFunctionType, int .... neuronsInLayers)
Step 7:	Provide implementation of specific neural network algorithm MomentumbackpropogationLearningrule((MomentumBackPropogation) neuralNet. getLearningRule())
Step 8:	Set learning algorithm for this network public void setLearningRule(L learningRule)
Step 9:	Perform calculation on the whole network public Calculate()
Step 10:	Return network output vector public double[] getOutput() here Output vector is an array collection of double value.

the prediction process that is carried out by the four algorithms as mentioned in above sections.

Those are as mentioned:

- Sopen
- Shigh
- Slow
- Sclose
- Sadjclose

where

Sopen   opening price of company stocks for the day.
Shigh   highest price of company stocks for the day.
Slow   lowest price of company stocks for the day.
Sclose   closing price of company stocks for the day.
Sadjclose   is the normalization value which is adjacent to the closing value.

This is how the data set of the companies looks like, accumulating additional three attributes called sdate, svolume and symid which store date of storage, quantity of stocks and the symbol identifier of company, respectively. The table deposits actual stock values which are used by algorithms to perform mathematical calculation to generate next day's opening price and to predict whether to sell/hold/ buy the stock (Fig. 3).

? ...	sdate	sopen	shigh	slow	sclose	svolume	sadjclose	symbid
1	2016-02-18	117.00	120.35	112.00	114.10	752400	114.10	16
2	2016-02-17	112.15	115.60	108.40	114.85	1023800	114.85	16
3	2016-02-18	1097.00	1126.44995	1097.00	1122.69995	90700	1122.69995	3
4	2016-02-16	121.60	123.70	112.60	113.50	654900	113.50	16
5	2016-02-17	1082.00	1101.40002	1077.00	1095.44995	388300	1095.44995	3
6	2016-02-18	284.00	289.00	281.65	284.90	84200	284.90	14
7	2016-02-18	833.00	846.75	825.10	843.80	50400	843.80	6
8	2016-02-15	113.40	122.60	113.40	120.10	1074500	120.10	16
9	2016-02-18	53.60	54.50	52.35	53.60	488500	53.60	7
10	2016-02-17	279.65	284.00	271.80	281.90	90600	281.90	14
11	2016-02-16	1090.00	1092.00	1075.15002	1081.00	290800	1081.00	3
12	2016-02-18	2280.05	2319.45	2279.00	2315.30	31700	2315.30	11
13	2016-02-16	284.10	288.00	275.20	276.70	101800	276.70	14
14	2016-02-17	827.90	834.95	816.15	824.60	55900	824.60	6
15	2016-02-12	119.00	120.00	105.60	110.15	1337800	110.15	16
16	2016-02-18	983.00	990.75	977.00	989.60	190400	989.60	5
17	2016-02-17	52.25	53.75	50.35	53.00	660200	53.00	7
18	2016-02-18	531.50	542.75	531.09998	541.04999	48800	541.04999	13
19	2016-02-17	975.20	980.00	959.10	975.20	97700	975.20	5
20	2016-02-15	1090.00	1099.94995	1086.80005	1091.34998	1511000	1091.34998	3
21	2016-02-15	278.55	289.00	278.55	285.05	129500	285.05	14

Fig. 3 Snapshot of database table

In the equation given below, *Ai* is the actual value in the data fed to the algorithms, whereas *Pt* is the result value or the predicted value for the same day's opening prices of the particular company's stock and is the mean value of the actual values. Mean square error is the value showing the error rate or we call it as efficiency of the algorithms. Calculation of efficiency is based on comparing the actual output and the predicted output of the algorithm. Efficiency is calculated every time the algorithm runs.

$$\text{Mean Square Error} = \frac{\sum_{i=1}^{N1} (Ai - Pt)^2}{\sum_{i=1}^{N1} (Ai - \acute{A}t)^2}$$

Mean square error is the value showing the error rate or we call it as efficiency of the algorithms. Calculation of efficiency is based on comparing the actual output and the predicted output of the algorithm. Efficiency is calculated every time the algorithm runs.

Efficiency of each algorithm varies according to the variation in data which is shown in Fig. 4. Out of four algorithms, the one which is intelligent (adaptive to situation) named neural network algorithm has the highest efficiency (in case of ups and downs in the stock indices). Other three, the statistical ones perform well in case when parameters have stable graph, i.e. when stock indices are not discrete (thus the system is hybrid in a way combining statistical and adaptive approaches). Approximately, efficiency values are as stated below:

- Regression algorithm is 37%,
- Moving averages is 41%,
- Forecasting is 38% and
- Neural network is 47%.

Figure 5 is the example showing performance graph of the algorithms in case of particularly selected company. It depicts comparison of results produced by algorithm and the actual value which was supposed to be generated. Whenever both these values match, efficiency rises.

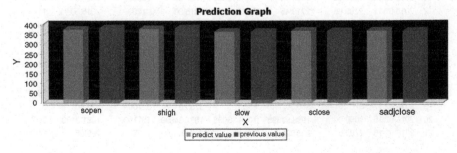

Fig. 4 Prediction graph

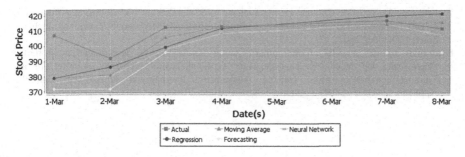

Fig. 5 Analytical graph of results

8 Conclusion

A lot of predicting softwares are constantly developed but generally employ a single approach. Having trained a neural network with optimal amount of data, we thus find the combinative approach of all the four algorithms best suited to cover all possible market fluctuations and predict with maximum efficiency. Our proposed Web-based application has a user-friendly interface with stock updates and has very minimal latency.

References

1. Gabriel Iuhasz, Monica Tirea, Viorel Negru, "Neural Network Predictions of Stock Price Fluctuations", 14th International Symposium on Symbolic and Numeric Algorithms for Scientific Computing, 2012.
2. D. Ashok Kumar, S. Murugan, "Performance Analysis of Indian Stock Market Index using Neural Network Time Series Model", Proceedings of the 2013 International Conference on Pattern Recognition, Informatics and Mobile Engineering (PRIME) February 21–22.
3. Decision support system for investing in stock market by using OAA-Neural Network Sabaithip Boonpeng; Piyasak Jeatrakul 2016 Eighth International Conference on Advanced Computational Intelligence (ICACI).
4. A comparative study of predictive algorithms for business analytics and decision support systems: Finance as a case study Meryem Ouahilal; Mohammed El Mohajir; Mohamed Chahhou; Badr Eddine El Mohajir 2016 International Conference on Information Technology for Organizations Development (IT4OD).
5. Gitansh Khirbat, Rahul Gupta and Sanjay Singh, "Optimal Neural Network Architecture for Stock Market Forecasting", International Conference on Communication Systems and Network Technologies, 2013.
6. Saima Hassan, Jafreezal Jaafar, Brahim B. Samir, Tahseen A. Jilani, "A Hybrid Fuzzy Time Series Model for Forecasting", Advance online publication: 27 February 2012.

7. Efthymia V. Tsitsika, Christos D. Maravelias, John Haralabous, "Modeling and forecasting pelagic fish production using univariate and multivariate ARIMA models", Article in Fisheries Science October 2007.
8. Osman Ahmed Abdalla, Abdelrahman Osman Elfaki, Yahya Mohammed AlMurtadha, "Optimizing the Multilayer Feed-Forward Artificial Neural Networks Architecture and Training Parameters using Genetic Algorithm", International Journal of Computer Applications (0975–8887) Volume 96–No. 10, June 2014.

Enhanced Prediction Model for Customer Churn in Telecommunication Using EMOTE

S. Babu and N.R. Ananthanarayanan

Abstract Customer churn is the term that refers to the customers who are in threat to leave the company. A growing number of such customers are becoming critical for the telecommunication sector, and the telecom sector is also in situation to retain them to avoid the revenue loss. Prediction of such behavior is very essential for the telecom sector, and classifiers proved to be the most effective one for the same. A well-balanced data set is a vital resource for the classifiers to yield the best prediction. All existing classifiers tend to perform poorly on imbalanced data set. An imbalanced data set is the one, where the classification attribute is not evenly distributed. Like the other real-time applications, the telecommunication churn application also has the class imbalance problem. So it is extremely vital to go for fine-balanced data set for classification. In this paper, an empirical method enhanced classifier for telecommunication churn analysis model (EC_for_TELECAM) using enhanced minority oversampling technique (EMOTE) has been proposed to improve the performance of the classifier for customer churn analysis in telecom data set. To evaluate the proposed method, experiments were done with various data sets. The experimental study shows that the proposed method is able to produce well-balanced data set to improve the performance of the classifier and to produce the best prediction model.

Keywords Telecommunication · Customer churn · Classifier
Imbalanced data set · Oversampling

1 Introduction

The telecom services are accepted all over the world as an important source of socioeconomic growth for a nation. Particularly, the Indian telecom has attained a phenomenal growth during the last few years and is expected to take a positive growth in the future also. This rapid growth is possible due to the different positive

S. Babu (✉) · N.R. Ananthanarayanan
SCSVMV University, Kanchipuram, Tamil Nadu, India
e-mail: babulingaa@gmail.com

© Springer Nature Singapore Pte Ltd. 2018
S.S. Dash et al. (eds.), *International Conference on Intelligent Computing and Applications*, Advances in Intelligent Systems and Computing 632,
https://doi.org/10.1007/978-981-10-5520-1_43

465

and proactive decisions of the government and the contribution of both by the private sector and the public. Due to the increasing competition, telecom sectors are facing the issue of customer churn. Customer churn is the term that refers to the customers who are in risk of leaving the company. Churn is a very critical issue in telecom because of its association with loss of revenue and the high cost of attracting the new customers. Prediction of such behavior is very essential for the telecom sector, and classifiers proved to be the effective one for the same.

Classification is widely accepted data mining technique for mining the data and predict about the future. By building the pertinent classifier, it is able to predict well about which class the new instance is [1]. In general, classifiers presume that the data set instances are uniformly distributed among different classes. The classifier is able to perform better on the data set whose distribution among the class is even but poor on the imbalanced data set. On the contrary, the real-world data sets are imbalanced among the distribution of the class attribute. The issue of class imbalance arises when there are many more instances in one class (majority class) and very less instances in another class (minority class) in the training data set. The performance of the classifiers built based on such imbalanced data set was extremely good at majority class but very poor on minority class [2]. Conversely, in many of the real cases, the most essential one for the prediction are minority class instances.

2 Related Work

As the churn is a critical issue in telecom, recent researchers showed more interest in churn analysis in telecom and proposed several methods to predict the same not by considering imbalance issue. The classifiers may lead to perform poorly on the imbalanced data set. To obtain the best prediction system, the imbalance problem needs to be resolved.

Ionut Brandusoiu, Gavril Toderean have proposed a churn prediction model using support vector machine learning algorithm with four kernel functions, namely RBF, LIN, PO, and SIG [3]. The churn data set is considered for experiments and results analyzed in technical and practical point of view. From the analysis, it is proved that the prediction model that employs with polynomial kernel function performs best. But in a practical point of view, the other three models like LIN, POL, and RBF perform best in the prediction of churn.

Nitesh V. Chawla has proposed the technique called synthetic minority over-sampling technique (SMOTE) [4]. In the proposed work, oversampling was used for minority class and under-sampling was used for majority class for balancing the data set. For oversampling, the samples of minority class are populated by creating artificial synthetic examples instead of replicating the real data. Based on the required amount of oversampling, five nearest neighbors using k-NN algorithm are taken randomly. From which two are chosen and the new instance was created in the direction of each. To analyze the performance, various data sets and the classifiers like C4.5, naive Bayes, and ripper are used. The obtained results prove that the proposed method performed well than the other re-sampling techniques.

Maisarah Zorkeflee has presented a new under-sampling method which is a combination of fuzzy distance-based under-sampling (FDUS) and SMOTE [5]. The process divides data set into two classes, namely majority (Ai) and minority (Bi) class. Using FDUS method re-sampling data set is repeated to produce the balanced data set. During the process, SMOTE is used to balance the data set if Ai becomes lesser than Bi. *F*-measure and *G*-mean are the measures used to analyze the performance of FDUS+SMOTE using the data set Pima, Haberman, and Bupa. The result of the analysis shows that the proposed method performed best on balancing the data set than the other techniques.

Piazza Jeatrakul has introduced a model by combining both the SMOTE and the complementary neural network (CMTNN) [6]. In the proposed method for under-sampling, CMTNN is applied and, for oversampling, SMOTE is applied. By combining these two methods, four techniques are developed to deal with imbalance problem. German, Pima, Spect, and Haberman's are the data sets considered for evaluation. To evaluate the proposed balanced data set, classifiers ANN, SVM, and k-NN are executed on the same. The comparison result of the measures like *G*-mean and AUC suggests that proposed method performs well.

3 Proposed Work

In the imbalanced data set, classifiers do not give high level of priority in classifying minority class instances. While constructing the tree also, the minority class instances are treated in the lower branches of the tree, and such a treatment may lead to the increase of the misclassification rate in the minority class instances. Ultimately, the sensitivity becomes very poor and the overall performance too. So, classifiers need to give more importance for the minority class instances also. With the above brief description, the proposed work is designed as follows:

Algorithm 1: Enhanced Classifier_for_TELEcommunication Churn Analysis Model (**EC_for_TELECAM**)

1: **Procedure** EC_for_TELECAM (*Ads*)
  Input: Imbalanced Dataset (*Ads*),
  Output: *Dt* - Decision Tree, *Dr* - Decision Rule
2: Get the Actual dataset *Ads*
3: **T_Dr** = call **EMOTE** (*Ads*, 5, '**True**') //To improve the Sensitivity of Dataset
4: **TR** = Extract the Decision Rule for '**True**' class from **T_Dr**
                            // Decision Rule for True Class
5: **F_Dr** = call **EMOTE** (*Ads*, 5, '**False**') //To improve the Specificity of Dataset
6: **FR** = Extract the Decision Rule for '**False**' class from **F_Dr**
                            // Decision Rule for False Class
7: Construct the Predicting Rule from **TR & FR**
8: Classify the Actual Dataset (*Ads*) with the Predicting Rule from Step 7
9: **end procedure**

An enhanced model EC_for_TELECAM is proposed with the main motivation to improve the classification accuracy of the classifier. The core idea behind the proposed model is that fine-tuning the misclassified instances into correctly classified instances using their nearest neighbors. The flow starts first by improving the sensitivity (true-positive rate) by calling the procedure EMOTE through which decision rule for true class is extracted. Then by calling EMOTE, the specificity (false-positive part) of the data set is also improved, by which decision rule for false class is extracted. Then to improve the overall performance of the classifier, the new classification rule is framed using the extracted decision rules of true and false classes.

Algorithm 2: Enhanced Minority Oversampling TEchnique (EMOTE)

1: **procedure** EMOTE (*Ai, k*)
  Input: Imbalanced Dataset (*Ai*), Nearest Neighbor (*k*)
  Output: *Dt* - Decision Tree, *Dr* - Decision Rule
2: 4:   Set the classifier and build the classifier
5:    Classify the dataset *Ai* and compute the measures
6:    Compute $acc \leftarrow (TP+TN) / (TP+FN+FP+TN)$
7:    set $imp\_acc \leftarrow acc$
8:    Compute *Sensitivity* $\leftarrow TP / (TP+FN)$
9:    Compute *Specificity* $\leftarrow TN / (TN+FP)$
11:   *repeat*
12:       *Partition* the Actual Dataset (*Ai*) into Misclassified instances as (*Mi*)
              and Correctly Classified instances (*CCi*)
13:           **for** each instance *i* in *Mi* do
14:               **if** *i = c* **then**   // c – **Class Value**
15:                   Populate *i* into *Ai*
16:                   find the *k* nearest neighbor (*Nn*) of *i* from *CCi*
                                                    //k- number of neighbor
17:                       **for** each instance *j* in *Nn* do
18:                           **if** *j = c* **then** // c – **Class Value**
19:                               Populate *j* into *Ai*
20:                           **end if**
21:                       **end for**
22:               **end if**
23:           **end for**
24:           Set acc $\leftarrow imp\_acc$
25:           rebuild the classifier with improved *Ai* and computer accuracy
26:           Compute $imp\_acc \leftarrow (TP+TN) / (TP+FN+FP+TN)$
27:   *until* (acc > imp_acc)
28:   return (**Dr**) // Returns Decision Rule
29: **end procedure**

The flow of the procedure EMOTE begins by obtaining the imbalanced data set (Ai) as input data set. In addition to this, the procedure also takes two additional parameters, namely number of nearest neighbors (k) for k-NN processing and class label (c) to define about, for which class (true of sensitivity or false for specificity) the data set has to be balanced to enhance the classification accuracy of the classifier. By setting the class attribute, the classifier is built on the imbalanced data set.

As a primary step, various performance measures like overall accuracy, sensitivity (true-positive rate) and specificity (false-positive rate) are computed from the results of the classifier. If the accuracy of the required class label (C) is not optimal then the actual data set (Ai) is partitioned into two different data sets, namely data set (CCi) with correctly classified instances and data set (Mi) with misclassified instances. As the secondary step, the instance from the misclassified instances (Mi) whose class label identical to c (true or false given through a parameter) is considered for tuning. To fine-tune the selected misclassified instance into correctly classified instances, the nearest neighbors of the misclassified instance from the correctly classified instances are retrieved. Then the selected misclassified instance and its equivalent (based on class label) nearest neighbors which as identical class labels are populated on the actual data set (Ai). The above step is repeated for each misclassified instance whose class label belongs to c. As a final step with the enhanced data set, the classifier is rebuilt, and the above said primary and secondary steps are repeated till the optimal accuracy is attained on the class label denoted by c.

4 Results and Discussion

To evaluate the performance of proposed work, the C# application has been developed. WEKA [7] is used for data mining process. To utilize the classes of WEKA in the C# code, the source file WEKA.jar is transformed to WEKA.dll using IKVM [8]. Classifier C4.5 is used in the developed code to build the classification model. Different data sets, namely yeast, adult, phoneme, thoracic surgery, churn from the UCI machine leaning repository, and oil spill data given by Kubat [9], are used to test the performance of the classifier.

As part of the work is pertained with descriptive research, the primary data were collected. Survey method was adopted using the structured questionnaire for the collection of primary data. The attributes of the questionnaire are defined with 5-point scale. In India, Tamil Nadu was the second largest in mobile subscription, so the same was selected as a sampling framework for this study. Totally 235 samples were collected and considered for processing.

To test the efficiency of the proposed method, different types of test have been carried out. They are,

1. Performance evaluation of the proposed model using C4.5 on various data sets.
2. Performance evaluation of the proposed model using various classifiers on the thoracic surgery data set.

3. ROC analysis of the proposed model using various classifiers on churn telecom data set.
4. Evaluation of the churn prediction model defined using the proposed work.

4.1 Evaluation of Proposed Model Using C4.5

To assess the proposed method, the classifier C4.5 was executed on the imbalanced actual data set using WEKA. From the results of the classifier, various performance measures are calculated. To balance the actual data set, the proposed model was executed on the actual data set. After balancing, the classifier is executed yet again on the balanced data set. The calculated values of both executions on various data sets are recorded and presented in Table 1.

The obtained experimental result given in Table 1 reveals that the performance of the classifier C4.5 on the actual data set is not fine on the prediction of true (sensitivity) class and fine on the prediction of false (specificity) class. However, the classifier is able to perform well in both the classes in the data set which is produced by EMOTE. This is because the proposed method fine-tune the misclassified instance into correctly classified instance by replicating the nearest neighbors.

4.2 Evaluation of Proposed Model Using Various Classifiers

In contrast to the comparison mentioned above, various classifiers like NB tree, random forest, simple cart, k-NN, and multi-layer perceptron (MLP) are executed

Table 1 Performance comparison of proposed method on various data sets

Measures		Data sets				
		Churn telecom	Yeast	Adult	Oil spill	Phoneme
Sensitivity	Actual	30.3	47.44	63.56	31.71	79
	Proposed	**96.97**	**95.81**	**99.23**	**95.12**	**99.24**
Specificity	Actual	99.62	86.07	93.32	97.43	90.36
	Proposed	**99.91**	**97.91**	**99.73**	**99.48**	**99.53**
Overall accuracy	Actual	97.72	74.88	86.15	94.56	87.03
	Proposed	**99.83**	**97.31**	**99.61**	**99.57**	**99.44**
G-mean	Actual	54.94	63.9	77.01	55.58	84.49
	Proposed	**98.43**	**96.86**	**99.48**	**97.52**	**99.39**
F-measure	Actual	42.11	52.24	68.85	33.77	78.14
	Proposed	**96.97**	**95.37**	**99.2**	**95.12**	**99.06**
AUC	Actual	0.8461	0.8571	0.8872	0.5146	0.8992
	Proposed	**0.9988**	**0.9763**	**0.9773**	**0.9999**	**0.9858**

Table 2 Performance comparison of proposed method on various classifiers

Measures		Classifiers				
		NB tree	Random forest	Simple cart	k-NN	MLP
Sensitivity	Actual	4.23	9.86	2.82	5.63	16.9
	Proposed	**90.29**	**100**	**99.03**	**97.09**	**98.06**
Specificity	Actual	96.25	93.75	92.5	94.5	89.5
	Proposed	**99.75**	**100**	**98.5**	**96.5**	**99.25**
Overall accuracy	Actual	82.38	81.1	78.98	81.1	78.56
	Proposed	**96.53**	**100**	**98.84**	**96.7**	**98.84**
G-mean	Actual	20.17	30.4	16.14	23.07	38.89
	Proposed	**94.9**	**100**	**98.89**	**96.79**	**98.65**
F-measure	Actual	6.74	13.59	3.88	8.25	19.2
	Proposed	**94.66**	**100**	**98.31**	**95.24**	**98.3**
AUC	Actual	0.5773	0.6536	0.4937	0.561	0.5901
	Proposed	**0.9682**	**1**	**0.988**	**0.9423**	**0.974**

on the actual thoracic surgery data set. The results of the experiments are recorded and found not fair. To enhance the performance of the classifier, the proposed model was executed on the actual data set to balance the same. After balancing, the classifiers stated above are again executed on the balanced data set. The results of the executions are presented in Table 2.

The results of the experiments confirm that all the classifiers are able to prove their efficiency only on the majority classes but fail to prove in minority classes on actual data set. The results as well reveal that the proposed model is an efficient one to solve such an issue. In specific, the performance of simple cart is very poor when compared to other classifiers on minority class. Still simple cart proves that the proposed model balances the data set in a better manner to enhance the performance of the classifier.

4.3 Roc Analysis of the Proposed Model

The receiver operating characteristic (ROC) curves are the graphical approach for summarizing and displaying the performance trade-offs between true-positive and false-positive error rates of the classifiers [10]. In the ROC curve, the y-axis indicates the sensitivity (true-positive rate) and x-axis indicates the specificity (false-positive rate). The point (0, 1) in the ROC curve would be the ideal point, i.e., it represents that all positive instances are correctly classified as positive and no negative classes are incorrectly classified as positive. In an ROC curve, the following are the various important points: (0, 0)—states all as a negative class, (1, 1) —states all as a positive class, (0, 1)—ideal.

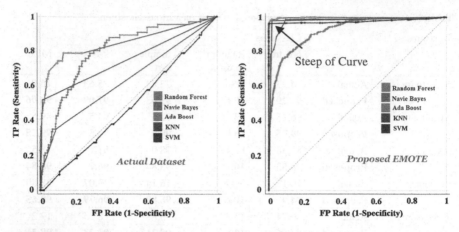

Fig. 1 ROC comparison of proposed method on various classifiers

The ROC curves depicted in Fig. 1 are defined using various classifiers like random forest, naïve Bayes, ADA boost, k-NN, and SVM as a base classifier on churn telecom data set. Initially, curves created by executing classifiers on the actual data set and next on the data set, which is balanced by the proposed model. The depicted ROC curves indicate that the performance of the classifier has huge variance between the actual imbalanced and the proposed balanced data set. This shows that the number of positive correctly classified instances is less and the number of negative incorrectly classified instances is high on the actual imbalanced data set. However, after balancing by proposed method, the classifiers are able to improve the result by attaining very few misclassifications on both positive and negative instances. The steep of the curve toward the point (0, 1) on the proposed method proves the same. This is the significant evidence to prove that the proposed model significantly enhances the performance of the classifier.

4.4 Evaluation of Proposed Churn Prediction Model

To define the enhanced churn prediction model, the primary data set is collected through questionnaire (235 instance: 89 True-class instances and 146 false-class instances) are considered. In order to extract the prediction rule, the proposed method **EC_for_TELECAM** was executed on the data set. The method primarily improves the true-positive rate (**sensitivity**) of the data set and extracts rule for true class. As a secondary step, false-positive rate (**specificity**) of the data set was improved, and the rule for false class is extracted. The prediction rule for the churn predicting system was framed using the rules extracted from primary and secondary step. To check the consistency of the defined predicting system, the cross-validation has been carried out with the actual data set. The results of the predicting system have obtained predictive accuracy of 97.55%, which is high about 18.35% than actual, and the error rate is 2.45%. In addition, it also has the true-positive rate (sensitivity) as 94.92% and false-positive rate (specificity) as 98.93%.

The statistical test is the most effective one to prove the significance of the model over the data. In this view, the statistical test is focused to analyze whether the prediction rules of the predicting system have significant over the predicting class variable or not. Discriminant function analysis is useful to deal with such an analysis. It is a statistical test to predict the dependent variable through the independent variable. It is also useful in finding whether a set of variable is effective in the prediction of the dependent variable or not [11]. The discriminant function looks like the following:

$$fk_m = u_0 + u_1X_1k_m + u_2X_2k_m + \cdots + u_pX_pk_m \tag{1}$$

There are three important variables in the discriminant factor analysis for analyzing the significance of the rule. They are eigen value, canonical correlation, and Wilk's lambda value. To investigate the significance of the prediction rules, the null hypothesis is tested using discriminant function analysis.

Ho: The variables involved in the prediction rule have no significance over the dependent variable.

To perform the test, the instances are filtered based on prediction rule and the variables that involved in this rule are tested against the filtered instances. The process is repeated on the randomly selected rules, and the results of the test are presented in Table 3.

The values of Table 3 shows that the eigenvalues of all the rules except the last two are >1, the canonical correlation $r_c > 0.35$, and Wilks lambda values are all less values (<0.7). The above values indicate that the rules are a good fit to the predictor. In addition to this, p-value <0.005, hence the null hypothesis **Ho** is rejected. This proves that the variables involved in the prediction rule are statistically significant to the dependent variable. As a summary, it is proved that the proposed method is able to define a best predicting system for customer churn in telecommunication.

Table 3 Results obtained from the discriminant factor analysis of decision rule

Eigen values			Wilks' lambda		
Function	Eigen value	Canonical correlation	Wilks' lambda	Chi-square	Sig.
1	1.810	0.803	0.356	12.914	0.005
1	1.008	0.709	0.498	27.196	0.000
1	3.000	0.866	0.250	4.852	0.028
1	1.585	0.783	0.387	25.645	0.000
1	0.600	0.612	0.625	12.455	0.006
1	0.591	0.609	0.629	18.804	0.000

5 Conclusion

The issues with imbalanced data set are inherent when used in the process of classification. It impacts the overall performance of the classifier. Many earlier studies focused on various approaches to improve the performance of the classifiers, not by considering the imbalance issues. Hence, the classifiers are not able to shine on the prediction of minority class instances. In this study, an enhanced model **EC_for_TELECAM** is proposed to handle the issue of imbalance in the data set. To evaluate the proposed method, various data sets are considered. The experimental results show that the proposed method well balances the data set by which it also improves the performance of the classifier. Hence, it is concluded that the role of the nearest neighbors of the misclassified instances is more vital in tuning misclassified instances into correctly classified instances. It is also concluded that the proposed method is more precious in such a data set where uniform distribution over the class attributes is not present. In order to define the prediction model for customer churn, the primary data which is collected through questionnaire was used on the proposed method. To test the significance of the rules involved in the prediction model statistically, Discriminant factor analysis using SPSS is carried out. The results of the test show that the rules of the prediction model are most significant with the related attributes. As a summary, the results of the experiments show that the proposed method EC_for_TELECAM outperformed and defined the best predicting system for customer churn. Through the predicting system, it has been concluded that billing, offers, accessibility, mobile number portability (MNP), and tariff plan are the most significant factors which influence the customer churn in telecommunication. Through this work, it is also suggested that these five factors are more valuable factors which needs to be focused more by the service providers in order to reduce the customer churn rate over the telecom industry.

References

1. S. J. Yen and Y. S. Lee, "Cluster-based under-sampling approaches for imbalanced data distributions," Expert Systems with Applications, vol. 36, pp. 5718–5727, (2009).
2. Qiang Wang, "A Hybrid Sampling SVM Approach to Imbalanced Data Classification", Hindawi Publishing Corporation, (2014).
3. Ionut Brandsoiu, "Churn Prediction in the Telecommunications Sector using Support Vector Machine", Annals of the Oradea university, May (2013).
4. Chawla, N. V, Bowyer, K. W., & Hall, L. O. "SMOTE: Synthetic Minority Over-sampling Technique", 16, 321–357, (2002).
5. Maisarah Zorkeflee, Aniza Mohamed Din, and Ku Ruhana Ku-Mahamud, "Fuzzy And Smote Resampling Technique For Imbalanced Data Sets", Proceedings of the 5th International Conference on Computing and Informatics, ICOCI2015 11–13 August, Istanbul, Turkey Universiti, Utara, Malaysia, (2015).
6. Piyasak Jeatrakul, Kok Wai Wong, and Chun Che Fung, "Classification of Imbalanced Data by Combining the Complementary Neural Network and SMOTE Algorithm", Springer-Verlag, (2010).

7. https://en.wikipedia.org/wiki/Weka_(machine_learning).
8. https://sourceforge.net/projects/ikvm/files.
9. Kubat, M., Holte, R., & Matwin, S. "Machine Learning for the Detection of Oil Spills in Satellite Radar Images. Machine Learning", 30, 195–215, (1998).
10. Tom Fawcett, "ROC Graphs: Notes and Practical Considerations for Data Mining Researchers", Intelligent Enterprise Technologies Laboratory, HP Laboratories Palo Alto, HPL-2003–4, January, (2003).
11. Benjamin Oghojafor, Godson Mesike, "Discriminant Analysis of Factors Affecting Telecoms Customer Churn", International Journal of Business Administration Vol. 3, No. 2; March (2012).

An Inter-Test Cube Bit Stream Connectivity-Optimized X-Filling Approach Aiming Shift Power Reduction

Sanjoy Mitra and Debaprasad Das

Abstract Transitions in scan cells bear much impact on the power consumption in scan-based VLSI test systems. X-filling approaches aim to fill don't care bits of test cube by typically assigning binary values (i.e. either 1s or 0s) in such a way that the mean switching activity gets lessened. We propose here a new X-filling approach named as bit stream connectivity optimization-based X-filling technique, called BSCO—a technique to decrease average shift-in transitions crop up during scan-based testing. In our approach, we have not only considered the shift-in switching activity specific to test vectors but also minimized inter-test cube switching activity by applying BSCO approach. The experimental outcomes attained from the benchmark ISCAS'89 clearly shows that the method is effective for reducing average transitions during scan shift operation.

Keywords BSCO · X-filling · X-bit · Switching activity · Shift power

1 Introduction

Like other testing schemes, scan-based testing also dissipates power at a higher extent in comparison with normal mode operation [1, 2]. This causes the following difficulties that peril the reliability of the circuits under test (CUTs).

S. Mitra (✉)
Department of Computer Science and Engineering, Tripura Institute of Technology,
Agartala 799015, Tripura, India
e-mail: mail.smitra@gmail.com

D. Das
Department of Electronics and Communication Engineering, TSSOT, Assam University,
Silchar 788011, Assam, India
e-mail: dasdebaprasad@yahoo.co.in

© Springer Nature Singapore Pte Ltd. 2018
S.S. Dash et al. (eds.), *International Conference on Intelligent Computing
and Applications*, Advances in Intelligent Systems and Computing 632,
https://doi.org/10.1007/978-981-10-5520-1_44

1. Thermal weight rises with the increase in mean power consumption and owing to its potential to cause structural harm to the silicon chip; this mean power consumption is needed to be moved away from the CUT.
2. Owing to shrinking of functional power, the test power keeps increasing, and their gap has become spacious from about 2X to 5X [1].
3. Excessive voltage drop occurring from disproportionate peak power dissipation may result in faulty data transfer during testing, especially during capture stage of at-speed testing; hence, the testing process becomes nullified and unneeded test yields loss [1, 3, 4] occurs.

A shift register is formed by functional linking of flip-flops in scan mode of testing. Three prime operating modes of scan-based testing are shift mode, capture mode and normal mode. During functional mode or normal mode, test signals are made 'OFF' and this enables the circuit to operate in its actual configuration. Capture and shift modes together are known as test mode. Scan flip-flops are toggled at a larger extent during test mode which causes switching activity raise in contrast to normal functional mode.

Here, we put forward a novel BSCO-based X-filling method to drop down shift-in switching in scan tests. In this approach, only shift power is taken into account for switching activity reduction. Consumption of capture power occurs at the last part of the edge of capture cycle for a period of CUTs-rated clock duration [5]. The flip-flops are scanned-in with bit stream data during shift operation. The shift operation during scan-based testing triggers significant raise in switching activity and hence considered as the important cause of power dissipation [6]. This proposal suggests a unique X-bit filling method to cut down average shift power consumption by lowering the shift-in transitions of the applied test cube. Here, test relaxation is applied to figure out non-essential bits in the input vector.

2 Overview of Prior Work on Low-Power Testing

Alterations are made to the circuit under test (CUT) for test power optimization, and this was researched by quite a lot of research groups, and their works include scan enable-disabling [7], virtual circuit partitioning [8], combinational logic division [9–15], toggling suppression through circuitry insertion [16, 17] or scan chain segmentation, power optimization for built-in self-test (BIST) application [18–20] and clock gating [21]. Although above approaches are quite useful in reducing test power, they are DFT cost-expansive and inconvenient for customary IC design and test flow. The gating scan architecture suggested in [22] reduces average power and also decreases peak capture transitions. The said method lowers average power by 28.17% when compared with usual scan architecture devoid of clock gating.

Usually, the X-filling techniques perform the job of filling unspecified (X) bits present in test vector with suitable binary values in such a way so that excessive switching activity during testing gets lessened. A much cited shift power reduction technique, adjacent filling [23], suffers from capture power violation. Besides this, MTR-fill [24], MT-fill [25], also aims to reduce shift power. MTR-fill is basically a combination of random filling and fill-adjacent techniques. In the random fill, don't care bits are filled up arbitrarily with 0s or 1s. In case of LC-filling [26] method, filling of X-bits is steered by the effect of filling individual don't care bits by '0' or '1'. The influence of every unspecified bits on the capture and shift power is considered, and accordingly, don't care bits are filled [27]. The sequence of filling these unspecified bits contributes to power reduction percentage. In this context, the prime challenge lies in the selection of target X-bit and further on the selection of a suitable value to be filled in the opted target bit position.

3 Proposed Technique of X-Filling

Here, we describe the proposed bit stream connectivity optimization-based X-filling technique, called BSCO—a X-filling technique aiming reduction in average shift power. Our focal concern is to lower transitions in test cubes during shift operation in order to optimize test power. In our approach, significant reduction in shift power is achieved by filling unspecified bits in the test cubes in such a way that starting bit stream and ending bit stream of each test cube (vector) are maximized in order to establish homogeneous bit connectivity to the maximized starting or ending bit stream of another test cube. Initially, starting and ending bit stream maximization corresponding to each test vector may be achieved with the following simple X-bit filling heuristic.

Firstly, in a test cube, fill all the don't care bits with the value of the left nearest care bit.

After filling in such fashion, if any X-bit still remains to fill, then fill this with the closest care bit on the right part of the cube.

Thus, the transition among successive bits will be lessened from the intra-vector perspective and also inter-vector perspective. The majority of the research work in this domain is focused towards filling X-bits to reduce intra-vector transition activity resulting in shift power reduction (Fig. 1).

The shift-in transitions are measured by means of a weighted transition metric (WTM) [28] which is shown in Eq. 1. Assuming scan vector $T_k = T_{k,1}, T_{k,2}, T_{k,3}, T_{k,l}$ with $T_{k,1}$ scanned-in before $T_{k,2}$, etc., and scan chain length to be l. The WTM for the scan-in test vector T_k can be expressed by

$$\text{WTM}_k = \sum_{i=1}^{l-1} (l - i) \left(T_{k,i} \oplus T_{k,i+1} \right). \tag{1}$$

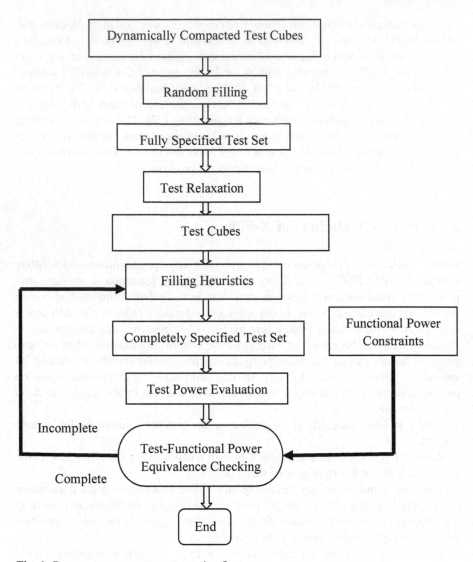

Fig. 1 Power-aware test pattern generation flow

Scan-in average shift power, P_{ave} corresponding to a test set $T_D = \{T_1, T_2, T_3, \ldots T_n\}$, can be estimated as follows:

$$P_{\text{ave}} = \frac{\sum_{k=1}^{m} \sum_{i=1}^{l-1} (l-i) \left(T_{k,i} \oplus T_{k,i+1}\right)}{n} \qquad (2)$$

In our approach, our objective is not limited to only reducing average power by reducing shift power consumption of each test cube (vectors). We have also

Fig. 2 Resultant test vector matrix

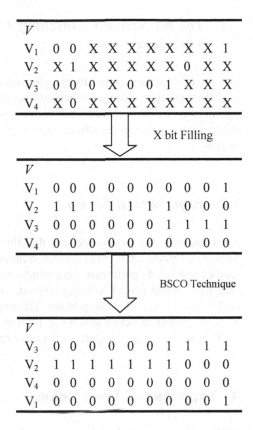

emphasized to drop down the switching activity arising out of different vector shift into the scan chain by using bit stream connectivity optimization (BSCO). We explain the BSCO approach with the following example.

In Fig. 2, V_1, V_2, V_3, V_4 initially represent incompletely specified test cubes generated after test relaxation process. These cubes (vectors) comprise don't care (X) bits along with other care bits. In the second stage of this figure, X-bits are filled up using our proposed method in such a manner that the end stream bit sequence and start stream bit sequence are optimized. In the final stage of the above figure, BSCO technique is applied among all the test vectors belonging to a test set (i.e. in this figure it is 4) in order to optimize inter-vector homogeneous bit stream length, and in this figure, it is shown as homogeneous bit stream of 1s for stream length '11' and homogeneous bit stream of 0s for stream length '22'.

3.1 The Bit Stream Connectivity Optimization (BSCO) Technique

The novel bit stream connectivity optimization technique is put forwarded here to find the optimum chain of homogeneous bits from different scan vectors. This optimization helps to reduce inter-vector switching activity to a great extent. One initial data structure in the tabular form is given in Table 1 for further computation in BSCO.

3.2 Experimental Set-Up

As mentioned in previous sections that the tools used are Design Compiler and Synopsys TetraMAX ATPG [29] tool. Synopsys TetraMAX was applied for getting relaxed test set. Experiments were conducted on ISCAS'89 [30] for verifying the efficacy of novel BSCO X-filling method. The algorithm is programmed in C and GCC *v 3.4.5* is used for compilation. The trials are done in core i3 machine having 3.2 GHz processor speed and 4 GB primary memory. Weighted transition metric (WTM) is utilized to compute shift power as depicted in Eq. 2.

4 Experimental Observation

In Table 2, a set of ISCAS'89 benchmarks on which experiment was conducted is described. Here, the basic circuit information like the number of gates that are equivalent within the circuit, 'scan cell quantity' and fault coverage is described. The findings on shift transitions from BSCO fill algorithm are recorded in Table 3.

Table 1 Bit stream vector table

Fully specified test vector	Starting bit stream			Ending bit stream		
	Prefix	Length	Code	Prefix	Length	Code
0000000001	0	9	09	1	1	11
1111111000	1	7	17	0	3	03
0000001111	0	6	06	1	4	14
0000000000	0	10	010	0	10	010

Table 2 Experimented benchmark circuit details

S. No.	ISCAS'89 benchmark	No. of gates	No. of scanned flip-flops	Fault coverage (%)
01	s27	30	3	100
02	s349	154	15	100
03	s420	142	16	100
04	s510	227	6	100
05	s713	235	19	100
06	s838	279	32	100
07	s1196	529	18	100
08	s1238	508	18	98.2
09	s1423	657	74	100
10	s1488	653	6	88
11	s5378	2836	179	100
12	s9234	5597	228	100
13	s13207	7479	638	99.1
14	s15850	9775	534	100

Table 3 Average shift-in switching comparison

S. No.	Benchmark circuit	Average shift-in switching				
		0 fill	1 fill	Adjacent fill [10]	Random fill	Proposed method
01	s27	1.40	1.45	0.89	1.80	1.14
02	s349	5.72	5.89	4.31	6.87	5.35
03	s420	3.83	3.77	3.36	5.41	3.32
04	s510	2.84	2.89	2.82	2.94	2.91
05	s713	6.10	5.53	5.78	8.63	5.51
06	s838	4.21	4.27	3.73	4.49	3.68
07	s1196	5.19	3.61	4.34	5.48	3.39
08	s1238	5.42	4.19	3.53	7.17	3.41
09	s1423	16.27	5.34	6.83	14.22	6.12
10	s1488	15.27	5.78	6.92	12.68	5.82
11	s5378	36.68	27.86	36.90	38.31	26.14
12	s9234	62.43	64.41	63.70	101.93	72.15
13	s13207	97.95	133.31	146.64	227.12	93.53
14	s15850	103.42	201.41	156.81	164.89	98.67

Table 4 Comparison of percentage reduction in average shift-in switching with respect to random filling approach

S. No.	Benchmark circuit	Average shift-in switching			
		0 fill	1 fill	Adjacent-fill	Proposed method
01	s27	22.00	19.44	50.55	36.67
02	s349	16.73	14.20	37.26	22.12
03	s420	29.20	30.31	37.89	38.63
04	s510	3.00	1.70	4.08	1.02
05	s713	29.30	5.53	33.08	36.15
06	s838	6.23	35.92	16.92	18.04
07	s1196	5.29	43.43	20.38	38.14
08	s1238	24.40	41.56	50.76	52.44
09	s1423	−14.36	62.44	51.96	56.96
10	s1488	−11.47	54.41	45.43	54.10
11	s5378	4.25	27.27	3.69	31.77
12	s9234	38.75	36.80	63.70	29.21
13	s13207	56.87	41.30	37.50	58.81
14	s15850	37.27	−16.07	4.90	40.16

Algorithm BSCO

---

1. Load the vector table into the memory
2. Sort the code values in starting bit stream and ending bit stream
3. Connect the highest ending bit stream code value with highest compatible starting bit stream code value and scan the corresponding ending bit stream code.
4. Repeat step 3 for presently scanned ending bit stream code value.
5. Repeat step 3 to 4 until all the entries in the first column are connected.
6. End

In Table 3, a shift-in switching data obtained with our approach is contrasted with random fill, 1-fill, adjacent-fill [22], 0-fill, etc. Data in Table 3 shows that consumed shift power is nearly analogous to the adjacent-fill technique which is renowned for its efficiency of lowering the shift power. Also, the proposed scheme attains comprehensive fall down in shift transition quantity compared with other X-filling schemes. Usually, it is observed that random fill generates lofty increase in switching activity in contrast to 0-fill and 1-fill. The obtained results of the experimented method are judged against other relevant filling techniques. A discernible reduction in switching activity average is observed in the case of our method. In many cases, it is found that average capture and shift switching decline gradually with the circuit size enlargement. A 36.67% cut-down for the least sized circuit (s27) and a 58.81% cut-down of average shift transitions for the larger circuit (s13207) are observed. Maximum drop in transitions for an average shift is observed in the s13207 circuit as compared to other filling approaches is shown in Table 4.

5 Conclusion

One of the prime concerns of silicon testing is power consumption. The approach is perhaps the first ever reported method in this low-power testing domain to give emphasis on inter-test cube switching activity minimization through efficient bit filling of the partially specified test cubes. This technique exhibits a satisfactory level of performance as compared to the other approaches. We have performed the experiment on a set of ISCAS'89 full-scan benchmarks. However, the same approach can be done on partial scan or multiple scan designs in order to get a higher percentage of reduction in average shift power. Besides the above, this can be integrated with SOC test data compression schemes to concurrently lessen test power as well as a large volume of test data.

References

1. Girard P (2002) Survey of low-power testing of VLSI circuits. In IEEE Des. Test Computing, vol. 19, no. 3, pp. 82–92, doi:10.1109/MDT.2002.1003802
2. Girard Patrick, Wen Xiaoqing, Touba Nur(2007). Low Power Testing. Morgan Kaufmann. System on-Chip Test Architectures: Nanometer Design for Testability, pp. 207–350, 978-0-12-373973-5
3. Wang Jing., Walker Duncan. M. H, Majhi A., Kruseman B., Gronthoud G., Villagra L. E., Wiel P. V. D. and Eichenberger S (2006) Power supply noise in delay testing. IEEE Design & Test of Computers. Volume: 24, No. 3, pp 226–234 doi:10.1109/MDT.2007.76
4. Saxena Jaayasree, Butler Keneth. M., Jayaram Vinay. B. and Kundu S. (2003). A case study of IR-drop in structured at-speed testing, In Proceedings of IEEE International Test Conference (ITC), pp. 1098–1104 doi:10.1109/TEST.2003.1271098
5. Wu F., Dilillo L., A. Bosio, Girard P., Pravossoudovitch S., Virazel A., J. Ma, W. Zhao, Tehranipoor M. and Wen X (2010) Analysis of Power Consumption and Transition Fault Coverage for LOS and LOC Fault Testing Schemes In Proc. of 13th IEEE Symposium on Design and Diagnostics of Electronic Circuits and Systems, pp. 376–381 doi:10.1109/ DDECS.2010.5491748
6. Butler Keneth M., Saxena Jaayasree., Jain A., Fryars T., Lewis J. and Hetherington G. (2004) Minimizing power consumption in scan testing: Pattern generation and DFT techniques, In proc. IEEE International Test Conference (ITC), pp. 355–364 doi:10.1109/ TEST.2004.1386971
7. Sankaralingam R., Pouya B. and Touba N. A. (2001) Reducing power dissipation during test using scan chain disable, in Proc. IEEE VLSI Test Symposium (VTS), 2001, pp. 319–324 doi:10.1109/VTS.2001.923456
8. Xu Q., Hu D., and Xiang D (2007) Pattern-directed circuit virtual partitioning for test power reduction, in Proc. IEEE International Test Conference (ITC), pp 1–10 doi:10.1109/TEST. 2007.4437633
9. Lee K. J., Huang T. C. and Chen J. J (2000) Peak-power reduction for multiple-scan circuits during test application, in Proc. IEEE Asian Test Symposium (ATS), pp. 453–458 doi:10. 1109/ATS.2000.893666

10. Rosinger P. M., Al-Hashimi B.M. and Nicolici N (2004) "Scan architecture with mutually exclusive scan segment activation for shift- and capture-power reduction," IEEE Trans. Computer.-Aided Design Integration, Circuits Systems, vol. 23, no. 7, pp. 1142–1153 doi:10. 1109/TCAD.2004.829797

11. Lee K. J., Hsu S. J., and Ho C. M (2004) Test power reduction with multiple capture orders. In Proceedings of IEEE Asian Test Symposium (ATS), 2004, pp. 26–31 doi:10.1109/ATS. 2004.82

12. Bonhomme Y., Girard P., Landrault C., and Pravossoudovitch S. (2002) Power driven chaining of flip-flops in scan architectures, in Proc. IEEE International Test Conference (ITC), pp. 796–803 doi:10.1109/TEST.2002.1041833

13. Bonhomme Y., Girard P, Guiller L., Landrault C. and Pravossoudovitch S (2003) "Efficient scan chain design for power minimization during scan testing under routing constraint," in Proc. IEEE Int. Test Conf. (ITC), Vol 1, pp. 488–493 doi:10.1109/TEST.2003.1270874

14. Li J., Hu Y. and Li X (2006) A scan chain adjustment technology for test power reduction. In Proc. 15th IEEE Asian Test Symposium (ATS) pp. 11–16 doi:10.1109/ATS.2006.260986

15. Naeini M. M. and Ooi C. Y. (2015) A Novel Scan Architecture for Low Power Scan-Based Testing, VLSI Design, Vol. 2015, Article ID 264071, doi:10.1155/2015/264071

16. Sankaralingam R. and Touba N. A. (2002) Inserting test points to control peak power during scan testing. In Proceedings of 17th IEEE International Symposium Defect Fault Tolerance VLSI Systems (DFT), pp. 138–146 doi:10.1109/DFTVS.2002.1173510

17. Sharifi S., Jaffari J., Hosseinababy M., Afzali-Kusha A. and Navabi Z (2005) Simultaneous reduction of dynamic and static power in scan structures. In Proc. of Design Automation Test Eur. (DATE), pp. 846–851 doi:10.1109/DATE.2005.270

18. Gerstendorfer S. and Wunderlich H. J. (1999) Minimized power consumption for scan-based BIST. In Proceedings of IEEE International Test Conf. (ITC), pp. 77–84 DOI 10.1109/TEST.1999.805616

19. Girard P., Guiller L., Landrault C. and Pravossoudovitch S (1999) Circuit partitioning for low power BIST design with minimized peak power consumption. In Proc. of IEEE Asian Test Symposium (ATS), pp. 89–94 doi:10.1109/ATS.1999.810734

20. Basturkmen N. Z., Reddy S. M., and Pomeranz I (2002) A low power pseudo-random BIST technique, in Proc. Int. Conf. Comput. Des. (ICCD), pp. 468–473 doi:10.1109/ICCD.2002. 1106815

21. Bhunia S., Mahmoodi H., Ghosh D., Mukhopadhyay S. and Roy K. (2005) Low-power scan design using first-level supply gating. IEEE Transactions on Very Large Scale Integration (VLSI) Systems, vol. 13, no. 3, pp. 384–395 doi:10.1109/TVLSI.2004.842885

22. Whetsel L.: Adapting scan architectures for low power operation, in Proc. IEEE International Test Conference (ITC), pp. 863–872 doi:10.1109/TEST.2000.894297

23. Butler K. M., Saxena J., Fryars T., Hetherington G., Jain A., and Lewis J (2004) Minimizing power consumption in scan testing: Pattern generation and DFT techniques. In Proceedings International Test Conference, pp. 355–364 doi:10.1109/TEST.2004.1386971

24. Song D., Ahn J., Kim T., and Kang S (2008) MTR-fill: A simulated annealing-based X- filling technique to reduce test power dissipation for scan-based designs, IEICE Transactions on Information and Systems, vol. E91-D, no. 4, pp. 1197–1200 doi:10.1093/ietisy/e91-d.4.1197

25. Ghosh S., Basu S. and Touba N.A. (2003) Joint minimization of power and area in scan testing by scan cell reordering, In Proceedings. IEEE Computer Society Annual Symposium on VLSI, pp 246–249 doi:10.1109/ISVLSI.2003.1183485

26. Li J. Xu Q., Hu Y., and Li X (2008) On reducing both shift and capture power for scan-based testing, in Proceedings of the Asia and South Pacific Design Automation Conference, ASP-DAC pp. 653–658 doi:10.1109/ASPDAC.2008.4484032

27. Wen X., Yamashita Y., Morishima S., Kajihara S., Wang L., Saluja K. K. and Kinoshita K. (2005) Low-capture-power test generation for scan-based at-speed testing, in Proceedings International Test Conference, pp. 1019–1028. doi:10.1109/TEST.2005.1584068

28. Sankaralingam R., Oruganti R. R. and Touba N.A (2000) Static compaction techniques to control scan vector power dissipation, in Proc. IEEE VLSI Test Symposium. (VTS), pp. 35–40 doi:10.1109/VTEST.2000.843824
29. Synopsys Inc.: TetraMAX ATPG user Guide, 2006.
30. F. Brglez; D. Bryan and K. Kozminski (1989) Combinational profiles of sequential benchmark circuits, In IEEE International Symposium on Circuits and Systems, Vol 3, pp 1929–1934 doi:10.1109/ISCAS.1989.100747

Studying the Role of Patient and Drug Attributes on Adverse Drug Effect Manifestation Using Clustering

Dipali, Yogita and Vipin Pal

Abstract Adverse drug reactions represent the unwanted or undesired effects of drugs. Timely extraction of such effects is highly required so that early warnings can be raised against if any serious beforehand to save patients from any further loss. It also helps in framing alternate treatment plan. There are two different ways for identifying the side effects of drugs. First is premarketing trials, which are conducted before floating drugs into markct. But this approach is not as effective as these trials are carried out on a restricted population for restricted time. That is why another approach called as postmarketing surveillance is used. Under this approach, data mining techniques have been applied frequently for finding adverse reactions of drugs. But most of the existing techniques are based on the assumption that all attributes are equally responsible for drug side effects which may not be applicable for all real-life cases. In this paper, we study the role of different patients and drug attributes in manifestation of adverse drug reactions using clustering technique.

Keywords Adverse drug reactions · Clinical trials · Data mining
Clustering · Pharmacovigilance

1 Introduction

Drugs/medicines are prescribed to patients for improving their health conditions, but unfortunately sometimes instead of improvement in health, these medicines induce some side effects in patients; these side effects are called adverse drug

Dipali (✉) · Yogita
UIET, Panjab University, Chandigarh, India
e-mail: bisladipali68@gmail.com

Yogita
e-mail: thakranyogita@gmail.com

V. Pal
Manipal University, Jaipur, India
e-mail: vipinrwr@yahoo.com

© Springer Nature Singapore Pte Ltd. 2018
S.S. Dash et al. (eds.), *International Conference on Intelligent Computing and Applications*, Advances in Intelligent Systems and Computing 632,
https://doi.org/10.1007/978-981-10-5520-1_45

effects/adverse drug reactions [1]. In clinical trials, positive effects of drugs become evident easily, but there is not always necessary information available to find out the negative effects of drugs. Also, these clinical trials are performed only on a restricted population for a restricted amount of time, and it is practically not possible to cover different age groups, drug dosage, etc., during this time [2]. That is why a lot of research is going on in the area of finding out the adverse reactions of drugs after they have been marketed for use. The activity of monitoring the impacts of medicines after they have been allowed for use focusing on the aim of identifying and assessing unintended and unreported responses beforehand comes under pharmacovigilance [3].

Timely extraction of such effects is highly required so that early warnings can be raised against if any serious beforehand to save patients from any further loss who which have not yet taken the drug [4, 5]. It also helps the medical practitioners in framing out any treatment for the adverse effects timely. A lot of factors are there which affect the occurrence of ADRs like age, weight, gender, pregnancy, drug dosage, alcohol intake [6]. Out of these factors, patient attributes like age, weight, and drug factors have a significant effect on adverse drug effect occurrence [7, 8].

Adverse drug reactions can be detected in two ways: First is through clinical trials before the drug is floated into market for use also called premarketing clinical trials. These trials are carried out so that the adverse effects of drugs if any could be found out before supplying the drug into market. But these trials are not so effective because they are performed on a restricted number of persons and for a restricted time so it is not possible to cover all possible cases to detect adverse effects of drugs [9, 10].

That is why another method called postmarketing surveillance is used for detecting harmful effects of drugs after they have been approved for use. Nowadays, reports on adverse effects of drugs are filed by doctors, pharmacists, consumers, manufacturers, etc., and this complete data is digitally available. One example of this is the spontaneous reporting system of US Food and Drug Administration (FDA) Adverse Event Reporting System (AERS) database [11]. Data mining techniques have been explored in postmarketing surveillance. Data mining is defined as a process of discovering useful patterns from the large amounts of data [12]. It has the potential of unearthing the hidden relationships of different drugs and their side effects. Most of the existing works in this area utilized association, classification, and clustering techniques of data mining. But these works assume all attributes equally responsible for occurrence of ADRs [2, 13–15]. But this can be possible that some attributes have more effect as compared to others in occurrence of ADRs.

In this paper, our aim is to study the role of different patients and drug attributes in occurrence of adverse effects of drugs. The effect of different attributes has been studied using clustering technique, and the contribution of different attributes in causing adverse effects of drugs has been analyzed.

The rest of the paper is organized as follows: In Sect. 2, literature survey of different data mining techniques used in the area of adverse drug effect detection is presented. In Sect. 3, methodology for the complete work is given. In Sect. 4, parameter setting for clustering is done. In Sect. 5, result analysis is given. In Sect. 6, the present work has been concluded.

2 Literature Survey

Adverse reactions of drugs are a serious problem. Detection of ADRs is necessary for alerting the healthcare professionals so that early warnings can be raised against these ADRs, and any effective solution could be framed out.

There are two different ways for identifying the side effects of drugs as follows:

Premarketing Clinical trials: These types of clinical trials are performed before a new drug is launched to check how effective drug for a disease is and to detect adverse effects if any so that can be addressed before starting the supply of the drug to customers [16]. One advantage of these trials is that the positive effects of drugs become evident easily and some of the adverse reactions are also identified. But there are some disadvantages of these trials too. Since these trials are conducted on a restricted population and for a restricted time frame, it is practically not possible to cover different age groups, drug dosage, etc. That is why it is not always possible to detect the rare adverse effects of drugs.

To overcome these limitations, postmarketing surveillance methods are used.

Postmarketing Surveillance: Postmarketing surveillance is performed after the drug is launched into market for use. Data mining techniques have been applied in this area for uncovering meaningful associations of medicines and their side effects.

As our work falls in the postmarketing surveillance, a brief summary of related work for this category has been discussed next.

2.1 Association Mining

It is a technique for finding the correlation between different data attributes from a large volume of data [12].

In paper [7], the author has extracted relationships between the adverse drug reactions, adverse event outcomes, and the patient demographics using Apriori algorithm, and the quality of results has been examined on bases if support, confidence, and accuracy measure. From the results, it is observed that diarrhea, anemia, and arthralgia are highly occurring in patients having age between 44 and 64 years. The work of paper presented in [17] has focused on analyzing the bleeding complications instigated on antiplatelet administration and to rank the relationships if any on basis of their frequency. As a result, total of 736 associations are detected out of which 147 were bleeding complications. On examining these, it is concluded that the association of hemorrhage is strong with drug clopidogrel than aspirin. On the other hand, for gastrointestinal bleeding complications, it is just opposite. The author in [18] has generated list associations of drug events with sex differences which can be helpful in investigating the sex difference in pharmaco-dynamics and pharmacokinetics. It also gave evidence for tailored medication instruction and prescription. For this purpose, chi-square test is used, and reporting odds ratio is considered to quantify the sex signal differences for drug-event combinations.

2.2 Clustering

It is an unsupervised data mining activity that deals with the generation of different groups of data in such a way that similar data instances come under the same group [12].

The author in [19] has identified those instances where different drug intakes lead to similar adverse reactions and intake of same drug has resulted in different reactions depending upon the patients' states. For this, the author has relied on the biclustering technique for detection of subsets of the drug adverse effects and of drugs for which frequencies of adverse reactions persistently high for different drugs. As a result of biclustering, total of 163 clusters are generated and about 89% of them showed the Pearson correlation coefficients smaller than 0.5 which signify large similarity in drug reactions and comparatively large variation in drug indi-cations which supports the likelihood of occurrence of similar drug reactions on taking same drug with different medical conditions. The author of paper [20] has applied k-means clustering method on data of such children who have given antibiotics. For this analysis, different attributes have considered, namely age of first antibiotic use, child age, type of antibiotic, age when adverse reaction occurs. On the analysis of clusters, it is found that the adverse effects in children occur mainly during the late preschool age and year of birth.

2.3 Classification

It is defined the learning the features of different classes in such a manner that this can be used for segregating future data instances in corresponding classes [12]. In [8], the author studies hidden relationships between patient information and Fosamax adverse events using association and classification techniques. Patient's age, gender and adverse event attributes, drugs reported for the event are used. Apriori algorithm is used to perform association analysis and single-label classifi-cation is performed in WEKA and multi-label classification is performed in MEKA software. Confidence in association rules is found between 90 and 96% which is greater as compared to previous studies. *F*-score is used for single-label classification evaluation. Accuracy and Hamming score are used for multi-label classification. The performance of multi-label classification is found best among all because a patient can have multiple adverse events at a time.

Most of the existing works which we have reviewed in the area of detecting adverse effects of drugs have utilized association mining technique. Also, these works assume all attributes equally responsible in occurrence of ADRs. But some attributes can contribute more as compared to others in generating ADRs. In this paper, we study the effect of different patients and drug attributes in manifestation of adverse effects of drugs using clustering technique.

3 Methodology

The methodology of the complete work is shown in Fig. 1.

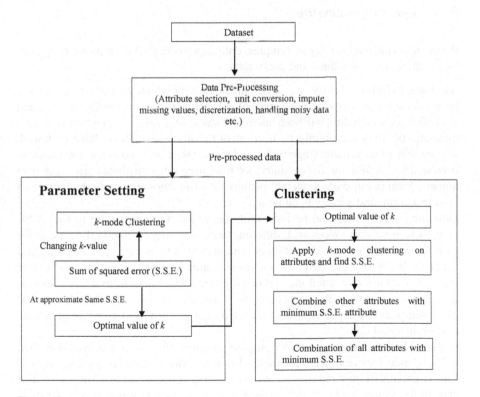

Fig. 1 Complete methodology flowchart

3.1 Dataset Selection

Dataset used in our study is publicly available on US Food and Drug Administration Web site [11]. For our study, we have chosen 2014 quartile 2 data which consists of total of 685,430 records. The dataset contains 7 data files including patient demographic and administrative information, report sources, patient outcomes, drug information, adverse events, drug therapy, and indications for use. Out of these, we have chosen 4 data files for our study DEMO, DRUG, REAC, and INDI files. There are a number of drugs in the dataset, but for our study, we have chosen aspirin and its variants, for example, cardioaspirin, bayaspirin. We have chosen aspirin particularly for this study because in our dataset there are a lot of missing values, but in case of aspirin, missing values are less comparatively and number of records is more as compared to other drugs. With larger number of missing values, the quality of data degrades; that is why we have chosen aspirin for our study because the quality of results depends on data quality.

3.2 Data Preprocessing

Before applying data mining techniques, data is preprocessed to improve its quality by handling missing values and noisy data.

Attribute Selection: Firstly, we select attributes with which we have to work. We have selected five attributes dose amount, patient age, patient weight, gender, and indication for which drug has been taken and studied the result of these attributes on incidence of adverse drug effects. In most of the studies which we have reviewed, we see that mainly patient age and gender are taken as a contributing factor in occurrence of ADRs, but in our study, we have used other attributes also. Age is in numeric form in our dataset, weight is numeric, dose amount is numeric, indication is in text form, and gender is numeric.

Unit conversion: Then, we perform unit conversion. Corresponding to each attribute, there were data values with different units. We select a standard data unit for each attribute and convert the different data units into their corresponding standard data unit. We have converted all dose amount units including g, μg, ng, and mg/kg into mg. We have converted the age of all patients into years from decade, month, day, and week. For weight attribute, we have chosen kg as standard unit and converted pound and grams into kg. If we do not perform unit conversion, it may lead to distorted data.

Impute Missing Values: Then, we impute missing values in our dataset. For that, we have used k-nearest neighbor method (k-NN). Most of the works done on this dataset previously just remove the records having missing values, but removal of data from dataset leads to information loss; as a result, the quality of results degrades. That is why we prefer to impute missing values rather than removing them.

Discretization of attributes: After that, we discretize our attributes using discretization by user specification method.

Handling noisy data: For dealing with noisy data instances, we have used outlier detection method given in [21]. When we applied this method on the entire dataset, the outliers are not detected properly, because the value may be an outlier for one attribute and may not be for another. That is why we applied the method on every individual attribute and then removed outliers from the entire dataset.

3.3 Parameter Setting

In this section, we set parameters for clustering. We have used k-mode clustering technique for our study. k-mode clustering requires the count of clusters to be generated; as the parameter k needs to be mentioned a priori, we have to set this parameter before applying clustering. We varied the value of k and find S.S.E. at each point; then, we select that optimal value of k at which the value of S.S.E. starts becoming approximately same. Then with that optimal value of k, we perform clustering.

3.4 Clustering

We take the optimal value of k and apply k-mode clustering by taking all attributes individually and find S.S.E. Then, we select attribute for which S.S.E. is minimum and add other attributes one by one with this attribute. We then repeat this process again and again until we get combination of attributes with minimum S.S.E. After that, we analyze the effect of attributes on clustering.

4 Parameter Setting

We have used k-mode clustering technique for grouping the data. This technique requires number of clusters to be predefined before applying clustering technique. So we cluster the data by varying the count of total clusters generated, that is, the value of k from 5 to 70, and accordingly finding the S.S.E. of clustering for each value of k. The value of S.S.E. varies from 0.442 to 0.030. Then, we plot a graph between number of clusters k and S.S.E. by taking k along x-axis and S.S.E. along y-axis.

As shown in Fig. 2, the value of S.S.E. decreases as the number of clusters is increasing, and after a certain point, the S.S.E. starts becoming approx. same around $k = 50$. So we chose 50 as the number of clusters or optimal value of k for clustering our data.

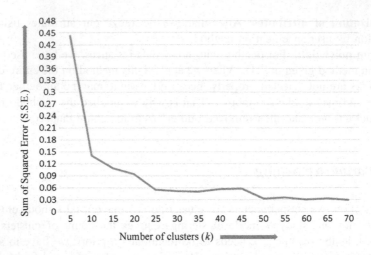

Fig. 2 Parameter setting

5 Result Analysis

k-mode clustering technique has been used in our study for analyzing the effects of different patients and drug attributes in causing ADRs.

k-Mode Results: Initially, all attributes have been considered individually and k-mode clustering technique is applied using the optimal value of k found in parameter setting section.

Table 1 shows all attributes taken individually with their corresponding S.S.E. In Fig. 3, we have plotted individual attributes along x-axis and S.S.E. along y-axis. Table 1 and Fig. 3 show that the S.S.E. is minimum for weight attribute. From this, it can be concluded that grouping or clustering is better in case of weight attribute as compared to other attributes, and weight attribute is more effective in grouping ADR instances.

Further, we combine attribute having minimum S.S.E. with all other attributes one by one and find S.S.E. for two-attribute combinations using clustering. Table 2 shows the combinations of all attributes with weight attribute, and the same is plotted in Fig. 4. In Table 2 and Fig. 4, WD stands for weight and dose, WA represents weight and age, WG represents weight and gender, and WI represents weight and indication.

Table 1 Individual attributes and their S.S.E

Attributes	S.S.E.
Dose(D)	0.678
Age(A)	0.515
Gender(G)	0.501
Weight(W)	0.214
Indication(I)	0.308

Fig. 3 Individual attributes
and their S.S.E.

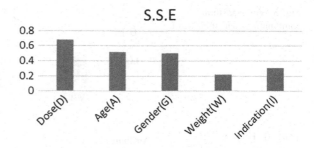

Table 2 Two-attribute
combinations and their S.S.E

Attributes	S.S.E.
WeightDose	0.151
WeightAge	0.119
WeightGender	0.112
WeightIndication	0.071

Fig. 4 Two-attribute
combinations and their S.S.E

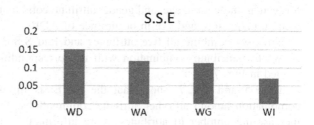

Table 2 and Fig. 4 show that after adding other attributes with weight attribute, S.S.E. is the lowest for weight and indication combination. So we can say that after weight, indication is a significant factor in manifestation of adverse drug effects.

After that, we add remaining attributes with the minimum S.S.E. attributes, that is, weight and indication, and find out S.S.E. for three-attribute combinations. Table 3 shows all three-attribute combinations with their corresponding S.S.E., and the same is plotted in Fig. 5. In Table 3 and Fig. 5, WID stands for weight, indication, and dose; WIA represents weight, indication, and age; and WIG represents weight, indication, and gender.

From Table 3 and Fig. 5, we can see that after adding other attributes with weight and indication attributes, in all three-attribute combinations S.S.E. is minimum for weight, indication, and age attribute. So we can say that after weight and indication, age is an important factor in manifestation of ADRs.

Table 3 Three-attribute
combinations and their S.S.E.

Attributes	S.S.E.
WeightIndicationDose	0.064
WeightIndicationAge	0.049
WeightIndictionGender	0.056

Fig. 5 Three-attribute
combinations and their S.S.E

Table 4 Four-attribute
combinations and their S.S.E

Attributes	S.S.E.
WeightIndicationAgeDose	0.064
WeightIndicationAgeGender	0.049

Then, with three-attribute combinations we combine two remaining attributes and find S.S.E. Table 4 shows all four-attribute combinations with their corresponding S.S.E., and the same is plotted in Fig. 6.

Table 4 and Fig. 6 show that in four-attribute combinations, S.S.E. is minimum for weight, indication, age, and gender attribute combination. So gender comes after weight, indication, and age in occurrence of ADRs.

Now, we combine all five attributes and find S.S.E. using clustering. Table 5 shows five-attribute combination with its corresponding S.S.E., and the same is plotted in Fig. 7.

Table 5 and Fig. 7, show that the S.S.E. is minimum for all five-attribute combination among all other attribute combinations. This happens because as we increase the number of attributes, more information becomes available for clustering or grouping, and clustering performance improves.

After analyzing the above results, we can say that the order of significance of different patients and drug attributes in manifestation of ADRs is as follows:

Weight > Indication > Age > Gender > Dose

Fig. 6 Four-attribute
combinations and their S.S.E.

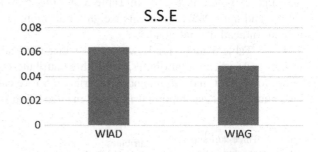

Table 5 Five-attribute
combinations and their S.S.E.

Attributes	S.S.E.
WeightIndicationAgeGenderDose	0.033

Fig. 7 Five-attribute combinations and its S.S.E.

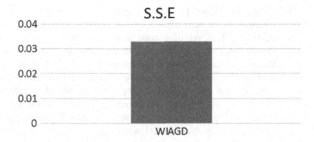

6 Conclusion

The effect of patients and drug attributes in causing adverse drug reactions using clustering technique has been studied in the present work. Our results show that weight is most relevant among other attributes in causing adverse drug reactions. After that, indication for which the drug is being given to a patient is significant because two patients can have same weight, age, etc., but if they are taking a drug for different diseases, then the adverse reaction occurred can be different in both cases. After that, age, gender, and dose amount play significant roles in causing adverse drug reactions, respectively, in that order.

Another observation from our results is that the S.S.E. of individual attributes is quite high, but as we keep on increasing the number of attributes, S.S.E. decreases correspondingly. This happens because more amount of information becomes available for clustering and as a result of which the clusters becomes more compact and similar kinds of records collect in same clusters, so clustering performance improves correspondingly. The range of S.S.E. in case of individual attributes varies from 0.214 to 0.678. After adding other attributes, S.S.E. reduces to 0.033 at the end when all attributes are combined.

References

1. K. Amery, "Why there is a need for pharmacovigilance," Pharmacoepidemiology and Drug Safety, vol. 8, no. 1, pp. 61–64, 1999.
2. Ji, Yanqing, Hao Ying, Peter Dews, Ayman Mansour, John Tran, Richard E. Miller, and R. Michael Massanari. "A potential causal association mining algorithm for screening adverse drug reactions in postmarketing surveillance." IEEE Transactions on Information Technology in Biomedicine, vol. 15, no. 3, pp. 428–437, 2011.
3. Liu, Mei, Michael E. Matheny, Yong Hu, and Hua Xu. "Data mining methodologies for pharmacovigilance." ACM SIGKDD Explorations Newsletter, vol. 14, no. 1, pp. 35–42, 2012.
4. Onakpoya, Igho J., Carl J. Heneghan, and Jeffrey K. Aronson. "Delays in the post-marketing withdrawal of drugs to which deaths have been attributed: a systematic investigation and analysis." BMC medicine, vol. 13, no. 1, pp. 26, 2015.

5. Wysowski, Diane K., and Lynette Swartz. "Adverse drug event surveillance and drug withdrawals in the United States, 1969–2002: the importance of reporting suspected reactions." Archives of internal medicine, vol. 165, no. 12, pp. 1363–1369, 2005.
6. Alomar, Muaed Jamal. "Factors affecting the development of adverse drug reactions (Review article)." Saudi Pharmaceutical Journal, vol. 22, no. 2, pp. 83–94, 2014.
7. Yildirim, Pinar, Ilyas Ozgur Ekmekci, and Andreas Holzinger. "On knowledge discovery in open medical data on the example of the fda drug adverse event reporting system for alendronate (fosamax)." In Human-computer interaction and knowledge discovery in complex, unstructured, big data, pp. 195–206. Springer Berlin Heidelberg, 2013.
8. Ibrahim, Neveen, Nahla Belal, and Osama Badawy. "Data Mining Model to Predict Fosamax Adverse Events." International Journal of Computer and Information Technology, vol. 3, no. 05, pp. 936–941, 2014.
9. Liu, Yihui, and Uwe Aickelin. "Detect adverse drug reactions for the drug Pravastatin." In Biomedical Engineering and Informatics (BMEI), 2012 5th International Conference on, pp. 1188–1192. IEEE, 2012.
10. Liu, Yihui, and Uwe Aickelin. "Detect adverse drug reactions for drug Pioglitazone." arXiv preprint arXiv: 1308.5144 (2013).
11. http://www.fda.gov/Drugs/GuidanceComplianceRegulatoryInformation/Surveillance/AdverseDrugEffects/ucm082193.htm.
12. Jiawei Han, Micheline Kamber and Jian Pei Data Mining: Concepts and Techniques, 3rd ed. The Morgan Kaufmann Series in Data Management Systems Morgan Kaufmann Publishers, July 2011.
13. Wang, Chao, Xiao-Jing Guo, Jin-Fang Xu, Cheng Wu, Ya-Lin Sun, Xiao-Fei Ye, Wei Qian, Xiu-Qiang Ma, Wen-Min Du, and Jia He. "Exploration of the association rules mining technique for the signal detection of adverse drug events in spontaneous reporting systems." PloS one 7, no. 7 (2012): e40561.
14. Ngufor, Che, Janusz Wojtusiak, and Jyotishman Pathak. "A Systematic Prediction of Adverse Drug Reactions Using Pre-clinical Drug Characteristics and Spontaneous Reports." In Healthcare Informatics (ICHI), 2015 International Conference on, pp. 76–81. IEEE, 2015.
15. Yang, Fan, Xiaohui Yu, and George Karypis. "Signaling adverse drug reactions with novel featurebased similarity model." In Bioinformatics and Biomedicine (BIBM), 2014 IEEE International Conference on, pp. 593–596. IEEE, 2014.
16. Ogwueleka, Francisca N., and Timothy Moses. "Predicting Risk of Direct-to-Customer Drug Prescription using K-Mean Clustering Technique." International Journal of Computer Applications, vol. 121, no. 17, 2015.
17. Tamura, Takao, Toshiyuki Sakaeda, Kaori Kadoyama, and Yasushi Okuno. "Aspirin-and clopidogrel-associated bleeding complications: data mining of the public version of the FDA adverse event reporting system, AERS." Int J Med Sci 9, no. 6, pp. 441–6, 2012.
18. Yu, Yue, Jun Chen, Dingcheng Li, Liwei Wang, Wei Wang, and Hongfang Liu. "Systematic Analysis of Adverse Event Reports for Sex Differences in Adverse Drug Events." Scientific reports 6 (2016).
19. Mizutani, Sayaka, Yousuke Noro, Masaaki Kotera, and Susumu Goto. "Pharmacoepidemiological characterization of drug-induced adverse reaction clusters towards understanding of their mechanisms." Computational biology and chemistry, vol. 50, pp. 50–59, 2014.
20. Yildirim, Pinar, Ljiljana Majnarić, Ozgur I. Ekmekci, and Andreas Holzinger. "Knowledge discovery of drug data on the example of adverse reaction prediction." BMC bioinformatics, vol. 15, no. Suppl 6: S7, 2014.
21. Ramaswamy, Sridhar, Rajeev Rastogi, and Kyuseok Shim. "Efficient algorithm for mining outliers from large data sets." In ACM SIGMOD Record, vol. 29, no. 2, pp. 427–438. ACM, 2000.

FPGA Implementation of a Passive Filter for Harmonic Reduction in Inverter Outputs in PV-Based Renewable Energy Systems

Meenakshi Jayaraman and V.T. Sreedevi

Abstract Passive filters are considered as the most inexpensive solution to reduce harmonics. This paper focuses on the design and analysis of inverter output LC-type passive filter for harmonic reduction in PV-based renewable energy systems. The design procedure is based on resonant frequency characteristics, and Bode plots are presented to validate the effectiveness of the filter design. Simulation and experimental results are projected to validate the filter design with respect to total harmonic distortion reduction, improved output waveforms and reduced resonant peaking. A FPGA-based pulse-width-modulated inverter prototype is implemented using SPARTAN 3E-XCS250E processor. Experimental results prove the utility of the designed LC filter for PV applications.

Keywords Photovoltaic (PV) · Pulse-width modulation (PWM)
Inverters · Field-programmable gate array (FPGA) · Harmonics
Total harmonic distortion (THD)

1 Introduction

Solar energy is dominant amid renewable sources as its utilisation is ecologically friendly. Solar photovoltaic (PV) systems are divided into grid-connected and stand-alone systems [1–4]. Grid-connected PV systems fitted in household communities and commercial establishments utilise a PV module that produces DC power which is transformed to AC power through an inverter to the grid [4, 5]. In contrast, stand-alone systems widely known for their capability to electrify off-grid rural areas are energised by PV panels to operate independent of the electric grid, and are usually designed and sized to supply residential loads [2, 3]. Figure 1 shows a general representation of a PV system.

M. Jayaraman (✉) · V.T. Sreedevi
School of Electrical Engineering, VIT University, Chennai, India
e-mail: jayaraman.meenakshi@gmail.com

© Springer Nature Singapore Pte Ltd. 2018   501
S.S. Dash et al. (eds.), *International Conference on Intelligent Computing
and Applications*, Advances in Intelligent Systems and Computing 632,
https://doi.org/10.1007/978-981-10-5520-1_46

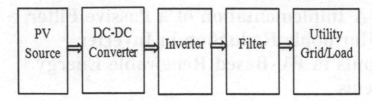

Fig. 1 General representation of a PV system

Inverters are a crucial part in most PV systems mounted for grid-connected or stand-alone applications. They use pulse-width modulation (PWM) strategy which involves high-speed switching of semiconductor switches to generate an AC output which produces high-frequency harmonics and noise [5–7]. These drawbacks end up with a distorted inverter output waveform which can be harmful to the stand-alone load reducing its lifetime and efficiency or while connecting to the grid. The output voltage of an inverter should follow a pure sine wave to exhibit good power quality. There are various methods to improvise the inverter output waveform quality to decrease its harmonic content and level of electromagnetic interference generated during the switching process [8–12].

Traditionally, passive filters have been used to obtain good quality output waveforms from PWM inverters [13–15]. Passive filters are considered as an effective and economical solution to reduce harmonics. Various topologies of passive filters are projected in the literature like single-tuned filter, first-order filter, second-order high-pass filter [13–16]. PWM inverters use first-order L filters on the output side to attenuate current ripples arising due to inverter switching process. Conversely, they carry limitation due to their huge size. PWM inverters utilise LC filters on its output side to attenuate voltage ripples [16]. A higher order LC filter provides better harmonic suppression at lower switching frequencies with a reduction of overall filter size [17]. The LC filter as an alternative of L filter is prominent as it offers harmonic attenuation with the same inductance value. Also, it allows the inverter to function in stand-alone and grid-connected modes, making it suitable to distributed generation systems [18–20]. Higher order filters like LCL, LCCL, LTCL are well-known and emerging configurations for grid-connected systems. Though they attenuate harmonics significantly, the design of their parameters is quiet complex [11, 21, 22]. Though such passive filters carry disadvantages like resonance, tuning issues, they are cost-effective for low-power stand-alone solar photovoltaic systems.

The design of inverter output passive filters is very crucial for PV-based systems. This article comprehensively discusses the design consideration of a simple LC-type passive filter for grid-connected/stand-alone PV systems. The design procedure considers resonant frequency, filter parameter values and inverter switching frequency. The usefulness of the considered filter is verified using frequency response characteristics and fast Fourier transform (FFT) analysis of the inverter output waveforms. Further, to assess the performance of the filter, experiments have been conducted on an inverter prototype with PV as input. The gating pulses for the inverter are generated using SPWM technique and implemented using a SPARTAN-3E-XCS250E field-programmable gate array (FPGA) processor. It is found that the designed passive filter improves the high-frequency harmonic attenuation, reduces peaking in the resonance and gives a lesser harmonic distortion on the inverter output waveforms.

This article is described in the following fashion: Sect. 2 shows analysis and design of LC-type passive filter for inverter output in grid-connected/stand-alone PV systems. Simulation and experimental outputs are displayed in Sect. 3. Section 4 gives the conclusion.

2 Analysis and Design of Inverter with Output Passive Filter

The power circuit of a single-phase full-bridge inverter with output filter is shown in Fig. 2.

Sinusoidal pulse-width modulation (SPWM) technique is employed to produce gating signals for the inverter switches S_1, S_2, S_3 and S_4. The switching pulses are created by matching a sinusoidal signal and a triangular carrier signal. The sinusoidal waveform frequency decides the inverter output frequency. The intersection of carrier and reference waves determines the switching instants [5–7]. Dead times are inserted into PWM pulses to protect inverter switches from shoot-through. Dead times being smaller than actual PWM period are injected at the start of each PWM pulse. This will help to delay the turn-on period of a switch when another switch on the same inverter leg finishes its turn-off process. This dead time may produce a serious distortion on the inverter output waveform. In this work, harmonics generated due to PWM switching action is taken into consideration for the design of passive filters.

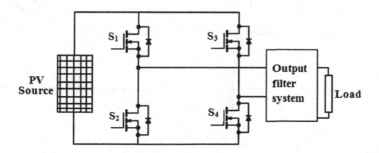

Fig. 2 Power circuit of full-bridge inverter with output filter

2.1 Constraints and Procedure on LC Filter Design

A simple technique to mitigate harmonics from an inverter is the use of passive LC filter. The general LC filter structure applied to an inverter is depicted in Fig. 3a. The filter contains an inductor and a capacitor to pass a selected portion of the harmonics generated because of PWM. Harmonics, voltage drop, reactive power absorption and power loss are key parameters to be considered when designing the filter. The reactive elements of the filter should provide high-frequency harmonic attenuation and pass the fundamental component. Further, the elements of the filter should be as small as possible and the fundamental voltage drop across them should be small. Over and above, the cost of the filter system attracts attention.

The PWM inverter output spectrum consists of harmonics at the switching frequency, multiples of switching frequency and around them posing as sidebands. Therefore, the frequency response of the filter requires to be widespread to display

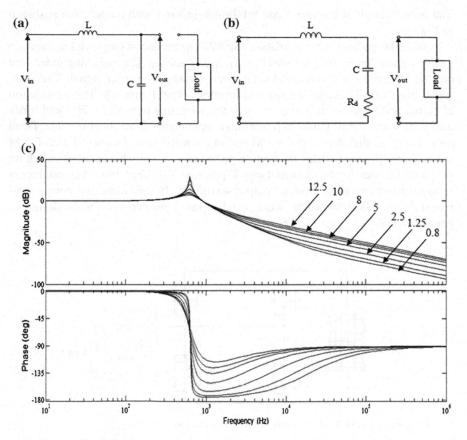

Fig. 3 LC filter system. **a** System model without damping resistance, **b** system model with damping resistance, **c** frequency response characteristics

negligible resistance at all these frequencies. The passive LC filter is chosen by the ratio of filter impedance to the load impedance at cut-off frequency. The LC filter impedance is given in Eq. (1).

$$Z_{\text{filter}} = Ls + \frac{1}{Cs}. \tag{1}$$

The total impedance (Z_{system}) of the filter system considering the load (Z_{load}) as shown in Fig. 3a is given in Eq. (2).

$$Z_{\text{system}} = Ls + \frac{Z_{\text{load}}}{1 + Z_{\text{load}}Cs}. \tag{2}$$

The transfer function in the frequency domain is given in Eq. (3).

$$G(j\omega) = \frac{1}{1 + LC(j\omega)^2}. \tag{3}$$

The frequency at which resonance occurs is given in Eq. (4).

$$f_r = \frac{1}{2\pi\sqrt{LC}}. \tag{4}$$

Equation (4) decides the frequency to attain required filter impedance at the switching frequency. The resonant frequency is usually kept below the inverter switching frequency in Eq. (4). Lower values of resonant frequency provide better attenuation of higher frequency components.

The value of Eq. (4) should be such that it is greater than the fundamental frequency value and lesser than half the PWM carrier frequency value. This would ensure sufficient attenuation in and around the switching frequency and its multiples. This would also ensure that the fundamental component is not filtered. For high-frequency PWM converters, the resonant frequency can be selected to be between (1/10) and (1/5) times of the carrier frequency [23]. The size of the elements decreases with a higher cut-off frequency value. Hence, LC filter values must be chosen such that it produces broader bandwidth at the fundamental, switching and resonant frequencies. Moreover, at rated frequency, the voltage drop across the inductance should be ensured to be lesser than 5% of the system voltage. Also, the capacitor branch current should not exceed 10% of the output current [18] to ensure good filter design.

The selection of L value depends on the inverter switching frequency. It is chosen using Eq. (5), where 'f_{sw}' is switching frequency, V_{dc} is the input DC voltage and Δ_{ripple} is ripple current chosen to be between 15 and 20% of the peak value of the rated current [16].

$$L = \frac{1}{8} \frac{V_{dc}}{\Delta_{ripple} f_{sw}}. \tag{5}$$

The inductance L is usually kept small due to its huge volume. Moreover, smaller value of 'L' helps to keep the impedance at a lower value.

The determination of the capacitor size is a significant factor in the filter design. The capacitor in a passive LC filter reduces harmonics by creating a low impedance path to ground. The capacitance is selected on the basis of reactive power absorption of the LC filter [14, 16]. It should be able to offer a good power factor at output frequency. The capacitance is judiciously selected to be low in order to restrict currents that would harm the power switches. Equation (6) governs the choice of 'C' for the passive filter connected to inverter output.

$$C = \frac{\alpha P}{2\pi f_o (V_{out})^2}, \tag{6}$$

where α represents reactive power absorption assumed to be below 5% [16], 'P' is the system power, 'f_o' is the output frequency. V_{out} is the inverter output voltage. However, these passive filters introduce potentially unstable dynamics (that appears as resonant peaks in the frequency response characteristics) that should be damped. A direct passive damping technique is implemented by connecting a resistor R_d in series with 'C' as shown in Fig. 3b. Considering damping resistance R_d, the transfer function is modified as given in Eq. (7).

$$G(j\omega) = \frac{j\omega \frac{R_d}{L} + \frac{1}{LC}}{(j\omega)^2 + j\omega \frac{R_d}{L} + \frac{1}{LC}}. \tag{7}$$

The selection of LC filter damping resistance is another design aspect. The damping resistance value is based on quality factor (QF) [24], which is given in Eq. (8),

$$QF = \frac{Z_{nat}}{R_d}, \tag{8}$$

where Z_{nat} is the impedance of the filter at natural resonance and given in Eq. (9)

$$Z_{nat} = \sqrt{\frac{L}{C}}. \tag{9}$$

The quality factor is the degree of tuning sharpness which governs the damping resistance value. The more it is, the more distinct is the valley at the resonant frequency. For lower values, the response at resonant frequency vanishes. The value of the damping resistance may be obtained by selecting an appropriate value of quality factor in the range of 0.5–30 [13].

A passive LC filter is designed based on above considerations. The Bode plot of the designed LC filter for different damping resistance values is shown in Fig. 3c. It is visible that the resonant peak decreases with increased damping resistance value; however, the harmonic attenuation decreases. Thus, an appropriate value of damping resistance that exhibits a good harmonic attenuation and, at the same time, reduces the peak of resonance is selected.

3   Simulation and Experimental Results

The inverter with output LC-type passive filter is simulated using MATLAB (2012b). Table 1 depicts the design specifications used for the simulation. SPWM technique is employed for the inverter as described in Sect. 2.

Figure 4a shows the simulated waveforms without using filter. Figure 4b, c gives the harmonic spectra of the corresponding waveforms. It is observed from Fig. 4b, c that the THD content on the voltage waveform is 54.22% and the load current waveform is 7.04%. It is clear from Fig. 4 that the magnitude of voltage and current harmonics is higher near the switching frequency of 5 kHz and its multiples.

The simulated inverter output waveforms with passive LC filter is depicted in Fig. 5a. The harmonic spectra of the corresponding waveforms are pictured in Fig. 5b, c, respectively. It is visible that the output voltage and load current waveforms are more towards sinusoidal and the THD content on the output waveforms is greatly reduced. The THD is 3.57% on the output voltage and 2.60% on the load current waveform.

To validate the simulations, a laboratory model of an inverter is built with PV as input. To obtain the input DC voltage, a series connection of four PV panels (NEPC solar panel) is done. The output waveforms and harmonics are measured using Tektronix digital storage oscilloscope and Fluke-43B power quality analyzer. A picture of experimental arrangement is shown in Fig. 6. The output obtained from one solar PV panel is shown in Fig. 7.

In this work, SPWM technique for the inverter is implemented using SPARTAN-3E XCS250E field-programmable gate array (FPGA) controller. VHDL which means VHSIC (Very High Speed Integrated Circuit) Hardware Description

Table 1 System design parameters

Parameter	Specification
DC voltage (from PV)	120 V
Switching frequency	5 kHz
Switches: S_1 to S_4 and driver IC	MOSFET IRFP460 and TLP 250
Load	RL load: 50 Ω, 10 mH
Inverter output voltage	85 V

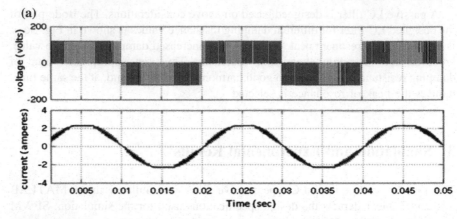

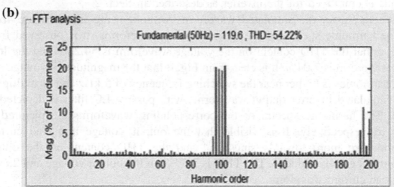

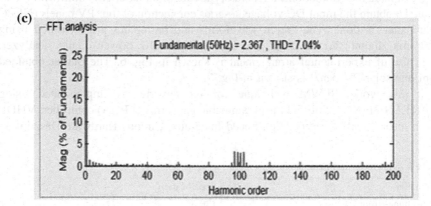

Fig. 4 Simulated waveforms without any filter. **a** Output voltage and current, **b** voltage spectrum, **c** current spectrum

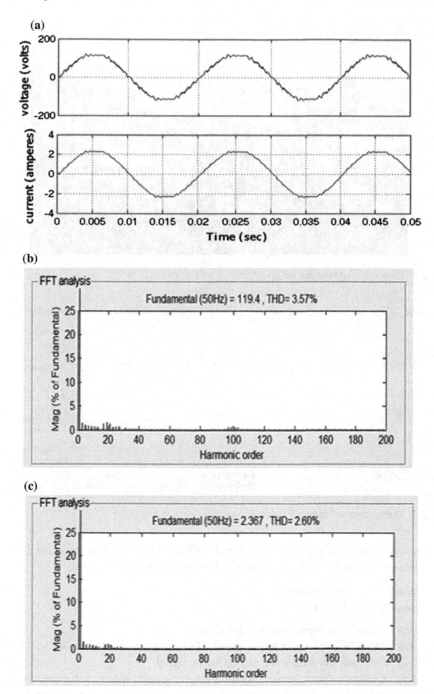

Fig. 5 Simulated outputs with LC filter. **a** Load voltage and current, **b** voltage harmonic spectra, **c** current harmonic spectra

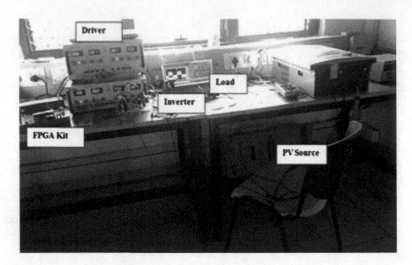

Fig. 6 Picture of experimental set-up

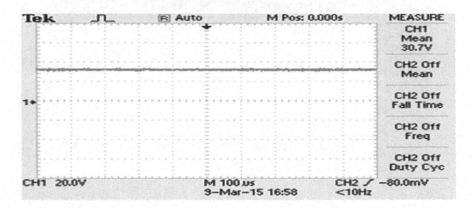

Fig. 7 Output obtained from a PV panel

Language is used to generate code for the FPGA kit. The steps used for FPGA implementation of SPWM technique for the inverter is given below:

1. Initialise system constraints.
2. Execute VHDL code to:

   (i)   Generate reference sinusoidal signal.
   (ii) Determine the modulation index (The modulation index is set as 1).
   (iii) Generate the triangular carrier wave signal.
   (iv) Compare the reference signal with the carrier signal for generation of pulses to S_1. Generate pulses for S_2 through a NOT operation on S_1.

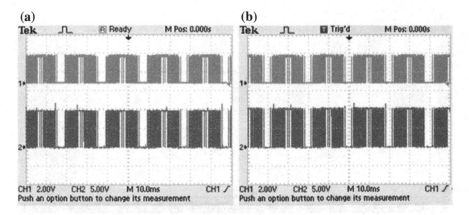

Fig. 8 Pulses generated for the inverter switches. **a** S_1, **b** S_2

 (v) Compare the phase-shifted reference signal with the triangular signal for generation of pulses to S_3 and generate pulses for S_3 through a NOT operation on S_4.

3. View the PWM waveforms through Xilinx.

The gating pulses generated using SPARTAN-3E XCS250E FPGA kit for the switches 'S_1' and 'S_2' are shown in Fig. 8.

Figure 9a, b displays the experimental waveforms for two different cases, namely without filter and with a passive LC filter.

It is observed that the designed passive filter has improved the output waveforms and they are more towards sinusoidal. The harmonic spectrum of the output waveforms using LC-type passive filter is presented in Fig. 10. With designed LC filter, the THD is reduced to 3.5% for the output voltage while that in the load current waveform, it is decreased to 2.6% which satisfies IEEE standards.

4 Conclusion

Effective design of passive filters is very important for PV applications. This paper has included a comprehensive parameter design steps and analysis of the performance of a passive LC filter in rejecting harmonics on the inverter output in a PV system. An experimental model of a PV-fed inverter with the designed passive filter is implemented using FPGA controller to judge the usefulness of the filter configuration. The results reveal that the designed filter structure provides a considerable amount of harmonic reduction in the high frequency band and reduces the peaking in the resonance. The THD content obtained with the LC-type filter is 3.5% for the output voltage and 2.6% for the output current. With the implementation of

(a)

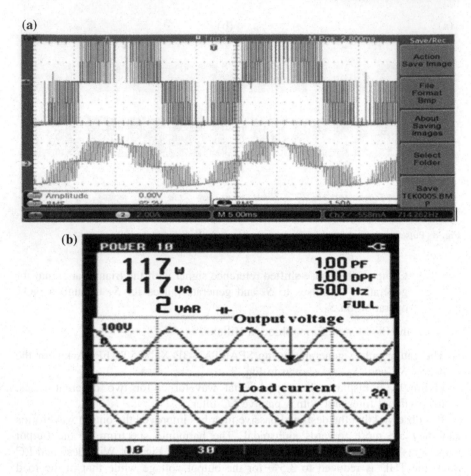

Fig. 9 Output voltage and load current waveforms. **a** Without any filter (50 V/div, 2 A/div), **b** with passive LC filter

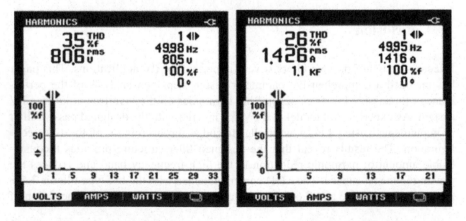

Fig. 10 Output voltage (*left*) and load current (*right*) harmonic spectrum with passive LC filter

the passive LC filter, the output waveforms are more towards sinusoidal. Further, the use of FPGA for the experimental implementation has provided rapid proto-typing, easy computational approach and simple hardware/software design. The filter design techniques described in this paper can be retrofitted for PV-based small-scale renewable energy conversion systems.

References

1. Akikur, R.K., Saidur, R., Ping, H.W., Ullah, K.R.: Comparative study of stand-alone and hybrid solar energy systems suitable for off-grid rural electrification: A review. Renew Sust Energ Rev. 27, 738–752 (2013).
2. Turkay, B.E., Telli, A.Y.: Economic analysis of standalone and grid connected hybrid energy systems. Renew Energ. 36(7), 1931–1943 (2011).
3. Kamalapur, G.D., Udaykumar, R.Y.: Rural electrification in India and feasibility of photovoltaic solar home systems. Int J Elec Power. 33(3), 594–599 (2011).
4. Yang, B., Wi, L., Zhao, Y., He, X.: Design and analysis of a grid-connected photovoltaic power system. IEEE T Power Electr. 25(4), 992–1000 (2010).
5. Kjaer, S.B., Pedersen, J.K., Blaabjerg, F.: A review of single-phase grid-connected inverters for photovoltaic modules. IEEE T Ind Appl. 41(5), 1292–1306 (2005).
6. Gow, J.A., Manning, C.D.: Photovoltaic converter system suitable for use in small scale stand-alone or grid connected applications. IEE P-Elect Pow Appl. 147(6), 535–543 (2000).
7. Wai, R.J. Wang, W.H.: Grid-connected photovoltaic generation system. IEEE T Circuits-I. 55 (3), 953–964 (2008).
8. Yang, S., Lei, Q., Peng, F.Z., Qian, Z..: A robust control scheme for grid-connected voltage-source inverters. IEEE T Ind Electron. 58(1), 202–212 (2011).
9. Mohamed, A., Elshaer, M., Mohammed, O.: Control enhancement of power conditioning units for high quality PV systems. Electr Pow Syst Res. 90, 30–41 (2012).
10. Hamza, D., Qiu, M., Jain, K.P.: Application and stability analysis of a novel digital active EMI filter used in a grid-tied PV microinverter module. IEEE T Power Electr. 28 (6), 2867–2874 (2013).
11. Anzalchi, A., Moghaddami, M., Moghaddasi, A., Sarwat, A.I., Rathore, A.K.: A new topology of higher order power filter for single- phase grid-tied voltage-source inverters. IEEE T Ind Electron. 63(12), 1464–1472 (2016).
12. Chinnaiyan, V.K., Jerome, J., Karpagam, J.: An experimental investigation on a multilevel inverter for solar energy applications. Int J Elec Power. 47, 157–167 (2013).
13. Akagi, H.: Modern active filters and traditional passive filters. Bull Pol Acad Sci-TE. 54(3), 255–269 (2006).
14. Rivas, D., Moran, L., Dixon, L.W., Espinoza, J.R.: Improving passive filter compensation performance with active techniques. IEEE T Ind Electron. 50(1), 161–170 (2003).
15. Das, J.C.: Passive Filters – Potentialities and Limitations. IEEE T Ind Appl. 40(1), 232–241 (2004).
16. Ahmed, K.H., Finney, S.J. Williams, B.W.: Passive filter design for three-phase inverter interfacing in distributed generation. in Proc. IEEE 5th International Conference-Workshop on Compatibility in Power Electronics, May 2007, 1–9.
17. Calzo, G.L., Lidozzi, A., Solero, L., Crescimbini, F.: LC filter design for on-grid and off-grid distributed generating units. IEEE T Ind Appl. 51(2), 1639–1650 (2015).
18. Kim, S.H., Kim, Y.H.: Design and analysis of an LC trap/LCR output filter for a single-phase NPC three-level inverter. Int J Electron. 95(12), 1279–1292 (2008).

19. Jayaraman, M., Sreedevi, V.T., Balakrishnan, R.: Analysis and Design of Passive Filters for Power Quality Improvement in Standalone PV Systems. In: 2013 Nirma University International Conference on Engineering, pp. 1–6 (2013).
20. Wang, T.C.Y., Ye, Z., Sinha, G., Yuan, X.: Output filter design for grid-interconnected three-phase inverter. In: IEEE 34th Annual Power Electronics Specialists Conference, pp. 779–784 (2003).
21. Liu, T., Hao, X., Yang, X., Zhao, M. Xiong, L.: A novel grid voltage feed forward control strategy for three-phase grid-connected VSI with LCCL filter. In: 2012 IEEE International Symposium on Industrial Electronics, pp. 86–91 (2012).
22. Ji, J., Wu, W., He, Y., Lin, Z., Blaabjerg, F., Chung, H.S.-H.: A simple differential mode EMI suppressor for the LLCL-filter-based single-phase grid-tied transformerless inverter. IEEE T Power Electr. 62(7), 4141–4147 (2015).
23. Li, N., Zhi, N., Zhang, H., Wang, Y.: A novel output LC filter design method of high power three-level NPC converter. In: IEEE 2014 International Power Electronics and Application Conference and Exposition, pp. 68–71 (2014).
24. Channegowda, P., John, V.: Filter optimization for grid interactive voltage source inverters. IEEE T Ind Electron. 57(12), 4106–4114 (2010).

Domain Ontology Graph Approach Using Markov Clustering Algorithm for Text Classification

Rajinder Kaur and Mukesh Kumar

Abstract Text categorization means dividing a set of input documents into the two or more classes to which these documents belong. Because of increase in availability of data in digital form in large amount, it becomes necessary to organize it. Feature extraction is the crucial step in text classification. Most of the existing text classifiers are lacking in finding out the relations among the terms. We proposed a probabilistic approach for text classification in which the nonlinear relations among the terms are also considered. This model uses the domain ontology graph (DOG) with Markov clustering (MCL) algorithm. Here, ontology graph is constructed using DOG model and then clustering of ontology graph is done by MCL algorithm. This approach is scalable to huge dataset also and its classification power is not affected if relations among terms are large. Experimental results have shown that our system is 91% accurate for 8 categories and decreases, as we increase the classes from 8 to 10 and then to 12, from 91 to 88% and then to 85%, respectively. We have compared our classifier with existing Naive Bayes and k-Nearest Neighbor classifiers. Experimental results show that our proposed model is more accurate than these two classifiers. The better results demonstrated that our presented system is developed effectively.

Keywords Domain ontology graph · Markov clustering algorithm
Text mining · Text classification

R. Kaur (✉) · M. Kumar
Computer Science & Engineering, University Institute of Engineering & Technology,
Panjab University, Chandigarh, India
e-mail: mrajinder14@gmail.com

M. Kumar
e-mail: mukesh_rai9@yahoo.com

© Springer Nature Singapore Pte Ltd. 2018   515
S.S. Dash et al. (eds.), *International Conference on Intelligent Computing
and Applications*, Advances in Intelligent Systems and Computing 632,
https://doi.org/10.1007/978-981-10-5520-1_47

1 Introduction

Text mining also known as text data mining or knowledge discovery in text (KDT) is the base of extracting high-quality information from raw data or text. High quality means the information should be according to user's need. Text classification is an active research field of text mining. As computers take text as sequence of strings, they can't extract useful information. So, specific algorithms and techniques should be used for preprocessing of raw data in order to get useful information or patterns [1].

Text (document) classification is the active research area of text mining in which assigning of text documents into classes or categories is done [2, 3]. These text documents include letters, newspapers, articles, blogs, proceedings, journal papers, etc. Text categorization means dividing a set of input documents into the two or more classes to which these documents belong. Because of increase in availability of data in digital form in large amount, it becomes necessary to organize it.

Text classification techniques can be divided into two categories: supervised document classification and unsupervised document classification (or document clustering). Supervised classification is one in which for defining the classes and classifying the documents, an external mechanism (e.g., human feedback) provides the information. Supervised machine learning techniques like Support Vector Machine, k-Nearest Neighbors, Naive Bayes, Decision Tree are applied frequently in text classification [1].

In unsupervised classification, the system doesn't have any pre-defined classes and it works without any external reference. Classification mode can also be semi-supervised in which some documents are pre-classified by external means for better learning of classifier. k-means, hierarchical clustering, etc., are commonly used as unsupervised learning techniques in text classification.

Text classification is divided into two phases: training phase and testing phase as shown in Fig. 1. Set of pre-classified or labeled documents $D = \{d_1, d_2, d_3, \ldots, d_n\}$ as training set is belonging to set classes $C = \{c_1, c_2, c_3, \ldots, c_p\}$. The training set is used for machine learning, i.e., to train the classifier. Depending upon the features selected, classifier is trained and classification algorithm is defined. The set of unlabeled documents referred as test set is used to test the classifier's accuracy by comparing the result driven by classifier for known label of document in the test set.

In this paper, we propose a probabilistic approach for text classification. It generates the domain ontology for each pre-defined class using training dataset. This model needs no human intervention in the process of ontology learning. Here, DOG is generated using MCL algorithm to train the classifier. The rest of the paper is composed of background, detailed methodology used for generating the DOGs and text classification algorithm, observations, conclusions and future scope, and limitations.

Fig. 1 Text classification steps

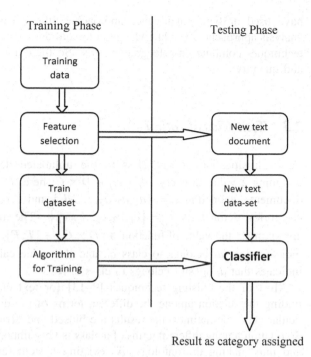

Result as category assigned

2 Background

2.1 Ontology

Ontology is basically a representation of real world's knowledge. Ontology defines a set of representative parameters for designing the model of domain of knowledge. These representational primitives are in machine readable format and are understandable by the human beings also [4, 5]. These formats are composed of attributes (properties), classes, and relationship among them. Ontology helps to develop knowledge-based systems, like Web search engines, text classification systems, content management systems, very effectively and efficiently. Ontology helps in real-time applications also. So we can conclude that ontology can be widely used as standard for semantic-based Web systems.

2.2 Ontology Learning

Ontology learning from textual data is very useful method as text data is the real source of human knowledge. Analyzing textual data requires some natural language processing approach [6, 7]. In recent years, for ontology learning most researchers have used artificial intelligence approaches like machine learning or statistical analysis approaches. The knowledge in textual data is implicit and vague, but these techniques compute knowledge explicitly. So there are difficulties in both quality and quantity.

2.3 Text Classification

Text Classification assigns class to the unlabeled documents. It is a task of assigning a value to every $(d_i \times c_j) \in D \times C$; here, D is the set of all unlabeled documents, defined as $D = \{d_1, d_2, d_3, \ldots, d_n\}$ and C is domain of all pre-defined categories defined as $C = \{c_1, c_2, c_3, \ldots, c_p\}$. The main target of TC is to approximate the value of function $\emptyset : D \times C \rightarrow \{T, F\}$. The value $\{d_i \times c_j\} \rightarrow T$ indicates that d_i belongs to class c_j, and the value calculated as $\{d_i \times c_j\} \rightarrow F$ indicates that d_i doesn't belongs to class c_j [8].

Most of the existing techniques [9–13] for text classification are lacking in finding the relation among the different terms of the document belonging to particular class. Sometimes, the results are biased and give error while classification. So relation among different terms of a class is very important point for classification and thus making the ontology. As existing system for text classification is not considering the term relation and treating every term as a unique identity for classification, error rate is high in them. If some systems have used the ontology for relation of terms, they are very complex and not much efficient. Ontology-based text classification improves the traditional system performance in terms of accuracy and also reduces the problem of over fitting. In this paper, we propose a probabilistic approach for text classification. It generates the domain ontology for each pre-defined class using training dataset. This model needs no human intervention in the process of ontology learning. Here, DOG is generated using MCL algorithm to train the classifier.

3 Methodology

Text classification process is divided into two phases: training phase and testing phase. In training phase, DOG is generated using feature extraction and MCL algorithm. During testing phase, text classification is done for unlabeled documents.

To model the ontology of knowledge in domain, ontology graph approach is used by the knowledge seeker system. Ontology graph is made up of four levels of conceptual units (CUs), linked together by different types of associations. The four CUs can be defined as:

Term T: the smallest conceptual unit extracted from the text which is relevant to the user's need.

Concept C: grouping up of related terms together with conceptual relation (CR) build the concepts, these are the basic units for concept graph (CG).

Concept Cluster CC: group of related clusters form a concept cluster CC. It tightly binds up the clusters to form hierarchy of knowledge.

Ontology Graph: grouping up of all CC forms a big and largest cluster of knowledge, termed as ontology graph.

3.1 Ontology Learning

At this stage, the domain of knowledge in the form of DOG is created. Graph creation is a knowledge-extraction process. A bottom–up approach is defined for extracting the features and designing the DOG in the form of CU (cluster unit) and CR (cluster relations). Bottom–up means the extraction is started from the smallest unit, i.e., term T and it ends up with the highest level, i.e., DOG. The five learning sub-processes are defined for ontology learning [14, 15]. These are the following:

I. **Term Extraction**: It is the process in which all the relevant terms are extracted from the dataset. A candidate term list T: $\{t_1, t_2, t_3, \ldots, t_n\}$ is extracted by eliminating the irrelevant terms from the text corpus. Stop word removal, stemming and lemmatization are done at this step.

II. **Term-to-Class Relationship Mapping**: The next step is term-to-class relationship mapping. The term-to-class mapping is done by using the nonlinear relation among the term and classes mutual information and information entropy is used for mapping. The information entropy for each term t and class c is calculated using Eq. 1.

$$H(X) = -\sum_{x \in X} P(x) \log p(x) \tag{1}$$

Then, mutual information among the term and class is calculated using Eq. 2.

$$I(t|c) = -\sum_{c \in C} \sum_{t \in T} P(t, c) \log \frac{p(t|c)}{p(t)p(c)} \tag{2}$$

Then, $R(t, c)$ relationship factor is calculated as:

$$R(t, c) = \frac{2I(t|c)}{H(t) + H(c)}. \tag{3}$$

If,

$R(t, c) < 1$: term a is negative dependence and not considered in the class c.

$R(t, c) > 1$: then term t is positive dependence and considered in the class c.

All the terms having negative dependence on class are ignored. And the terms having positive dependence are considered for class. So term lists are prepared for all the pre-defined classes/categories in this step.

III. **Term-to-Term Relationship Mapping**: Further interrelationship among the different terms in class is measured by term-to-term relationship mapping. Similarly, the term-to-term mapping is done by using the R-factor value as given by Eq. 4.

$$R(t_a, t_b) = \frac{2I(t_a|t_b)}{H(t_a) + H(t_b)} \tag{4}$$

Here, we will get the relationship factor among the terms extracted at previous step for each class. The R-factor visualization for two classes as an example for medical and space class is as shown in Fig. 2. In this figure, only first 11 terms are considered for each class.

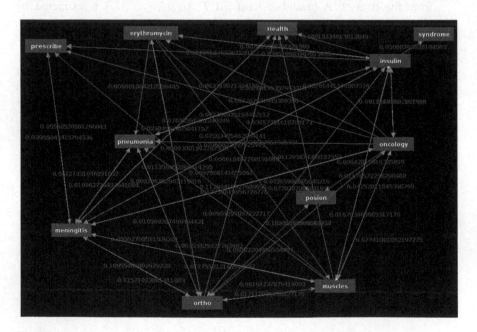

Fig. 2 Term-to-term relationship mapping for medical class

IV. **Concept Clustering using MCL algorithm**: at this step, the graph generated at previous level is clustered into tight semantic-related group. In this paper, Markov clustering algorithm [16] defined as Algorithm 1 for graph clustering is used.

Algorithm 1: The MCL algorithm for graph clustering

```
  I.   Input is an un-directed graph, power parameter e, and inflation parameter
       r.(by default e = r =2)
 II.    Creation of associated matrix;
III.    Normalization of matrix;
 IV.    Expanding the matrix by taking up to eth power;
  V.    Inflation of resulting matrix with parameter r;
 VI.    Repetition of steps 4 and 5 until a steady state is reached;
VII.    Interpretation of resulting matrix to find clusters.
```

V. **DOG Generation**: using the different concepts and relations defined in previous levels, DOG is generated. The node having maximum relation value among all terms in a cluster is selected as a label for that cluster.

Algorithm 2: DocOG generation algorithm

```
  I.  Input an unlabelled text document;
 II.  Obtain the term list of document as: T= {t1, t2, t3,..., ti};
 II.  Compute the term and relation set for document by comparing with all DOGs as
      T_d and R_d;
 IV.  Calculate the weight of every term in T_d by using
```
$$wi = tfi/n; \quad \text{n=number of terms in document}$$
```
  V.  For every pair of related terms calculate the weight of edges as
```
$$Wti,tj = wi * wj \; ;$$
```
 VI.  DocOG is generated using the created relation set and term set.
```

3.2 Text Classification

Text classification is achieved by finding the similarity of unlabeled document with the DOG. For this, text classification process is comprised of three steps: (1). Generation of DocOGs for the unlabeled document corresponding to each pre-defined class DOG. (2) Deriving the score vector of document for each class. (3) Select the class having highest score as category for the document.

3.2.1 Generation of DocOG

DocOG will be created by using the DOGs generated at previous level. An unlabeled document is input to the system, and then DocOGs of this document corresponding to all pre-defined classes are generated. Algorithm 2 is used to generate DocOG

3.2.2 Score Vector Calculation

Here the score vector for the document as given by Eq. 5.

$$\text{Score}(\text{doc}, \text{DOG}_j) = \text{score}(\text{DocOG}, \text{DOG}_j) \tag{5}$$

$S = \{S_1, S_2, S_3, \ldots, S_n\}$ is calculated for all the n-pre-defined categories. These scores are calculated by finding the number of nodes matching of all DocOGs with corresponding DOG.

3.2.3 Category Selection

After obtaining the score vector of all DocOGs here, comparison among the different scores for classes is done and the document is assigned to the class having highest score, i.e., the highest scored DocOG is selected as classified domain.

4 Software and Dataset

This proposed approach is implemented in Linux operating system using java, python, and C. Twenty Newsgroups dataset is downloaded from Jason Rennie's page. This dataset is a collection of 20,000 newspaper documents approximately, partitioned in 20 categories. This dataset is freely available. We have filtered the documents of only 12 classes, i.e., Advertisement, Automobile, Computers, Cryptography, Electronics, Games, Medical, Politics, Religion, Science, Graphics, and Windows for our research from these 20 categories dataset. Each file contains on an average of 70 words. We have used different number of classes for comparison and for checking the efficiency of classifier. First we had taken 8 categories then, we took 10 categories and then we had taken 12 categories for text classification. This variation of classes is done in order to check the effect of number of categories on classification power of classifier (Fig. 3).

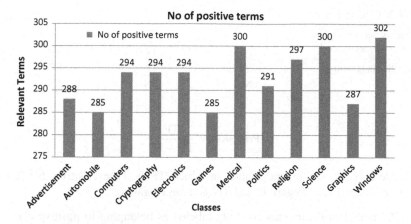

Fig. 3 Number of positive terms for each class

Table 1 Distribution of dataset

S. No.	Class	No. of documents	Used for learning	Used for testing	No of positive terms	No of negative terms
1	Advertisement	250	220	30	288	1423
2	Automobile	300	245	55	285	1444
3	Computers	249	215	34	294	1396
4	Cryptography	251	219	32	294	1414
5	Electronics	250	215	35	294	1398
6	Games	250	210	40	285	1405
7	Medical	250	210	40	300	1389
8	Politics	250	215	35	291	1379
9	Religion	252	215	37	297	1412
10	Science	250	220	30	300	1300
11	Graphics	250	220	30	287	1362
12	Windows	250	220	30	302	1317

4.1 Performance Measures

Error rate is the performance measure used to evaluate the classifier efficiency and accuracy. In this precision, recall and *f*-measure [8] are the basic performance measuring parameters. These can be defined as following (Table 1).

4.1.1 Precision

Precision is also known as positive predictive value. It calculates the accuracy by finding the percentage of documents correctly retrieved to the total retrieved documents.

Table 2 Contingency table

Category set $C = \{c_1, c_2, c_3, \ldots, c_j\}$		Expert judgments	
		Yes	No
Classifier judgments	Yes	TP	FP
	No	FN	TN

$$\text{Precision} = \frac{TP}{TP + FP} \tag{6}$$

Table 2 is the contingency table, in which TP is True Positive, FP is False Positive, FN is False Negative, and TN is True Negative.

- TP: number of documents correctly labeled as belonging to positive class.
- FP: number of documents incorrectly labeled as belonging to the class.
- FN: number of documents which are not labeled as belonging to the positive class but should have been.
- TN: number of documents which are correctly labeled as not belonging to the class.

4.1.2 Recall

Recall is also known as sensitivity. It calculates the ability of the classifier by measuring the percentage of correctly classified documents to the total classified documents.

$$\text{Recall} = \frac{TP}{TP + FN} \tag{7}$$

4.1.3 *F*-measure

It is the measure of harmonic mean of precision and recall. It gives the closeness between precision and recall. It is defined by as mentioned in Eq. 3.

$$F\text{-measure} = \frac{2 \times \text{precision} \times \text{recall}}{\text{precision} + \text{recall}} \tag{8}$$

4.2 Experimental Results

The proposed algorithm for text classification is implemented and compared with Naive Bayes and k-Nearest Neighbor classifier. Naive Bayes and k-Nearest Neighbor classifiers are implemented in Python using the inbuilt library "sklearn." In k-Nearest Neighbor algorithm, ten nearest neighbors are considered for measuring the distances in classification.

The three classifiers are implemented using 8, 10, and 12 classes or categories to measure the performance of classifiers effectively. It is done to evaluate and compare the effect of number of categories on the classification power of the classifier. To evaluate the power of classifiers, the comparison is done using precision, recall, and f-measure.

4.2.1 Experiment 1

In Experiment 1, we have considered the number of categories $N = 8$ for text classification. These are Automobile, Electronics, Religious, Sports, Medical, Cryptography, Science, and Politics. Then, the performance is evaluated using precision, recall, and f-measure. Table 3 shows the values of precision, recall, and f-measure for different models using number of classes $N = 8$. Figure 4 gives the representation for comparison of f-measure for different classifiers for different classes.

The accuracy power for DOG is 91%, while those of Naive Bayes are 84% and that of k-NN is 77%. This f-measure value shows that the DOG proposed model performs better than other two classifiers. k-NN has lowest accuracy. And k-NN has lower classification power as compared to others. Proposed model shows maximum of 97% accurate results for class Electronics and minimum of 81% for class Science. This result shows that proposed model gives better result as compared to

Table 3 Precision, recall, and f-measure for $N = 8$

Class	Precision			Recall			f-measure		
	NB	k-NN	DOG	NB	k-NN	DOG	NB	k-NN	DOG
Automobile	0.89	0.75	0.93	0.96	0.85	0.95	0.92	0.79	0.94
Electronics	0.95	0.84	0.94	0.97	0.88	0.99	0.96	0.86	0.97
Religious	0.58	0.71	0.92	0.98	0.9	0.96	0.73	0.8	0.94
Sports	0.93	0.79	0.89	0.7	0.62	0.87	0.8	0.69	0.88
Medical	0.92	0.87	0.93	0.88	0.62	0.87	0.9	0.72	0.9
Cryptography	0.88	0.87	0.92	0.94	0.82	0.96	0.91	0.84	0.94
Science	0.93	0.64	0.9	0.61	0.73	0.74	0.73	0.68	0.81
Politics	0.98	0.76	0.79	0.51	0.78	0.88	0.67	0.77	0.83
Average	0.88	0.78	0.91	0.84	0.77	0.91	0.84	0.77	0.91

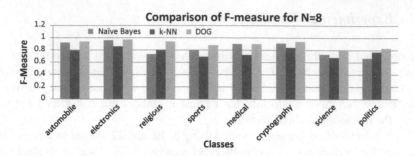

Fig. 4 Representation for comparison of *f*-measure for $N = 8$

Table 4 Precision, recall, and *f*-measure for $N = 10$

Class	Precision			Recall			*f*-measure		
	NB	*k*-NN	DOG	NB	*k*-NN	DOG	NB	*k*-NN	DOG
Automobile	0.83	0.61	0.84	0.86	0.73	0.9	0.84	0.67	0.87
Electronics	0.94	0.61	0.84	0.74	0.63	0.91	0.83	0.62	0.87
Religious	0.83	0.76	0.91	0.95	0.77	0.93	0.89	0.77	0.92
Sports	0.94	0.83	0.92	0.97	0.84	0.98	0.95	0.84	0.95
Medical	0.56	0.72	0.91	0.98	0.89	0.96	0.71	0.79	0.94
Cryptography	0.85	0.72	0.9	0.65	0.54	0.7	0.74	0.62	0.79
Science	0.9	0.86	0.93	0.88	0.58	0.87	0.89	0.69	0.9
Politics	0.87	0.85	0.9	0.94	0.81	0.96	0.9	0.83	0.93
Advertisement	0.92	0.64	0.91	0.6	0.71	0.72	0.72	0.67	0.8
Computer	0.98	0.75	0.78	0.5	0.76	0.87	0.66	0.76	0.82
Average	0.86	0.74	0.89	0.82	0.73	0.89	0.82	0.73	0.88

the other two techniques. This comparison can also be expressed using graphical representation. Figure 4 shows the graphical representation for comparison of the three techniques such as proposed model, Naive Bayes, and *k*-NN algorithm for text classification.

4.2.2 Experiment 2

In Experiment 2, we have considered the number of categories $N = 10$ for text classification. These are Automobile, Electronics, Religious, Sports, Medical, Cryptography, Science, Politics, Advertisement, and Computer. Then, the performance is evaluated using precision, recall, and *f*-measure. Table 4 shows the values of precision, recall, and *f*-measure for different models using number of classes $N = 10$. Figure 5 gives the representation for comparison of *f*-measure for different classifiers for different classes. The accuracy power for DOG is 88%, while those of Naive Bayes are 82% and that of *k*-NN is 73%.

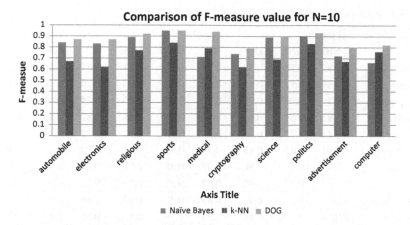

Fig. 5 Representation showing comparison of *f*-measure for $N = 10$

This *f*-measure value shows that the DOG performs better as compared to other two classifiers. *k*-NN has lowest accuracy. And *k*-NN has lower classification power as compared to others. Proposed model shows maximum of 95% accurate results for class Sports and minimum of 79% for class Cryptography. This result shows that proposed model gives better result as compared to the other two techniques. This comparison can also be expressed using graphical representation. Figure 5 shows the graphical representation for comparison of the three techniques such as proposed model, Naive Bayes, and *k*-NN algorithm for text classification.

4.2.3 Experiment 3

In Experiment 3, we have considered the number of categories $N = 12$ for text classification. These are Automobile, Electronics, Religious, Sports, Medical, Cryptography, Science, Politics, Advertisement, Computer, Graphics, and Windows. Then, the performance is evaluated using precision, recall, and *f*-measure. Table 5 shows the values of precision, recall, and *f*-measure for different models using number of classes $N = 10$. Figure 6 gives the representation for comparison of *f*-measure for different classifiers for different classes.

The accuracy power for DOG is 85%, while those of Naive Bayes are 79% and that of *k*-NN is 69%. This *f*-measure value shows that the DOG proposed model performs better than other two classifiers. *k*-NN has lowest accuracy. And *k*-NN has lower classification power as compared to others. Proposed model shows maximum of 92% accurate results for class Medical and Cryptography, and minimum of 76% for class Politics. This result shows that proposed model gives better result as compared to the other two techniques. This comparison can also be expressed using graphical representation. Figure 6 shows the graphical representation for comparison of the three techniques such as proposed model, Naive Bayes, and *k*-NN algorithm for text classification.

Table 5 Precision, recall, and *f*-measure for $N = 12$

Class	Precision			Recall			*f*-measure		
	NB	*k*-NN	DOG	NB	*k*-NN	DOG	NB	*k*-NN	DOG
Automobile	0.85	0.52	0.88	0.7	0.71	0.81	0.77	0.6	0.84
Electronics	0.87	0.57	0.83	0.67	0.6	0.83	0.76	0.58	0.83
Religious	0.7	0.59	0.78	0.8	0.64	0.76	0.75	0.62	0.77
Sports	0.95	0.61	0.83	0.73	0.54	0.91	0.82	0.57	0.87
Medical	0.82	0.77	0.91	0.95	0.73	0.93	0.88	0.75	0.92
Cryptography	0.91	0.82	0.87	0.97	0.82	0.98	0.94	0.82	0.92
Science	0.47	0.7	0.86	0.97	0.88	0.96	0.63	0.78	0.91
Politics	0.84	0.73	0.89	0.63	0.52	0.66	0.72	0.61	0.76
Advertisement	0.9	0.86	0.92	0.88	0.58	0.87	0.89	0.69	0.89
Computer	0.84	0.87	0.86	0.93	0.79	0.95	0.88	0.83	0.91
Graphics	0.93	0.62	0.89	0.56	0.7	0.7	0.7	0.66	0.78
Windows	0.98	0.73	0.74	0.45	0.77	0.86	0.62	0.75	0.8
Average	0.83	0.7	0.86	0.79	0.69	0.86	0.79	0.69	0.85

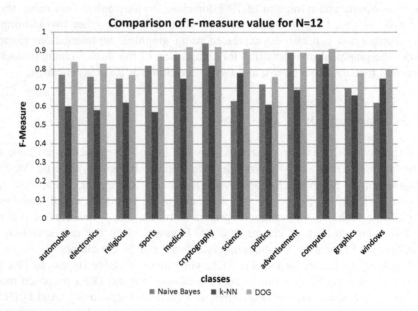

Fig. 6 Representation showing comparison of *f*-measure for $N = 12$

Table 6 Average value of *f*-measure for all classifiers

f-measure			
	Naïve Bayes	*k*-NN	DOG
Classes = 8	0.84	0.77	0.91
Classes = 10	0.82	0.73	0.88
Classes = 12	0.79	0.69	0.85

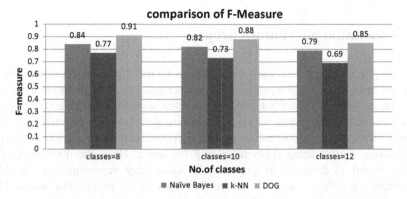

Fig. 7 Comparison of average *f*-measure value

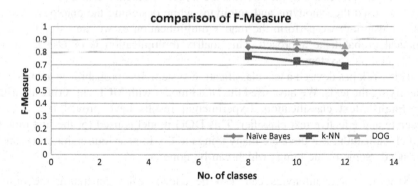

Fig. 8 Line-graph representations for *f*-measure comparison

4.2.4 Experiment 4

Experiment 4 shows the average value of precision, recall, and *f*-measure for all the three classifiers (Table 6).

Figures 7 and 8 show the graphical representation of average *f*-measure for all the three classifiers. These show that with increase in number of categories, the accuracy of the classifier decreases. As we can see that the *f*-measure value decreases for every classifier with increase in value of *N*. By comparing the average values, it is also proved that the proposed classifier is more accurate than traditional Naive Bayes and k-NN approaches for text classification.

Figure 8 shows using line-graph representation how the accuracy power of different classifiers decreases with increase in number of categories. From this figure, it is also concluded that proposed classifier is the more accurate among the three classifiers.

5   Conclusion and Future Scope

This proposed scheme, to classify English texts by using probabilistic approach, is a fully automatic system. We just give the dataset and pre-defined classes as input to our system. DOG with hierarchical clustering was used for Chinese text. It is for the first time that DOG model with MCL clustering is used for English text. DOG model increases the classification power. Effective feature extraction is done by considering the nonlinear relations among the terms. Here, the domain ontology graph model is designed to generate the knowledge representation and MCL clustering algorithm is used to cluster the terms of the graph. The use of MCL algorithm makes the system efficient as it is mathematical approach, so is more accurate. This approach is scalable to huge dataset also and its classification power is not affected if relations among terms are large. But there are limitations also. We have not used the synonyms and antonyms while designing the ontology. Also, in MCL clustering, we perform the matrix multiplication as we are using the mathematical probabilistic approach. This matrix multiplication is of $O(n^3)$. So it is highly complex system.

This work has devised a text classification system. It is probabilistic approach for classifying the texts. We have used the DOG model with MCL clustering algorithm for English text classification. Experimental results have proved that it is an accurate and effective text classifier. This DOG model is used for the first time for text classification. But there are many things to do. In near future, this work can be extended. Some of the things which can be done are:

- Synonyms and antonyms can also be added while generating the domain ontology graph.
- Semantic-based learning approach can also be used in future for improving the system.
- It can also be applied to other languages.

References

1. Hotho, Andreas, Andreas Nürnberger, and Gerhard Paaß. "A brief survey of text mining." *Ldv Forum*. Vol. 20. No. 1. 2005.
2. W. Fan, L. Wallace, S. Rich, and Z. Zhang, "Tapping into the Power of Text Mining", Journal of ACM, Blacksburg, 2005.
3. C. Aggarwal, and C. Zhai, "A survey of text classification algorithms". In Mining text data. Springer. 2012. pp 163–222, 2012.
4. James N. K. Liu, Yu Lin He, Edward H.Y. Lim, "A New Method For Knowledge and Information Management Domain Ontology Graph Model" IEEE Transactions On Systems, Man, and Cybernetics: Systems, VOL. 43, No.1, January 2013.
5. P. Buitelaar and P. Ciomiano, Ontology Learning and Population: Bridging the Gap between Text and Knowledge. Amsterdam, The Netherlands: IOS Press, 2008.
6. M. Y. Dahab, H. A. Hassan, and A. Rafea, "TextOntoEx: Automatic ontology construction from natural English text," Expert Syst. Appl., vol. 34, no. 2, pp. 1474–1480, Feb. 2008.
7. R. Gacitua, P. Sawyer, and P. Rayson, "A flexible framework to experiment with ontology learning techniques," Knowl.-Based Syst., vol. 21, no. 3, pp. 192–199, Apr. 2008.
8. F. Sebastiani. Machine learning in automated text categorization. ACM Computing Surveys. Vol. 34, No. 1, pp. 1–47, March 2002.
9. F. Sebastiani, "Text categorization", In Alessandro Zanasi (ed.), Text Mining and its Applications, WIT Press, Southampton, UK, pp. 109–129, 2005.
10. J. Han, M. Kamber, "Data Mining Concepts and Techniques", Morgan Kaufmann publishers, USA, 70–181, 2001.
11. S. Rasane, D. V. Patil, "Handling Various Issues In Text Classification: A Review", International Journal of Emerging Trends in Technology, Vol. 3, 2016.
12. J. Han, M. Kamber, "Data Mining: Concepts and Techniques," Elsevier, Second Edition, 2006.
13. Y. Yang "An Evolution of statistical Approaches to Text Categorization" Information Retrieval Vol. 1, pp. 69–90, 1999.
14. M. Y. Dahab, H. A. Hassan, and A. Rafea, "TextOntoEx: Automatic ontology construction from natural English text," Expert Syst. Appl., vol. 34, no. 2, pp. 1474–1480, Feb. 2008.
15. P. Soucy, and G. Mineau, "Feature Selection Strategies for Text Categorization", AI 2003, LNAI 2671, pp. 505–509, 2003.
16. Van Dongen, S. "Graph Clustering by Flow Simulation". PhD Thesis, University of Utrecht, The Netherlands, 2000.

An Embedded System for Color Point Control of LEDs Against Ambient Temperature Variations

R. Srividya, Ciji Pearl Kurian, C.R. Srinivasan and R. Sowmya

Abstract Dynamic variation and stabilization of color temperature according to user preferences, using multichip Red-Green-Blue LEDs, is a challenging task. Color point can vary due to variation in junction temperature caused due to self-heating, variation in ambient temperature, and device aging. In this paper, we present a closed-loop system that can tune and control the color points from 2700 to 6500 K from the variations of ambient temperature. Open-loop results obtained using forward voltage technique clearly show the effect of temperature on forward voltage, total flux, tristimulus values, color point, and peak wavelength. The effect of PWM dimming factor on tristimulus values is also studied. Nonuniform decay rates of Red-Green-Blue LEDs demand the design of separate control loops. Thus, a cost-effective hardware-in-loop simulation system with individual color control, compensating the temperature by instantly detecting the diode forward voltage with minimum number of components, is designed and validated.

Keywords Tristimulus values · Hardware-in-loop simulation · PWM dimming Peak wavelength

R. Srividya · C.P. Kurian · R. Sowmya
E&E Department, Manipal Institute of Technology, Manipal, India
e-mail: srividya.r@manipal.edu

C.P. Kurian
e-mail: ciji.pearl@manipal.edu

R. Sowmya
e-mail: sowmya.r@manipal.edu

C.R. Srinivasan (✉)
ICE Department, Manipal Institute of Technology, Manipal, India
e-mail: cr.srinivasan@manipal.edu

1 Introduction

The demands for mood lighting and smart lighting are increasing day by day. People try to prefer cool white sources during the daytime in offices and warm white sources during the evening time for their day-to-day activities at home. Retailers would like to display and advertise their products under different lighting conditions. Conventional light bulbs using fluorescent, incandescent, or tungsten filaments can produce only one color temperature which cannot be varied. Hence, for dynamic lighting one lamp of each color temperature needs to be installed which would result in increased installation cost and wastage of space [1]. Advancements in solid state lighting technology have given a smarter, efficient, and environmental-friendly solution for this problem. Long life, small size, flexibility, and nontoxic nature make LEDs more energy efficient. The two most generalized methods for producing white light with LEDs are (i) combining a short-wavelength LED with single or multiple phosphors and (ii) mixing of monochromatic LEDs in appropriate proportions [2]. These methods have their own advantages and disadvantages. Phosphor LEDs are available as a single LED package but the low and varied absorption rates of phosphors cause light loss. Color LEDs mixing technique allows the user to vary proportion externally thereby producing much more efficient white light making it suitable for dynamic lighting. The recent introduction of multichip multicolored LEDs in the market has removed the need of using color filters and has simplified the job of color mixing. Thus, multichip Red-Green-Blue (RGB) LEDs are selected for the work presented here. Stabilizing the colors of RGB LEDs to maintain a color point is a challenging task, which includes luminous flux control using switch-mode power converters and color point maintenance due to variation in junction temperature and ambient temperature [3]. Junction temperature can be controlled within 150 °C as specified by several manufacturers by designing a proper heat sink focusing on its thermal resistivity, size, and shape of heat sink and uniform spacing of LEDs on the heat sink. There are many papers on junction temperature control where the effect of ambient parameter is neglected as it is not significant [4–7]. But for accurate color point control, this parameter needs to be taken into account. Thus, in this paper, we have analyzed ambient temperature effect on optical and electrical characteristics of LEDs. Forward voltage technique which is widely adopted for the measurement of junction temperature is used in this paper to measure ambient temperature. At constant current, junction temperature is linearly related to forward voltage of LEDs. When there is rise in temperature, the drop in forward voltage is nonuniform for RGB LEDs due to the fact that forward voltage depends on color of LEDs. Thus, this nonuniform variation in forward voltage causes drop in total luminous flux and nonuniform variation in tristimulus values finally causing a shift in the color point. Similarly, ambient temperature is also proportional to forward voltage when there is no self-heating in the LEDs. Apart from these major factors, color point can also vary due to variation in the dimming factors or current levels. Due to this reason, PWM driving technique is preferred over AM [2, 8]. Hence, we focus on the design of a system which measures the optical and electrical parameters of the LEDs during the on condition of PWM current for the control of color point.

This paper is organized as follows: Sect. 2 gives an overview of RGB color mixing. Section 3 gives a detailed study of various parameters affecting color point. Section 4 explains the system design with validation and Sect. 5 concludes the paper.

2 RGB Color Mixing

The CIE 1931 chromaticity space diagram gives the distribution of colors ranging from 380 to 780 nm on an x-y-plane as shown in Fig. 1. Correlated color temperatures (CCTs) ranging from 1600 to 10,000 K lie on the Planckian locus and are used to indicate a color from a black body radiator [9, 10]. The black point located on the locus, at the center of the diagram, is the equal energy point representing white light. Reference [11] shows the color mixing algorithm using which the tristimulus or mixing proportions of Red, Green, and Blue (Yr Yg Yb) or (X Y Z) were calculated theoretically for CCTs from 2700 to 6500 K by knowing the x-y coordinates of the desired CCTs and x-y coordinates of individual red, green, and blue LEDs. Figure 1 shows the color point of 6500 K located below the Planckian locus obtained practically using spectro-radiometer when appropriate wavelengths of Red, green, and blue LEDs were mixed. RGB LEDs having peak wavelengths of 625, 528 and 464 nm, respectively, were placed inside an integrating sphere and were powered with PWM driving current levels (Dr Dg Db) according to the theoretical mixing proportionality, and the corresponding CCTs were verified. Ambient temperature (T_o) was maintained at 25 °C inside the sphere, and care was

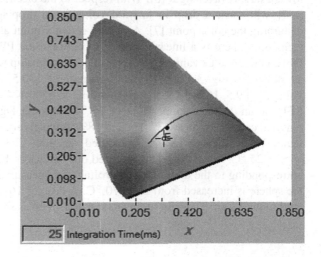

Fig. 1 CIE 1931 chromaticity diagram for 6500 K

Fig. 2 SPD of RGB LED
with peak wavelengths of
625, 528, 464 nm for 6500 K

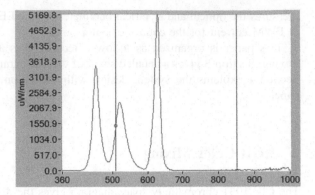

taken to avoid the junction temperature rise such that $T_o = T_j$. The results of the
same are shown in [11].

Similarly, the tristimulus values ($X\ Y\ Z$) for all other CCTs were practically
verified. The total lumen output of the lamp consisting of 5 RGB LEDs [12] placed
on an aluminum heat sink of dimension (10.5 × 10.5) cm and thermal resistivity of
1.62 °C was fixed at 420 lm [11]. Figure 2 shows the spectral power distribution of
RGB LEDs under study obtained using spectro-radiometer for a CCT of 6500 K.

3   Experimental Study of Factors Affecting Color Point

After producing the color points, next step is to stabilize them against variations in
ambient temperature and driving techniques. Considering the optical characteristics
of LEDs as the ambient temperature increases the total target lumen output
decreases, color point shifts toward shorter wavelength and peak wavelength shifts
toward larger wavelength [8]. With respect to the electrical characteristics, changes
in T_o will cause a drop in forward voltage and drop in forward current thereby
disturbing the color point [7]. Accuracy of color point also depends on the driving
technique. There is a linear relationship between the PWM driving levels (Dr Dg
Db) and tristimulus values ($X\ Y\ Z$) [4]. This relationship was experimentally verified
as shown in Fig. 3a–c. A forward current of 136.5, 301, 63 mA and forward
voltages of 9.5, 16.5, 13.3 V corresponding to 6500 K CCT are applied to the RGB
LEDs at an ambient temperature of 25 °C. The (Dr Dg Db) values to obtain the
required current levels are 39, 86, and 18%, respectively, for the selected RGB LED
lamp. The values of ($X\ Y\ Z$) obtained are 100, 293, and 24 lm, respectively,
@ t_o = 25 °C. Normalizing the required duty cycles as 100%, the tristimulus values
corresponding to the diode forward voltage are measured as the temperature inside
the sphere is increased from 25 to 70 °C. A drop of 16 and 14 lm was observed in
tristimulus X and Y corresponding to red and green as shown in Fig. 3a, b, whereas

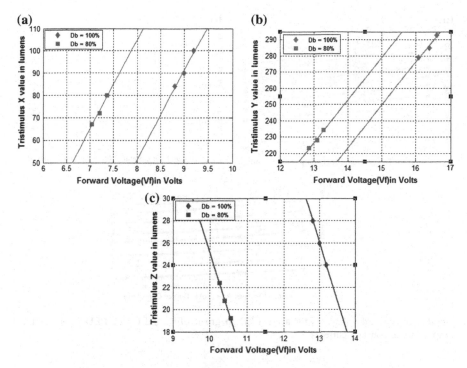

Fig. 3 a–c Experimental tristimulus values X, Y, Z versus diode forward voltage of **a** red LED, **b** green LED, and **c** blue LED at different duty cycles. The data points on the graph are measured with arise in ambient temperature from 25 to 70 °C

a gain of 4 lm was observed in Z corresponding to blue over the span of temperature rise as shown in Fig. 3c. The drop in forward voltage was 0.4 V for red, 0.5 V for green, and 0.38 V for blue. Red LED shows the largest lumen depreciation and less forward voltage decay compared to green. Green LED shows the largest forward voltage depreciation and less lumen output decay compared to red. Blue LED gives an increased lumen output and very less forward voltage depreciation compared to other two [13, 14]. This is because of the color characteristics of blue, and it will produce an increased lumen output or maintain 100% lumen output for the first 500 h of operation and then will tend to decrease. The results also clearly show that tristimulus $(X\,Y\,Z)$ with diode voltages linearly change with temperature under stable duty cycle [4]. Because of this linearity, the $(X\,Y\,Z)$ for other duty cycles can be derived as shown in Fig. 3a–c.

At constant current, the temperature and voltage are linearly related to each other with negative slope [15]. Experimental results shown in Fig. 4 establish the relation between ambient temperature and forward voltage V_f of the LED lamp. The experiment was performed inside a humidity chamber where the ambient temperature was controlled in steps and the corresponding voltages were noted down.

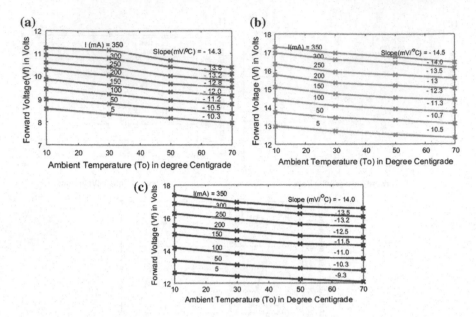

Fig. 4 **a–c** Forward voltage V_f for **a** red LED, **b** green LED, and **c** blue LED versus ambient temperature at constant currents

The same procedure was repeated for several pulsed current levels of 5 to 350 mA. Care was taken in ensuring that there is no rise in junction temperature by self-heating such that $T_j = T_o$. Enough time was given between the readings for the LEDs to adjust to the new ambient temperature, and short current pulses were supplied for the duration of 50 μs. As seen from Fig. 4, the slope $\Delta V_f/\Delta T_o$ becomes steeper as current is varied from 5 to 350 mA. The slopes reduce with increasing current from −10.3 to −14.3 mV/°C for red, −10.5 to −14.5 mV/°C for green, and −9.3 to −14.0 mV/°C for blue. Blue decays at a faster rate compared to other two colors which can be clearly seen by comparing their slopes @ 350 mA current [5, 13]. As seen from Fig. 5 a strong temperature dependence can be clearly seen in the *V–I* characteristics of the LED luminaire. When the ambient temperature is changed from 10 to 70 °C corresponding to fixed driving currents, drop in forward voltage is notified. This approach also confirms that the forward voltage degradation is more for green compared to red and blue.

When forward voltage reduces, the *x-y* coordinates shift resulting in a shift in the color point. Because of the nonuniform distribution of colors in CIE 1931, it is recommended to use CIE 1976 *u'-v'* coordinates diagram for calculating color difference. If the color point variation is less than 0.002, it will be indistinguishable [3, 7, 16, 17]. After converting *x-y* to *u'-v'*, color variation $\delta(u'v')$ of 0.02 was observed for 6500 K when T_o was raised from 25 to 70 °C. A similar significant shift was also observed for other color points.

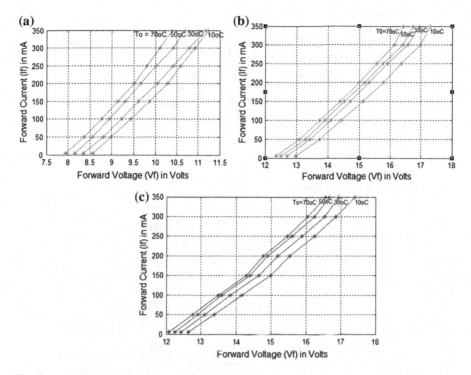

Fig. 5 **a–c** Experimental values of forward voltage versus forward current for different ambient temperatures of 25–70 °C

4 System Description with Evaluation

From the above experiments performed with the open-loop system, it was clear that due to the variations of ambient temperature there will be shift in the color point. Hence, it is necessary to design a closed-loop control so that the drop in V_f can be compensated by providing increased forward current [6]. A clear view of the designed embedded system for color point control of an RGB LED lamp is shown in Fig. 6. The algorithm to calculate the tristimulus values with corresponding PWM current levels for user preferred CCTs (Dr Dg Db) is designed and executed using LabVIEW. Three individual PWM signals at a switching frequency of 300 Hz were generated using the counter channels of NI X series USB 6356 DAQ. 350 mA current-controlled floating buck converters LM3407 were used to drive the RGB LEDs.

Ambient temperature variation was sensed by a temperature sensor and was sent as feedback to LabVIEW where this sensed voltage was compared with the reference voltage. A simple proportional controller was designed on LabVIEW to automatically sense the drop in voltage every 100 ms from the sensor and accordingly increase the driving current levels by increasing the levels of (Dr Dg Db) and thereby compensate for the drop in voltage to maintain the color point.

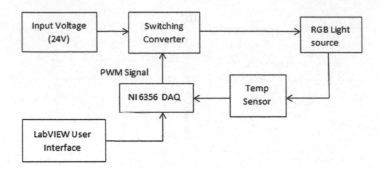

Fig. 6 Block diagram of the system

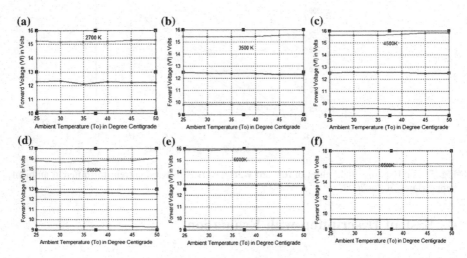

Fig. 7 a–f Closed-loop response for each CCT showing constant forward voltage against variations from ambient temperature

The closed-loop experimentation was carried in two steps: (1) calibration of the sensor and (2) actual measurement. In calibration step, sensor and RGB lamp were placed inside the humidity chamber. Placement of the sensor was near to the lamp but not on the heat sink to avoid unnecessary resistance adding up on the heat sink. The sensor temperature was compared with the humidity panel temperature to check the accuracy of the sensor. Later, a digital thermometer was also inserted along with the setup inside the chamber. All the three temperatures were compared to ensure accuracy of the input given to the controller from the feedback. After calibration of the sensor, actual measurement was performed outside the humidity chamber in a dark room. The variation in the ambient temperature was produced by a soldering iron, and the performance of the system was validated. Figure 7a–f shows the results of the closed-loop action on some CCTs. The gain (K) that was set inside the controller was determined after many iterations of the closed-loop system

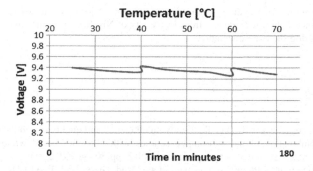

Fig. 8 Closed-loop response evaluation

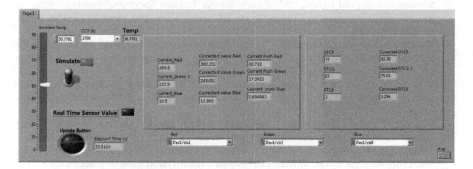

Fig. 9 Screenshot of the designed GUI for process control

to achieve accuracy. The selected K values were common for all CCTs. Accordingly, the system could efficiently maintain a constant forward voltage for variations in ambient temperature from 25 to 50 °C.

These results were again validated by considering the open-loop response of the system and instead of continuous closed-loop control; control was done only at temperatures of 40 and 60 °C as shown in Fig. 8. Figure 9 shows the screenshots of the developed VI on LabVIEW.

5 Conclusion

Making RGB LEDs suitable for dynamic white lighting with accurate color control is a challenging task. The nonuniform variation in the electrical and optical parameters of red-green-blue LEDs has been verified using experimental data. The method of measuring the drop in forward voltage at the on condition of PWM and thereby providing increased duty ratios for color point control is proposed. The designed system can effectively reduce the variation in color by changes in ambient temperature with reduced complexity.

Acknowledgements The authors wish to acknowledge Djordje Velickovic, Electrical and Computer Science Engineer graduated from University of Nis, Serbia for his constant support and contribution during the experimentations.

References

1. Niko Rolamo, Diwakar Bista et al., "Study on efficiency and quality of solid state light source by the combination of monochromatic sources with phosphor based white light-emitting diode", Rentech Symposium Compendium, Vol 2, pp 30–36, 2012.
2. Subramanian muthu, Frank j. schuurmans et al., "Red, Green, and Blue LEDs for White Light Illumination", IEEE journal on selected topics in quantum electronics, vol. 8, no. 2, pp 333–338, 2002.
3. Srividya R, Ciji pearl Kurian, "White Light Source Towards Spectrum Tunable Lighting- A Review" IEEE International Conference on Advances in Energy Conversion Technologies, pp 203–208, 2014.
4. Xiaohui qu, Siu-chung wong et al., "Temperature measurement technique for stabilizing the light output of RGB LED lamps", IEEE transactions on instrumentation and measurement, vol. 59, no. 3, pp 661–670, 2010.
5. S. K. Ng, K. H. Loo et al., "Color Control System for RGB LED With Application to Light Sources Suffering From Prolonged Aging", IEEE Transactions on Industrial Electronics, vol. 61, pp 1788–1798, 2014.
6. Folkert d, Roscam abbing et al., "Light-emitting diode junction-temperature sensing using differential voltage/current measurements", IEEE sensors conference, pp 861–864, 2011.
7. Xiaohui Qu, Siu-Chung Wong, Chi K. Tse, "Temperature Measurement Technique for Stabilizing the Light Output of RGB LED Lamps", IEEE Transactions on Instrumentation and measurement, vol. 59, no. 3, pp 661–670, 2010.
8. Subramanian muthu, James Gaines et al., "Red, Green and Blue LED-based White Light Source: Implementation Challenges and Control Design", IEEE Industry applications conference, vol. 1, pp 515–522, 2003.
9. William Dotto Vizzotto, Guilherme Gindri Pereira et al., "Electrothermal Characterization Applied to the Study of Chromaticity Coordinates In RGB LEDs", Power Electronics Conference, pp 1146–1152, 2013.
10. Ingo Speier, Marc Salsbury, "Color Temperature Tunable White Light LED System", International Conference on Solid State Lighting, Proc. of SPIE, Vol. 6337, pp 1–12, 2006.
11. R. Srividya, Ciji.Pearl Kurian et al., "Implementation of a Tunable RGB LED Light Source", International Jol. of Control theory and applications, vol 8, no. 3, pp 1251–1260, 2015.
12. Edixeon RGB LEDs, Edixeon Lighting Company, 2015.
13. Narendran, L Deng et al., "Performance characteristics of Light emitting diodes", International conference on Solid state lighting, proceedings of SPIE, pp 267–275, 2004.
14. Lukas Lohaus, Emanuel Leicht et al., "Advanced Color Control for Multicolor LED Illumination Systems with Parametric Optimization", IEEE Conference, pp 3305–3310, 2013.
15. KR Shailesh, CP Kurian et al., "Measurement of junction temperature of light-emitting diodes in a luminaire", Lighting Research Technology, pp 1–13, 2014.
16. Rodrigo G. Cordeiro, Alexandre S. Cardoso et al., "Indirect Control of Luminous Flux and Chromatic Shift Methodology Applied to RGB LEDs", IEEE Conference, 2014.
17. Saijo Prathap, Sonia Sunny, Aju S Nair, "Colour temperature tuning to improve efficacy of white light", Elsevier Procedia Technology, Vol 24, pp 1186–1193, 2016.

Enhance Incremental Clustering for Time Series Datasets Using Distance Measures

Sneha Khobragade and Preeti Mulay

Abstract Incremental clustering is a prevalent task associated with time series dataset analytics. Distance measures play important roles in incremental clustering to form, update, and append the clusters. In this research, a system is proposed to recommend suitable distance measure for a time series dataset. The system comprises of four distance measures, two incremental clustering algorithms, and mapping outcome of these techniques is based on time series datasets from varied domains. The choice of suitable distance measure is crucial to the incremental clustering process due to their diverse characteristics. With the objective of simplifying this task, genetic algorithm-based process provides the suitable distance measure and incremental clustering algorithm for particular time series dataset(s). This proposed system is an amalgamation of four distance measure techniques and enhanced fuzzy C-means, X-means clustering algorithms.

Keywords Incremental clustering · Distance measures · Time series datasets
Mapping · Validation

1 Introduction

Latest research in data mining focuses to find valuable informatics from temporal patterns. Enhancement in Information gathering strategies has facilitated access to huge quantity of temporal information, frequently termed as time series information. This information has generated a novel kind of datasets in which each instance consists of whole time series. The major distinctiveness of this category of information is high dimensionality, its dynamism, its auto-association, and noisy values

S. Khobragade (✉) · P. Mulay
Symbiosis Institute of Technology, Symbiosis International University, Lavale,
Pune 412115, India
e-mail: sneha.khobragade@sitpune.edu.in

P. Mulay
e-mail: preeti.mulay@sitpune.edu.in

© Springer Nature Singapore Pte Ltd. 2018   543
S.S. Dash et al. (eds.), *International Conference on Intelligent Computing
and Applications*, Advances in Intelligent Systems and Computing 632,
https://doi.org/10.1007/978-981-10-5520-1_49

that obscure analysis on deeper extent. In vision of above problem scenario, numerous scholars have addressed on discovering new techniques or on acclimatizing present data-mining procedures to derive useful statistics from these dataset. Incremental learning via incremental clustering have been effectively tailored to time series information in numerous application domains.

Distance measure factor is crucial in finding degree of similarity or dissimilarity in time series information via cluster formation process. Common distance measure like Euclidean using in data-mining algorithms fails to handle noisy information in time series dataset and is not suitable for incremental clustering on time series dataset. In order to overcome this, research scholars have presented large measure for specific kind of information.

Practical results propose that not every distance procedures are suitable for dataset. This is most likely due to exact individuality of every dataset that makes certain distance procedures extra apt compared to other. However, option of similarity measure is not petty as to find association amid uniqueness of temporary datasets and properties of diverse distance procedures. The integrated approach presented in this research work, which is mapping incremental clustering algorithm, time series datasets and distance measures, with enhancements in two algorithms, is not very widely presented in previous research. This work reduced "trial and error" efforts while selecting specific distance measure suitable for time series data and incremental clustering approach. In practical scenario, work with large temporal dataset with multiple distance measures is impossible and achieving best trainable machine is out of scope. Multilabel classification technique automatically selects the most suitable distance measure for clustering a time series dataset [1]. Different time series datasets as image data, GIS data, etc. will use multilabel classification. Multilabel classification will not give proper solution for particular time series dataset.

This research paper presents incremental clustering technology and different distance procedures to achieve best optimal results. The present paper focuses on the selection of incremental clustering algorithms, distance measures for particular time series dataset. The research carried out maps the best suitable incremental clustering algorithm and distance measures for given time series dataset. Mapping among those techniques gives better performance, execution time, and more attention to incremental data because of its effectiveness with large data. It compares results between time series input data with two incremental clustering algorithms and four distance measures. A genetic algorithm is used to take decision over the methods. The symmetric tree-mapping algorithm selects the multiple paths. Quick execution time and minimum path will select the incremental clustering algorithm and distance measures. At any stage of implementation for better execution anytime, algorithm plays an important role. Validation simulates important role to improve the cluster quality. Splitting the cluster in the different cluster can take decision for better performance. In this situation, validation takes an important role. Validation improves the execution quality. In this paper, validation is checked over clusters. It checks train test validation technique using Pearson correlation method. As incrementally data is added, then it again checks the validation condition and creates incremental clusters.

2 Literature Review

Mobile medical recommendation system using multi-label modeling is discussed in [2]. The recommendation system gives limited user satisfaction and performance. It is further suggested that with incorporating incremental clustering in [2] the performance can be enhanced further.

To improve searching result of a Web site, it used technique like clustering algorithm and cross-validation technique [3]. Cross-validation technique is accepted for partial examination and recital factors as precision, relevance speed, and user feedback are considered for evaluation. Future scope is to develop system with user feedback consideration. In the present paper for particular time series dataset, the decision is taken by genetic algorithm to find out best result over incremental clustering algorithm and distance measure technique. Cross-validation is checked using Pearson correlation technique for improving the quality of cluster.

A distributed k-means clustering algorithm is used for the wireless sensor that gives the idea of clustering in networking [4]. Multi-agent theory is used to develop algorithm for data exchange. Proposed algorithm is optimal k-means solutions. Major drawback in research is it is simulation-based and result outcomes are similar to any centroid-based clustering. Distributed clustering in networking gives limited performance over the network data. In the present paper, time series dataset is used. For increment number of cluster estimation, x-means incremental clustering algorithm gives suitable result for particular time series dataset.

RBF network kernel-based fuzzy clustering algorithm is compared with several other clustering techniques like k-means, fuzzy c-means, and x-means clustering algorithm by authors in paper [5]. RBF networks' future direction is to focus on the improved quality of KFCM-based clustering, and designing learning algorithm. The present paper uses the concept of fuzzy-c incremental clustering on varied distance measure techniques.

Computing cluster centers at minimum computational time with validity indexes is discussed by authors in paper [6]. Authors states that most research work do not consider number of clusters used in clustering process. Therefore, it is very vital to consider cluster formation structure and number of clusters to be formed. Additionally, data outburst innumerous requests, several issues such as grouping for Web information, imbalanced data, and high attribute data, are future scope of the work. The research proposed in this paper uses symmetric tree clustering for calculating suitable group of clusters. It also gives result in minimum time period and improves clustering quality using incremental clustering algorithm.

Tree-based incremental overlapping clustering gives the multiple paths for a query. Search tree technique is used to show multiple paths [7]. Static dataset is only considered for cluster formation process, and in real-time scenario, however, cluster generation process needs to consider dynamic datasets. Research presents a tree data structure incremental clustering technique with three-way decision support system. Tree data structure group's data points having increasing relevance. Overlapping cluster generation is achieved and gives three-way decision support in

[7]. The present paper uses the symmetric tree based technique for comparing incremental clustering algorithm and distance measure technique.

Enormous increase in use of web information and related retrieval is discussed by authors of [8]. Authors said that cluster formation is a challenging task, especially when quick search results are expected from specialized field like healthcare. In [8] user trust is measured while clusters are formed to facilitate users to get required information. Euclidean measure is used which give limited number of searching paths. Hence in this research work varied distance measure techniques are used by replacing static clustering approach to incremental.

In time series clustering, previous scholar work concludes that distance measures largely affect cluster output select this factor is not easy and remains challenge [1]. Work presents multilabel classification framework to select most suitable distance measure dynamically. Experimental analysis on five attributes has been carried out, and observation suggests that system simplifies time series cluster formation process. Future definition analysis and relation mapping to database would even simplify clustering process.

Performance of clustering algorithm for high dimensional dataset uses clustering techniques like k-means, agglomerative hierarchical clustering [9]. The evaluation has been tested for k-values of input clusters. The empirical evaluation concludes that hierarchical clustering is better in terms of time complexity compared over k-means clustering. In the present paper, for particular time series dataset, a genetic algorithm will give suitable distance measure and incremental clustering algorithm.

In single medical dataset, seven distance measures are compared with clustering algorithm. It checks impact of different distance measures on different clustering algorithms [10]. It examines the effect of different distance measures on clustering. Only Euclidean and Minkowski have similar results for tested dataset. Future complexity needs to be studied for all distance measures not covered in this paper. Present paper will compare the four time series data with two incremental clustering algorithms and four distance measures to check the impact on the techniques.

Image clustering issue is been addressed in this article that has majorly k-means and similar procedures fail for image cluster generation process. Author has implemented Gaussian-based hierarchical cluster generation process for high-value X-ray images. k-measures are used to group similar features with Minkowski and fractional distance measures [11]. Gaussian test technique gives validation over clusters. In the present paper for incremental cluster validation, the Pearson correlation technique will be used.

3 Issues of Multilabel Classification

Multilabel classification gives a response to search multiple paths [12]. Multilabel classification technique is a challenging problem in time series datasets like music. This technique will not handle the music-categorized data. Multilabel classification technique has chance of poorly ordered chain, and it will access limited data. For large data, incremental clustering will find the best way to create incremental clusters.

Multilabel classification only checks similarity and minimum distance between the data but it cannot deal with incremental clustering and mapping between the datasets. Experiment in the present paper will use to find out the best incremental clustering algorithm. Our focus is to take the future performance into account complexity and accuracy of incremental clustering in this system. The focus of this research work is to map an incremental clustering algorithm to distance measures suggested for time series dataset from specific domain.

4 Proposed System

4.1 Architecture of the System

Figure 1 shows the architecture of the research work presented in this paper. The time series data is given as input for preprocessing. The use of self-learning genetic algorithm is for taking a decision over incremental data and distance methodology. Next step is to select a distance measure including Mahalanobis, Manhattan, Chessboard, and Minkowski etc. Incremental clustering techniques i.e x-means incremental clustering algorithm or fuzzy-c incremental clustering algorithm is enhanced for generating quality clusters and appropriate mapping results. Symmetric tree mapping is used to find out different paths over incremental clustering data and distance methodology. A train, test validation technique using Pearson correlation, helps to compute cluster performance. For incremental data, on arrival of new input data series either the alforithm decides to update the cluster or form a new cluster. This way the entire effectual mapping of distance measures and time series domain is obtained by enhancing two incremental clustering algorithms.

4.2 Mind Map of Proposed System

Figure 2 shows mind map of the present paper. The mind map consists of the following subsection:

1. Input,
2. Data analysis,
3. Symmetric tree mapping,
4. Validation,
5. Output.

Input given to this proposed research system is time series datasets from varied domains. These input datasets are first preprocessed as the pre-requisite for clustering. Principal components are extracted from preprocessing. Data Analysis is carried out primarily on these principal components.

Time series dataset is an input component, along with four distance measures to two different incremental clustering algorithms. This integrated approach generates different paths like a tree structure. Finally different paths are compared to select the minimum path suggesting probable mapping.

Person correlation algorithm is used for validation of clusters. On arrival of new data, genetic algorithm is invoked by the system to calculate clusters to append already existing cluster database.

5 Methodology

5.1 Enhance Incremental Clustering Algorithms

Incremental clustering is a technique to form a group based on measuring the similarity between the incremental data. The enhanced incremental clustering analysis will select the cluster based on the different incremental clustering algorithms. The present paper uses two incremental clustering algorithms. In the algorithm, the BOLD point shows enhancement of incremental clustering.

5.1.1 Enhance Fuzzy c-Means Incremental Clustering

The fuzzy algorithm has a fuzzy approach clustering. In the present paper, fuzzy-c incremental clustering algorithm will be based on abstract classifier. Fuzzy-c is an interference engine. Steps of fuzzy-c incremental clustering algorithm are as follows:

Input: Select preprocessed dataset attributes from different distance measures.
Process:
Step 0: Start

Step 1: Randomly initialize the membership matrix using Eq. (1)

$$\sum_j c = 1 \mu_j(x) = 1 \quad i = 1, 2, \ldots k.$$ (1)

Step 2: Calculate centroid using Eq. (2) shown below

$$c_j = \sum i [\mu_j(x_i)] \frac{m_{xi}}{\sum_i [\mu_j^{x_i}]^m}.$$ (2)

Step 3: Calculate dissimilarity between data points and centroid using Euclidean distance Eq. (3) shown below

$$D_i = \sqrt{(x_2 - x_1)^2 + (y_2 - y_1)^2}.$$ (3)

Step 4: Update new membership matrix using Eq. (4) shown below

$$\mu_j(x_i) = \frac{N}{D},$$

$$\mu_j(x_i) = \frac{\left[\frac{1}{d_{ji}}\right]^{\frac{1}{m-1}}}{\sum c_{k=1} \left[\frac{1}{d_{ki}}\right] \frac{1}{m-1}}.$$ (4)

For creating clusters incrementally, update the membership matrix function.
Step 5: Compute membership matrix numerator factor using Eq. (5)

$$N(\text{value}) = \left[\frac{1}{d_{ji}}\right]^{\frac{1}{m-1}}.$$ (5)

This extension is implementing for fitting the time series dataset size.
Step 6: Compute membership matrix denominator value factor using Eq. (6) as shown below

$$\mu_j(x_i) = \sum c_{k=1} \left[\frac{1}{d_{ki}}\right] \frac{1}{m-1}.$$ (6)

Here, m is fuzzy parameter. The range of m is always [1.25–2].
This extension is implementing for fitting the time series dataset size if the size is not fitted in numerator.

Step 7: Go back to Step 2 unless the centroid is not changed.
Output: Enhanced clusters.

5.1.2 Enhance X-Means Incremental Clustering

The X-means incremental clustering algorithm is an extension of the k-means algorithm to find out a number of cluster centers. However, in our approach, our approach cannot use k-means. In this research work, X-means is used for auto-iteration of the time series dataset because it has a low computational cost and reduces time complexity. It has to select the coverage of cluster, respectively. It is a rule-based classifier. It automatically creates a number of clusters. In this paper, our plan is to apply the following steps of x-means clustering algorithm.

Input: Select preprocessed dataset attributes using genetic algorithm.
Process:
Step 0: Prepare p-dimensional data whose sample size is n.
Step 1: Set an initial number of clusters to be k_0, which should be sufficiently small.
Step 2: Apply k-means to all data with setting $k = k_0$. We name the divided clusters C_1, C_2, ...C_{k0}.
Step 3: Repeat the following procedure from Step 4 to Step 9 by setting $i = 1$, 2, ..., k_0.
Step 4: For a cluster of C_i, apply k-means by setting $k = 2$. We name the divided clusters C_{i1}, C_{i2}.
Step 5: We assume the following p-dimensional normal distribution for the data x_i contained $C_i = f(\emptyset_i; x) = (2\pi) - p/2|V_i| - 1/2 \times \exp\left[\frac{1}{2}(x - \mu_i)^t V_i^{-1}(x - \mu_i)\right]$.
Step 6: Recursive distance identification. N is number of attributes. N! = N * (N − 1)*...*2 * 1 & that 0! = 1.
This step is processed for incremental data.
Step 7: Apply distance, density estimation change for number of cluster estimation. C_n is cluster number, min is minimum distance, A is set of attributes. $Cn = \int_0^{min} \sum |A_i - A_{i+1}|$.
For neighbor-based cluster creation approach.
Step 8: Euclidean distance calculation. E_d is Euclidian distance

$$E_d = \sqrt{\sum_{i=1}^{n}(A_i - A_{i+1})^2}.$$

To calculate exact distance between two point.

Step 9: Cluster formulation, where C is cluster set and C_i is ith cluster

$$C = \int_0^n A_i \in E_d C_i.$$

Output: Different cluster formulations.

6 Genetic Algorithm

Multi-optimization has been a major challenge to be dealt in algorithm evolutions, and previously scalar functions have been generated to solve multi-objective problem. Genetic algorithm is a comparative process that learns over time to compute multi-objective problem and has current scope of research. In this present paper, genetic algorithm is used to take the decision for selection of incremental clustering and distance measure. Genetic algorithm takes input like incremental clustering algorithm list, distance measures techniques, dataset selected attribute, and complexity parameter. For processing first, it calculates complex evaluation. After the evaluation, analysis factor analyzed the data from the attribute list. Next step is to identify the fitness function. Finally, genetic algorithm takes decision over the techniques like incremental clustering and distance measure. Then, the output is advising perfect incremental clustering, one-distance measure technique between two incremental clustering algorithms, four distance measure techniques for a specific domain of time series.

Input:

(A) Selection of time series dataset from the time series datasets attribute list (wine, weather, cancer, sports, indoor signal)
(B) Selection of distance measures techniques:

    (i) Mahalanobis distance measure

$$Dm = \sqrt{(x - \mu)^T S^{-1}(x - \mu)} \qquad (1)$$

(ii) Manhattan distance measure

$$Md = |x2 - x1| + |y2 - y1| \tag{2}$$

(iii) Minkowski distance measure

$$Mid = \left(\sum_{i=1}^{n} |x - y|^p \right)^{1/p} \tag{3}$$

iv. Chessboard distance measure

$$D_c = \max(|x_2 - x_1|, |y_2 - y_1|) \tag{4}$$

(C) Selection of incremental clustering algorithms:

   (a) Enhance fuzzy c-means incremental clustering.
   (b) Enhance x-means incremental clustering.

(D) Select complexity parameter.
    Pearson correlation(r) for cluster validation.

Process:
Step 0: Start
Step 1: Complexity evaluation
Step 2: Analysis factor for population
Step 3: Fitness function identification
Step 4: Decision making over techniques
Output: For particular time series dataset suitable

   I. Incremental clustering algorithm among enhanced fuzzy-c and x-means incremental clustering.
   II. Distance measure technique among Mahalanobis, Manhattan, Minkowski, and Chessboard distance measure.

7 Results and Discussion

In this research paper the focus is more on mapping of most widely used distance measure, i.e. Mahalanobis distance measure from the steps. Table 1 shows computations of cluster validations including Standard Deviation (SD) and covariance Table 2 shows Mahalanobis distance (MD), time in milliseconds for the. "Indoor User Movement Prediction" from RSS dataset time series dataset. Graph 1 plot indicates the average time taken for the calculation on Mahalanobis distance verses size of selected dataset.

Table 1 Standard deviation and covariance using Mahalanobis distance measure

Dataset rows	Mean1	Mean2	SD1	SD2	SD1^2	SD2^2	Covariance	MD	Time in ms
300	−0.2361	−0.085	0.2159	0.1834	0.0466	0.0336	0.0105	0.9625	114
400	−0.2348	−0.0816	0.2128	0.1793	0.0452	0.0321	0.0096	1.001	155
550	−0.237	−0.0811	0.2134	0.1773	0.0454	0.3168	0.0097	1.0218	153
700	−0.237	−0.0806	0.2124	0.1768	0.0451	0.0312	0.0093	1.0317	156
753	−0.235	−0.0827	0.2115	0.1756	0.0447	0.0308	0.0095	1.0085	164
SOS	−0.2363	−0.0814	0.2119	0.1755	0.0449	0.0308	0.0093	1.074	161
856	−0.2371	−0.0824	0.2121	0.1762	0.0449	0.031	0.0092	1.0201	165
905	−0.2375	−0.0823	0.2123	0.1761	0.045	0.031	0.0093	1.0275	166
977	−0.2308	−0.0845	0.2135	0.1765	0.0455	0.0311	0.0098	1.0107	173
1153	−0.2391	−0.0841	0.2129	0.176	0.0453	0.0309	0.0097	1.0209	171
1170	−0.2401	−0.0862	0.2141	0.1763	0.0458	0.0312	0.0102	1.004	172
1213	−0.2407	−0.0856	0.2141	0.1761	0.0457	0.031	0.0102	1.011	173
1240	−0.2398	−0.0846	0.2132	0.175	0.0454	0.03	0.0099	1.021	174

Table 2 Mahalanobis
calculation

Dataset Rows	Time in Milliseconds
300	114
400	155
550	153
700	156
753	164
808	161
856	165
905	166
977	173
1153	171
1170	172
1218	173
1240	174

Graph 1 Mahalanobis
calculation graph

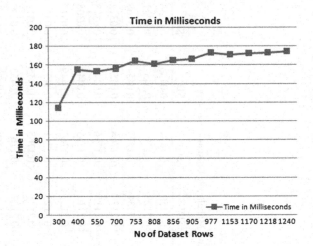

8 Conclusion and Future Direction

Incremental clustering improves the cluster quality and execution time. In this paper, the incremental clustering algorithm is used along with distance measures to overcome the drawbacks of multilabel classification for mapping over time series datasets [1]. Here, the incremental clustering algorithm decides the length of the time series datasets. In addition, the various distance measures are used to find out the minimum distance and similarity in the objects. Using distance measures techniques is easy to calculate outliers, covariance, grid-based path, ratio for absolute zero values in the attribute selection. Survey analysis on more than ten articles concludes that incremental clustering output is highly influenced by distance measure. Such as, selection of distance measure challenge has overcome in the incremental clustering. Genetic procedure is promised to better option for selection of distance measure techniques and incremental clustering techniques for particular time series dataset. Future direction of the present paper is numerous distance measures that are compared with numerous time series dataset and incremental clustering, and it enhances with number of incremental clustering.

Appendix

See Figs. 1 and 2.

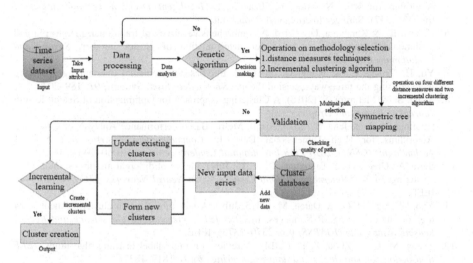

Fig. 1 Architecture of the present research paper

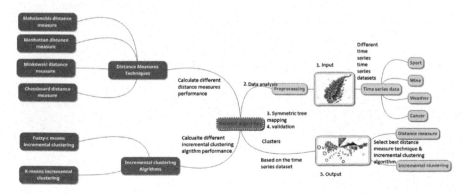

Fig. 2 Mind map of the present research paper

References

1. Usue Mori, Alexander Mendiburu, and Jose A. Lozano (2015), Similarity Measure Selection for Clustering Time Series Databases, IEEE Transactions on Knowledge and Data Engineering, 2015.
2. Guo, L., Jin, B., Yu, R., Yao, C., Sun, C., & Huang, D. (2016). Multi-Label Classification Methods for Green Computing and Application for Mobile Medical Recommendations.
3. Mehrotra, S., & Kohli, S. (2016). Application of Clustering for Improving Search Result of a Website. In *Information Systems Design and Intelligent Applications* (pp. 349–356). Springer India.
4. Liu, Q., Fu, W., Qin, J., Zheng, W. X., & Gao, H. (2016, March). Distributed k-means algorithm for sensor networks based on multi-agent consensus theory. In *2016 IEEE International Conference on Industrial Technology (ICIT)* (pp. 2114–2119). IEEE.
5. Czarnowski, I., & Jędrzejowicz, P. (2016). Kernel-Based Fuzzy C-Means Clustering Algorithm for RBF Network Initialization. In *Intelligent Decision Technologies 2016* (pp. 337–347). Springer International Publishing.
6. Hancer, E., & Karaboga, D. (2016). A comprehensive survey of traditional, merge-split and evolutionary approaches proposed for determination of cluster number. *Swarm and Evolutionary Computation*.
7. Yu, H., Zhang, C., & Wang, G. (2016). A tree-based incremental overlapping clustering method using the three-way decision theory. *Knowledge-Based Systems*, *91*, 189–203.
8. Kohli, S., & Mehrotra, S. (2016). A Clustering Approach for Optimization of Search Result. *Journal of Image and Graphics*, *4*(1).
9. Chormunge, S., & Jena, S. (2015, April). Metric Based Performance Analysis of Clustering Algorithms for High Dimensional Data. In *Communication Systems and Network Technologies (CSNT), 2015 Fifth International Conference on* (pp. 1060–1064). IEEE.
10. Serra, A., Greco, D., & Tagliaferri, R. (2015, July). Impact of different metrics on multi-view clustering. In *2015 International Joint Conference on Neural Networks (IJCNN)* (pp. 1–8). IEEE.
11. Yao, W., Loffeld, O., & Datcu, M. (2015, July). A hierarchical patch clustering method for high-resolution TerraSAR-X images. In *2015 IEEE International Geoscience and Remote Sensing Symposium (IGARSS)* (pp. 2370–2373). IEEE.
12. Zhang, M. L., & Zhou, Z. H. (2014). A review on multi-label learning algorithms. *IEEE transactions on knowledge and data engineering*, *26*(8), 1819–1837.

BCH/Hamming/Cyclic Coding Techniques: Comparison of PAPR-Reduction Performance in OFDM Systems

Pankaj Kumar, Amit Kumar Ahuja and Ram Chakka

Abstract OFDM (orthogonal frequency division multiplexing) is a spectrum-efficient digital modulation scheme. It is widely used in various applications involving today's as well as next-generation network (NGN). Despite many benefits, it suffers from the drawback of high *PAPR* (peak-to-average power ratio) of the transmitted OFDM signal. Channel codes are known to reduce *PAPR* (Rajasekhar et al. in PAPR reduction performance in OFDM systems using channel coding techniques. International Conference, IEEE, pp 1–5, 2014; Tsai et al. in IEEE Trans Wireless Commun 7:84–89, 2008; Jiang and Wu in IEEE Trans Broadcast 54:257–268, 2008; Chen and Liang in IEEE Trans Wireless Commun 6:3524–3528, 2007) [1–4]. In (Rajasekhar et al. in PAPR reduction performance in OFDM systems using channel coding techniques. International Conference, IEEE, pp 1–5, 2014) [1], PAPR-reduction performance of OFDM signal using Hamming and cyclic codes was evaluated and compared. BCH codes are widely used in mobile communication networks (Agrawal and Zeng in Introduction to wireless and mobile systems, 2015; Hazan and Ran in Study of indoor LTE green small-cells using mobile front haul architecture over hybrid fiber-wireless channels, IEEE, pp 185–189, 2016) [5, 6]. Objective of this paper is to evaluate PAPR-reduction performance of BCH-coded OFDM signal and compare the same with the case when Hamming and cyclic codes are used. Simulation results are evaluated using MATLAB® R2015(a). *CCDF* versus *PAPR* curves are plotted and used for comparing the performances of the three channel coding schemes.

P. Kumar
Meerut Institute of Technology (MIT), Meerut, Uttar Pradesh, India
e-mail: pku.kiet@gmail.com

A.K. Ahuja (✉)
Meerut Institute of Engineering and Technology (MIET), Meerut, Uttar Pradesh, India
e-mail: amit123ahuja@gmail.com

R. Chakka
RGM College of Engineering and Technology, Nandyal, Andhra Pradesh, India
e-mail: ramchakka@yahoo.com

© Springer Nature Singapore Pte Ltd. 2018
S.S. Dash et al. (eds.), *International Conference on Intelligent Computing and Applications*, Advances in Intelligent Systems and Computing 632,
https://doi.org/10.1007/978-981-10-5520-1_50

Keywords OFDM · Wireless communication · PAPR · Cyclic code
Hamming code · BCH code

1 Introduction

With increasing usage of communication technologies, the demand for services with higher data rates such as multimedia, voice, data over wireless links has also increased [7]. The received signal is usually distorted due to time-varying characteristics of the channel, also termed as fading [8]. MIMO (multi-input-multi-output) systems use space diversity to reduce the effect of fading [9, 10]. Alternatively, the multi-carrier modulation techniques are used [10, 11]. The basic idea of multi-carrier modulation technique is to divide the single high data rate stream into several low data rate streams to overcome the effect of multi-path fading channel [12]. A multi-carrier system such as FDM (frequency division multiplexing), as shown in Fig. 1a, divides the total available bandwidth into subcarriers and transmits the data simultaneously [12]. To avoid inter-carrier-interference (ICI) in FDM, the carrier frequencies are spaced sufficiently far apart (refer to Fig. 1a). This results in lower overall data rate at the output of FDM system. OFDM (orthogonal frequency division multiplexing), on the other hand, is a spectrum-efficient digital modulation scheme in which the complete channel bandwidth is divided into multiple subcarriers that are mutually orthogonal to each other [7, 12], as shown in Fig. 1b. Orthogonality gives the subcarrier a valid reason to be closely spaced, even partially overlapped, without interference (refer to Fig. 1b).

Due to its various advantages like high spectral efficiency, robustness to channel fading, immunity to interference etc., the OFDM is widely used in digital audio and video broadcasting, referred to as 'DAB' and 'DVB,' respectively [13]. Further, most of the wireless LAN standards like IEEE 802.11a or IEEE 802.11g use OFDM as main multiplexing scheme for better spectrum efficiency [14, 15]. However, despite various benefits, OFDM suffers from the drawback of very high *PAPR* (peak-to-average power ratio). This is due to large number of subcarriers that are used in OFDM signals, as illustrated in Fig. 2.

Fig. 1 **a** FDMA spectrum,
b OFDM spectrum

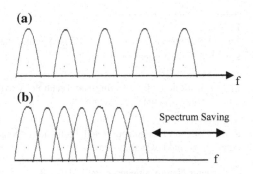

Fig. 2 High peaks in OFDM signal amplitude after adding multiple sinusoidal signals

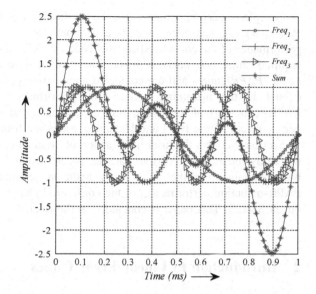

As can be seen from Fig. 2, the OFDM signal, which is sum of signals at different frequencies ('Freq$_1$', 'Freq$_2$', 'Freq$_3$') has large dynamic range of amplitude as compared to that of the single-frequency components. Further, two peaks, with amplitude larger than that of the single-frequency components, can be seen at 0.1 and 0.9 ms (refer to Fig. 2) signifying increase in the *PAPR* of the OFDM signal.

PAPR is formally defined as, measure of variation in signal amplitude from its mean value [1]. In a given duration, *PAPR* is the ratio of 'peak amplitude squared' and 'average signal power,' in that duration. If *PAPR* is too high, the signal may be out of the scope of LPA (linear power amplifier) to process. This introduces distortion which is usually nonlinear in nature and changes the superposition of signal spectrum. The result is degradation in system performance. If no measures are taken to counter this, the MIMO-OFDM system could face serious restrictions on its use in practical applications [16]. This leads to motivation for identifying different techniques to control *PAPR* of the transmitted OFDM signal, without compromising its performance. A promising method, which has attracted considerable attention of researchers in past, is to use 'error control codes (ECC),' also referred to as 'channel codes' [1, 17].

Channel coding is known to reduce *PAPR* [1, 17]. In [1], the capability of cyclic and Hamming codes, in achieving PAPR-reduction, has been demonstrated. BCH code is an important channel coding scheme, in today's as well as futuristic next-generation networks (NGNs). Thus, it becomes imperative to evaluate the PAPR-reduction performance of the BCH code and compare the same with the case when cyclic and Hamming codes are used. This is precisely the objective of this paper.

Complementary cumulative distribution function (*CCDF*), defined as $F_X(x) = P_r(X > x)$ [1], has been used as a parameter to compare the PAPR-reduction performance of the considered channel codes. MATLAB® R2015(a) is used for the required simulation analysis study. The evaluated results are plotted as *CCDF* versus *PAPR* curves and are compared appropriately. Using methodology, as discussed in the upcoming section, the performance superiority (in terms of PAPR-reduction) of the cyclic code is demonstrated over the other two, while performances of the BCH and Hamming codes are found to be almost similar.

Section 2 covers the brief introduction of channel codes that are considered for this study and analysis. Various parameters selected and methodology adopted have been explained in the Sect. 3. In addition, evaluation and comparison of PAPR-reduction performance, of the considered channel codes, have been carried out in Sect. 3. This paper is concluded in the Sect. 4.

2 Introduction to Linear Error Codes

As mentioned earlier, cyclic, Hamming, and BCH codes are considered for this study and analysis. Following is the brief description of each of these.

2.1 Hamming Codes

The Hamming code is defined as, for any positive integer 'm' with $(m \geq 3)$, there exists a Hamming code with the following parameters [18]:

- Code length $(n) = 2^m - 1$.
- Number of information symbols $(k) = 2^m - m - 1$.
- Number of parity-check symbols $(n - k) = m$.
- Error correcting capability $(t) = 1 (\text{or}, d_{\min} = 3)$.

More details about Hamming codes can be seen from [18].

2.2 Cyclic Code

The major advantage of cyclic codes is the easier implementation of its encoder and decoder structures. A binary code is said to be cyclic if it exhibits the following two fundamental properties [19, 20]: (a) sum of any two codewords is also a codeword. This is referred to as linearity property, (b) any cyclic shift to a codeword gives another codeword. This is referred to as cyclic property. If the components of a n-tuple, $v = (v_0, v_1, \ldots, v_{n-1})$ are cyclically shifted 'i' places to the right, the resultant n-tuple would be:

$$v^{(i)} = \left(v_{n-i}, v_{n-i+1}, \ldots, v_{n-1}, v_0, v_1, \ldots, v_{n-i-1}\right) \tag{1}$$

In Eq. (1), for binary system, parameter 'v' corresponds to either '0' or '1', while 'i' is nonzero positive integer. More details about cyclic codes can be seen from [19, 20].

2.3 BCH Codes

The Bose, Chaudhuri, and Hocquenghem (BCH) codes form a large class of random error correcting cyclic codes [17, 20]. This class of codes is a remarkable generalization of the Hamming code, for the purpose of multiple-error corrections. For any positive integers 'm' with $m \geq 3$, and, t with $t < 2^{m-1}$, there exists a binary BCH code with the following parameters [20]:

- Block length $(n) = 2^m - 1$.
- Number of parity-check digits $(n - k) \leq m.t$.
- Minimum distance $(d_{min}) \geq 2t + 1$.

We call this code a t-error-correcting BCH code. Further details regarding BCH code can be seen from [17–20].

3 Parameter Selected, Methodology Adopted, Results, and Discussions

The study and analysis have been carried out under following parametric assumptions:

OFDM parameters: Total data bits $= 52 \times 10^3$; number of frames $= 10^3$; number of subcarriers per frame $= 52$; frame length $= 80$ bits; cyclic prefix bits $= 16$; IFFT length $= 64$; pilot insertion bits $= 12$; Modulation technique $=$ binary phase-shift keying (BPSK).

Channel coding parameters: Message length $(k) = 4$; codeword length $(n) = 7$; parity-check bits $(m) = 3$.

Figure 3 shows the methodology which has been adopted for this study and analysis. Using this methodology, the *PAPR* is evaluated and the corresponding *CCDF*, as defined earlier, is evaluated for the considered channel coding schemes. The results are plotted as *CCDF* versus *PAPR* curves as shown in Figs. 4, 5, and 6.

Figure 4a displays the *CCDF* versus *PAPR* curve corresponding to cyclic code. In order to show the capability of channel codes, in reducing *PAPR*, the *CCDF* versus *PAPR* curve corresponding to the uncoded scheme (when transmitted data is sent without channel coding) is also plotted in the same plot (refer to Fig. 4a). The corresponding results when BCH and Hamming codes are used are shown in

Fig. 3 Flowchart depicting
methodology adopted

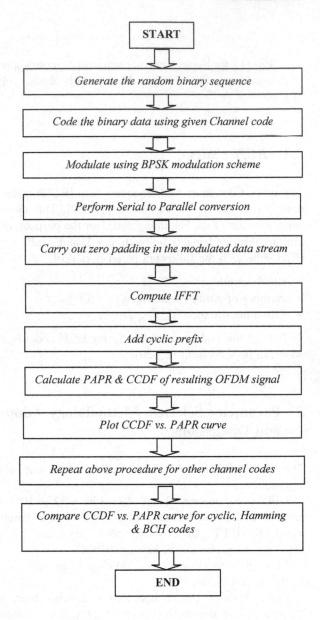

START

Generate the random binary sequence

Code the binary data using given Channel code

Modulate using BPSK modulation scheme

Perform Serial to Parallel conversion

Carry out zero padding in the modulated data stream

Compute IFFT

Add cyclic prefix

Calculate PAPR & CCDF of resulting OFDM signal

Plot CCDF vs. PAPR curve

Repeat above procedure for other channel codes

Compare CCDF vs. PAPR curve for cyclic, Hamming
& BCH codes

END

Figs. 4b and 5a respectively. In Fig. 5b, BCH, cyclic, and uncoded schemes are
compared, while Fig. 6a shows the comparison of cyclic, Hamming, and uncoded
schemes. Figure 6b depicts the comparison of BCH, Hamming, and uncoded
schemes, while comparison of all the considered channel codes has been shown in
Fig. 6c. It is to be noted here that, at a given *PAPR*, lower is the *CCDF*, better
would be the performance (in terms of PAPR-reduction) and vice versa.

From Figs. 4, 5, and 6, the following conclusions can be drawn.

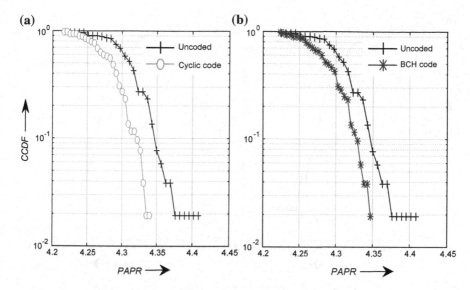

Fig. 4 *CCDF* versus *PAPR* plots for **a** cyclic code and uncoded scheme, **b** BCH code and uncoded scheme

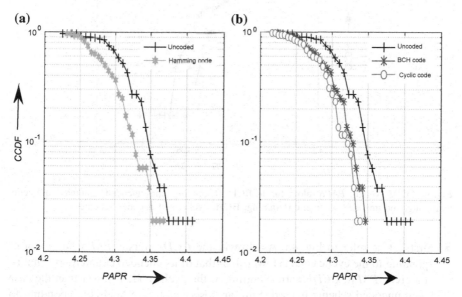

Fig. 5 *CCDF* versus *PAPR* plots for **a** Hamming code and uncoded scheme, **b** BCH, cyclic, and uncoded scheme

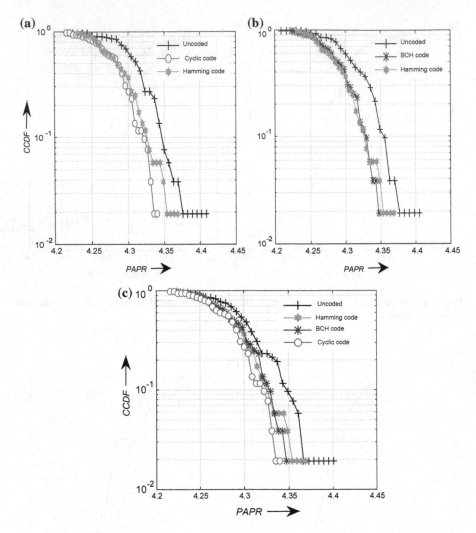

Fig. 6 *CCDF* versus *PAPR* plots for **a** BCH, Hamming and uncoded schemes, **b** cyclic, Hamming, and uncoded scheme, **c** Hamming, BCH, cyclic, and uncoded schemes

- First, let us refer to Fig. 4a. As shown, the *CCDF* versus *PAPR* curve corresponding to the cyclic code is lower than that of its uncoded counterpart. That is, at a given *PAPR*, *CCDF* corresponding to the cyclic code is lower than the case when uncoded scheme is used. Similar observations are held corresponding to other channel coding schemes, that is, BCH code (refer to Fig. 4b) and Hamming code (refer to Fig. 5a). These are expected results as it shows the capability of the channel coding schemes in reducing *PAPR*.
- As can be observed from Fig. 5b, at a given *PAPR*, the value of *CCDF* corresponding to cyclic code is small as compared to the case when BCH code is

used. This is observed for almost all the *PAPR* values in the range from 4.2 to 4.35 (refer to Fig. 5b). Similar observations are held when cyclic code is compared with Hamming code as can be seen from Fig. 6a. This shows the superiority of the cyclic code, in PAPR-reduction, over the other two, that is, Hamming and BCH.

- Refer to Fig. 6b in which BCH and Hamming codes are compared. As observed, performance of the BCH code is almost similar to that of the Hamming code. This is because of almost similar *CCDF* at a given *PAPR*, corresponding to both the schemes.
- All the above observations can be held from Fig. 6b in which *CCDF* versus *PAPR* curves corresponding to all the considered coding schemes have been displayed and compared.

Thus, it can be said that, among the three channel coding schemes that are considered here, the cyclic code introduces maximum linearity (minimum *PAPR*), while the linearity introduced by Hamming and BCH codes are almost similar.

4 Conclusions

This paper evaluates and compares the PAPR-reduction performance of the OFDM signal using three channel coding schemes namely, cyclic, Hamming and BCH codes. *CCDF* versus *PAPR* curves are plotted and used for comparing the performances of these channel coding schemes. From the obtained results, cyclic code is observed to introduce maximum linearity (minimum *PAPR*), while linearity introduced by the Hamming and BCH codes is found to be almost similar.

One of the major advantages of this study is that, in order to gain further improvements, it can be augmented with the selective mapping (SLM) technique.

References

1. Rajasekhar C, Raghava VY, & Hanith D (2014) PAPR reduction performance in OFDM systems using channel coding techniques. In: Electronics and Communication Systems (ICECS), 2014 International Conference, IEEE, pp 1–5.
2. Tsai YC, Deng SK, Chen KC & Lin MC (2008) Turbo coded OFDM for reducing PAPR and error rates. IEEE Transactions on Wireless Communications, 7(1) pp 84–89
3. Jiang T, Wu Y (2008) An overview: Peak-to-average power ratio reduction techniques for OFDM signals. IEEE Transactions on broadcasting, 54(2), pp 257–268.
4. Chen H, Liang H (2007) Combined selective mapping and binary cyclic codes for PAPR reduction in OFDM systems. IEEE Transactions on Wireless Communications, 6(10), pp 3524–3528.
5. Agrawal DP & Zeng QA (2015) Introduction to wireless and mobile systems. Cengage learning.

6. Hazan Y & Ran M (2016) Study of indoor LTE green small-cells using mobile front haul architecture over hybrid fiber-wireless channels. In: Networks and Communications (EuCNC), European Conference, IEEE, pp 185–189.
7. Furht B, Ahson SA (Eds.) (2016) Long Term Evolution: 3GPP LTE radio and cellular technology. CRC Press.
8. Simon MK & Alouini MS (2005) Digital communication over fading channels (Vol. 95). John Wiley & Sons.
9. Suarez-Casal P, Fresnedo O, Castedo L, Garcia-Frias J (2016) Transmission of analog correlated sources over MIMO fading channels. In: Smart Antennas (WSA 2016); Proceedings of the 20th International ITG Workshop on VDE, pp 1–7.
10. Bolcskei H (2006) MIMO-OFDM wireless systems: basics, perspectives, and challenges. IEEE wireless communications, 13(4), pp 31–37.
11. Xu K, Xie W, Ma WF, Zhang DM, Wang C, Li N & Xu YY (2016) On Max-SINR timing for multicarrier modulation system with hexagonal Time-Frequency lattice over doubly dispersive channel. In: Electronics, Electrical Engineering and Information Science: Proceedings of the 2015 International Conference on Electronics, Electrical Engineering and Information Science (EEEIS2015). World Scientific, pp 469–483.
12. Monk A, et al. (2007) Broadband network for coaxial cable using multi-carrier modulation, U. S. Patent No. 7, 29, 5518.
13. Poullin D (2005) Passive detection using digital broadcasters (DAB, DVB) with COFDM modulation. IEE Proceedings-Radar, Sonar and Navigation, 152(3), pp 143–152.
14. Joshi A, Saini DS (2010) Performance analysis of coded OFDM for various modulation schemes in 802.11 a based digital broadcast applications. In: Information Processing and Management. Springer Berlin Heidelberg, pp 60–64.
15. Burbank JL, Andrusenko J, Everett JS, Kasch WT (2013) Wireless networking: Understanding internetworking challenges. John Wiley & Sons.
16. Van Zelst A, Schenk TC (2004) Implementation of a MIMO OFDM-based wireless LAN system. IEEE Transactions on signal processing, 52(2), pp 483–494.
17. Gupta P, Kumar BA, Jain SK (2015) Peak to average power ratio reduction in OFDM using higher order partitioned PTS sequence and Bose Chaudhuri Hocquenghem Codes. In: Signal Processing and Communication Engineering Systems (SPACES), 2015 International Conference, IEEE, pp 443–447.
18. Hamming RW (1950) Error detecting and error correcting codes. Bell Labs Technical Journal, 29(2), pp 147–160.
19. Peterson WW, Weldon EJ (1972) Error-correcting codes. MIT press.
20. Imai H (Ed.) (2014) Essentials of error-control coding techniques. Academic Press.

Study of Segmentation Techniques for Cursive English Handwriting Recognition

Pritam S. Dhande and Reena Kharat

Abstract This paper aims to present a study of different segmentation techniques for optical character recognition of handwritten cursive English script. Optical character recognition is a very challenging research field. There are scanners with inbuilt OCR for printed documents but not for handwritten documents. Character recognition of handwritten cursive English script is a very challenging task. In cursive English handwriting, the characters in a word are connected to each other. So the segmentation and feature extraction of cursive English script are much difficult.

Keywords Feature extraction · Handwritten cursive English script
Optical character recognition · Segmentation

1 Introduction

Optical character recognition (OCR) is a method which identifies and recognizes the text which is stored in an image of the format JPEG, GIF, etc. The text in an image is converted into a machine-readable format viz. ASCII or Unicode. The pixel representation of a letter is converted into its equivalent character representation using OCR. It is the machine simulation of human reading. Character recognition is the study of how machines can observe the environment, distinguish character of interest from their background, and make the correct decisions about the characters. OCR has numerous applications in pattern recognition, computer vision, and artificial intelligence. The task of character recognition is broadly separated into two categories:

P.S. Dhande (✉) · R. Kharat
Department of Computer Engineering, Pimpri Chinchwad
College Engineering, Pune, India
e-mail: dhandespritam@gmail.com

© Springer Nature Singapore Pte Ltd. 2018
S.S. Dash et al. (eds.), *International Conference on Intelligent Computing and Applications*, Advances in Intelligent Systems and Computing 632,
https://doi.org/10.1007/978-981-10-5520-1_51

- Handwritten documents and
- Machine-printed documents.

The characters in machine-printed documents are straight with uniform alignment, and there is uniform spacing between the characters. Handwritten characters varies in shape and size because each and every person on the earth has different handwriting. In general, there are six steps in the process of character recognition of handwritten cursive script. These steps are as follows:

- Image acquisition,
- Preprocessing,
- Segmentation,
- Feature extraction,
- Classification, and
- Post-processing.

The flowchart for optical character recognition of handwritten documents is given in Fig. 1.

2   Related Work

In the middle of 1940s, the first character recognition system is appeared with the development of digital computers. The early work on the automatic recognition of characters has been concentrated either upon machine-printed text or upon a small set of well-distinguished handwritten text or symbols.

In the literature of cursive English handwriting recognition, in 2016, Abhishek Bala and Rajib Saha proposed a horizontal and vertical projection method for segmentation of cursive handwritten document. The proposed segmentation method can segment the text lines and words. This technique gives better result for multiple skew and overlapping characters. The samples from IAM database are taken for the experiment. The accuracy for the line segmentation is 95.65% and for the word segmentation is 92.56%. Proposed skew normalization method normalizes the skew up to 96% [1].

In 2016, Kanchan Keisham and Sunanda Dixit proposed the segmentation method based on the information energy. This energy is calculated for each and every pixel. Artificial neural network is used for recognition of characters. The accuracy of recognition for line segmentation is 95%, and for word and character segmentation is 94 [2].

In 2014, Subhash Panwar and Neeta Nain proposed a novel connectivity strength parameter. This parameter is used for deciding the groups of the components which belong to the same line. In this approach, over-segmentation is removed by using depth-first search method with the iterative use of the connectivity strength function. This technique is implemented with English, Hindi, and Urdu text images

Fig. 1 Flowchart for OCR

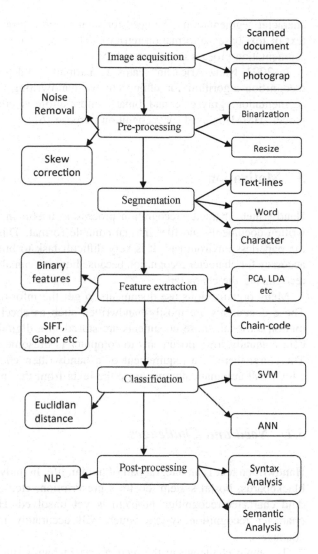

taken from benchmark database. The strength of this technique is that it is a language adaptive approach. The average accuracy of the proposed method is 97.30% [3].

In 2009, G. Louloudisa, B. Gatosb, I. Pratikakisb, C. Halatsisa proposed a segmentation methodology of handwritten documents for text line and word segmentation. Hough transform is used for text line segmentation [4].

In 2015, Namrata Dave discussed various methodologies to segment a text-based image such as pixel counting approach, histogram approach. Various levels of segmentation are also discussed viz. text line segmentation, word segmentation, character segmentation. The need for segmentation process is justified in the context of text-based information retrieval. Then, the various factors which affect the segmentation process are discussed [5].

In 2002, Nafiz Arica and Fatos T. Yarman-Vural proposed segmentation and recognition algorithm for offline cursive handwriting recognition. For character segmentation, grayscale and binary information is combined. Hidden Markov model (HMM) is used for recognition purpose [6].

3   Motivation

Handwritten character recognition process is useful in converting the old handwritten documents and files into an editable format. This will be helpful to create the paperless environment. It is very difficult task to build the system with 100% accuracy for character recognition because humans can also make the mistake while recognizing the characters.

Many organizations use documents to get the information from the customers. These documents are mostly handwritten such as checks, forms. For easy information retrieval, those documents are stored in a digital format. Filling the same data manually from documents to computer is tiresome and time-consuming task. Therefore, there is a requirement of a handwritten character recognition system which will automatically recognize the texts from the image.

3.1   Need and Challenges

Handwritten text is an integral part of human life. In daily life whatever we write on blackboards, postal system, doctor's prescription, etc., everything is handwritten. And character recognition problem is yet unsolved. Hence, there is a need of character recognition system which will accurately recognize the handwritten characters.

The main challenge in the recognition of handwritten characters is that every person on the Earth has different handwriting. There are various other factors also which cause difference in handwriting such as multi-orientations, skewness of the text lines, overlapping characters, connected components, pressure points. Many scripts are there with their intrinsic variations. A single character can be written in many forms, so it is also a challenging task to recognize a particular handwritten character.

4 Flow of the System

A. **Preprocessing**:

In preprocessing, many steps will be carried out. They are as follows:

- **Noise removal**:

This is a process of removing noise from scanned image by using appropriate filter such as smoothing linear filter, order statistic filter. Smoothing process is used for blurring and reducing noise.

- **Binarization**:

In this, image is converted into a grayscale image Then, grayscale image is subjected to binarization. Binarization is performed by using Otsu's thresholding method. This method converts the grayscale image into black and white image, wherein the pixel values of the image are either 0 or 1. Such an image is called as a binary image.

- **Cropping and resizing**:

The extra portion which is present in the image other than the portion occupied by the character need to be eliminated so that only the character can be processed. This process is called as cropping. In the cropping process, initially, the top-leftmost black pixel of the character is first identified, and stored in a temporary variable. Similarly, the top-rightmost black pixel, bottom-leftmost black pixel, and bottom-rightmost black pixel of the character are identified and stored. These values are given to the cropping function in order to extract only the character from the image. After cropping the character image, the image is resized.

- **Thinning**:

The thinning process preserves the structure of original image and reduces the amount of data needed to process. Thinning helps to improve the extraction accuracy and efficiency.

B. **Segmentation**:

In segmentation, an image is decomposed into sub-images of individual character. Segmentation includes the following:

- Line segmentation which is a separation of line from paragraph,
- Word segmentation which is a separation of word from line, and
- Character segmentation which is a separation of character from words.

C. **Feature extraction**:

In this stage, essential information of the character in an image is extracted. This process plays an important role in the recognition process.

D. **Classification**:

In this stage, character recognition system uses the methodologies of pattern recognition, which assigns an unknown sample into a predefined class.

E. **Post-processing**:

In post-processing, output text is matched with the dictionary, so as to improve the recognition rate. If any character from the word is not correctly recognized, then matching word from dictionary will be picked up.

4.1 Mathematical Model

The mathematical model for the segmentation technique is as follows:

$$F(x) = F(I \rightarrow O|P),$$

where $F(x)$ is a function, mapping set of input images $\{I\}$ to set of output strings $\{O\}$, given a set of operations P.

The definitions of sets are as follows:

$$\{I\} = \{F(I1) = \{F(x1) : x1 \in A\},$$
$$F(I2) = \{F(x2) : x2 \in A\},$$
$$\cdot$$
$$F(In) = \{F(xn) : xn \in A\}$$
$$\}$$

where A is a set of possible pixel values, i.e., $\{0\ldots255\}$

$$\{O\} = \{F(O1) = \{F(y1) : y1 \in B\},$$
$$F(O2) = \{F(y2) : y2 \in B\},$$
$$\cdot$$
$$F(On) = \{F(yn) : yn \in B\}$$
$$\}$$

where B is a set of possible strings which themselves are formed by smaller set of characters, small case and upper case.

$$B = \{C1, C2\}$$
$$C1 = \{a, b, \ldots, z\}$$
$$C2 = \{A, B \ldots Z\}$$
$$\text{Now, } P = \{P1, P2, \ldots Pn\}$$

where $P1 \ldots Pn$ are the operations followed in a linear fashion.

1. $P1$ = Noise reduction/binarization

$$P1 = \{F(I) : I1 \rightarrow I2\}$$
$$I1 = \{F(x1) : x1 \in A\}$$
$$I2 = \{F(x2) : x2 \in B\}$$

where
$$A = \{0, 1, \ldots 255\}$$
$$B = \{0, 255\}$$
where mapping function is given by

- Histogram equalization and
- Weight and mean initialization.
- For all thresholds,

    i. Update w and u (weight and mean).
    ii. Find covariance.

2. P_2 = Cropping

$$P_2 = \{F(I) : I_1 \rightarrow I_2\}$$
$$I_1 = \{F(x_1) : x_1 \in A\}$$
$$I_2 = \{F(x_2) : x_2 \in B\}$$

where
$$A = \{0, 255\}$$
$$B = \{0, 255\}$$
where mapping function is given by

- $I1 = F(x, y)$
- $r\_\text{sum} = \sum_{i=0}^{n} F(x_i)$
- $c\_\text{sum} = \sum_{i=0}^{m} F(y_i)$
- $I_2(x, y) = I_1(x, y)$ if $(r\_\text{sum} > 0)$ or $(c\_\text{sum} > 0)$

3. P_3 = Segmentation (line)

$$P_3 = \{F(I) : I_1 \rightarrow \{I_n\}\}$$
$$I_1 = \{F(x_1) : x_1 \in A\}$$
$$\{I_n\} = \{I_1^1, I_1^2, I_1^3, \ldots I_1^n,\}$$

where
$$I_I^i = \{F(x_1) : x_1 \in A\}$$
$$A = \{0, 255\}$$

I_I^i = Images of cropped lines
where mapping function is given by

- Horizontal and vertical projection methods.

5 Algorithms Used

For segmentation, horizontal and vertical projection methods are used [1].

A. **Algorithm for line segmentation**:

Line segmentation algorithm is carried out by horizontal projection as follows:

Step 1: Read a handwritten document image as a multi-dimensional array.
Step 2: Check whether the image is a binary image or not. If binary image, then store it into a 2-D array IMG[][] with size $M \times N$ and go to Step 4; otherwise, go to Step 3.
Step 3: Convert the image to binary image and store into a 2-D array IMG[][].
Step 4: Construct the horizontal projection histogram of the image IMG[][] and store into a 2-D array HPH[][].
Step 5: Measure the height, starting and ending row number of each rising section present in horizontal projection histogram and store these values in 3-D array LH[][][] sequentially.
Step 6: Count the number of rising sections by counting the rows of the 3-D array LH[][][]. Then, measure the threshold (T_i) value by calculating average height of rising sections from the 3-D array LH[][][].
Step 7: Select each rising section from 3-D array LH[][][] and check whether the height of that rising section is less than the threshold or not. If yes, then this rising section is not considered as a line and go to Step 9; otherwise, rising section is treated as a line and go to Step 8.
Step 8: Find the rising section's starting and ending rows number from the array LH [][][]. Let starting and ending rows be $r1$ and $r2$, respectively. Extract the line segment between $r1$ and $r2$ from the original binary image denoted by IMG[][].
Step 9: Go to Step 7 for next rising sections till all rising sections are not under consideration; otherwise, go to next step.

Step 10: End.

B. Algorithm for word segmentation:

Word segmentation algorithm is carried out by vertical projection as follows:

Step 1: Read a segmented binary line as 2-D binary image LN[][].
Step 2: Construct the vertical projection histogram of the line LN[][] and store into a 2-D array LVP[][].
Step 3: From the vertical projection histogram (LVP[][]), measure width of each inter-word and intra-word gaps and store the width into 1-D array GAPSW[].
Step 4: Count the total number of gaps as TGP by calculating the size of GAPSW[]. Add width of all gaps by adding the elements of GAPSW[] and store into TWD.
Step 5: Calculate the threshold (T_i) as follows:

$$T_i = \text{TWD}/\text{TGP} \tag{1}$$

In equation, T_i is the threshold value denoting average width of inter-word gaps, TWD denotes total width of all gaps, and TGP denotes the total number of gaps.
Step 6: For each $i(1 \leq i \leq \text{sizeof(GAPSW[])})$, if $\text{GAPSW}[i] \geq T_i$, then these gaps are treated as inter-word gaps; otherwise, gaps are treated as intra-word gaps. Depending on inter-word gaps width, words are segmented from the line.
Step 7: End.

6 Conclusion

This work deals with the study of segmentation techniques for recognition of cursive handwritten English characters. The samples used are of high quality to reduce the complexities in the recognition process. In this work, a horizontal and vertical histogram projection-based approach has been used as a segmentation strategy for offline cursive handwriting recognition. The experiments using the proposed approach on CCC benchmark dataset have been conducted.

References

1. Abhishek Bala and Rajib Saha, "An Improved Method for Handwritten Document Analysis using Segmentation, Baseline Recognition and Writing Pressure Detection", 6th International Conference On Advances In Computing & Communications, ICACC 2016, 6–8 September 2016, Cochin, India, Elsevier-2016.
2. Kanchan Keisham and Sunanda Dixit, "Recognition of Handwritten English Text Using Energy Minimisation", Information Systems Design and Intelligent Applications, Advances in Intelligent Systems and Computing, Bangalore, India, Springer-2016.

3. Subhash Panwara & Neeta Naina, "A Novel Segmentation Methodology for Cursive Handwritten Documents", IETE JOURNAL OF RESEARCH|VOL 60|NO 6|NOVDEC 2014
4. G. Louloudisa, B.Gatosb, I.Pratikakisb, C.Halatsisa, "Text line and word segmentation of handwritten documents" doi:10.1016/j.patcog.2008.12.016, Elsevier -2009
5. Namrata Dave, "Segmentation Methods for Hand Written Character Recognition", International Journal of Signal Processing, Image Processing and Pattern Recognition Vol. 8, No. 4 (2015), pp. 155–164.
6. Nafiz Arica, Student Member, IEEE, and Fatos T. Yarman-Vural, Senior Member, IEEE, "Optical Character Recognition for Cursive Handwriting", IEEE transactions on pattern analysis and machine intelligence, vol. 24, no. 6, June 2002

Wide Band Triangular Patch Textile Antenna with Partial Ground Plane

Kunal Lala, Archana Lala and Vinod Kumar Singh

Abstract A triangular patch microstrip textile antenna with partial ground is estimated for wide band application. The antenna resonates at 7.43 GHz frequency having the bandwidth of 36.35% ranging from 6.06 to 8.75 GHz frequency band. The simulated results of proposed triangular patch textile antenna are carried by CST software. The antenna is very compact (50 mm × 50 mm × 1.0) in size and simple to design. The reflection coefficient, gain, right polarization, directivity and left polarization of the proposed antenna remain valid for operating frequency band.

Keywords Return loss · Polarization · Directivity and CST software

1 Introduction

Nowadays the fast development of modern communication systems is required for transportable devices for some important features which include easy designing, lightweight, small in size, compatible with microwave, millimetre-wave integrated circuits, less production cost and easy fabrication of microstrip antennas. The microstrip antenna has numerous useful properties which include minute size, low cost of the fabrication, lightweight, ease of installation, but the main limitations of

K. Lala
Department of Electronics & Communication Engineering, RKGIT,
Ghaziabad, UP, India
e-mail: kunalec3@gmail.com

A. Lala
Department of Computer Science Engineering, SRGI, Jhansi, UP, India
e-mail: archanalala2006@yahoo.com

V.K. Singh (✉)
Department of Electrical Engineering, SRGI, Jhansi, UP, India
e-mail: singhvinod34@gmail.com

© Springer Nature Singapore Pte Ltd. 2018
S.S. Dash et al. (eds.), *International Conference on Intelligent Computing and Applications*, Advances in Intelligent Systems and Computing 632,
https://doi.org/10.1007/978-981-10-5520-1_52

577

printed antennas remain their narrow bandwidth features which limit the range of frequency over which the antenna can work effectively. Microstrip antenna comprises three most important parts which is substrate, patch and ground [1–5]. A large bandwidth capability of supporting numerous applications has become the unavoidable feature of any modern wireless communication system. Therefore, there remains continuous increasing demand of the performance objectives of the printed antenna mostly used in such systems. In this regard, antennas which support circular polarization are being preferred and their properties are continuously investigated by researchers around the world [6–9]. Microstrip antenna plays major role in wireless communication system and is used in high-performance aircrafts, radar, missiles and other spacecraft. It has many advantages such as its lightweight, simple structure, ease of addition and less cost. Microstrip antenna requires very less space for installation as these are simple and small in size. The only space that required is the space for the feed line [10–12].

The anticipated article is compact with partial ground which is applicable for wide band application. The proposed triangular patch gives wide band of 36.35% covering the frequency range from 6.06 to 8.75 GHz.

2 Antenna Configuration

The radiating element of the proposed antenna is triangular patch which comprises triangular slots. The rectangular dielectric substrate with the overall dimensions of 50 × 50 mm and several changes have to be made to attain the essential return loss, gain and directivity. Firstly, basic configuration is made using the triangles, and the triangular slots are inserted to achieve the desired reflection coefficient and the gain which is necessary for the wide band application. Copper is served as the conductive material for the slots of the antenna in order to achieve the desirable conductivity. The design parameters are given in Table 1 (Fig. 1).

Table 1 Design parameters of triangular patch textile antenna

Design parameters	Value
Inner triangle patch dimension	4 × 4 mm
Inner triangle slot dimension	10 × 10 mm
Outer triangle patch dimension	25 × 25 mm
Microstrip feed line (L × W)	2 × 24
Substrate dimension (Ls × Ws)	50 × 50
Partial ground plane (Lg × Wg)	50 × 15

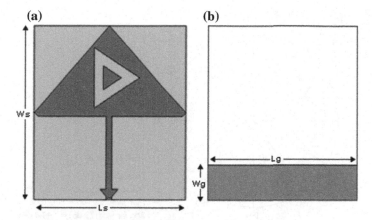

Fig. 1 Triangular patch textile antenna geometry: **a** front view and **b** back view

Fig. 2 Simulated reflection coefficient versus frequency of proposed triangular patch textile antenna

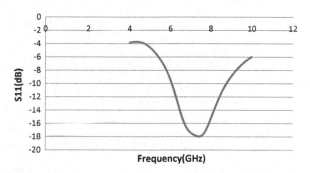

3 Result and Discussion

Figure 2 shows the simulated reflection coefficient verses frequency plot of presented antenna. The optimized antenna resonates at 7.43 GHz and gives the −18 dB return loss at resonating frequency. The reflection coefficient verses frequency plot in the figure below clearly shows that the presented triangular patch textile antenna has wide band width suitable for C-band application. Figure 3 depicts two-dimensional radiation pattern of anticipated textile antenna at 7.43 GHz which describes the main lobe direction = 143°, angular width (3 dB) = 44.8° and main lobe magnitude = 2.6 dBi at φ = 90. Also it gives the main lobe direction = 34° and

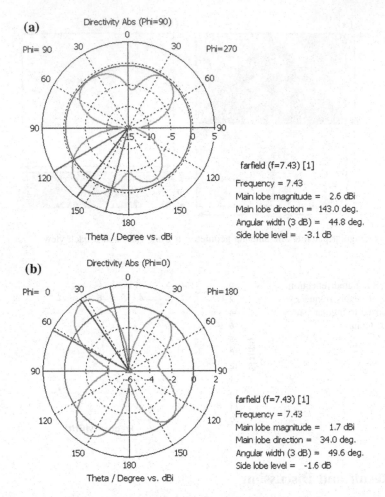

Fig. 3 Two-dimensional radiation pattern of presented triangular patch textile antenna at 7.43 GHz

main lobe magnitude = 1.7 dBi at $\varphi = 0°$. Figure 4a shows three-dimensional radiation pattern of proposed textile antenna at 7.43 GHz which gives total efficiency of about −6.393 dB and directivity of 4.453 dBi.

(a)

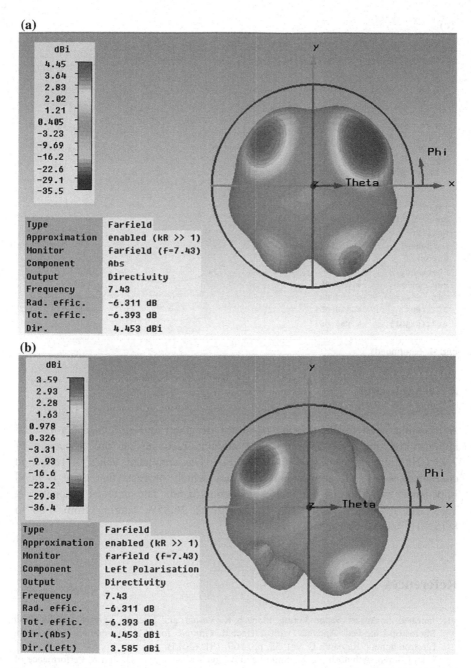

(b)

Fig. 4 Simulated three-dimensional radiation pattern of proposed triangular patch textile antenna at 7.43 GHz

(c)

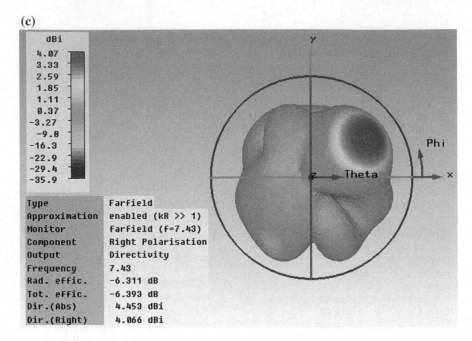

Type	Farfield
Approximation	enabled (kR >> 1)
Monitor	farfield (f=7.43)
Component	Right Polarisation
Output	Directivity
Frequency	7.43
Rad. effic.	-6.311 dB
Tot. effic.	-6.393 dB
Dir.(Abs)	4.453 dBi
Dir.(Right)	4.066 dBi

Fig. 4 (continued)

4 Conclusion

A triangular path textile antenna has been designed and simulated for C-band application using CST software. The antenna resonates at the frequency of 7.43 GHz ranging from 6.06 to 8.75 GHz. The simulated value of reflection coefficient is −18 dB at resonant frequency which suggests that there is good impedance matching in the entire frequency band. The antenna resonates at 7.43 GHz frequency having the bandwidth of 36.35% ranging from 6.06 to 8.75 GHz frequency band.

References

1. Jamshed A. Ansari, Sapna Verma, Mahesh K. Verma, and Neelesh Agrawal,: Wide Band Microstrip-Line-Fed Antenna with Defected Ground for CP Operation, Progress In Electromagnetic Research C, vol. 58, pp. 169–181, (2015).
2. Nikhil Singh, Ashutosh Kumar Singh, and Vinod Kumar Singh: Design & Performance of Wearable Ultra Wide Band Textile Antenna for Medical Applications, Microwave and Optical Technology Letters, vol. 57, no. 7, pp.1553–1557, (2015).
3. Stuti Srivastava, Vinod Kumar Singh, Zakir Ali, and Ashutosh Kumar Singh: Duo Triangle Shaped Microstrip Patch Antenna Analysis for WiMAX lower band Application,

International Conference on Computational Intelligence: Modelling Techniques and Applications(CIMTA-2013), Procedia Technology Elsevier 10, pp. 554–563, (2013).

4. Raghupatruni V., Ram Krishna, and Raj Kumar: Design of Temple Shape Slot Antenna for Ultra Wideband Applications, Progress In Electromagnetic Research B, vol. 47, pp. 405–421, (2013).

5. Balanis CA: Antenna Theory: Analysis and Design, John Wiley and Sons, New York, 2004.

6. Amit Kumar Rawat, Vinod Kumar Singh, and Shahanaz Ayub, Compact Wide band Microstrip Antenna for GPS/WLAN/WiMax Applications, International Journal of Emerging Trends in Engineering and Development, no. 2 vol. 7, pp. 140–145, (2012).

7. Rishabh Kumar Baudh, Ranjan Kumar, and Vinod Kumar Singh: Arrow Shape Microstrip Patch Antenna for WiMax Application, Journal of Environmental Science, Computer Science and Engineering & Technology, vol. 3, no. 1, pp. 269–274, (2013).

8. Janabeg Loni, Shahanaz Ayub, and Vinod Kumar Singh: Performance analysis of Microstrip Patch Antenna by varying slot size for UMTS application, IEEE Conference on Communication Systems and Network Technologies (CSNT-2014), pp. 01–05, (2014).

9. N. M. Din, C. K. Chakrabarty, A. Bin Ismail, K. K. A. Devi, and W. Y. Chen: Design of RF Energy Harvesting System For Energizing Low Power Devices, Progress In Electromagnetics Research, vol. 132, pp. 49–69, (2012).

10. Mai A. R. Osman, M. K. A. Rahim, M. Azfar N. A. Samsuri, F. Zubir, and K. Kamardin: Design, implementation and performance of Ultra-Wideband Textile Antenna, Progress In Electromagnetics Research B, vol. 27, pp. 307–325, (2011).

11. Vinod Kumar Singh, and Nikhil Kumar Singh: Compact Circular Slotted Microstrip Antenna for Wireless Communication Systems, Journal of Microwave Engineering & Technologies, vol. 1, no. 1, pp. 07–14, (2015).

12. Rajat Srivastava, Shahanaz Ayub, and V. K. Singh: Dual Band Rectangular and Circular Slot Loaded Microstrip Antenna for WLAN/GPS/WiMax Applications, IEEE Conference on Communication Systems and Network Technologies (CSNT-2014), pp 45–48, (2014).

A Survey on Image Enhancement Techniques Using Aesthetic Community

Priyanka Chaudhary, Kailash Shaw and Pradeep Kumar Mallick

Abstract Nowadays, digital imaging devices and social network provide prevalence of sharing images through social media like Facebook. Image sharing is nothing but artistic enhancement of images by various applications such as Instagram, Microsoft Office Picture Manager, Adobe Photoshop as images are the source of information for interpreting and analyzing data. For enhancements, an image is converted from original image to new modified image, so the result is better image which is obtained from collection of techniques which give improved visual appearance of an image. This technique is widely used in printing industry, graphic design, cinematography, forensic purpose, etc. In particular, to enhance image attractiveness, the aesthetic appearance of an image is used where latent Dirichlet allocation (LDA) can be used.

Keywords Image enhancement · Aesthetic · Community detection
LDA · Topic model

1 Introduction

Image enhancement is one of the techniques which is widely used in image research, computer graphics. It is the subareas of image processing. The main aim of image enhancement techniques is processing an image so the result is improved than the

P. Chaudhary (✉) · K. Shaw
D.Y. Patil College of Engineering, Akurdi, Savitribai Phule Pune University,
Pune, India
e-mail: priyanka.sherekar@yahoo.co.in

K. Shaw
e-mail: kailash.shaw@gmail.com

P.K. Mallick
Vignana Bharathi Institute of Technology (VBIT), Ghatkesar, Aushapur (V),
Hyderabad, Ranga Reddy District, Telangana, India
e-mail: Pradeepmallick84@gmail.com

© Springer Nature Singapore Pte Ltd. 2018
S.S. Dash et al. (eds.), *International Conference on Intelligent Computing
and Applications*, Advances in Intelligent Systems and Computing 632,
https://doi.org/10.1007/978-981-10-5520-1_53

previous image. It improves the quality of images with respect to human views. It is also used to remove blur, noise. Most of the time it is required to enhance the values of images and therefore some image enhancement techniques are used.

Generally, image enhancement techniques are distributed in either of one spatial domain class or it can be frequency domain class. Spatial domain method works to manipulate the image pixels. The pixels values are calculated and manipulated to get the desired enhancement. Partial domain techniques are logarithmic transforms, power law transforms, histogram equalization which are related to direct use of pixels in the image. It is also categorized into two types, i.e., point processing operation and spatial filter operation [1]. Point processing operations (or intensity transformation function) are the simplest spatial domain operations, and it is performed on only single pixel value. The point processing approaches can be divided into four categories, i.e., image negatives, log transformations, inverse log transformations, and power law transformations [1]. Spatial filters are categorized as linear spatial filters and nonlinear spatial filters. In the linear spatial filtering, a convolution mask used with an image and is a weighted mask over the entire image.

The enhanced image is not linearly related to pixels in the neighborhood of original image in linear spatial filters. Frequency domain techniques are calculated by manipulating the orthogonal transform of the image. It is used to compute 2-D discrete unitary transform of the image [1]. The conventional approaches for image enhancement consists of the problem like an ideal image enhancement framework is built by professional photographers but most of the time, due to lack of experience photographers cannot find particular aesthetic topic, some photographers can upload thousands of images on image hosting site like Flickr, whereas some upload very few, so it is also difficult task to describe aesthetic interest of photographers similarly as topic of an images are retrieving from tags associated with images, some images tag are noisy or even missing. Fully automatic image enhancement algorithm for big scale application is little bit hard task [2].

To solve the above problem, a fully automatic framework can be used to enhance image visual appearance by discovering aesthetic communities using a tag-regularized LDA model can have advantage. Different image enhancement tasks are combining into a framework using a probabilistic model [2].

2   Related Work

Over the past few years, image enhancement in image processing has become a popular research topic because of the increased use of social media. The image aesthetic enhancement algorithm can be mainly related to image analysis and data mining. Tag regularized topic model i.e., Latent Dirichlet Allocation (LDA) is used to detect aesthetic topic which is understandable and those are similar topics of image users are discovered for making aesthetic communities which is a main part of image enhancement. The related work is divided into two parts: aesthetic model and community detection [2] are as follows.

A. *Aesthetic Communities*

Aesthetic model gives the visual features extracting from the image. Many conventional aesthetic models are researched in present days. Multimedia application [3] which most of the time users use for assessing the aesthetic quality of a photographers using geometric rule of composition. The spatial structure, i.e., local and global between image region from multiple channels are used to assess image aesthetics [4]. Semantic relations are considered for optimization between foreground objects [5] which is the image enhancement method for optimizing image composition.

In human scenery aesthetic composition, [6] author mainly implements the system on an Android platform and produces souvenir images as a result. Aesthetic classifier [7] trained on compositional, content, and sky-illumination attributes provides good results for human quality judgments for images. Content based image quality assessment [8] proposed by regional and global features. Working with large database and efficiently extracting feature have impact over various images.

The aesthetic-driven image recomposition techniques survey [9] explains the number of image recomposition problem, objectives, complete analysis of effectiveness of each technique for obtaining the recomposition objectives. Semantically rich photos can be retargeted [10] by finding image meaning by its tags, by a multilable SVM. Latent stability discovery (LSD) is a key technique used as a generative model.

B. *Discovering Community*

The objective of discovering community is to identify structure of community in network. Community is group of closely connected nodes which are loosely connected to others. The community detection issue, detection of overlapping communities are mentioned in data mining [11, 12], bipartite graphs clusters [13], and cluster detection [14].

Image annotation approach [15] explained labels latent semantic community with multikernel learning. Constructing a label concept graph indicates the concepts relationship. Semantic communities are discovered using an automatic community detection method. The generative probabilistic modeling [16] is used to detect community in which every community is a combination of semantic topics. Author explains entropy filtering Gibbs sampling [16].

Topic models such as LDA [2, 17] and its different variants [18, 19] are used to detect social communities. A social link graph is created from topic models into a generative model. It categorizes users into different communities by sampling process.

Multiassignment clustering (MAC) combines Boolean vectorial data among multiple clusters [2, 20]. Two community detection schemes, i.e., modularity optimization and Potts model clustering, on a set of benchmark graphs are evaluated in [21]. Regularized probabilistic latent semantic analysis [22] is used for learning aesthetic communities from Flicker site.

3   Materials and Methods

A. Material

Material refers to the datasets used in different papers. There are many datasets for evaluating photo aesthetics which are mentioned below.

The MIRFLICKR

The MIRFLICKR-25000 is an open evaluation project contains 25,000 images which are extracted from the social photography site Flickr. These images are stored with extra information called as tags which is a new feature.

It also consists of EXIF (Exchangable image file format) is metadata representing a number of properties and settings of the camera. This explains information about camera, its settings, image settings, etc. [23].

The Photo.net

The Photo.net is online community which provides a platform for those people who are having photography as a hobby. It consists of information about the hottest gear, tutorials to help people on journey, inspirational interviews of people, etc. It provides a robust and peer to peer educational platform where anyone can connect to person and post their photos. This site also gives suggestions on improving photography. It consists of more than 4.8 million photographs [24]. This dataset is also used in [4, 7].

The AVA

AVA, Aesthetic Visual Analysis, is a new bulky database. It consists of more than 200,000 images with a variety of metadata, aesthetic scores for image. It includes well-formed labels for over 60 categories and also labels based on photographic approach [24]. This dataset is used in [4, 9, 25].

The DPChallenge

The DPChallenge photography is league and site for images hosting which provides a way for photographers to stay together as a team and compete against other teams using their collective vote averages. It also gives platform for photographers to show their talent through image. The DPL is addition to DPChallenge consists of group of people which helps to learn new things and new techniques. People can share, browse images through it. This dataset is used in [7].

Flickr

Flickr is one of the photo, video hosting websites with web services. Flickr is online community allows users to exchange photographs. Image researchers and bloggers used to host images in blogs and social network also; academic researchers download images as a dataset and can be used for research. Flickr provides images according to categories with tags [1, 7].

B. Methods

1. **Tag-Wise Regularized LDA** [2]
   A tag-wise regularized topic model is used to explore topic which is aesthetic image user. It gives systematic and understandable topics from the visual features and tags of images. LDA is used on images with similar aesthetic values share similar latent aesthetic topics [2].
   Modeling the aesthetic topic of each Flickr user based on the conventional LDA suffers from two problems. Different Image users upload numbers of the images. Some of the users might have uploaded 50,000 photos, while others have fewer than 10 images, thus the learned LDA model is susceptible to over fitting. Second, the tag channel is informative for reflecting a human semantically aesthetic perception of an image, but it cannot be integrated into the conventional LDA conveniently in the same way as color and textural features, since many Flickr images do not have tags [2].
   To solve the above problems, a tag-wise regularizer on the conventional LDA. It uses Flickr images as a dataset and according to tags and latent aesthetic topics they found good topics over a dataset. The regularizer is a structured prior which is based on all the photo tags, which constitute an auxiliary dataset. The structured prior depends on the relationships of tag histograms calculated from the auxiliary dataset [2].

2. **Block-LDA** [26]
   Generating text documents with entities with external connection between entities pairs which allows for using supplementary annotated text for influencing and for improving link modeling. It uses latent topics stochastic block models blocks. The joint modeling allows information sharing of latent topics among the network structure with text, giving result as coherent topics. In entities, co-occurrence patterns as well as words related to them gives modeling of links in the graph [26]. The topics in the text are extracted by entity–entity links. Author proposed a method for performing relative inference using a collapsed Gibbs sampler. The author uses The Munich Institute for Protein Sequencing (MIPS) database [26, 27]. MIPS includes a hand-crafted protein interactions collection including 8000 protein complex associations in yeast. They found the result which shows that joint modeling gives good. Recovery is necessary for the joint model which is internally evaluated by perplexity in two datasets and externally by protein functional class prediction in the yeast dataset [26].

3. **Mixed Membership Stochastic Block (MMSB)** [28]
   Mixed membership stochastic block models are a member in relational data at latent variable. The mixed membership model contains unit of observation with number of clusters by membership probability. The different cluster of concurrent membership contain words constituting each document contains different underlying topics. The objects multiple latent roles give relationships to others. Mixed membership approach for relational data explains the interaction

between objects of multiple roles. For example, various proteins' intercommunication may be led by different function. The MIPS protein intercommunication dataset and interpretations of the blocks in MMB of biological functions are used [28].

In results, model analyzes applications in which observations giving collection of graph is unipartite. The nested variation inference algorithm is able to be parallel and permits fast relative inference on large graphs [28].

The following methods are for community detection where different author tries to detect the community with nodes; similarly, overlapping community is also focused.

4. **Community-Affiliation Graph Model** [29]

Network communities are main principles, which give nodes sets arrange into densely linked clusters or groups. Communities contains many nodes which can be repeated and may be overlap in network [29]. The research elaborates communities overlaps are less closely connected than the non-overlapping portion. By validating this assumption on six huge networks categories such as social, collaboration, information networks in which nodes state community memberships. There are four online social networks: the LiveJournal blogging community, the Friendster online network, the Orkut social network and the Youtube social network. The information network is Amazon product co-purchasing network is used which is fifth and last is the collaboration network of DBLP where nodes represent authors/actors and edges connect nodes that have co-authored a paper. By ground-truth communities, they found result that the community overlaps are more closely connected, which is opposite to the conventional wisdom that community overlaps are more sparsely connected than the communities themselves [29].

Existing community discovery methods fail for detecting dense overlaps communities. In [29], community-affiliation graph model is proposed which is a community detection model-based method and uses bipartite node community-affiliation networks. This method accurately captures overlapping, non-overlapping, and hierarchically nested communities which identifies relevant communities more accurately [29]. Author uses dataset which is a collection of six huge social, collaboration with information networks where community memberships, first four are online social networks, the LiveJournal blogging community [29], the Friendster online network [29], the Orkut social network [29], and the YouTube social network where in each of these networks persons create their groups where others can join. They focused on interests, styles, hobbies, and geographical regions; second type is Amazon product which is a co-purchasing network [29] where community is different. The network nodes show products with edges link among co-purchased products. Each product (i.e., node) of one or more hierarchically arranged product categories and products from the same category show a cluster which view as a ground-truth community. The last dataset is DBLP [29] which is an online computer science publications dataset, providing the title, author list, and conference of publication for more than 350,000 papers. In it, nodes show authors/actors and edges connect nodes that have co-authored a paper. They found result that AGM performs good in discovering network communities and the overlaps between communities [29].

5. **Cluster-Overlap Newman Girvan Algorithm Optimized (CONGO)** [11]
Many networks consist of a community format, like vertices form densely connected groups, which are sparsely linked to another groups. Communities discovering in networks is a demanding part, in term of overlapping. The first algorithm, i.e., CONGA detects overlapping communities using split betweenness. They improved the algorithm known as CONGO by adding betweenness of local form, which is good and faster [11]. The CONGO algorithm contains steps, where it tries to removes an edge from the network or separate a vertex into two vertices by using local betweenness. The CONGO algorithm can be derived from [11]. Execution time of CONGOs is built upon network structure. They run CONGO algorithm on several real-world networks like netscience, cond-mat-2003. They are collaborating networks of co-authorships and having different size. In result, they found that CONGO is an powerful, fast algorithm for detecting overlapping communities in networks than CONGA [11].

6. **Community Detection with Propinquity Dynamics** [12]
Most of time networks representation models complex systems. Detection of community structure is found to be demanding task in large network data. In [12], proposes a novel community detection algorithm, is dynamic process by opposition of the network topology and the topology propinquity. The vertices which are overlapped shared between communities can be determined by simple post-processing. The propinquity is incrementally calculated for achieving better efficiency. This algorithm is implemented on 1000 of machine. Community detection algorithm is necessary due to abundant community structures hidden in real networks. Social network is the implicit interpersonal communities which are constantly as well as automatically formed by the collective general decision of each personal. The community structure gives natural results with local form which is a dynamic process called as propinquity dynamics and its incremental form [12]. For performance and scalability of the algorithm on large graphs uses Wikipedia linkage graph dataset. In this graph, nodes are Wikipedia pages and hyperlinks in Wikipedia pages which are pointing to other normal Wikipedia pages. If accessed, some of them will be redirected to another normal page. The other dataset is Edinburgh Associative Thesaurus (EAT), a word association network, which includes large volume of data word norm stimulus and responds. The results show that the algorithm correctly places overlapping community over real graph in dynamic scenario.

Table 1 Comparative BER scores of different communities discovery algorithms [2]

Flickr Group	MMSB	BLDA	KC	HC	LC	CP	LRE	MAC	Gregory	Zhang	Yang	Ours (ICL)	Ours
The light fan	0.565	0.533	0.633	0.412	0.545	0.523	0.445	0.502	0.4765	0.488	0.512	**0.6085**	0.599
The noir mood	0.678	0.645	0.645	0.508	0.634	0.575	0.534	0.608	0.5765	0.588	0.632	0.7089	**0.715**
Graphic designers	0.517	0.489	0.467	0.455	0.476	0.504	0.467	0.481	0.4876	0.499	0.512	**0.5926**	0.583
Aesthetics failure	0.754	0.722	6789	0.7	0.666	0.674	0.734	0.688	0.6335	0.654	0.698	0.7889	**0.8**
Green is beautiful	0.448	0.456	0.412	0.466	0.434	0.47	0.493	0.504	0.4876	0.489	0.504	**0.5683**	0.545
Colors	0.822	0.804	0.756	0.779	0.781	0.764	0.733	0.765	0.7786	0.79	0.811	0.8376	**0.835**
Closer	0.703	0.713	0.67	0.685	0.666	0.684	0.676	0.692	0.6774	0.695	0.712	**0.7872**	0.774
Less is more	0.633	0.622	0.605	0.621	0.589	0.608	0.587	0.606	0.6231	0.577	0.622	**0.6763**	0.665
Field guide	0.745	0.723	0.711	0.72	0.678	0.689	0.701	0.712	0.7111	0.714	0.734	0.7487	**0.754**
Night lights	0.703	0.724	0.69	0.703	0.688	0.689	0.701	0.665	0.6923	0.692	0.712	**0.7941**	0.786
Black and white	0.346	0.305	0.31	**0.433**	0.345	0.312	0.299	0.355	0.3122	0.308	0.317	0.3345	0.333
Stick figure	0.722	0.701	0.669	0.677	0.694	0.685	0.677	0.682	0.6783	0.691	0.705	0.7521	**0.753**
Writing mach.	0.856	0.833	0.767	0.788	0.822	0.767	0.811	0.775	0.7869	0.792	0.822	**0.8773**	0.864
Through glass	0.754	0.722	0.712	0.734	0.689	0.703	0.713	0.756	0.7589	0.761	0.76	0.7721	**0.777**
Fog and rain	0.434	0.403	0.388	0.382	0.395	0.412	0.423	0.432	0.4376	0.442	0.448	**0.4732**	0.466
Architecture	0.757	0.732	0.718	0.723	0.701	0.689	0.688	0.694	0.7032	0.708	0.713	0.7387	**0.739**
Window seat	0.834	0.801	0.766	0.793	0.809	0.795	0.795	0.813	0.8121	0.805	0.812	0.8435	**0.846**
Movement	0.505	0.489	0.494	0.465	0.522	0.513	0.467	0.489	0.4937	0.498	0.517	**0.5511**	0.544
Orange and Blue	0.545	0.521	0.534	0.534	0.503	0.512	0.488	0.433	0.4786	0.49	0.516	0.5867	**0.595**
Jump project	0.668	0.621	0.642	0.623	0.595	0.611	0.603	**0.621**	0.5673	0.588	0.592	0.5767	0.605
Average	0.643	0.631	0.613	0.631	0.613	0.597	0.546	0.621	0.6271	0.631	0.677	0.7378	**0.75**

4 Result Analysis

In [2], author compares their method with eight popular clustering algorithms, containing those that consider only the graph/network structure, explore information about profile and combine the both. The eight algorithms are: (1) mixed membership stochastic block (MMSB) [2, 28]; (2) block-LDA [2, 26]; (3) K-means clustering (KC) [2]; (4) hierarchical clustering (HC) [2]; (5) link clustering (LC) [2]; (6) clique percolation (CP) [2]; (7) low-rank embedding (LRE) [2]; and (8) multiassignment clustering (MAC) [2]. They also compare their method with three communities detection algorithms: Gregory's [2, 11] and Zhang et al. [2, 12]'s algorithms support overlapping communities, and Yang and Leskovec [2, 29]'s algorithm model nodes hard memberships into multiple overlapping communities. The number of aesthetic communities for all three algorithms is determined by the BIC criterion.

In [2] compared approach with the above 11 clustering and communities detection algorithms. Fixing the cluster/community number at 20 for each algorithm and the algorithms multimodal visual features are same. In Table 1 [2], the observations are shown. For 17 from 20 Flickr communities, their technique outperforms for all its oppositors, as the BER scores are the highest. Tag-wise regularized topic model gives advantage by optimally capturing user aesthetic topic. Aesthetic communities can be more accurately detected like for architecture. The method in [2] performs better on Flickr groups that contain very less users. This again demonstrates need to incorporate a regularized term on the conventional LDA. The Block-LDA [26] jointly models connection among entities and associated text annotated that allows co-occurrence information in text to influence link modeling and vice versa. The methods results show that joint modeling performs good by using only one source of information in used dataset. Improvements are given spatially when the joint model is calculated internally and externally.

The topics convinced by the model examined effective in understanding type of the data not only for topics discussed but also for the connectivity characteristics between entities [26]. The Mixed membership stochastic block [28] learned information regarding the mixed membership of objects to latent groups as well as the patterns of connectivity between them. For describing the functional information of the MIPS collection of protein interactions, estimates are essential part of it. For this, using MMB as a dimensionality reduction approach which is effective for model driven denoising of collections of interactions [28].

5 Performance Metrics

In [2], the performance is deriving from the BER score where they set different parameters for aesthetic communities detection that is the weight of the regularizer and parameter in the affinity matrix. The BER score reflects the accuracy of aesthetic communities discovery. They [2] believe that accurately mining aesthetic communities can greatly improve image enhancement operation.

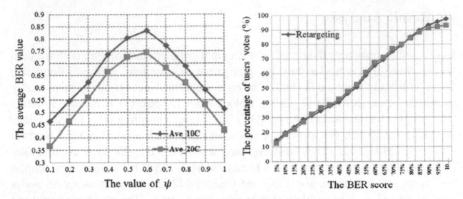

Fig. 1 *Left* the average BER value under different values of ψ; *Right* The percentage of users voting our algorithm as the best performer under different BER scores [2]

To validate this, they calculate the percentage of users who vote their method as the best, under different BER scores. As shown on the right of Fig. 1 [2], the percentage of users mostly preferring their method increases stably by increasing the BER score. This reflects the need to tune parameter setting carefully to achieve a high BER score.

6 Conclusion

This paper is a survey of the current research activities of image enhancement by using aesthetic community's detection. It has been generally accepted that aesthetic appeal of an image can be very useful for the image enhancement. A new approach is shown where aesthetic community using tags of the images are used to form the communities and it will become important part of image enhancement. A tag-wise regularized topic model developed to detect the aesthetic topic of each Flickr user. By discovering relationship between image users with aesthetic topic different communities can be formed. Image can be enhanced from the visual features and using operations like retargeting and recomposition.

Acknowledgements We would like to thank our reviewer for giving positive comment for which our paper came to this standard. We would also like to thank the publishing house for accepting our paper. We thank our college authority for providing infrastructure to complete this work.

References

1. Mudigonda, Shanmukha Priya, and Koustubha Priya Mudigonda. "Applications of Image Enhancement Techniques—An Overview," in MIT International Journal of Computer

Science and Information Technology, Vol. 5, No. 1, January 2015, pp. 17–21. ISSN 2230-762.

2. Hong, Richang, Luming Zhang, and Dacheng Tao. "Unified photo enhancement by discovering aesthetic communities from Flickr," IEEE Transactions on Image Processing 25.3 (2016): 1124–1135. Digital Object Identifier 10.1109/TIP.2016.2514499.

3. S. Bhattacharya, R. Sukthankar, and M. Shah, "A framework for photoquality assessment and enhancement based on visual aesthetics," in Proc. 18th ACM Int. Conf. Multimedia, 2013, pp. 271–280. DOI:10.1145/1873951.1873990.

4. L. Zhang, Y. Gao, R. Zimmermann, Q. Tian, and X. Li, "Fusion of multichannel local and global structural cues for photo aesthetics evaluation," IEEE Trans. Image Process., vol. 23, no. 3, pp. 1419–1429, Mar. 2014. Digital Object Identifier 10.1109/TIP.2014.2303650.

5. F.-L. Zhang, M. Wang, and S.-M. Hu, "Aesthetic image enhancement by dependence-aware object recomposition," IEEE Trans. Multimedia, vol. 15, no. 7, pp. 1480–1490, Nov. 2013.

6. Y. Wang et al., "Where2Stand: A human position recommendation system for Souvenir photography," ACM Trans. Intell. Syst. Technol., vol. 7, no. 1, 2015, Art. ID 9. DOI:http://dx.doi.org/10.1145/2770879.

7. Dhar, S., Ordonez, V., & Berg, T. L. (2011). High level describable attributes for predicting aesthetics and interestingness. In *2011 IEEE Conference on Computer Vision and Pattern Recognition, CVPR 2011* (pp. 1657–1664). [5995467] DOI:10.1109/CVPR.2011.5995467.

8. Luo, X. Wang, and X. Tang, "Content-based photo quality assessment," in Proc. IEEE Int. Conf. Comput. Vis., Nov. 2011, pp. 2206–2213. DOI: 978-1-4577-1102-2/11/$26.00.

9. Islam, Md. Baharul, Wong Lai-Kuan, and Wong Chee-Onn. "A survey of aesthetics-driven image recomposition." Multimedia Tools and Applications (2016): 1–26. DOI:10.1007/s11042-016-3561-5.

10. Zhang, Luming, et al. "Retargeting semantically-rich photos." IEEE Transactions on Multimedia 17.9 (2015): 1538–1549. DOI:10.1109/TMM.2015.2451954.

11. S. Gregory, "A fast algorithm to find overlapping communities in networks," in Machine Learning and Knowledge Discovery in Databases, vol. 5211. Berlin, Germany: Springer, 2008, pp. 408–423. DOI:10.1007/978-3-540-87479-9_45.

12. Y. Zhang, J. Wang, Y. Wang, and L. Zhou, "Parallel community detection on large networks with propinquity dynamics," in Proc. 15th ACM SIGKDD, 2009, pp. 997–1006. DOI 10.1145/1557019.1557127.

13. S. Papadimitriou, J. Sun, C. Faloutsos, and P. S. Yu, "Hierarchical, parameter-free community discovery," in Machine Learning and Knowledge Discovery in Databases, vol. 5212. Berlin, Germany: Springer, 2008, pp. 170–187. DOI 10.1007/978-3-540-87481-2_12.

14. T. Yang, R. Jin, Y. Chi, and S. Zhu, "Combining link and content for community detection: A discriminative approach," in Proc. 15th ACM SIGKDD, 2009, pp. 927–936. DOI: 10.1145/1557019.1557120.

15. Y. Gu, X. Qian, Q. Li, M. Wang, R. Hong, and Q. Tian, "Image annotation by latent community detection and multikernel learning," IEEE Trans. Image Process., vol. 24, no. 11, pp. 3450–3463 Nov.2015. DOI:10.1109/TIP.2015.2443501.

16. Zhou, E. Manavoglu, J. Li, C. L. Giles, and H. Zha, "Probabilistic models for discovering e-communities," in Proc. 15th Int. Conf. World Wide Web, 2006, pp. 173–182. DOI: 10.1145/1135777.1135807.

17. M. Blei, A. Y. Ng, and M. I. Jordan, "Latent Dirichlet allocation," J. Mach. Learn. Res., vol. 3, pp. 993–1022, Mar. 2003.

18. G. Costa and R. Ortale, "A Bayesian hierarchical approach for exploratory analysis of communities and roles in social networks," in Proc. IEEE/ACM Int. Conf. ASONAM, Aug. 2012, pp. 194–201. DOI:10.1109/ASONAM.2012.42.

19. E. Yao et al., "Probabilistic text modeling with orthogonalized topics," in Proc. 37th Int. ACM SIGIR Conf. Res. Develop. Inf. Retr., 2014, pp. 907–910. DOI:10.1145/2600428.2609471.

20. M. Frank, A. P. Streich, D. Basin, and J. M. Buhmann, "Multiassignment clustering for Boolean data," J. Mach. Learn. Res., vol. 13, no. 1, pp. 459–489, 2012.

21. Lancichinetti, S. Fortunato, and F. Radicchi, "Benchmark graph for testing community detection algorithms," Phys. Rev. E, vol. 78, p. 046110, Oct. 2008. DOI:https://doi.org/10.1103/PhysRevE.78.046110.

22. H. Zhang, R. Edwards, and L. Parker, "Regularized probabilistic latent semantic analysis with continuous observations," in Proc. 11th ICMLA, Dec. 2012, pp. 560–563. DOI:10.1109/ICMLA.2012.102.

23. The MIRFLICKR Retrieval Evaluation. Available: http://press.liacs.nl/mirflickr/#sec_introduction.

24. Photo.net Available: http://photo.net/ AVA: A Large-Scale Database for Aesthetic Visual Analysis Available: http://www.lucamarchesotti.com/.

25. N. Murray, L. Marchesotti, and F. Perronnin, "AVA: A large-scale database for aesthetic visual analysis," in Proc. CVPR, 2012, pp. 2408–2415. DOI:10.1109/CVPR.2012.6247954.

26. R. Balasubramanyan and W. W. Cohen, "Block-LDA: Jointly modeling entity-annotated text and entity-entity links," in Proc. SDM, 2011, pp. 450–461. DOI:http://dx.doi.org/10.1137/1.9781611972818.39.

27. Mewes, Hans-Werner, et al. "MIPS: analysis and annotation of proteins from whole genomes." Nucleic acids research 32. suppl 1 (2004): D41–D44. DOI:https://doi.org/10.1093/nar/gkh092.

28. E. M. Airoldi, D. M. Blei, S. E. Fienberg, and E. P. Xing, "Mixed membership stochastic blockmodels," J. Mach. Learn. Res., vol. 9, pp. 1823–1856, 2008.

29. J. Yang and J. Leskovec, "Community-affiliation graph model for overlapping network community detection," in Proc. IEEE 12th Int. Conf. Data Mining, Dec. 2012, pp. 1170–1175. DOI:10.1109/ICDM.2012.139.

Parameter Optimization for Medical Image Modality Classification

B. Sundarambal and K. Bommanna Raja

Abstract With increasing usage of medical images for the diagnosis in healthcare sector, the size of the image repository grows enormously. Image retrieval becomes a critical task with increasing size of repository. To address this problem, this article deals with the design of an automated system to predict the modality of medical image. This work then can be incorporated into image retrieval system with a large collection of medical images. Six modalities such as CT (computed tomography), XR (X-ray), PET (positron emission tomography), US (ultrasound), MR (magnetic resonance imaging) and PX (photograph) are considered in this experiment. Dense SIFT (scale-invariant feature transform) features, sampled at regular intervals, are extracted from the images, represented with bag-of-words histogram and classified by SVM (support vector machine). This paper explores three directions to improve the classification accuracy—usage of increasing number of training images, preferring spatial histogram rather than simple histogram and extending kernel map from linear to hellinger in SVM classifier. The obtained results are compared with existing complicated approaches and proved that better classification results are obtained with proposed simple approaches.

Keywords Medical Image Modality Classification · SIFT · BoVW · SVM

B. Sundarambal (✉)
Department of Computer Science and Engineering, AKT Memorial College
of Engineering and Technology, Kallakurichi, Tamil Nadu, India
e-mail: sundariprabu@gmail.com

K. Bommanna Raja
Department of Electronics and Communication Engineering, KPR Institute
of Engineering and Technology, Coimbatore, Tamil Nadu, India
e-mail: dr.k.bommannaraja@gmail.com

© Springer Nature Singapore Pte Ltd. 2018
S.S. Dash et al. (eds.), *International Conference on Intelligent Computing
and Applications*, Advances in Intelligent Systems and Computing 632,
https://doi.org/10.1007/978-981-10-5520-1_54

1   Introduction

Over the last three decades, healthcare sector is ruled by medical imaging where the physician mainly depends on various medical imaging modalities to diagnose and treat the diseases. Based on the type of diseases, different modalities are preferred for different organs. For example, X-rays are suitable for diagnosing lung disease and bone fractures, CT for tumour detection in head and abdominal disease, ultrasound during pregnancy, etc. With the day-to-day invention of latest medical equipment, medical images are also acquired at an increasing rate. These bulk volumes of images are stored in a centralized repository and accessed frequently for diagnosis and study purposes. Retrieving image from such a large repository poses a difficult task, and hence, an effective and efficient computerized system is required to retrieve such images. The survey shows that modality is used as one of the filters to reduce the search space [1]. Hence, an automated system to identify and classify the modality of medical images becomes an emerging area of the research.

To promote the research in this sector, ImageMedCLEF—a forum, organized contest for modality classification task from the year 2010 [1]. Modality classification is the important task of ImageMedCLEF till 2013. Evaluation of the contest is based on the percentage of correctly classified images. Many research groups registered for the contest and submitted their promising results. The latest task proposed by ImageMedCLEF in 2016 such as compound figure separation also requires the results of modality classification task. Hence, the research is continued in modality classification task and explored in multidimensions to outperform the classification results obtained so far.

This paper is organized as follows: literature review is given in Sect. 2. The proposed work is discussed in Sect. 3, and experimental results are reported in Sect. 4. At the end, Sect. 5 concludes the paper and explores the way to extend the work in future.

2   Related Work

Several research groups performed experiments on modality classification tasks and submitted their results in ImageCLEFmed2013 competition [2]. IBM, the research group stood first in modality classification task, adopted sophisticated multimodal fusion techniques and obtained 81.68% classification accuracy [3]. FCSE group ranked second in modality classification task extracted densely sampled SIFT features and employed spatial pyramid [4]. The medGIFT group ranked fourth in modality classification performed feature fusion from many features descriptors and to name a few, colour and edge directivity descriptor (CEDD), bag of visual words (BoVW) using SIFT, fuzzy colour and texture histogram (FCTH) [5].

The fifth position bagged by Image and Text Integration (ITI) group adopted flat and hierarchical classification strategies with SVM [6]. The best classification accuracy in the modality classification task obtained by DEMIR research team was 64.60%, and they preferred mixed approach by combining CEDD, FCTH and colour layout descriptor (CLD) features along with textual information [7]. MIILab (Medical Image Information Laboratory) participated in ImageCLEFmed2013 modality classification task and submitted the results [8]. They extracted features using the fast filtering techniques and SURFContext with classical BoF (bags of features) approach. The overall classification accuracy is around 65%. Dimitrovski I et al. [9] evaluated classification results from different combinations of visual and textual descriptors and obtained 87.10% accuracy which is the best classification result reported so far. In [10], authors extracted different visual and textual features and employed a strategy called joint kernel equal contribution (JKEC) to give equal weightage to all the features used. Kalpathy-Cramer et al. developed neural network-based, hierarchical classifier and achieved greater than 95% classification accuracy with greyscale image [11].

Csurka et al. [12] used Fisher vector representation of the images from visual aspect and the image captions from textual aspect for classification. The authors in [13] used BoVW, bag of colours (BoC), CEDD, FCTH and fuzzy colour histogram (FCH) descriptors to represent the image. Thus, the detailed survey proves that SIFT features are used in almost many modality classification tasks giving the best classification results. Hence, experiments are conducted to optimize some parameters in the existing SIFT feature extraction and also in the classification methodologies to improve the overall accuracy still better.

3 Proposed Methodology

On seeing the frequent application of SIFT-based BoVW (bag of visual words) representation of images particularly for the classification tasks, we intended to extract dense SIFT features from the image and converted to BoVW histogram but along with some modifications in the parameters normally employed. The changes are introduced based on the contributions from the three works as follows:

- Akata et al. [14] suggested different ways to improve the classification accuracy with large-scale images. Among them, one suggestion is to have good number of training images.
- Vedaldi and Zisserman [15], in their assignment on Image Classification Practical, 2011, suggested to include spatial histogram to improve the classification accuracy.
- Swathi Rao [16] proved that hellinger kernel outperformed linear kernel.

The proposed method combined the advantages of the above-stated three approaches and tested for experiments. The proposed system consists of extraction

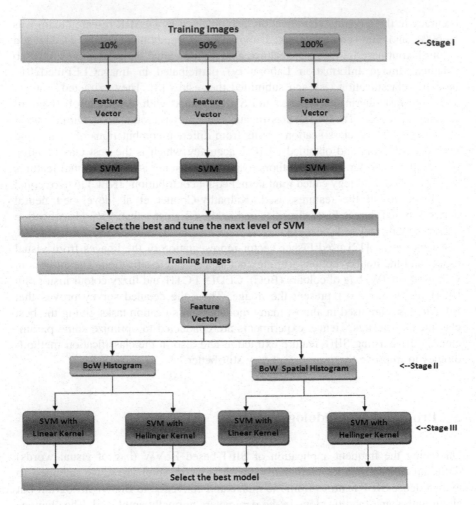

Fig. 1 Architecture of the proposed system for modality classification

of densely sampled SIFT descriptors of reasonable size of training images, inclusion of spatial histogram from bag-of-words representation of images and comparison of classification results with SVM classifier using linear and Hellinger kernel mapping. The architecture of the proposed system is illustrated in Fig. 1. The classifier performance is evaluated at three stages, stage I with the number of images, stage II with inclusion of spatial histogram and stage III with the comparison of linear and Hellinger kernels.

The various stages of the proposed system are discussed briefly in subsequent subsections.

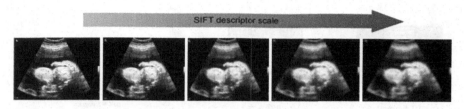

Fig. 2 Scale space of one image from training set. With increasing scale, it is observed that resolution of image decreases

3.1 Dense SIFT Feature Extraction

Bag of visual words formed with SIFT features is used traditionally in many classification problems. SIFT keypoints can be extracted in three modes, key point detection, dense sampling and random sampling. SIFT keypoint represents a circle with its centre depicting x and y coordinates, the radius of the circle depicting scale and the angle depicting its orientation.

To obtain keypoints at multiple scales, Gaussian scale space is constructed. The scale space is a collection of images obtained by smoothing the input images progressively. Such a scale space is shown in Fig. 2. Smoothing the image results in reducing the resolution of images.

The keypoints are then extracted at four different scales (sigma = 0.6, 1, 1.3 and 1.6 for the Gaussian filter) and sampled densely with an interval distance of 4 pixels in an image grid. For each keypoint, 128-dimensional descriptor is obtained. To reduce the large dimension of descriptors, the obtained descriptors are then mapped to a codebook containing say 1000 codewords. Then, histogram containing the proportion of the descriptors to that specific codeword is constructed.

3.2 Bag of Visual Words

The origin of BoVW is based on the regular text analysis. Normally any text document is interpreted as the collection of words and to analyse the document; we identify the frequency of occurrences of those words. Similarly, the image can also be interpreted as the collection of visual words and to analyse the image, we identify the frequency of occurrences of those visual words.

Among three modes of SIFT feature extraction, dense sampling approach provides more keypoints as the features are extracted from the whole grid image with an interval of normally 2–4 pixels. Hence, much feature will be obtained with this approach when compared with the other two modes, keypoint detection and random sampling. Thus, to reduce the feature descriptor size appropriately, feature quantization is done by simply running k-means on the obtained descriptors. The centroids of the k-means represent the visual words of the image.

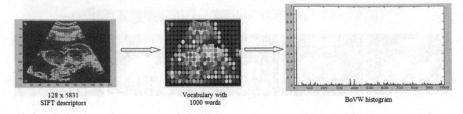

128 x 5831
SIFT descriptors

Vocabulary with
1000 words

BoVW histogram

Fig. 3 Formation of BoVW histogram

The various steps in forming the visual words of an image are as follows:

1. Dense SIFT features are extracted from the training images.
2. Each feature has its descriptors in 128 dimensions. k-means with say 1000 centroids is run on the obtained SIFT descriptors to end up with 1000 words.
3. To represent a particular image using the visual vocabulary, again dense features are extracted from it and assigned to the visual vocabulary. The assignment is based on calculating the Euclidean distance (L2 distance) between a word and a given descriptor.
4. Finally, a histogram of visual words is built to represent that particular image.

The procedure for representing BoVW histogram for one image is visually summarized in Fig. 3.

3.3 Bag of Visual Words with Spatial Information

Another approach to improve classification accuracy is incorporating spatial information on the existing plain BoVW histogram containing 1000 words. To achieve this, the given image is divided into 2×2 subregions and the histogram is computed for each subregion. Thus, 4 histograms with 1000 words are obtained and they are then stacked to form an array of single dimension of size 4000 (1000×4). Figure 4 shows the partition of an image into 2×2 subregions.

3.4 SVM Classifier with Linear and Hellinger Kernel

The support vector machine (SVM) introduced by Boser, Guyon and Vapnik in 1992 is used as the classifier along with kernel trick to maximize the margin of hyperplanes [17]. This algorithm just plots the feature in feature space, and using hyperplane, it identifies the boundary of each class. Kernel trick is employed to identify the best hyperplane segregating the different classes. Two SVM classifiers one with linear and other with Hellinger kernel are used for classification. Square root of the histogram is considered for implementing Hellinger kernel.

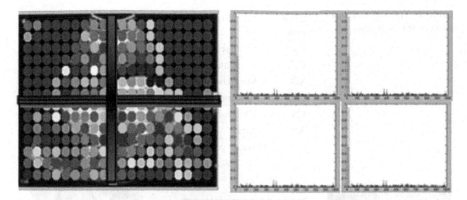

Fig. 4 BoVW with 2 × 2 spatial tiling on left and its histogram for each tile on right

To classify the images of multiple classes, two flavours of SVM, one-versus-one and one-versus-all approaches, can be used. We preferred one-versus-all approach in which a classifier is built for each modality/class. The examples pertaining to that class are assigned positive labels and the remaining examples are assigned as negative labels. SVM with linear and Hellinger kernel mapping is used.

The one-vs-all SVM classifier classifies the feature vector as positive or negative using the Eq. 3.1.

$$
\begin{aligned}
w^t x + b \geq 0 \text{ for positive classification} \\
w^t x + b < 0 \text{ for negative classification}
\end{aligned}
\tag{3.1}
$$

where x, w and b are the feature vector to be classified, weight vector and bias, respectively. The values of w and b are determined during training process and the equations are then used to obtain decision hyperplane which classify the images as positive or negative. The crucial aspect is to find a set of weight and bias such that the margin is maximized. *Kernel tricks* are employed to obtain the best margin. The kernel makes the data linearly separable.

4 Results and Discussion

Data set

The experiments are carried out on 780 images of six different modalities. The training set consists of 50% of images while the testing set forms 50% of images in the data set. Table 1 contains the detailed split up of the images into training and testing set.

Table 1 Training and testing samples for each class

Class code	Modality/class	Train	Test
PX	Photograph	65	65
CT	Computed tomography	65	65
MR	Magnetic resonance imaging	65	65
PET	Positron emission tomography	65	65
US	Ultrasound	65	65
XR	X-ray	65	65
	Total	390	390

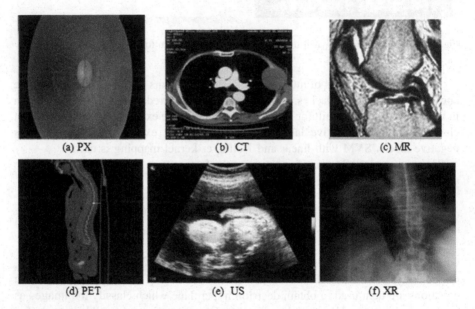

(a) PX   (b) CT   (c) MR

(d) PET   (e) US   (f) XR

Fig. 5 Sample image from each modality/class in training set

The images are collected from open-i biomedical image search engine filtered by image modality and PubMed collections [18]. Examples of images from the training data set are shown in Fig. 5.

The images obtained are of different size, and it is resized not to exceed 480 pixels in the row, and the column is adjusted automatically such that image aspect ratio is preserved. In all experiments, densely sampled SIFT features on the whole image grid with an interval of 4 pixels are extracted at 4 scales with sigma of 0.6, 1, 1.3 and 1.6. k-means with 1000 centroids is then applied on the extracted features.

As a next level, the image is partitioned into 2×2 subregions and again the histogram is computed separately for each subregion.

The visual words of training image from each modality are formed and their histograms are constructed as shown in Fig. 6.

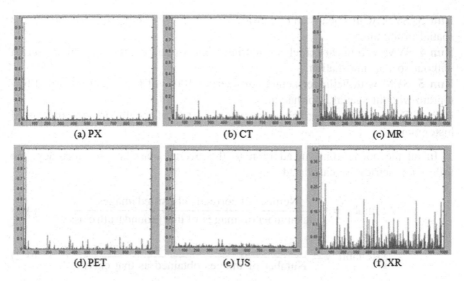

Fig. 6 Sample histogram from each modality/class in training set

This histogram is the signature of the image, and because of its uniqueness for each modality, the classifiers are trained with different histograms. One-vs-all SVM classifiers for all modalities are tested for all the test images with two variants of SVM classifiers—linear and hellinger kernels.

The proposed system is evaluated by identifying the overall classification accuracy. The overall accuracy of the system is the ratio of the number of correctly classified images to the number of all images. This is the commonly used evaluation strategy for any classification problem.

The results are tabulated as the confusion matrix for the test set and the main diagonal depicts the number of images correctly classified.

Evaluation with similar training and testing sets was performed for the following choices:

1. Varying number of training images.
2. BoVW histogram and BoVW spatial histogram.
3. SVM with linear and Hellinger kernel.

In this section, the results of the proposed system for automatic classification of medical imaging modalities are reported. Six runs are performed for modality classification task. The classification result of all runs for each modality classifier is shown as confusion matrix.

Run 1: SVM with linear kernel considering 10% of training images and 2×2 spatial histogram.
Run 2: SVM with linear kernel considering 50% of training images and 2×2 spatial histogram.

Run 3: SVM with linear kernel considering 100% of training images and 2×2 spatial histogram.

Run 4: SVM with linear kernel considering 100% of training images and histogram without spatial information.

Run 5: SVM with hellinger kernel considering 100% of training images and histogram without spatial information.

Run 6: SVM with hellinger kernel considering 100% of training images and spatial histogram.

In all the above runs, in addition to the overall classification accuracy, the following metrics are calculated:

$$\text{Accuracy} = \frac{\text{Number of correctly classified images}}{\text{Total number of images of that ground truth class}} \qquad (4.1)$$

$$\text{Reliability} = \frac{\text{Number of correctly classified images}}{\text{Number of images obtained as that class}} \qquad (4.2)$$

$$\text{Overall accuracy} = \frac{\text{Total number of correctly classified images}}{\text{Total number of test images}} \qquad (4.3)$$

The *kappa* is another metric that is also used to evaluate the classifiers. It compares observed accuracy with expected accuracy from a random classifier. It is calculated using the formula

$$\text{Kappa} = \frac{\text{observed accuracy} - \text{expected accuracy}}{1 - \text{expected accuracy}} \qquad (4.4)$$

The classifier for each modality is trained with 65 images of each modality. The entire test image set consisting of 65 images of each modality is given to all the classifiers to classify the corresponding modality images. The confusion matrix for the six runs is tabulated as shown in Tables 2, 3, 4, 5, 6, 7. The various metrics are calculated to assess the performance of the classifier as given in Eqs. 4.1–4.4.

Table 2 Confusion matrix of Run 1—SVM with linear kernel, 10% of training images and 2×2 spatial histogram

Modality	XR	US	PX	PET	MR	CT	Producer accuracy (precision)
XR	**47**	1	2	3	8	4	72.3
US	1	**29**	13	10	6	6	44.6
PX	9	7	**38**	5	4	2	58.4
PET	7	14	8	**21**	10	5	32.3
MR	7	5	8	6	**27**	12	41.5
CT	0	3	3	12	15	**32**	49.2
User accuracy (recall)	66.1	49.1	52.7	36.8	38.5	52.4	

Table 3 Confusion matrix of Run 2—SVM with linear kernel, 50% of training images and 2 × 2 spatial histogram

Modality	XR	US	PX	PET	MR	CT	Producer accuracy (precision)
XR	56	0	4	1	3	1	86.1
US	1	48	7	4	4	1	73.8
PX	3	3	45	5	0	9	69.2
PET	7	6	4	32	4	12	49.2
MR	2	8	9	7	29	10	44.6
CT	1	2	3	18	11	30	46.1
User accuracy (recall)	80	71.6	62.5	47.7	56.8	47.6	

Table 4 Confusion matrix of Run 3—SVM with linear kernel, 100% of training images and 2 × 2 spatial histogram

Modality	XR	US	PX	PET	MR	CT	Producer accuracy (precision)
XR	55	0	5	0	4	1	84.6
US	1	53	5	2	3	1	81.5
PX	0	1	48	4	3	9	73.8
PET	8	4	1	40	2	10	61.5
MR	1	7	6	6	38	7	58.4
CT	0	2	3	12	13	35	53.8
User accuracy (recall)	84.6	79.1	70.5	62.5	60.3	55.5	

Table 5 Confusion matrix of Run 4—SVM with linear kernel, 100% of training images and simple histogram

Modality	XR	US	PX	PET	MR	CT	Producer accuracy (precision)
XR	56	0	4	0	3	2	86.1
US	1	52	4	4	3	1	80
PX	1	1	47	4	2	10	72.3
PET	2	0	3	48	2	10	73.8
MR	0	9	5	2	40	9	61.5
CT	0	0	8	9	15	33	50.7
User accuracy (recall)	93.3	83.8	66.1	77.6	61.5	50.7	

The overall classification accuracy and kappa for the six runs are tabulated in Table 8. According to Fliess, kappa > 0.75 is the best classifier, 0.40–0.75 is as fair as good and <0.40 is the worst classifier [19].

608 B. Sundarambal and K. Bommanna Raja

Table 6 Confusion matrix of Run 5—SVM with Hellinger kernel, 100% of training images and simple histogram

Modality	XR	US	PX	PET	MR	CT	Producer accuracy (precision)
XR	59	0	3	0	3	0	90.7
US	1	52	5	3	3	1	80
PX	2	3	49	3	3	5	75.3
PET	3	3	4	44	5	6	67.6
MR	0	8	4	4	38	11	58.4
CT	0	0	4	9	13	39	60
User accuracy (recall)	90.7	78.7	71	69.8	58.4	62.9	

Table 7 Confusion matrix of Run 6—SVM with Hellinger kernel, 100% of training images and spatial histogram

Modality	XR	US	PX	PET	MR	CT	Producer accuracy (precision)
XR	57	0	3	2	3	0	87.6
US	1	54	2	4	4	0	83.0
PX	3	4	48	3	3	4	73.8
PET	2	4	1	47	3	8	72.3
MR	1	5	7	6	40	6	61.5
CT	0	0	3	13	10	39	60
User accuracy (recall)	89.0	80.5	75	62.6	63.4	68.4	

Table 8 Overall classification accuracy and kappa

Run #	Overall classification accuracy (%)	Kappa
1	49.744	0.397
2	61.538	0.538
3	68.974	0.628
4	70.769	0.649
5	72.051	0.665
6	73.077	0.677

Table 8 shows that both classification accuracy and kappa keeps on increasing in the consecutive 6 runs. The 6th run, the combination of SVM with Hellinger kernel, spatial histogram and 100% training images gives the better classification accuracy of 73.077% and kappa of 0.677. Even though the classifier cannot be rated as the best, it is as fair as good, according to Fliess. The overall classification accuracy and kappa for the different runs are plotted and shown in Figs. 7 and 8, respectively.

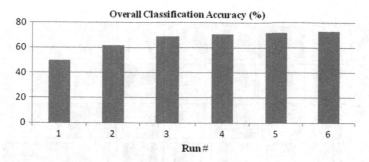

Fig. 7 Overall classification accuracy plot of six runs. Sixth run, the combination of SVM classifier with Hellinger kernel, spatial histogram and 100% training images, gives the best classification accuracy of 73.077%

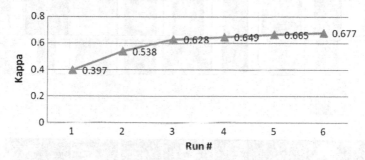

Fig. 8 Kappa plot of six runs. Sixth run, the combination of SVM classifier with Hellinger kernel, spatial histogram and 100% training images, gives the best value of 0.677. Kappa with the range of 0.40–0.75 is rated as good classifier

	Group Name	Ranking position	Accuracy
Table 9 Comparison of proposed work with existing research works submitted in ImageCLEF 2013 modality classification task	IBM	1st	80.79
	FCSE	2nd	77.14
	Proposed work	**3rd**	**73.077**
	MiiLab	4th	66.46
	medGIFT	5th	63.78
	ITI	6th	61.50
	CITI	7th	56.62
	IPL	8th	52.05

The comparison of the proposed work with the results submitted by the research groups in the conference organized by ImageCLEF 2013 for modality classification task is tabulated in Table 9.

The output of the best run for each class is shown in Fig. 9. It can be seen from the output that some images are misclassified in each class. XR and US classifiers perform much better compared with other modality classifier. The reason behind

(a)

(b)

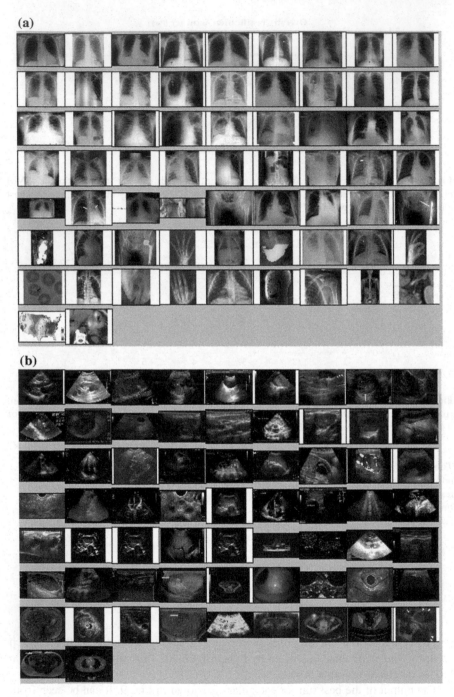

Fig. 9 **a** Output of XR classifier. **b** Output of US classifier. **c** Output of PET classifier. **d** Output of PX classifier. **e** Output of CT classifier. **f** Output of MR classifier

(c)

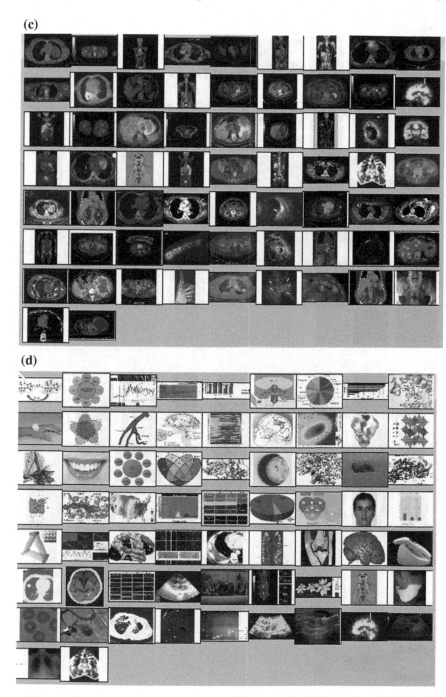

(d)

Fig. 9 (continued)

(e)

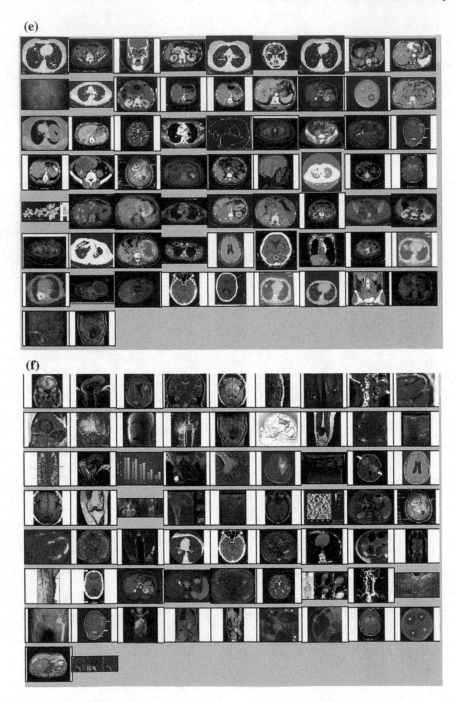

(f)

Fig. 9 (continued)

that is PET mostly comes in combination with CT which is misclassified as CT. As visual similarities among CT, MR and PET are confusing even for human, the system predicts many images from these groups in a wrong manner. Hence, further tuning of the parameters is still required to improve the classification accuracy still better. Perhaps if the training set is built strongly including similar type of images which are wrongly misclassified, the classification accuracy can be improved still better. But that approach also should not end up in overfitting. Hence, deep analysis of wrongly misclassified images should be taken into consideration and the changes in the parameters from multiple views can be performed to achieve the goal.

5 Conclusion

The experimental results are reported for the proposed system to classify the modalities of medical images. This work is mainly to integrate into medical image retrieval system where the medical images are retrieved based on its modality. Using a data set of 780 images, six approaches are evaluated and the approach combining densely sampled SIFT descriptors and bag-of-words spatial histogram along with Hellinger kernel mapping of SVM gives the best overall classification accuracy. The maximum overall classification accuracy obtained is 73.077%. In the experiments, we have shown that increasing training images, incorporating spatial histogram and extending linear to Hellinger kernel mapping of SVM produce good results. As an extension to existing work, we plan to tune other parameters in future to improve classification results.

References

1. http://www.imageclef.org/2010/medical (2010).
2. Alba Garcia Seco de Herrera, Jayashree Kalpathy-Cramer, Dina Demner-Fushman, Sameer Antani and Henning Muller, Overview of the ImageCLEF 2013 medical tasks, in: CLEF working notes 2013, Valencia, Spain, (2013).
3. Mani Abedini, Liangliang Cao, Noel Codella, Jonathan H. Connell, Rahil Garnavi, Amir Geva, Michele Merler, Quoc-Bao Nguyen, Sharathchandra U. Pankanti, John R. Smith, Xingzhi Sun, and Asaf Tzadok, IBM Research at ImageCLEF 2013 Medical Tasks, IBM Multimedia Analytics ImageCLEF (2013).
4. Ivan Kitanovski, Ivica Dimitrovski, and Suzana Loskovska, FCSE at Medical Tasks of ImageCLEF 2013, CLEF working notes (2013).
5. Alba G. Seco de Herrera, Dimitrios Markonis, Roger Schaer, Ivan Eggel, Henning Muller, The medGIFT Group in ImageCLEFmed 2013, CLEF working notes (2013).
6. Matthew S. Simpson, Daekeun You, Md. Mahmudur Rahman, Dina Demner-Fushman, Sameer Antani, and George Thoma, ITI's Participation in the 2013 Medical Track of ImageCLEF, CLEF working notes (2013).
7. Okan Ozturkmenoglu, Nefise Meltem Ceylan, Adil Alpkocak, DEMIR at ImageCLEFMed 2013: The Effects of Modality Classification to Information Retrieval, Procs. of ImageCLEFMed (2013).

8. Xin Zhou, Miaofei Han, Yanli Song, Qiang Li, Fast filtering techniques in medical image classification and retrieval, CLEF working notes (2013).
9. Ivica Dimitrovski, Dragi Kocev, Ivan Kitanovski, Suzana Loskovska, Saso Dzeroski, Improved medical image modality classification using a combination of visual and textual features, Computerized Medical Imaging and Graphics 39, pp 14–26, (2015).
10. Xian-Hua Han and Yen-Wei Chen, Biomedical Imaging Modality Classification Using Combined Visual Features and Textual Terms, International Journal of Biomedical Imaging, vol. 2011, Article ID 241396, 7 pages, doi:10.1155/2011/241396, (2011).
11. Jayashree Kalpathy-Cramer, William Hersh, Automatic Image Modality Based Classification and Annotation to improve Medical Image Retrieval, MEDINFO (2007).
12. Gabriela Csurka, Stéphane Clinchant and Guillaume Jacquet, XRCE's Participation at Medical Image Modality Classification and Ad-hoc Retrieval Tasks of ImageCLEF 2011, CLEF working notes (2011).
13. Jacinto Arias, Jesus Martinez-Gomez, Jose A. Gamez, Alba G. Seco de Herrara, Henning Muller, Medical images modality classification using discrete Bayesian Networks, Computer Vision and Image Understanding, Volume 151, Pages 61–71, October (2016).
14. Zeynep Akata, Florent Perronnin, Zaid Harchaoui, Cordelia Schmid, Good Practice in Large-Scale Learning for Image Classification. IEEE Transactions on Pattern Analysis and Machine Intelligence, Institute of Electrical and Electronics Engineers, 36 (3), pp. 507–520, (2014).
15. Andrea Vedaldi and Andrew Zisserman, Image Classification Practical, (2011).
16. Swathi Rao G., Effects of Image Retrieval from Image Database using Linear Kernel and Hellinger Kernel Mapping of SVM International Journal of Scientific & Engineering Research, Volume 4, Issue 5, May (2013).
17. B. E. Boser, I. M. Guyon, and V. N. Vapnik. A training algorithm for optimal margin classifiers. In D. Haussler, editor, 5th Annual ACM Workshop on COLT, pages 144–152, Pittsburgh, PA, ACM Press, (1992).
18. https://openi.nlm.nih.gov.
19. J. Fleiss, Statistical Methods for Rates and Proportions, 2nd ed. New York: Wiley, (1981).

Exploring the Architectural Entropy in Cloud Computing System

A.R. Manu, Shakil Akhtar, Vinod Kumar Agrawal, K.N. Bala Subramanya Murthy and V. Suma

Abstract Architecture entropy has an impact on the quality of the software and further on the level of customer satisfaction. Hence, every industry aims to reduce this architectural entropy such that it should not lead to degeneration and architecture decay. This might cause an increase in the architectural vulnerabilities at the root level resulting in the architecture erosion over time. This work aims to understand the architecture entropy and its impact. This is because architecture evolves over time due to various specification changes by the customers. The changes thus made may lead to increased architecture smells which in turn leads to architectural decay. This paper, hence, puts forth the rationale for some of the architectural decay in terms of entropy and has further suggested some steps to overcome the same thorough instances and case study. This work enables architects to understand the impact of entropy on value-added parameters such as consistency, reliability, integrity, and authenticity of the cloud computing architecture in particular.

A.R. Manu (✉)
CORI Lab, Department of ISE, PESIT, VTU, PES Institute of Technology,
100 Feet Ring Road BSK III rd Stage, Bangalore 560085, Karnataka, India
e-mail: manu.a.ravi@gmail.com; manu_ar@nitk.ac.in

S. Akhtar
GE Digital, Bangalore, Karnataka, India
e-mail: shakilsoz17ster@gmail.com; 2shakil.akhtar@gmail.com

V.K. Agrawal
CORI Lab, Department of Information Science and Engineering,
PES-University, Bangalore 560085, Karnataka, India
e-mail: vk.agrawal@pes.edu

K.N. Bala Subramanya Murthy
PES-University, BSK 3rd Stage, 100 Feet Ring Road, Bangalore, Karnataka, India
e-mail: vice.chancellor@pes.edu

V. Suma
Department of Information Science and Engineering, Research and Industry Incubation
Centre, Dayananda Sagar College of Engineering, Kumaraswamy layout,
Bangalore, Karnataka, India
e-mail: sumavdsce@gmail.com

© Springer Nature Singapore Pte Ltd. 2018
S.S. Dash et al. (eds.), *International Conference on Intelligent Computing
and Applications*, Advances in Intelligent Systems and Computing 632,
https://doi.org/10.1007/978-981-10-5520-1_55

Keywords Cloud computing system · Architectural degradation
Architectural debt · Architectural degradation · Architectural vulnerability
Architectural erosion · Architectural entropy · Architectural decay
Architectural smells · Architectural degeneracy · Architectural deprivation

1 Introduction

Each part or component of a system software or a software application gets
architecturally designed, developed, executes and validated in order to manage and
maintain it. Any architectural types, however, can be categorized to be either
dogmatic architecture or exploratory architecture and regulatory architecture.
Introducing the preamble design choices and resolution into the computing system's explanatory architecture which is not incorporated in, and/or covered by, or
else implied by the narrowed inflexible architecture leads to architectural drift while
architectural erosion happens due to changes in design choices of the computing
system's illustrative structural and architectural design choices that breach its
dogmatic and regulatory architecture [1].

Architecture decays, if the computing system does not meet the intended
stakeholder's business needs and if it is not documented as per industry benchmarks. Despite the architecture meeting the customer requirements, deviation from
its intended documentation may also lead to architecture decay causing the entropy.
Architecture entropy also increases with the clustered multilateral group of stakeholder needs to be fulfilled by the intended computing system transforms but not
with the architecture. As an outcome, it results in weakening and decreasing the
ability and available resiliency of computing systems software architecture to
congregate to business stakeholder requirements [2].

The quantity of entropy is also a gauge of the architectural disorder, or disambiguation or unpredictability, or uncertainty of randomness, or arbitrariness of a
computing system [3]. The notion of entropy gives profound insight into the way of
artless unplanned changes for various daily event occurrences. The estimation of
entropy affords an algebraic way to program the innate notion of which sequence is
not practical, although they would not violate the fundamental law of architectural
effort conservation—Convoy's law.

An irreversible architectural debt arises due to shortcuts used in crafting the
architecture. It is reversible to normal stage with the infinitesimal effort of work
required to change its route from progressive freezing to progressive attenuation of
entropy with compressed architectural refactoring. For reversible progression, the
computing system is stable in its balanced computing environment, while for
irrevocable processes it is not.

Architectural, structural smells arise due to code in the incorrect place, issues in
class, package, subsystem and layer relationships of the object-oriented and procedural packages/programs in computing systems, deficient decomposition, excessive
decomposition, obsolescence, and overgeneralization, etc. The shortcuts used to

devise the architecture and its design, using short paths, also bypassing enterprise architecure strategies and diluting the business and industry standards and guidelines, due to business pressure, and market competitiveness in terms of time period, working effort, and cost for implementing and executing auxiliary transform boost—efficiency yields but shrink the Quality-of-Service. It has turned out to be tougher to envisage the consequence of additional changes in the architectural design using shortcuts and diluting standards results in consuming more price, time schedule, value, and further amendments usually cause the as-is architecture to digress further from the as-anticipated architecture—worsening the condition of the architecture performance and security implementation leads to architecture decay and finally causing the architecture erosion. Figure 1 indicates the complexity involved in various types of architecture in the computing systems. From the figure, it can be inferred that when changes have to be included in such complex designs, it may lead to any of the architectural issues addressed above (Fig. 2).

Since entropy is an undesired feature, reduction or elimination of entropy becomes a necessary action. Elimination of entropy is commonly known as entropy cutback. This is yet again measured and quantified in the preparatory sketch,

Fig. 1 Hadoop architecture.
Source [4]

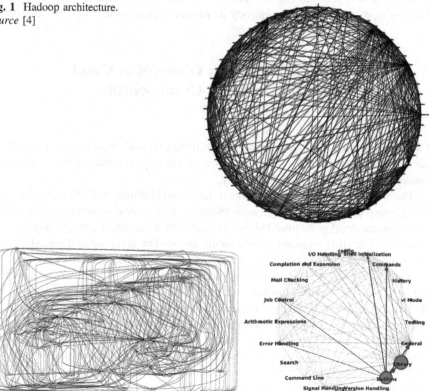

Fig. 2 Hadoop complete architecture and top-level architecture of Bash. *Source* [4]

reserved for tidying up the technical debt acquired by past progressive effort put forth during software development life cycle by using shortcuts.

Entropy measurement is the process which identifies and provides opportunity to reuse or refactor the architectural design and code using the post hoc reviews to clean up technical debt. It is used to measure the holes in upfront design and architecture to address the debt issues and vulnerabilities.

In IT organizations specifically in computing system architecture, the entropy disorder gains are due to various reasons such as computing system changes in terms of making alterations, due to business market changes by using the shortcuts, because of the mistakes in the architecture and so on. The architectural entropy in a closed computing system like cloud computing systems results in the loss of a data set or information message in transmission at rest and local storage which is due to disorder and randomness in the computing system. There is a penchant affinity for all material matters/substances in the space of computation to progressively move toward a state of inert uniformity along with the accrued uncertainty and complexity of the computing system. Hence, universal architecture tends and moves toward to the disordered state than another state. It suffices that it is inevitable to visualize the robust worsening (descent) of computing system heading to a state of disorder/randomness leading to architecture decay. Subsequent section explains various types of architectural entropy as observed in a cloud-based environment.

2 Architectural Entropy Types Observed in Cloud Computing Architecture and Cloud Security Architecture

Cloud architecture (including security architecture) can at times be experienced like an unacknowledged task. Various types of entropy are budgetary, technology, transitional, and time schedule entropy.

Developing the as-is or as-intended target architecture with cloud stakeholder organization approval and budget is problematic. It can consume several weeks of time to design and develop the intended architecture with organization consent, finances, and resources. The crisis increases rapidly as the architecture is agreed and signed off. The irked ink has desiccated and gets degenerated with time due to market pressure and technological changes. It starts to degrade in the course of a process of architectural entropy. Entropy is termed from the second law of thermodynamics. It means that the designed things are inclined to deteriorate over time. In the box of thermodynamics, the value of accessible amount of energy required to execute the related work and effort needed to implement architectural design changes, due to the evolving customer business requirements. On the other hand, this course of deprivation degeneracy state cannot be inverted or reversed to devoid of paying out more energy.

In the security architectural framework, we employ entropy to articulate the measured slowly degenerated architectural degradation under the market and business pressure of the identified constriants on the designed security architecture over time. The architecture design location is exceptionally defined using the uncertainty principle,

$$\Delta A_m \Delta xy \geq \hbar/2, \tag{1}$$

where ΔA_m is the insecurity and vulnerabilities in the existing architectural disorder measured at a particular momentum of architectural design implementation and Δxy is the uncertainty in architectural design in xy direction. This entails that the momentum of a highly condensed degenerated architecture is very uncertain, since the architecture design is located in cramped space. As a result, to reduce the entropy of architecture in a very small space, terrific working effort is required to manage its architecture degeneracy momentum. The objective of as-intended architecture on its possession is futile. Previously defined, one should spend time, cost, resources, and effort spinning it into realism.

By reworking on the design of the architecture arises, possibilities of architecture degradation may occur which results in the rise of the entropy. This happens due to compromising owing to cost pressures. Time schedule shortage results with increasing entropy with quality degradation of as-intended target architecture. Achieving the final end state in the purest form that outweighs the benefit of the architecture represents the golden standard solution to vendor landscape with potential technological solutions, helping to achieve the intended goal to devise the vulnerability free as intended architecture with security plugin at root level of the design. Architectural design decision compromising leads to architecture entropy rise due to judgemental calls. The technology and architecture must be tested with due diligence to pan out the solution as expected to eliminate the vulnerability rooted in architecture due to architectural dimension of technical debt.

However, with the business change as a concern, the conversion from 'A' to 'B' is viewed as transformations of the architecture to require changes in the mindset of the people, process, and technology which is already in place performing to accomplish definite outcomes. People's political and economical influence also has an impact on architecture transition with respect to integrating the existing capability and idiosyncrasies to architecture results in more entropy.

Due to the business changes, coercion and cyber terrorization becomes more advanced with the chance in technology support for new systems added to the computing ecosystem due to innovation and also due to regulatory pressures leading to architecture entropy. With upfront capital investment in architectural design due to evolving changes using shortcuts and dilution of the standards results in more time consuming, increase in cost and more effort investment, by the security architect to invest and revalidate critical elements of security architecture. He/She needs to make sure the security relevant aims, architectural design principles, guidance, and values are in place with effective vigilance and management control systems in place to improve, make secure robust and rendezvous sculpt between deliverance, process actions, and architecture design. Security architect needs to factor it in a proper way.

3   Current Literature Survey and Related Work

Architectural smells arise either by ignorance or knowingly or by accidentally using the shortcuts while making the design decisions. A variety of automation tools, techniques, related processes, and industry-benchmarked practices established a way to find and catalog the smells and vulnerabilities cropping up at the granularity of design and code [5]. Authors of [4, 6] have provided research directions for architecture smells cataloging [7]. Inventing some major automation tools to support for identifying and refactoring the architecture smells and its vulnerability, for example, SonarQube and Sonargraph, are existing in current literature [8, 9]. Authors of [10] have put forth a financial plan of architecture refactoring enumerating architecture smells from technical debt and vulnerabilities. The new dimension of research considering the technical debt and its relationship with software vulnerabilities is found in [11]. The same is a good basis for the research on the architecture vulnerabilities leading to security risk.

Various causes for architecture entropy are further provided in various researchers work. Authors of [12] discuss the system architecture entropy with its complexity using an illustration of architecture entropy, and they draw the relationship between the architecture entropy and system complexity for legacy systems. This part of the research therefore puts forth the comprehension of the architecture entropy using the various real-time instances. Authors of [1, 7, 13, 14] have listed commonly occurring architecture smells catalogs which include elementary block overload concern, brick usage of overload, block-level cyclic dependency, ambiguous and unused interfaces, duplicate and scattered functionalities, component and connectors envy's, chaining connectors, and irrelevant superfluous adjoining connectors are some of the commonlly occuring archiectural smells as presented by the authors of [1, 7, 13]. The authors of [15] establish the relationship of architectural vulnerabilities with the technical debt. The authors of [16] provide smells catalog, whereas the authors of [17–22] provide various factors causing architectural decay and degradation very effectively. Further, it can be found the authors of [23–27] present interrelation between technical debt's architectural roots and vulnerabilities. The authors of [28–33] propose architecture refactoring as a solution of architecture decay and architecture degradation factors in real practical instances. The work [33–38] discusses our work on multilateral security architecture and its application to the Docker container to resolve test debt the another dimension of the technical debt.

However, authors of [4] provide structure smells of architecture in addition to their associated issues in class, related packages, computing subsystem components, subpackages, components, and subcomponents, methods' superclasses, and its layered relationships [4]. Authors of [39] have indicated the modes through which architecture violations can be talked with appropriate solutions to improve the structure of dependency cycles [39].

Hence, it is clear that architecture entropy has a scope for continual improvement and hence needs research to progress in the area. Subsequent section thus illustrates few real-time instances where architecture evolution is discussed. Further, the evolution is mapped to a case study taken from a leading software industry.

3.1 Real-Time Instances Illustrating Architecture Evolution

Current computing software systems such as cloud computing are with varied coding standards having in-depth and unpredictable specifications from various stakeholders. As an instance, the MAC-OS code, Microsoft OS, Windows, and other Linux OS flavors have evolved to several millions of LOC since the cloud era. Architecture evolution styles and patterns are considered in this section as an example from micro-services and container service patterns [40, 41]. In addition to evolved architectural patterns and styles, cloud supporting features such as scalability and elasticity have an impact in terms of schedule and computing size which leads to architecture prone to vulnerabilities. This in turn affects structural, functional, and behavioral performances. Hence, refactoring the architecture becomes a need.

Another instance where refactoring is required can be observed when architecture evolved from the Microsoft OS, from Windows 2003 to Windows-NT, and from Windows XP version to the Microsoft OS Windows Vista. The failure of

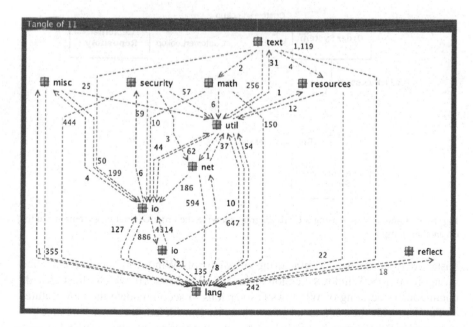

Fig. 3 Architecture disorder in Java language. *Source* [4]

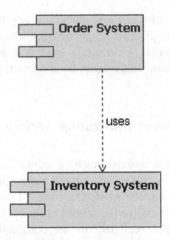

Fig. 4 Plain component: UML diagram showing the inventory system's order predicting generic dependency

Fig. 5 Substitute UML
diagram components with
required interface symbols

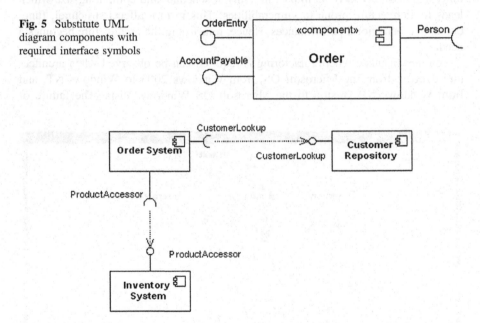

Fig. 6 Component drawn using a UML diagram showing the order computing system depending on another component

Vista led to complete OS refactoring of Microsoft Windows 7 and recent refactoring from MS Windows version 8 to MS Windows 10 which illustrates the continuous refactoring of Windows OS versions to accommodate the new features.

Further, with the code as an instance when considered, the Java language features and its development carried with bulk refactoring by developing language

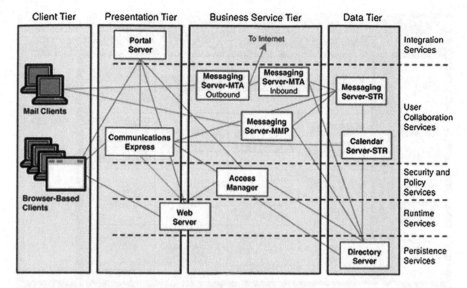

Fig. 7 Logical architecture diagram for typical enterprise communications setup. *Source* Oracle Corp

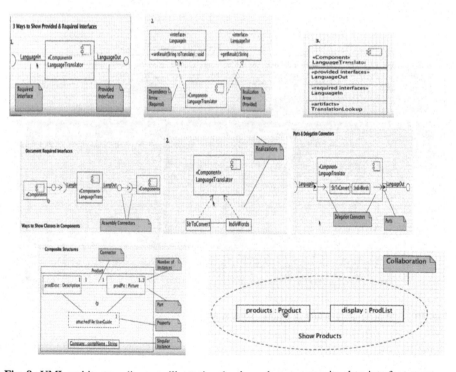

Fig. 8 UML architecture diagrams illustrating the dependency composite class interface structure and composite structures

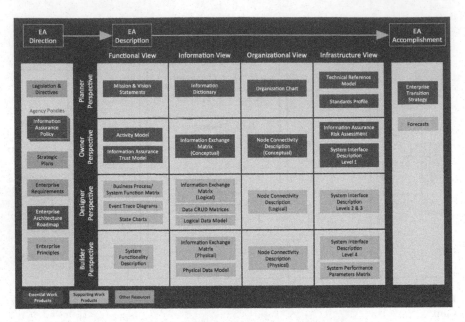

Fig. 9 Complex enterprise implementation architecture in cloud setup [42]

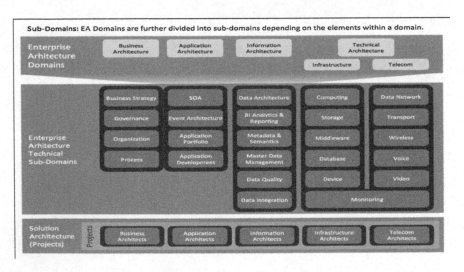

Fig. 10 Enterprise architecture domains diagram [42]

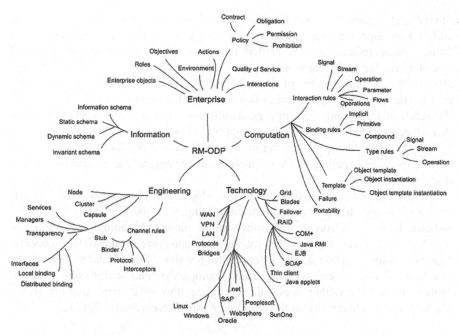

Fig. 11 Scale of complexity of the architecture viewing the communication dependency of RM-ODP [43] *Source* [4]

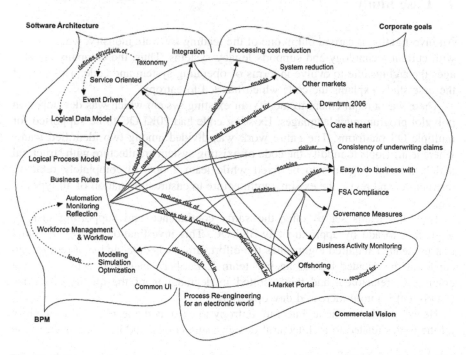

Fig. 12 Patterns and view of complex architectural consideration in the cloud computing ecosystem *Source* [4]

library and compiling features added to the modern virtualization features in addition to aspect-oriented programming language features support. Since the 1990s, major refactoring was carried out by evolving libraries with involved complexities due to the improvement with versions, releases of the Java, JDK, and JRE to the current version of Java 9 evolutions. Also, another illustration is refactoring the C language features to object C and the innovation of Go language is the result of refactoring the language features to support object-oriented features and aspect-oriented language features support. Figure 3 shows the complex tangle in Java language library illustrating the complex over dependency cycles between various packages leading to smell, thereby increasing in the architecture decay diagram (Source [4]).

Some more instances are depicted below which shows how architecture entropy increases in complexity with modern computing systems. Figures 4, 5, 6 and 7 indicate increase in complexity from simple module to complex modules as architecture evolves due to the market demands and stakeholders concern. However, with complete design of computing system, Figs. 8 and 9 indicate the structure of the system with increased complexity and architecture entropy. Figures 10 and 11 further depict the complete flow of system structure in the Hadoop and its complexity in the cloud ecosystem respectively (Fig. 12).

4 Case Study

An investigation is carried out in one of the leading software industry, which works with current technology and supports legacy systems. Thus, the company encourages the architecture to evolve in terms of operating system and code. This part of the case study explains the code which has led to entropy.

There was a requirement to modify an existing system which was developed in polyglot programming languages. Existing code had 20KLOC to be supported for multiple OS platforms. The entire work was divided among two identified teams where team 1 opts to modify the code of existing system architecture with structural and behavioral bugs too to be fixed while team 2 opts to completely refactor, restructure, and revise the existing architecture to ensure elimination of all types of bugs.

Both the teams set to offer the same continuance and supporting task of upgrading to add by approximately 3KLOC. This investigation, however, found that the team 1 required eight times the effort, time, schedule, and cost required to carry out the decided task as against team 2. Further, since architecture was not completely refactored, and this resulted in degradation of the quality-of-service (QOS) and it inturn increased dev-ops time.

Hence, this led to the architecture entropy to exist as the team failed to identify all the issues related to architectural design using corresponding state of the art in technology. This negligence toward architectural design decisions further led to the entry and subsequently decay in architectural design. Due to this fact, the productivity of the company too reduced.

Below, stated, is the equation which indicates the data system of the computing business processes based on the summation of all use-cases/user stories to indicate the complexity involved in making design decisions.

$$\text{Data systems} = \sum_{i=1}^{x} \text{Assorted Job mix of Use cases}_i \tag{2}$$

where an information system is defined by

$$\text{Information systems or data system} = \sum_{i=1}^{x} \text{Jobs Processess}_j \tag{3}$$

Thus, the work process is the set of performance actions executed by any persons either by individual, entity, or clustered section unit of individuals using the automation tools and protocol technique also defined by

$$\text{Information system or data systems} = \sum_{i=1}^{x} \text{Jobs effort}_k \tag{4}$$

Hence, it is clear that for information systems or data systems where evolution happens in architecture involves more amount of job processes and also more effort in addition to it also needs to pay attention on architectural design decisions. Failure to follow the standards and guidelines to design the architecture with evolving needs of the customer business which may lead to residual architectural entropy.

5 Conclusion

Architecture entropy (AE) is used to depict the slower architectural design erosion (wearing down) away from a prepared, planned, structured, ordered, controlled, and organized solution state moving toward more disordered messy, disrupted state as the architectural with structural and behavioral integrity, veracity, reliability of the computing system getting eroded. This paper provides details for architecture entropy, and its impact is further put forth using some instances in real-time applications and a case study where entropy is further illustrated. This disorder is dreadful in architecture as disorder drives, coerce impels with cost outlay. As a result, it influences the quality of the system in terms of reliability, veracity, and integrity over time. As systems evolve, however, entropy also exists and has the tendency to increase with increased complexity in the architecture of the system structure. In general, the architecture entropy gain cannot be avoided, but the state levels/phases of the entropy gain can be reduced with proper governance, control, dominance, monitoring, measuring the disorder, and proper budgeting.

Acknowledgements Authors would like to thank Ganesh Samarthyam for timely advice and needful suggestion to complete this work in a timely manner.

References

1. Nenad Medvidović: "Architectural Degradation The Plague of Maturing Software Systems", University of Southern California (accessed on 30th Nov 2016).
2. https://accu.org/content/conf2008/Dalgarno-WhenGoodArchitectureGoesBad.pdf (accessed on 30th Nov 2016).
3. http://stal.blogspot.in/2008/01/architectural-entropy.html (Accessed on 30th Nov 2016).
4. http://www.designsmells.com/articles/refactoring-for-software-architecture-smells-potential-research-directions/ (accessed on 10th Nov 2016).
5. Bass, L.; Clements, P.; & Kazman, R.: *Software Architecture in Practice, Second Edition.* Boston, MA: Addison-Wesley, (2003).
6. Martin Lippert and Stephen Roock: Refactoring in Large Software Projects: Performing Complex Restructurings Successfully. John Wiley & Sons. (2006).
7. Joshua Garcia, Daniel Popescu, George Edwards and Nenad Medvidovic: "Identifying Architectural Bad Smells", IEEE, European Conference on Software Maintenance and Reengineering, (2009).
8. Sonargraph. https://www.hello2morrow.com/products/sonargraph, 2016. (Online; accessed 11-Nov-2016).
9. http://www.sonarqube.org/?s=refactoring, (accessed on 30th Nov 2016).
10. L. Xiao, Y. Cai, R. Kazman, R. Mo, and Q. Feng. "Identifying and quantifying architectural debt. In Proceedings of the 38th International Conference on Software Engineering, pp 488–498, (ACM, 2016).
11. Robert L. Nord, Ipek Ozkaya, Robert L. Nord, Ipek Ozkaya, Edward J. Schwartz, Forrest Shull, Rick Kazman, "Can Knowledge of Technical Debt Help Identify Software Vulnerabilities?" (2016), (accessed on 30th Nov 2016).
12. Robert Cloutier, Mary Bone, Dinesh Verma, Kim Sommer, "System architecture entropy", (INCOSE2009), (accessed on 30th Nov 2016).
13. Gustavo Vale; Eduardo Figueiredo; Ramon Abílio; Heitor Costa, "Bad Smells in Software Product Lines: A Systematic Review, Bad Smells in Software Product Lines: A Systematic Review, Software Components, Architectures and Reuse (SBCARS), Eighth Brazilian Symposium on SBCARS (2014).
14. Ran Mo, Joshua Garcia Y, Yuanfang Cai, Nenad Medvidovicy, "Mapping Architectural Decay Instances to Dependency Models", IEEE-MTD 2013, San Francisco, CA, USA, (2013).
15. Robert L. Nord, Ipek Ozkaya, Robert L. Nord, Ipek Ozkaya, Edward J. Schwartz, Forrest Shull, Rick Kazman, "Can Knowledge of Technical Debt Help Identify Software Vulnerabilities?" (2016), (accessed on 30th Nov 2016).
16. Joshua Garcia, Daniel Popescu, George Edwards and Nenad Medvidovic, "Toward a Catalogue of Architectural Bad Smells", http://softarch.usc.edu/~daniel/publications/qosa_2009.pdf, (2009) (accessed on 30th Nov 2016).
17. http://stal.blogspot.in/2008/01/architectural-entropy.html (accessed on 30th Nov 2016).
18. Ran Mo, Joshua Garcia Y, Yuanfang Cai, Nenad Medvidovicy, "Mapping Architectural Decay Instances to Dependency Models", IEEE-MTD 2013, San Francisco, CA, USA, (2013).

19. Alessandro Gurgel, Isela Macia, Alessandro Garcia, Arndt von Staa, Mira Mezini, Michael Eichberg, Ralf Mitschke, "Blending and Reusing Rules for Architectural Degradation Prevention", MODULARITY'14, Switzerland, (April 22–26, 2014, Lugano).
20. http://www.ivencia.com/index.html?/softwarearchitect/chapter17/chapter17.html (accessed on 30th Nov 2016).
21. Ran Mo, Joshua Garcia Y, Yuanfang Cai, Nenad Medvidovicy, "Mapping Architectural Decay Instances to Dependency Models", IEEE - MTD 2013, San Francisco, CA, USA, (2013).
22. Alessandro Gurgel, Isela Macia, Alessandro Garcia, Arndt von Staa, Mira Mezini, Michael Eichberg, Ralf Mitschke, "Blending and Reusing Rules for Architectural Degradation Prevention", MODULARITY'14, Switzerland, (April 22–26, 2014, Lugano).
23. Rick Kazman, Yuanfang Caiz, Ran Moz, Qiong Fengz, Lu Xiaoz, Serge Haziyevy, Volodymyr Fedaky, Andriy Shapochkay, "A Case Study in Locating the Architectural Roots of Technical Debt", (accessed on 30th Nov 2016).
24. Hugo Andradey, Eduardo Almeida and Ivica Crnkovic, "Architectural Bad Smells in Software Product Lines: An Exploratory Study", (April 2014).
25. Isela Macia, Roberta Arcoverde, Alessandro Garcia, Christina Chavez, Arndt von Staa, "On the Relevance of Code Anomalies for Identifying Architecture Degradation Symptoms", (2012).
26. O. Zimmermann. "Architectural refactoring: A task-centric view on software evolution". IEEE Software, 32(2):26–29, (Mar 2015).
27. http://ibm.simongreig.com/2015/11/architecture-entropy.html (accessed on 30th Nov 2016).
28. Kim, Miryung, Thomas Zimmermann, and Nachiappan Nagappan. "An Empirical Study of Refactoring Challenges and Benefits at Microsoft." IEEE Transactions on Software Engineering: 1–1, 7 (2014).
29. Qi Mo, Fei Dai, Rui Zhu, Jian Da, Leilei Lin, Tong Li.: "A Distributed Business Process Collaboration Architecture Based on Entropy in Cloud Computing", Springer International Publishing, Cloud Comp (2014), LNICST 142, pp 126–134, (2015).
30. Y. Cai and R. Kazman, "Software architecture health monitor", In Proceedings of the 1st International Workshop on Bringing Architectural Design Thinking into Developers' Daily Activities, BRIDGE '16, pages 18–21, New York, NY, USA, ACM, (2016).
31. Manu A R, Shivanand M Handigund, Manoj Kumar M, Dinesha H A, V K Agrawal, K N Balasubramanya Murthy, Nandakumar A N, "Towards an Ameliorated Approach for Design and Maturity of Cloud Service Technical Activities and Cloud Project Management by Overcoming the Service Scope Creep", Journal of Computer Networks, 2017, Vol. 4, No. 1, 30-47 Available online at http://pubs.sciepub.com/jcn/4/1/4.
32. Manu A R, Manoj Kumar M, Dinesha H A, V. K Agrawal, K N Bala Subramanya Murthy "An Approach to Enhance Security of Cloud Computing Services using Software Engineering Model", International Journal of Computer Applications (0975–8887) Volume 67–No.11, April 2013.
33. Manu A R, V K Agrawal, K N Balasubramanya Murthy, Manoj Kumar M, "Towards realizing the Secured Multilateral Cooperative Computing Architectural framework", IEEE International Conference on Cloud Computing for Emerging Markets 2014, 15 & 17 OCTOBER 2014, BANGALORE, INDIA, IEEE Catalog Number: CFP14CCM-ART ISBN: 978-1-4799-6141-2
34. Manu A R, V K Agrawal PhD, K N Balasubramanya Murthy PhD, "A User Identification Technique to Access Big Data Using Cloud Services", ACM, DEBS, Doctoral Symposium, May, 2014, held at IITB, Powai, Mumbai, Maharastra, India, Published in International Journal of Computer Applications (0975–8887) Volume 91–No 1, April 2014.
35. Manu A R, Jitendra Kumar Patel, Shakil Akthar, V K Agrawal, K N Balasubramanya Murthy, "Docker Container Security via Heuristics-Based Multilateral Security- Conceptual and Pragmatic Study", 2016 IEEE - International Conference on Circuit, Power and Computing Technologies [ICCPCT], Tamil Nadu, India.

36. Manu A R, Jitendra Kumar Patel, Shakil Akthar, V K Agrawal, K N Balasubramanya Murthy, "A Study, Analysis and deep dive on Cloud PAAS security in terms of Docker Container security", 2016 IEEE- International Conference on Circuit, Power and Computing Technologies [ICCPCT], Tamil Nadu, India
37. Manu A. R, V. K Agrawal, K N Bala Subramanya Murthy, "An empirical hunt for ally co-operative cloud computing utility", IEEE - 2017 11th International Conference on Intelligent Systems and Control (ISCO), 2017, Tamil Nadu, India.
38. Manu A.R., Vinod Kumar Agrawal, K.N. Bala Subramanya Murthy, Suma V, "Exploring Cloud Computing Technological Test Debt", I J C T A, (International Journal of Control Theory and Applications (IJCTA) [Scopus Indexed]), 10(9), 2017, pp. 327-351, International Science Press.
39. Kim, Miryung, Thomas Zimmermann, and Nachiappan Nagappan. "An Empirical Study of Refactoring Challenges and Benefits at Microsoft." IEEE Transactions on Software Engineering: 1–1, 7 (2014).
40. S. Newman. Building Microservices. O'Reilly, (2015).
41. M. Stal. Software architecture refactoring. In Tutorial in The International Conference on Object Oriented Programming, Systems, Languages and Applications, (2007).
42. http://www.conceptdraw.com/solution-park/management-enterprise-architecture-diagrams (accessed on 30th Nov 2016).
43. https://accu.org/content/conf2008/Dalgarno-WhenGoodArchitectureGoesBad.pdf (accessed on 30th Nov 2016).

Analysis of Fronto-Temporal Dementia Using Texture Features and Artificial Neural Networks

N. Sandhya and S. Nagarajan

Abstract Fronto-temporal dementia patients show severe dis-functionality in executive, interpersonal, behavioral traits with impaired cognitive skills and decreased memory capabilities. This study classifies the given MRI brain images into "healthy" or "demented" using artificial neural networks. The features are extracted from FTD images using GLCM. We make use of back propagation network.

Keywords FTD (Fronto-temporal dementia) · GLCM (Gray level co-occurrence matrix) · Back propagation network (BPN)

1 Introduction

Fronto-temporal dementia patient shows emphatic changes in personality and has problems in copying, calculations, memory skills. The patient becomes tactless, impulsive, has speech and language problems, commits spelling mistakes, initially reduces talking, and later shows mutism. The patient generally fails to recognize photos, cannot remember, or name his/her acquaintances. The vision and hearing are also affected along with other executive dis-functionalities.

N. Sandhya (✉)
Department of MCA, Hindustan Institute of Technology & Science,
Padur, Chennai 603103, Tamil Nadu, India
e-mail: Sandhya.N.Deepak@gmail.com

S. Nagarajan
Department of Mechanical Engineering, Hindustan Institute of Technology & Science,
Padur, Chennai 603103, Tamil Nadu, India
e-mail: snagarajan1960@gmail.com

© Springer Nature Singapore Pte Ltd. 2018
S.S. Dash et al. (eds.), *International Conference on Intelligent Computing and Applications*, Advances in Intelligent Systems and Computing 632,
https://doi.org/10.1007/978-981-10-5520-1_56

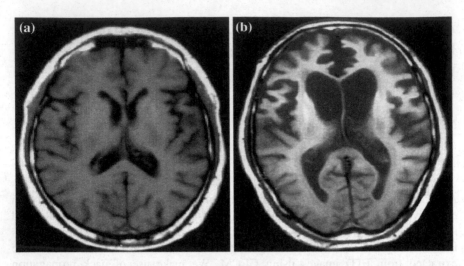

Fig. 1 Sample of FTD brain images. *Source* Images from the Study of Dementia, PubMed Central, US National Library of Medicine, National Institute of Health

Table 1 Brain images data

Brain images	Male	Female	Age group
20	12	8	50–68 Year

The patient does not take any interest in self-grooming, is stubborn, and does not care for personal health, lacks persistence and not tractable. The patient may exhibit changes in food intake. Depressions, lack of insight, dullness are some of the common symptoms. The patient fails to conduct properly with his/her own self, with others as well as with the society properly thus shows behavioral problems [1].

MRI is a method of getting images used in clinics to produce images of high quality of the human body. A horizontal tube called "bore of the magnet" runs all through. The patient lies flat into bore on a horizontal table. Depending on the type of examination to be executed patient goes in with head inside in the beginning or feet in side in the beginning. With the body part to be scanned lies in the isocenter of magnetic field, scanning starts. MRI gives a better picture of the internals of the human body. MRI provides good image information for diagnosing many injuries and situations like brain cancer, brain tumor, and dementia analysis (Fig. 1).

1.1 Materials and Methods: Patients for MR Imaging

20 brain images of patients are taken for this study (Table 1).

2 Proposed System Architecture

The main steps of the implementation are as follows:

1. Image acquisition (MR brain images)
2. Image preprocessing (converting RGB image to Gray scale image, resizing, edge detection using Sobel, binary dilation, histogram equalization, and thresholding)
3. Segmentation of image
4. Feature extraction (texture features are extracted using GLCM)
5. Classification (neural network) (Fig. 2).

2.1 Image Acquisition

The MRI brain images were obtained by 1.5 Tesla Siemens Magnetom-Avento SQ MRI scanner.

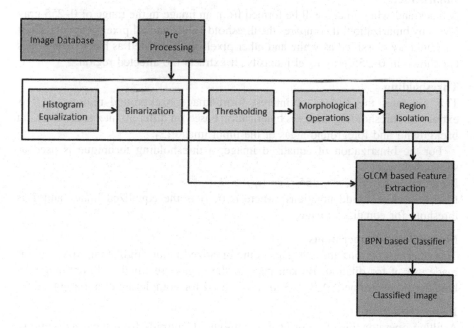

Fig. 2 Proposed system architecture

2.2 Image Preprocessing

Steps of Image preprocessing are stated as below.

Removal of Skull

Skull surrounds the brain in the outer part. Its non-cerebral tissues are to be removed. It poses a problem of segmenting non-cerebral and intracranial tissues owing to tissues' homogeneity intensities.
Steps in removal of skull are as follows:

(a) Find size and store the columns and rows in different variables
(b) Applying iteration for half of columns and all rows
(c) Half of image is processed by converting white pixels into black pixel and setting gray value to zero
(d) Repeat the same steps for another column and row.

Histogram Equalization

Histogram shows how frequently pixel occurs in the input image. The non-uniform image will be equalized to a uniform form of variation. It equalizes the image.

Binarization

A black-and-white image will be formed from an image in the range of 0–255 gray levels by binarization. It compares the threshold value, and all pixels standing about threshold are classified as white and other pixels are classified as black. The pixels are shown in 0–255 gray level intensity. It extracts the affected region.

Thresholding

This separates the regions of interest form image background regions. ROI is extracted. Thresholding achieves binarization based on different intensities/colors in background and foreground areas of the input image [2].

For the binarization of equalized image, a thresholding technique is used as shown below:
Binarized image Pm, $n = 255$ if eo $(i, j) > T$.
Else $P\ m$, $n = 0$ (add notations) where $o\ (i, j)$ is the equalized image and T is threshold for equalized image.

Morphological Operations

This sharpens regions and fills gaps in the binarized input. Imdilate in MATLAB is made use of for dilation. Broken gaps in the edges are filled and continuities at boundaries are formed. A 3×3 matrix is used for completing dilation operation.

Region Isolation

A filling operator fills close or nearby contours. Centroids form localized regions.

The extracted region will be used for eliciting massive region in the input image (Figs. 3 and 4).

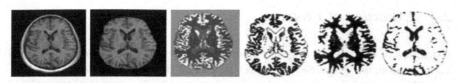

Fig. 3 Segmentation of normal healthy control. *Source* Sample images from MRI segmentation of the human brain, PubMed Central, US National Library of Medicine, National Institute of Health

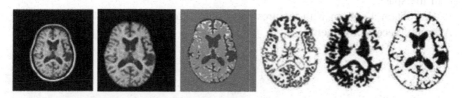

Fig. 4 Segmentation of a FTD brain. *Source* Sample images from MRI segmentation of the human brain, PubMed Central, US National Library of Medicine, National Institute of Health

2.3 Method for Feature Extraction

Segmentation divides the input image into many regions having various texture properties. Classification categorizes the regions which have undergone segmentation [3].

Texture has mutually related entities. A texture can be smooth, grained, rough depending on its structure. Tone is determined by the intensity characteristics of pixel, and structure is implied by spatial relationship between pixels. Texture refers to the spatial arrangement of texture primitives (a set of pixels constituting the simple or the most fundamental sub pattern), and texture elements called as Textone [4]. They are arranged in a systematic manner.

There are three methods of extracting text features.

1. Statistical methods
2. Syntactic structural methods
3. Spectral method.

Statistical Method—Statistical features which are extracted will be shown as vector in multidimensional space. They are emphasized on first-order, second-order, and higher-order statistics of an input image. These feature vectors are allocated to the class by a probabilistic decision algorithm [5].

Syntactic Method—Spatially arranged texture primitives placed according to placement rules form a full pattern. A similarity is derived from structural pattern and the syntax of the language [6].

Spectral method—Spatial frequencies define textures, and they are evaluated by autocorrelation function.

2.4 *Extraction of Features*

Extraction of features is a way of reducing the data for obtaining a set of needed variables from the image. Seven features are extracted. The four angles 0°, 45°, 90°, and 135° are used for extracting Haralick features [7]. The spatial relationship of pixel is determined.

The relative frequencies $P(m, n, k, a)$ determine a co-occurrence matrix, which shows how often pixels m occurs at a distance n at a distance k with an orientation a in the space.

We obtain a pattern with textures from input image.

Angular Second Moment

$$tf1 = \sum_{m,n=0}^{N-1} P_{m,n}^2$$

Contrast

$$tf2 = \sum_{m,n=0}^{N-1} P(m,n) * (m-n)^2$$

Inverse Difference Moment

$$tf3 = \sum_{m,n=0}^{N-1} \frac{P(m,n)}{1+(m-n)^2}$$

Dissimilarity

$$tf4 = \sum_{m,n=0}^{N-1} P(m,m) * |(m-n)|$$

Entropy

$$tf5 = \sum_{m,n=0}^{N-1} P(m,n) * [-\ln(P(m,n))]$$

Maximum Probability

$$tf6 = \max m, nP(m,n)$$

Inverse

$$tf7 = \sum_{m,n=0}^{N-1} \frac{P(m,n)}{(m-n)^2}$$

Energy: It measures the changing homogeneity.
Contrast: It measures clarity of the image and the texture of shadow depth. It is the diagonal near the moment of inertia.
Entropy: It is used to measure the randomness of textures in the image.
Correlation Coefficient: It calculates the probability of pixel pairs occurrences.
Homogeneity (Inverse Difference Momentum): It measures whether the elements are distributed along the diagonal of co-occurrence matrix and how close are they.

2.5 Classification Using ANNs

The study uses a back propagation neural network having input, one output, and hidden units. We extract the features and inputting it to a neural classifier.

Each connection of input–output unit has a weight associated with it. Weights are adjusted and the network is trained and accurate classification is obtained. The results expected would be "0" for healthy controls (HC) and "1" for FTD brains.

There are two phases where in a training phase features are elicited from input images. After training, the networks are made use of in the algorithm. When the input image is given as an input, it undergoes a simulation with the trained network which is stored and goes for data testing [8].

The classification process has training and testing Phases. In training phase, features are extracted from the images in which the diagnosis is known. After training, the trained networks are stored which will be used in the algorithm. When an image is taken as input in the algorithm, it is simulated with the trained networks and goes for testing the data [8].

2.5.1 Working of a Back Propagation Algorithm

For training a pattern (x, t) where x is the input and t is the desired output. "x" produces output, and real output is termed as "o." At the output layer, which is compared with the expected output t and algorithm finds the error $e = t - o$ (i.e., difference between expected and actual output) [9]. The rate of change of error with the activity level of the neuron is also found out by the BP algorithm. This if phase-I, as the transmission is in a forward direction and algorithm goes back to previous layer of the output, computes the weights (the weight between output, last layer) so that error is minimized; next algorithm steps back to the previous layer and computes the weights of last hidden layer and its previous layer (next to last layer).

Thus, the errors are found and new weight values are found marching backward from output layer to the input layer by the algorithm. When at the input layer, if there is no change, a steady state is reached and the algorithm repeats the process for next pair of (x, t). Response is in forward direction, but weights are in backward direction. Therefore, it gets the name BPN.

The error is based on the weights value. Weight updation depends on learning rate. The learning factors of BPN are initial weights, learning rate, momentum, generalization, hidden nodes, training data. Training has three levels—feed training patterns, calculating the error and sending them to previous layers, upgrading weights t.

Algorithm

The terminologies used are:

x is input pattern $(x_1, ...x_i..... x_n)$

t is output pattern $(t_1....t_k....t_m)$

α is parameter for learning

x_i is i^{th} input unit

v_{0j} is layer j^{th} bias

w_{0k} is layer kth bias z_j is a hidden unit

δ_k is weight adjusted for w_{jk} because of error at y_k which will be sent back to units hidden connected to y_k

δ_j is weight adjusted for v_{ij} because of error sent back to unit z_j

0: Initializing learning rate and weights.

  I. Repeat Steps II–IX when stopping condition is false
 II. Repeat Steps III–VIII for pair (x, t)
Phase 1:
III. Input x_i (i varies from 1 to n) is received by input unit and sent to hidden unit
 IV. Unit z_j (j varies from 1 to p) summates the weighted input and computes total input:

$$z_{inj} = v_{0j} + \sum_i x_i v_{ij}$$

Output is computed by activation function z_{inj} [(0, 1) or (−1, + 1)]

$$z_j = f(z_{inj})$$

Output propagated from hidden layer to output layer.

V. Compute total input yk (k varies from 1 to m)

$$y_{ink} = w_{0k} + \sum_{j=1}^{p} z_j w_{jk}$$

Output is calculated using activation function

$$y_k = f(y_{ink})$$

Phase 2:

VI. Error correction is calculated as δ_k (k varies from 1 to m)

$$\delta_k = (t_k - y_k) f'(y_{ink})$$

Based on error correction, bias and weights are changed and updated:

$$\Delta w_{jk} = \propto \delta_k z_j$$
$$\Delta w_{0k} = \propto \delta_k$$

Send the error correction δk to the previous layer in the backward direction.

VII. At hidden unit, inputs are summated:

$$\delta_{inj} = \sum_{k=1}^{m} \delta_k w_{jk}$$
$$\delta_j = \delta_{inj} f'(z_{inj})$$

Based on δ_j, weights, bias are updated:

$$\Delta v_{ij} = \propto \delta_j x_i$$
$$\Delta v_{0j} = \propto \delta_j$$

Phase 3:

VIII. Bias, weights are updated at y_k:

$$w_{jk}(\text{new}) = w_{jk}(\text{old}) + \Delta w_{jk}$$
$$w_{0k}(\text{new}) = w_{0k}(\text{old}) + \Delta w_{0k}$$

Updation of bias and weights happen in each hidden unit:

$$v_{ij}(\text{new}) = v_{ij}(\text{old}) + \Delta v_{ij}$$
$$v_{0j}(\text{new}) = v_{0j}(\text{old}) + \Delta v_{0j}$$

IX. Continue the process till actual output tallies with the desired output

2.6 Decision

Classification has error rate and at times fails to identify the demented brain or wrongly identifies a healthy brain as a demented brain.

Usually, error rate is described by true positive, false positive, true negative and false negative as follows:

True positive—correctly named as "demented" by the algorithm
False positive—incorrectly named as "demented" where they are normal
True negative—correctly named as "normal"
False negative—incorrectly named as "normal" where they are demented

The measures of performance being are sensitivity, specificity, accuracy.
Sensitivity (True-positive fraction) is defined as proportion of true positives which are correctly diagnosed.

$$\text{Sensitivity} = \text{TP}/(\text{TP} + \text{FN})$$

Specificity (True-negative fraction) is defined as the proportion of true negatives which are diagnosed as normal.

$$\text{Specificity} = \text{TN}/(\text{TN} + \text{FP})$$

Accuracy is the probability of rightly identifying images, i.e., it is the proportion of correct outcome either true positive or true negative.

$$\text{Accuracy} = \text{TP} + \text{TN}/(\text{TP} + \text{TN} + \text{FP} + \text{FN})$$

Sensitivity shows how well model identifies positive cases and specificity calculates how well it identifies the negative cases. Accuracy is expected to measure how well it recognizes both categories. Thus, if both sensitivity and specificity are high (low), accuracy will be high (low). If any one of the measures (sensitivity or specificity) is high and other is low, then accuracy will be inclined toward one of them. Hence, only accuracy cannot be a good performance measure.

3 Conclusion

This approach attempts to classify the given MRI brain images using texture-based feature extraction in artificial neural networks. At first, the MRI database is introduced which has healthy brains (HC), FTD brain images. The architecture of ANN is designed to classify the input images. It is implemented by MATLAB R2008b with image processing toolbox and neural network toolbox.

4 Scope for Future Study

The scope of the study is to improvise the architecture of ANNs by improving feature functions to get the segregated result.

Declaration We, the author(s) of the paper, hereby declare that the reference images used in this paper are the sample images taken from the PubMed Central, US National Library of Medicine, National Institute of Health database. These images are used in this paper to illustrate the concept and do not contain any information that is related to patient. The sample images referenced in this paper are part of the public domain research studies published in PubMed Central and may be used and reproduced without special permission. The sources of the images have been cited and acknowledged in the paper.

References

1. N. Sandhya, Dr. S. Nagarajan, Frontotemporal Dementia—A Supervised Learning Approach, Springer ERCICA 2015 Vol 3, DOI:10.1007/978-981-10-0287-8.
2. G. Vijay Kumar and Dr. G.V. Raju, "Biological Early Brain Cancer Detection using Artificial Neural Network", International Journal on Computer Science and Engineering, Vol. 02, No. 08, 2010.
3. R. C. Gonzalez and R. E. Woods, Digital Image Processing, Prentice Hall, New Jersy, 2002.
4. Jayashri Joshi and Mrs. A. C. Phadke, "Feature Extraction and Texture Classification in MRI", IJCCT, Vol. 2 Issue 2–4, 2010.
5. Fukunaga K. Introduction to statistical pattern recognition. 2nd ed. Academic Press; 1990.
6. Pavilidis T. Structural description and graph grammar. In: Chang S. K., Fu K. S., editors. Pictorial information systems. Springer Verlag Berlin: 1980. pp. 86–103.
7. Haralick R.M. Statistical and structural approaches to texture. Proc IEEE. 1979; 67:786–804.
8. S. Wadhwani, A. K. Wadhwani and M. Saraswat, "Classification of Breast Cancer Detection Using Artificial Neural Networks", Current Research in Engineering, Science and Technology (CREST) Journals, Vol. 01, Issue 03, pp. 85–91, 2013.
9. S. N. Sivanandam and S.N. Deepa, "Principle of Soft Computing", WILEY INDIA EDITION, Pages 74–83, 1993.

3. Conclusion

4. Scope for Future Study

References

Revisiting the Symptoms of Work-Life Balance: A Dependency Analysis of Employees in ICT Sector

G. Rajini

Abstract Background/Objectives: The study aims to analyse the influence of factors on work-life balance of employees in ICT sector, and to study the challenges faced by them. **Methods/Statistical analysis**: The data for this survey were collected from a sample of 150 employees in ICT sector and were analysed using multiple regression techniques to measure the combined effects of independent variables with SPSS 17.0. **Findings**: The studies found that majority of the employees do not share work-related issues with their families. This may lead to work-related stress and general dissatisfaction with work. Also the studies found that most employees are depending on technology for work purposes even outside working hours. Employees feel anxious when they cannot access technology to check work-related messages. With regard to health-related issues, it was found that there is a higher occurrence of health-related issues in female employees than in male employees. However, there is no effect of age on any of the factors of work-life balance. There is a positive effect of healthcare programs provided by the organization on health-related issues of the employees. As the experience in the organization increases, Managerial Support and Organization Support increase and commitment towards the organization increases. **Improvements/Applications**: It is found that long-term tenure of the employees depends highly on Organization commitment and Managerial Support. Also, as the experience in the organization increases, the desire for long-term tenure in the organization also increases.

Keywords Work-life balance · ICT companies · Organizational commitment and Support

G. Rajini (✉)
School of Management Studies, VELS University, Chennai, Tamil Nadu, India
e-mail: dr.rajini.g@gmail.com; rajini.sms@velsuniv.ac.in

© Springer Nature Singapore Pte Ltd. 2018
S.S. Dash et al. (eds.), *International Conference on Intelligent Computing and Applications*, Advances in Intelligent Systems and Computing 632,
https://doi.org/10.1007/978-981-10-5520-1_57

1   Introduction

Employees are the backbone of any organization. It is essential that organizations find ways to retain their employees and keep them satisfied. One effective method to achieve this is putting into effect work-life balance practices in the organization. Work-life balance is the process of attempting to balance the demands and expectations of one's career, personal life, interpersonal relationships, partnerships and family [1]. This balance is essential for them to function effectively both at work and in their personal lives. Work-life balance is not simply balancing work demands and personal life responsibility, but doing so in a way that is meaningful to him or her (the employee) [2]. It is a continuous process where the organization makes sure that the employee has enough time and opportunities to engage in their personal lives in addition to their work-life. It has been suggested that an individual's best interests are served by living a balanced life [3]. Work interference with family and family interference with work are shown to be negatively related to affective commitment [4]. This need for employees to give adequate attention to their work and other aspects of their life such as family, friends, personal health, spirituality, etc., is addressed in work-life practices implemented by the organization. Studies indicate that while work-life balance has conventionally been presumed to involve giving equal amounts of time to paid work and non-work roles, more recently the notion has been accepted as more difficult and has been expanded to integrate extra components [5]. When employees spend more and more time at their workplace or at work-related activities, they feel as though they are neglecting other areas of their life. Work-life practices work at removing this feeling. Such practices have been found to reduce the overall stress that the employees face, thus making them more motivated and satisfied, and ultimately making them more committed to the organization. Studies have shown that work-life balance has a positive effect on employees' affective commitment to their organizations [6]. Affective commitment is an emotional attachment to the organizations or the employers which can cause employees to want to remain with the organizations [7]. Work-life balance in the workplace has become a more important issue as it tends to exhibit positive results such as low turnover, work engagement, organizational citizenship behaviour, in-role performance, increased firm productivity, job satisfaction and organizational commitment [8]. This enhanced emotional attachment to the organization and led to increased outcomes [9]. Thus, work-life policies have benefited both the employee and by extension, the organization.

2   Materials and Methods

The study is of descriptive type because each and every item is clearly described. Research instrument that is used in this study is questionnaire. The author has obtained all ethical approvals from an appropriate ethical committee/institutional review board. The proposal of the study bearing the questionnaire was submitted to

Institutional Ethics Committee, VELS University, Chennai, India, and approval was accorded by the committee bearing ICE/ODS/2016/01 as official communication number. There are 39 items in this instrument, with five-point Likert-like scale [10]. Reverse scoring for few items has done [11].

1. Strongly Agree, Agree, Partially Agree, Disagree and Strongly disagree
2. Always, Most of the time, Sometimes, Seldom, Never
3. Extremely satisfied, somewhat satisfied, partially satisfied, somewhat dissatisfied, extremely dissatisfied.

Convenience sampling [12] in case of infinite population is a method of sample selection which gives each possible sample that was willing to participate in the study to be selected. Informed consent by all the participants was sought appropriately before data collection. The convenience sampling method has an advantage of voluntary participation of respondents, which paves way to research ethics protecting the privacy rights of individuals, and real data can be collected. The consents of the respondents have thus been taken to use the data in this research. In order to detect weaknesses in design and instrumentation and to provide proxy data for selection of a sample, a pilot test was conducted to answer the following questions:

- Whether the respondents understand the entire question?
- Whether certain words in the questions need explanation?
- Are there unnecessary questions?

The deficiencies on the questionnaire are corrected. SPSS Statistics 17.0 version is used for data analysis. Multiple regressions analysis [13] was applied which uses quite simple as well as multiple predictions to predict Y from X values.

3 Results

Cronbach's alpha value of all the 39 items is given as $\alpha = 0.813$. The questionnaire is reliable, and the items are internally consistent. The result is expressed in Table 1.

Multiple Regressions

3.1 Organizational Commitment

H_{01}: Organization commitment does not depend on the Experience of the employees, Co-workers Support, Managerial Support, Family Support and Organization Support.

Table 1 Reliability Statistics

Reliability statistics		
Cronbach's alpha	Cronbach's alpha based on standardized items	No. of items
0.813	0.805	39

A multiple regression was run for measuring the combined effects of indepen-dent variables (Experience X_1; Co-worker Support X_2; Managerial Support X_3; Family Support X_4; Organizational Support X_5) against dependent variable (Organization commitment).

The equation derived from previous theories is given as:

$$Y_1 = C_1 + b_1X_1 + b_2X_2 + b_3X_3 + b_4X_4 + b_5X_5$$

where b_1, b_2, b_3, b_4 and b_5 are beta coefficients.

The statistical significant of regression coefficients was worked out and tested. The model summary is exhibited in Table 2. The ANOVA model was employed to discern the significant variation among the five independent variables (Experience, Co-worker Support, Managerial Support, Family Support, Organization Support). The results are exhibited in Table 3.

The prediction ability of the regression model is denoted by R^2 with value 0.123 in which 12.3% of variance in the dependent variable belongs to independent variables with the F-value 4.044 in Table 3.

The beta weight of 0.263 and 0.256 for variable 5 and 1 (Organizational Support and Experience) in Table 4 meant that when other variable held constant, com-mitment towards the organization (Dependent variable) would increase by quarter the standard deviation (0.162 and 0.217). The order of importance or predictive importance was 0.217, 0.162, 0.149, 0.147 and −0.068 for variables 5, 1, 3, 4 and 2, respectively, exhibited in Table 4. The t test is to confirm that the results can be generalized to the total population by value 2.522, 2.066, 1.868, 1.573 and −0.719 in Table 4. From the ANOVA Table 3, it was found that the significant value was 0.02 which is less than 0.05 which concludes that the regression performed has generated a good model. By substituting the value in equation:

Table 2 Model[b] summary

Model	R	R square	Adjusted R square	Std. error of the estimate
1	0.351[a]	0.123	0.093	1.689

[a]Predictors: (constant), Organization Support, experience, family Support, co-worker Support, managerial Support
[b]Dependent variable: organization commitment

Table 3 ANOVA[b] table

Model		Sum of squares	Df	Mean square	F	Sig.
1	Regression	57.679	5	11.536	4.044	0.002[a]
	Residual	410.781	144	2.853		
	Total	468.460	149			

[a]Predictors: (constant), organization Support, experience, family Support, co-worker Support, managerial Support
[b]Dependent variable: organization commitment

Table 4 Coefficient[a]—t test

Model		Unstandardized coefficients		Standardized coefficients	T	Sig.
		B	Std. error	Beta		
1	(Constant)	4.738	1.401		3.382	0.001
	Experience	**0.256**	0.124	0.162	2.066	0.041
	Co-worker Support	−0.062	0.086	−0.068	−0.719	0.473
	Managerial Support	0.124	0.079	0.149	1.573	0.118
	Family Support	0.094	0.050	0.147	1.868	0.064
	Organization Support	**0.263**	0.104	0.217	2.522	0.013

[a]Dependent variable: organization commitment

$$Y_1 = C_1 + b_1 X_1 + b_2 X_2 + b_3 X_3 + b_4 X_4 + b_5 X_5$$

Organization commitment = −4.044 + **0.263** (**Organizational Support**) + **0.256** (**Experience**) + 0.124 (Managerial Support) + 0.094 (Family Support) − 0.062 (Co-worker Support).

3.2 Satisfaction of Policies

H_{02}: Satisfaction of policies does not depend on Organizational Support, Experience, Managerial Support.

Here, this multiple regression measures the combined effects of independent variables (Organizational Support X_{1a}; Experience X_{2a}; Managerial Support X_{3a}) against dependent variable (Satisfaction with policies).

The general form of equation derived from previous theory is given as:

$$Y_{1a} = C_{1a} + b_{1a} X_{1a} + b_{2a} X_{2a} + b_{3a} X_{3a}$$

where b_{1a}, b_{2a}, b_{3a} are beta coefficients.

The statistical significance of regression coefficients was worked out and tested by applying the t test, and the model summary is exhibited in Table 5. The ANOVA was employed to discern the significant variation among the three independent variables (Organizational Support, Experience and Managerial Support). The results are exhibited in Table 6.

Table 5 Model[b] Summary

Model	R	R square	Adjusted R square	Std. error of the estimate
1	0.322[a]	0.104	0.085	4.83899

[a]Predictors: (constant), managerial Support, experience, organization Support
[b]Dependent variable: satisfaction with policies

Table 6 ANOVA[b] table

Model		Sum of SQUARES	Df	Mean square	F	Sig.
1	Regression	394.870	3	131.623	5.621	0.001[a]
	Residual	3418.703	146	23.416		
	Total	3813.573	149			

[a]Predictors: (constant), managerial Support, experience, organization Support
[b]Dependent variable: satisfaction with policies

Table 7 Coefficient[a]—t test

Model		Unstandardized coefficients		Standardized coefficients	T	Sig.
		B	Std. error	Beta		
1	(Constant)	9.502	3.280		2.897	0.004
	Organization Support	0.516	0.292	0.149	1.765	0.080
	Experience	**1.333**	0.355	0.295	3.758	0.000
	Managerial Support	−0.077	0.201	−0.033	−0.384	0.701

[a]Dependent variable: satisfaction with policies

The prediction ability of the model is expressed by R^2 with value 0.104 in which only 10.4% of variance in the dependent variable belongs to independent variables with the F-value 5.621 in Table 6.

The beta weight for variable 2 is 1.333 (Experience) in Table 7 which means that when others independent variables held constant, Satisfaction of policies (Dependent variable) would increase by half the standard deviation (0.295). The order of importance or predictive importance was 0.295, 0.149 and −0.033 for variables 2, 1 and 3, respectively, shown in Table 7. The t test confirms that the results can be generalized to the total population by value 3.758, 1.765 and −0.384 in Table 7. The ANOVA Table 6 shows that the significant value was 0.001 which is less than 0.05 thereby concluding that the regression was a good model.

By substituting all values in the equation:

$$Y_{1a} = C_{1a} + b_{1a}X_{1a} + b_{2a}X_{2a} + b_{3a}X_{3a}$$

> Satisfaction with policies = 9.502 + **1.333** (**Experience**) + 0.516 (Organizational Support) + 0.077 (Managerial Support)

3.3 Long-Term Tenure

H_{03}: Long-term tenure does not depend on Organization Support, Co-workers Support, Managerial Support, Organization commitment or Satisfaction with policies.

Again this multiple regression is to measure the combined effects of independent variables (Organizational Support X_{1b}; Co-worker Support X_{2b}; Managerial Support X_{3b}; Organizational Commitment X_{4b}; Satisfaction of policies X_{5b}) on the dependent variable (Long-Term Tenure).

This equation derived from theory is:

$$Y_{1b} = C_{1b} + b_{1b}X_{1b} + b_{2b}X_{2b} + b_{3b}X_{3b} + b_{4b}X_{4b} + b_{5b}X_{5b}$$

where b_{1b}, b_{2b}, b_{3b}, b_{4b}, b_{5b} are beta coefficients.

The statistical significance of regression coefficients was derived and tested by applying the t test, and the model summary is displayed in Table 8.

The ANOVA was employed to discern the significant variation among the five independent variables (Organization Support, Co-workers Support, Managerial Support, Organization commitment and Satisfaction with policies). The results are exhibited in Table 9.

Table 8 Model[b] summary

Model	R	R Square	Adjusted R Square	Std. error of the estimate
1	0.702[a]	0.493	0.475	0.716

[a]Predictors: (constant), policies, managerial Support, organization commitment, organization Support, co-worker Support
[b]Dependent variable: long-term tenure

Table 9 ANOVA[b] table

Model		Sum of squares	Df	Mean square	F	Sig.
1	Regression	71.890	5	14.378	28.016	0.000[a]
	Residual	73.903	144	0.513		
	Total	145.793	149			

[a]Predictors: (constant), policies, managerial Support, organization commitment, organization Support, co-worker Support
[b]Dependent variable: long-term tenure

Table 10 Coefficient[a]—*t* test

Model	Unstandardized coefficients		Standardized coefficients	T	Sig.
	B	Std. error	Beta		
1 (Constant)	−1.324	0.562		−2.354	0.020
Organization Support	−0.051	0.045	−0.075	−1.127	0.262
Co-worker Support	0.056	0.036	0.110	1.545	0.125
Managerial Support	0.095	0.034	0.204	2.814	0.006
Organization commitment	0.341	0.035	0.610	9.790	0.000
Policies	0.009	0.012	0.048	0.798	0.426

[a]Dependent variable: long-term tenure

The model's prediction ability is denoted by R^2 with value 0.493 in which 49.3% of variance in the employee tenure belongs to independent variables with the F-value 28.01 in Table 9. The order of importance was 0.610, 0.204, 0.110, 0.048 and −0.075 for variables 4, 3, 2, 5 and 1, respectively, exhibited in Table 10. The *t* test confirmed that the results generalized to the total population by value 9.790, 2.814, 1.545, 0.798 and −1.127 in Table 10. ANOVA in Table 9 shows that the significant value was 0.00 and the regression was a good model.

By substituting the value in equation:

$$Y_{1b} = C_{1b} + b_{1b}X_{1b} + b_{2b}X_{2b} + b_{3b}X_{3b} + b_{4b}X_{4b} + b_{5b}X_{5b}$$

Tenure = −1.324 + **0.341 (Organizational Commitment)** + **0.095 (Managerial Support)** + 0.056 (Co-worker Support) + 0.009 (Satisfaction of policies) − 0.051 (Organizational Support)

4 Conclusion

This study has evaluated the degree of Support of each factor of work-life balance. Mainly three dependent variables Organizational commitment, Satisfaction on policies and employee tenure were analysed by the predictors Managerial Support, Organizational Support, Co-worker Support and Family Support (Figs. 1, 2 and 3). By these analyses it was found that the first and second model had dependency less than 50 %. The p-p plots are a check on normality and the plotted points should be in straight line. There is no Serious departures of the points from the line which means that normality assumption are met. The third model (multiple regression) which predicted about the tenure of employees significantly depends on Organizational commitment and Managerial Support is worthwhile statistically. The other two models showed less than 50% dependency on the predictors. The employees' positive perception towards Organizational commitment, Managerial Support leads to long-term tenure in the organization. Co-worker Support and Satisfaction of policies were not a matter of concern for long-term tenure.

Further descriptive analysis done reveals a thorough understanding of work-life balance as per the cognition of employees. Health is often neglected in the pursuit of career growth. Good work-life balance includes paying attention to personal health so that it is not negatively affected by the stresses of the workplace. There is a higher occurrence of health-related issues in female employees than in male employees in ICT Companies. However, there is no effect of age on any of the factors of work-life balance [14]. Time is insufficient for the employees for attention to personal health such as time for physical exercise. However, providing more healthcare programs to the employees was shown to have a positive effect on their health. Work-life balance is reflected in the ability of the employees to achieve a balanced lifestyle. The employees of ICT organizations were able only moderately to achieve a balanced lifestyle. Encouraging its employees to achieve a balanced lifestyle could be given importance by the organization. Most employees also do not share work-related issues with their families, and the presence of an on-site counsellor would help them share the problems and reduce their work-related stress. Technology is two-faced. It can help us stay connected and aid our career growth, but it can also be a cause for stress if overly dependent on it. Although technology [15] was found to have a positive effect on the stress levels of the employees, it was found that they become anxious when not able to access technology for work purposes.

Fig. 1 Normal P-P plot of regression Y standardized residual

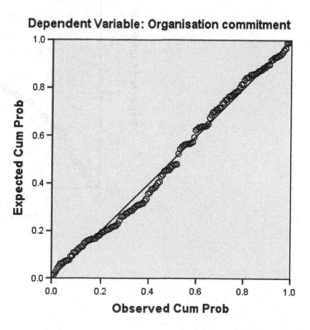

Fig. 2 Normal P-P plot of regression Y_{1a} standardized residual

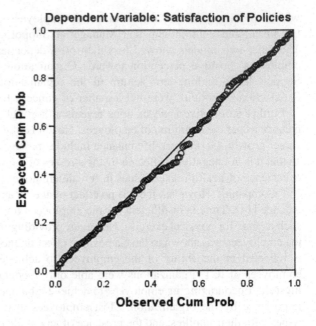

Fig. 3 Normal P-P plot of regression Y_{1b} standardized residual

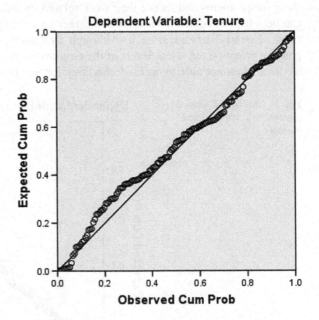

Acknowledgements My best wishes and gratitude to my MBA student of 2016 batch Smruthi Prabhakaran M.K. for her assistance in these studies.

Declaration As author of the paper, I affirm that all ethical approvals from appropriate ethical committee and consent from the individuals participated in the study were obtained. Neither the editors nor the publisher will be responsible on this accord.

References

1. Emslie C., and Hunt K., (2009) Live to work or work to live? A qualitative study of gender and work-life balance among men and women in mid-life. *Gender, Work and Organization*, Vol. 16 No. 1, pp. 151–172.
2. Reiter N., (2007) Work-life balance: What do you mean? *Journal of Applied Behavioural Science.* 43, 273–294. doi:10.1177/002188630730.
3. Kofodimos J.R. (1993) Balancing Act, San Francisco. CA: Jossey-Bass.
4. Netemeyer R.G, Boles J.S., and McMurrian R. (1996) Development and validation of work-family conflicts and work-family conflict scales. *The Journal of Applied Psychology.* Vol. 81 No. 4, pp. 400–410.
5. Parasuraman S., & Greenhaus J.H. (2002). Toward reducing some critical gaps in work-family research. *Human Resource Management Review*, 12, 3, 299–312.
6. Casper W.J., Harris C., Taylor-Bianco A., and Wayne J.H. (2011) Work-family conflict, perceived supervisor support and organizational commitment among Brazilian professionals. *Journal of Vocational Behavior.* Vol. 79 No. 3, pp. 640–652.
7. Allen T.D., Herst D.E., Bruck C.S., and Sutton M.C., (2000). Consequences associated with work-to-family conflict: A review and agenda for future research. *Journal of Occupational Health Psychology.* Vol. 5 No. 2, pp. 278–308.
8. Konrad A. M., and Mangel R. (2000). The impact of work-life programs on firm productivity. *Strategic Management Journal.* Vol. 21 No. 12, pp. 1225–1237.
9. Mayer R.C., and Schoorman F.D. (1992). Predicting participation and production outcomes through a two-dimensional model of organizational commitment. *Academy of Management Journal.* Vol. 35 No. 3, pp. 671–684.
10. G. Rajini and G. Madhumita. Exuberance Due to Celebrity Endorsement on Brands: A Product Categorical Study. Indian Journal of Science and Technology. Vol 9(32), DOI:10.17485/ijst/2016/v9i32/98664, August 2016, *ISSN (Print): 0974-6846, ISSN (Online): 0974-5645.*
11. G. Rajini (2013). Vital Strategies Discriminating Global and Local Organizations. *Global Business Review, 14*(2), Sage Publications, 225–241.
12. G. Rajini and Krithika M. (2016). A study about interdependence between attitude and perceived risk in online shopping. International Journal of Management and Social Science Research Review (IJMSRR). Vol. 1, No. 4, pp. 80–84.
13. G. Rajini and Krithika M. (2016). Factors influencing online shopping Intention: Impact of Perceived Risk. Advances in Natural and Applied Sciences. 10(10). Pages: 116–119.
14. G. Rajini and M. Krithika (2016). Online Purchase: Risk Cognizance Influencing Intention. Indian Journal of Science and Technology. Vol 9(32), DOI:10.17485/ijst/2016/v9i32/98660, August 2016.
15. Krithika M., and Rajini G. (2016). Comparative Review of Two Theoretical Frameworks in Technology based Consumer Acceptance Behavior. International Journal of Research in Management, Economics and Commerce. (Impact Factor: 6.384), Vol. 6, No. 4, pp. 16–26.

A Compact Tri-band Microstrip Patch Antenna for WLAN and Wi-MAX Applications

Namrata Singh, R.P.S. Gangwar and A.K. Arya

Abstract In this paper, a compact microstrip patch antenna with offset feeding technique is proposed for WLAN (2.4/5.2 GHz) and Wi-MAX (3.5 GHz) applications. The design includes defected ground structure (DGS) for bandwidth improvement. A single complimentary split ring resonator (CSRR) is introduced on the main patch. Along with a CSRR ring, the main radiator is also loaded with a rectangular slot. The proposed antenna is compact with a size of $24 \times 24 \times 1.6 \text{ mm}^3$. The design is validated and optimized by using Ansofts' HFSS 13.0, EM full wave software. Experimental and simulated results are compared, and a good agreement between them is observed.

Keywords WLAN · Wi-MAX · DGS · MSA · CSRR · Bandwidth improvement

1 Introduction

In recent years, there has been a considerable advancement in wireless communications field. Microstrip patch antennas are of main interest in this advancement of technology because of their low cost and simplicity in fabrication. Multiband MSA has attracted researchers as the single antenna can cover specific applications simultaneously [1]. As per IEEE standards, WLAN operates in 2.4 GHz band (2.4–2.48 GHz) and 5 GHz band (5.15–5.35 GHz), and Wi-MAX operates in 3.5 GHz band (3.4–3.6 GHz) [2]. Earlier, dual-band slot antennas for WLAN were

N. Singh (✉) · R.P.S. Gangwar · A.K. Arya
Department of Electronics & Communication Engineering, College of Technology,
G.B. Pant University of Agriculture & Technology, Pantnagar 263145,
Uttarakhand, India
e-mail: namrata1792@gmail.com

R.P.S. Gangwar
e-mail: profrpsgangwar@yahoo.com

A.K. Arya
e-mail: ashwiniarya.iitr@gmail.com

© Springer Nature Singapore Pte Ltd. 2018
S.S. Dash et al. (eds.), *International Conference on Intelligent Computing and Applications*, Advances in Intelligent Systems and Computing 632,
https://doi.org/10.1007/978-981-10-5520-1_58

655

developed [3, 4] that were quite large. Recently, MSA with defected ground structure (DGS) was developed for WLAN and Wi-MAX applications [5, 6]. Applying the DGS is one of the techniques to improve the antenna performance in terms of bandwidth and size.

Defect on the ground disturbs the shielded current distribution, and this disturbance depends upon the size of the defect.

Nowadays to achieve the multiband behaviour in the antennas, complimentary split ring resonators (CSRRs) are used [7]. CSRRs are couple of complementary split ring resonators (SRRs) and mainly used for synthesizing meta-material structures.

In this paper, the proposed antenna operates in 0–6 GHz frequency range for WLAN and Wi-MAX applications. The antenna has a defected ground plane and the main radiator patch of which is loaded with single CSRR ring and a rectangular slot for achieving tri-band operation that operates in 2.4, 3.5 and 5 GHz bands. Simulations are carried out by using Ansofts' HFSS. The proposed antenna is fabricated using photolithographic technique, and then its parameters in terms of reflection coefficient, gain and radiation patterns are measured using vector network analyzer (VNA) and anechoic chamber. The simulated results have been compared with the experimental results, and it is found that there is good agreement between them. The fabricated antenna is compact as compared to earlier reported works.

2 Materials and Methods

The proposed antenna is developed on FR-4 epoxy substrate having relative permittivity (ε_r) of 4.4 and tangent loss (δ) of 0.002. The patch is loaded with a single CSRR ring and a rectangular slot above it as shown in Fig. 1a. The design is developed as follows: first resonating band at 3.5 GHz is achieved because of the main patch; for second resonating band at 2.4 GHz, a CSRR ring is introduced on the patch; and for third resonating band at 5 GHz, a rectangular slot has been introduced above CSRR ring as shown in Fig. 1a. Further, the performance of the design is improved with the use of offset feed and DGS as depicted in Fig. 1. Dimensions of the patch are calculated using the following equations [6]:

$$\text{Wp} = c/2f_o[(\varepsilon_r + 1)/2]^{-1/2} \tag{1}$$

$$\text{Lp} = c/2f_o\sqrt{\varepsilon_{\text{reff}}} - 2\Delta L \tag{2}$$

where $\varepsilon_{\text{reff}}$ and ΔL are calculated from [6]. The values of the other parameters have been analysed after a series of optimization. Table 1 includes values of all design parameters used in design of the proposed antenna. A staircase defective ground structure has been used for bandwidth improvement, and the patch is slightly (0.5 mm) displaced from the centre axis of the microstrip antenna that contributes to impedance matching.

Fig. 1 Geometry of the proposed antenna

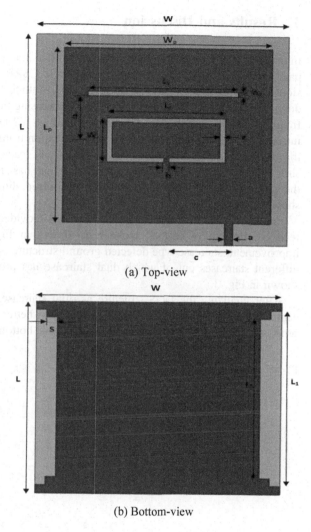

(a) Top-view

(b) Bottom-view

Table 1 Design parameters used in antenna geometry

Design parameter	Value (mm)	Design parameter	Value (mm)
L	24	a	0.8
W	24	b	0.5
Lp	20	c	6.4
Wp	20	d	5.5
Lr	11.4	e	0.5
Wr	2.5	L_1	20
Ls	14.5	L_2	22
Ws	0.5	S	1

3 Results and Discussion

In the proposed antenna design, resonating band at 3.5 GHz occurs because of basic patch. For resonating band at 2.4 GHz, a single CSRR ring is added. The dimensions of the ring affect resonance frequency [7]. With the increase of length of ring a downward frequency shift occurs, whereas decreasing the length of ring an upward frequency shift occurs. For resonating band at 5.2 GHz, a rectangular slot is introduced on the patch, and its length and location from centre of the patch affects the resonating band. Increasing the length of slot leads to a downward frequency shift in higher band, whereas decreasing the length has an opposite impact. Through the parametric study of all parameters, optimized dimensions are achieved and shown in Table 1.

From analysis, it is found that offset feed has provided better results. Thus, offset feed is selected for further analysis as shown in Fig. 2. Then for bandwidth improvement, staircase type defected ground structure is used. Analysis is done for different staircases out of which dual staircase has provided improved results as shown in Fig. 3.

After finalizing the DGS structure as dual staircase, the patch is shifted from centre towards the feed side (right) resulting in better impedance matching. The proposed antenna design is fabricated, and its top, bottom and perspective views are

Fig. 2 Feed comparison

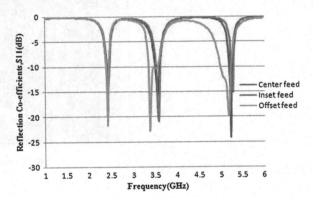

Fig. 3 Comparison in DGS structures

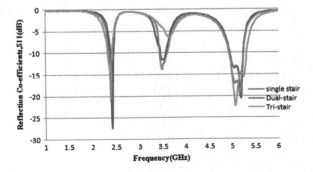

(a)						**(b)**						**(c)**

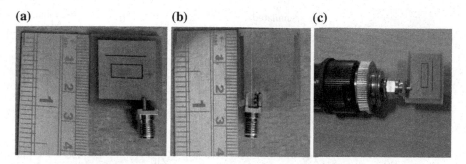

Fig. 4 Fabricated antenna: **a** top-view, **b** bottom-view, and **c** perspective view

Fig. 5 $|S_{11}|$ (dB) versus frequency

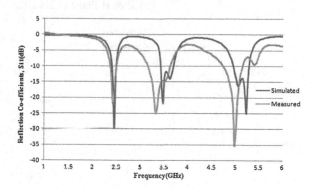

Table 2 Impedance bandwidth comparison

Analysis	Impedance bandwidth		
	2.4 GHz band	3.5 GHz band	5 GHz band
Simulation	2.3909–2.4873 (96.4 MHz)	3.4162–3.6853 (269.1 MHz)	4.9391–5.2843 (345.23 MHz)
Experimental	2.365–2.48 (115 MHz)	3.15–3.625 (475 MHz)	4.75–5.125 (375 MHz)

presented in Fig. 4 (a, b and c, respectively). Experimental results have been compared with the simulated ones.

The comparison between the simulated and experimental $|S_{11}|$ (dB) is presented in Fig. 5. It is found that these results are almost matched except some frequency shifts in upper bands. Impedance bandwidth for each frequency band in simulated and experimental analysis is presented in Table 2.

Radiation patterns for the above-mentioned bands are presented in Fig. 6 from which it is found that the simulated and measured patterns are similar. Figure 7 shows the measured peak gain at different resonant frequencies of interest. It is clear from this figure that the gain is quite good in these resonant frequencies.

In Table 3, the performance parameters of the fabricated antenna are compared with that of earlier reported works. It shows that antenna is quite compact.

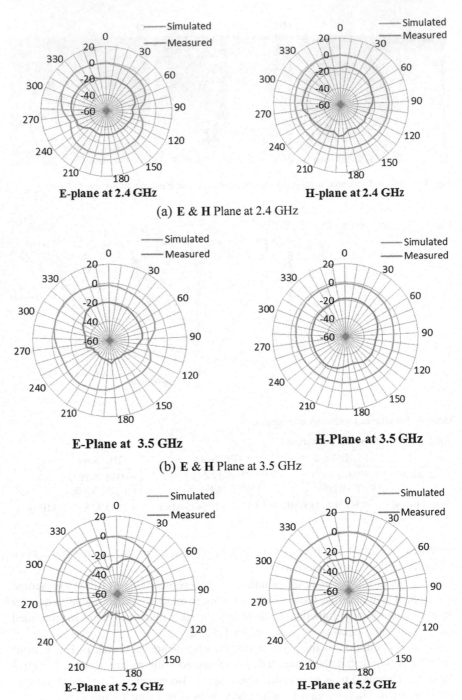

E-plane at 2.4 GHz   **H-plane at 2.4 GHz**

(a) **E & H** Plane at 2.4 GHz

E-Plane at 3.5 GHz   **H-Plane at 3.5 GHz**

(b) **E & H** Plane at 3.5 GHz

E-Plane at 5.2 GHz   **H-Plane at 5.2 GHz**

(c) **E&H** Plane at 5.2 GHz

Fig. 6 Radiation pattern at three resonating bands

Fig. 7 Measured peak gain in dB

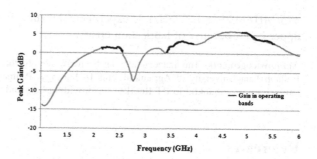

Table 3 Comparison with earlier reported works

Antenna design	Ground plane size (mm²)	Resonant band (GHz)	Band width (MHz)	Material used	
				ε_r	h (mm)
[1]	45 × 35	2.4	760	4.4 (FR-4)	1.57
		5.2	72,090		
[2]	100 × 45	2.4	90	2.2 (RT-5800)	0.127
		3.5	290		
		5.2	440		
[3]	260 × 200	2.4	90	0.999991 (Cu)	10
		5.2	39		
[4]	75 × 75	2.4	270	4.7	0.8
		5.2	1850		
[5]	50 × 50	2.4	82	4.4 (FR-4)	1.6
		3.5	175		
		5.2	546.9		
[6]	57 × 50	1.8	60	4.4 (FR-4)	1.6
		3.6	50		
[7]	40 × 46	2.4	100	4.4 (FR-4)	1.6
Fabricated antenna	24 × 24	2.4	115	4.4 (FR-4)	1.6
		3.5	375		
		5.2	475		

4 Conclusion

In this paper, a compact tri-band microstrip antenna is designed, simulated, fabricated and measured for applications of WLAN (2.4/5 GHz) and Wi-MAX (3.5 GHz) operations. The presented work is quite compact in design. Antenna is experimentally tested, and it is found that the simulated and experimental results are almost similar. Antenna works in 2.4 GHz band with impedance bandwidth of 115 MHz (2.365–2.48 GHz), in 3.5 GHz band with impedance bandwidth of 475 MHz (3.15–3.625 GHz) and in 5 GHz band with impedance bandwidth of

375 MHz (4.75–5.125 GHz). In experimental results, frequency shift is observed in upper bands because of fabrication errors.

Acknowledgements The financial help has been provided by the College of Technology, Govind Ballabh Pant University of Agriculture and Technology, Pantnagar, under TEQIP-II fund. The authors are very thankful to IIT Roorkee for providing help and support in the research work.

References

1. Smith, Rathore, A., Nilavalan, R., AbuTarboush, H. F. and Peter, T., "Compact Dual-Band (2.4/5.2 GHz) Monopole Antenna for WLAN Applications," IEEE International Workshop on Antenna Technology (iWAT) 2010.
2. Pazin, Lev, Telzhensky, Nikolay and Leviatan, Yehuda, "Multiband Flat-Plate Inverted-F Antenna for Wi-Fi/WiMAX Operation", *IEEE Antennas and Wireless Propagation Letters*, Vol. 7, 2008.
3. Su, Chih-Ming, Chen, Hong-Twu, Chang, Fa-Shian and Kin-Lu Wong, "Dual-Band Slot Antenna for 2.4/5.2 GHz WLAN operation," Microwave and Optical Technology Letters/Vol. 35, No. 4, November 20, 2002.
4. Wu, J.-W., Hsiao, H.-M., Lu, J.-H. and Chang, S.-H., " Dual broadband design of rectangular slot antenna for 2.4 and 5 GHz wireless communication," Electronics Letters 11th November 2004 Vol. 40 No. 23.
5. Zahraoui, Issam, Zbitou, Jamal, Errkik, Ahmed, Abdelmounim, Elhassane, Sanchez Mediavilla Angel, "A Novel Printed Multiband Low Cost Antenna for WLAN and WiMAX Applications," International Journal of Microwave and Optical Technology, Vol. 11, No. 2, March 2016.
6. Zahraoui, Issam, Zbitou, Jamal, Errkik, Ahmed, Abdelmounim, Elhassane, Sanchez Mediavilla Angel, "A New Design of a Microstrip Patch Antenna with Modified Ground for DCS and Wi-MAX Applications," International Journal of Microwave and Optical Technology, Vol. 11, No. 4, July 2016.
7. Pandeeswari, R. and Raghavan, S., "Microstrip Antenna with Complementary Split Ring Resonator Loaded Ground Plane for Gain Enhancement," Microwave and Optical Technology Letters/Vol. 57, No. 2, February 2015.

Statistical vs. Rule-Based Machine Translation: A Comparative Study on Indian Languages

S. Sreelekha, Pushpak Bhattacharyya and D. Malathi

Abstract In this paper, we present our work on a case study between statistical machine translation (SMT) and rule-based machine translation (RBMT) systems on English-Indian language and Indian to Indian language perspective. Main objective of our study is to make a five-way performance comparison: such as, (a) SMT and RBMT; (b) SMT on English–Indian language; (c) RBMT on English–Indian language; (d) SMT on Indian to Indian language perspective; (e) RBMT on Indian to Indian language perspective. Through a detailed analysis, we describe the rule-based and the statistical machine translation system developments and its evaluations. Further, with a detailed error analysis, we point out the relative strengths and weaknesses of both the systems. The observations based on our study are: (a) SMT systems outperform RBMT; (b) In the case of SMT: English to Indian language MT systems perform better than Indian to English language MT systems; (c) In the case of RBMT: English to Indian language MT systems perform better than Indian to English language MT systems; (d) SMT systems perform better for Indian to Indian language MT systems compared to RBMT. Effectively, we shall see that even with a small amount of training corpus SMT system has many advantages for high-quality domain-specific machine translation over that of a rule-based counterpart.

Keywords Machine translation · Statistical machine translation
Rule-based machine translation · English-Indian machine translation
Indian-Indian language machine translation

S. Sreelekha (✉) · P. Bhattacharyya
Department of Computer Science & Engineering,
Indian Institute of Technology Bombay, Mumbai, India
e-mail: sreelekha@cse.iitb.ac.in

P. Bhattacharyya
e-mail: pb@cse.iitb.ac.in

S. Sreelekha · D. Malathi
Department of Computer Science & Engineering, SRM University, Chennai, India
e-mail: malathi.d@ktr.srmuniv.ac.in

© Springer Nature Singapore Pte Ltd. 2018
S.S. Dash et al. (eds.), *International Conference on Intelligent Computing
and Applications*, Advances in Intelligent Systems and Computing 632,
https://doi.org/10.1007/978-981-10-5520-1_59

1 Introduction

Machine translation (MT) is a process of automating the translation process from one language to another using the techniques from linguistics, translation theory, and statistics. The difference between the source and target languages and their ambiguities are the major difficulties in MT. There are many ongoing attempts to develop MT systems for regional languages using various approaches [1, 20]. The approaches to MT are categorized as rule-based or knowledge-driven approaches, and corpus-based or data-driven approaches. The RBMT approaches are further classified into transfer-based MT, Interlingua MT, and dictionary-based MT, while the corpus-based approaches are classified into example-based MT and SMT. Many studies have been conducted in the case of English to Indian language and Indian to Indian language MT system development [2–8]. This paper discusses a comparative study on RBMT and SMT approaches used in English to Indian language and Indian to Indian language MT systems.

The organization of the paper is as follows: Sect. 2 starts with the discussion about rule-based and statistical-based MT approaches; Sect. 3 presents the experiments conducted, evaluations, and error analysis which convey the main components of the paper; Sect. 4 concludes the paper.

2 Rule-Based Versus Statistical

RBMT system requires a huge human effort to prepare the rules and linguistic resources, such as morphological analyzers, part-of-speech taggers, syntactic parsers, bilingual dictionaries, transfer rules, morphological generator, and reordering rules. In the case of English to Indian languages and Indian to Indian languages, there have been many attempts with all the four RBMT approaches [3–10]. Data-driven approaches, which provide an alternative to direct- and rule-based MT systems, have come to the fore of language processing research over the past decade. These approaches uses supervised or unsupervised statistical machine learning algorithms to build statistical models from the bilingual parallel corpora. There are three different statistical approaches in MT, word-based translation, phrase-based translation, and hierarchical phrase-based model. This paper discusses the phrase-based statistical approaches used against the rule-based approaches in the English-Indian language and the Indian-Indian language MT system developments to generate quality translations.

2.1 Rule-Based Machine Translation

Rule-based MT systems works based on the rules created for morphology, syntax, lexical selection, transfer, and generation. Collection of rules and a bilingual or multilingual lexicon is the resources used in RBMT. The transfer model involves three stages: analysis, transfer, and generation. Figure 1 shows the complete work flow of translation in the form of a pipeline.

During the analysis phase linguistic analysis is performed on the input source sentence in order to extract information in terms of morphology, parts of speech, phrases, named entity, and word sense disambiguation. During the lexical transfer phase: there are two steps namely word translation and grammar translation. In word translation, source language root word is replaced by the target language root word with the help of a bilingual dictionary and in grammar translation, suffixes are getting translated. In generation, phase genders of the translated words are corrected and it will be followed by short-distance and long-distance agreements performed by intra-chunk and the inter-chunk module. This ensures that the gender, number, and person of local groups of phrases agree as also the gender of the subject's verbs or objects reflects those of the subject.

2.2 Statistical Machine Translation

The statistical approach works by generating the statistical models from the parallel aligned bilingual text corpora with some probability [11–13]. The best translation

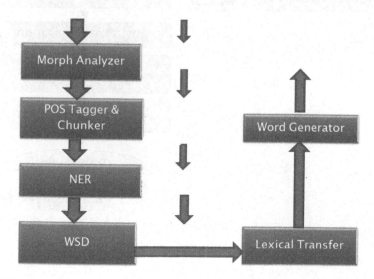

Fig. 1 RBMT work flow

will be extracted from the words which have the highest probability. Language divergence and reordering divergence are the major concerns while translating from a source language to target language [1, 8, 14]. Figure 2 shows the functional flow diagram of an SMT system. The major steps in SMT are: corpus preparation, training, decoding, and testing. Corpus preparation, alignment, and its cleaning will be done in the preprocessing step. Training is a process in which the statistical tables are extracted from the parallel corpora using supervised or unsupervised statistical algorithms [13]. In statistical machine translation, word by word and phrase-based alignment plays the major role during parallel corpus training. Translational model, language model, distortion table, phrase table, etcetera are modeled during the training. Decoding [11, 12, 15] is the most complex task in machine translation [15], where the trained models will be decoded. It is the major process in which the target language translations are being decoded using the generated phrase table, translation model, and language model. The two major concerns with SMT are the decoding complexity and the target language reordering [16].

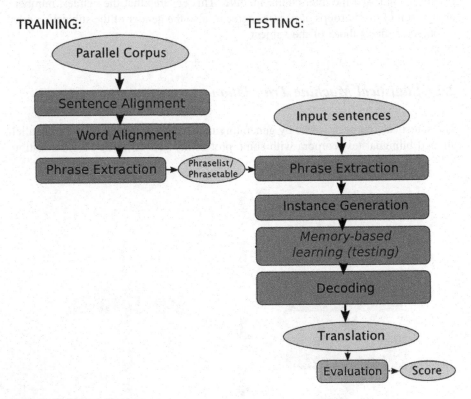

Fig. 2 SMT work flow

3 Experimental Discussions

We now describe the SMT system experiments and the comparisons with the results of the rule-based system described above. We use Moses [17] and Giza++ for the alignments and for the statistical model generation. Our experiments are focused on two research directions:

1. Indian-Indian language perspective[1,2];
2. English-Indian language perspective.[3,4]

For Indian-Indian language MT system case study, we have used Marathi–Hindi as the base language pairs, and for the English-Indian language MT system case study, we have used English–Malayalam as the base language pairs.

3.1 Statistical Machine Translation System Experiments

We manually cleaned a 90,000 sentence parallel corpus for both the Marathi-Hindi and the English–Malayalam language pairs. We have corrected the grammatical structure of the sentences and tokenized it, thereby making available a high-quality corpus for training. Table 1 describes the corpus resources we have used for training. We followed the training steps of Moses baseline system. In order to perform the tuning, we have used 500 sentence pairs. We observed that there was only slight improvement on the translation quality. Since the sentence pairs used for tuning had a number of stylistic constructions, and the BLEU based tuning tends to cause deterioration of quality. We have tested the translation system with a corpus of 1000 sentences taken from the 'ILCI tourism health' corpus as shown in Table 1. The added advantage in the case of Marathi–Hindi compared to English–Malayalam was the SOV ordering similarity between Marathi and Hindi. However, there were difficulties in handling inflected words.

Evaluation

To analyze the quality of translation, we have used both subjective evaluation and BLEU score [18] evaluation. We have used fluency as an indicator to evaluate the correct grammatical constructions present in the translated sentence and the adequacy as an indicator for the amount of meaning being carried over from the source to the target. Depending on how much sense the translation made and its

---

[1]http://tdil-dc.in/mt/common.php.

[2]http://www.cfilt.iitb.ac.in/SMT-System/.

[3]http://www.cfilt.iitb.ac.in/SMT-EM/.

[4]http://www.cfilt.iitb.ac.in/SMT-ME/.

Table 1 Corpus statistics

S. No.	Corpus source	Training corpus [manually cleaned and aligned]	Corpus size [sentences]
1	ILCI	Tourism	23,500
2	ILCI	Health	23,500
Total			47,000
S. No.	Corpus source	Tuning corpus [manually cleaned and aligned]	Corpus size [sentences]
1	ILCI	Tourism	250
2	ILCI	Health	250
Total			500
S. No.	Corpus source	Testing corpus [manually cleaned and aligned]	Corpus size [sentences]
1	ILCI	Tourism	1000
2	ILCI	Health	1000
Total			2000
S. No.	Corpus source	Testing corpus (subjective evaluation) [manually cleaned and aligned]	Corpus size [sentences]
1	ILCI	Tourism	250
2	ILCI	Health	250
Total			500

grammatical correctness, we assigned scores between 1 and 5 for each translation. The formula [19] used for computing the scores is shown in Eq. (1):

$$A/F = 100 * ((S5 + 0.8 * S4 + 0.6 * S3))/N. \qquad (1)$$

We considered the sentences with scores above 3 only. We penalize the sentences with scores 4 and 3 by multiplying their count with 0.8 and 0.6 respectively in order to make the estimate of scores much better. The results of our evaluations are given in Table 4.

3.2 English-Indian Language Case Study Results

In order to perform the English-Indian language case study, we have used English–Malayalam and Malayalam–English as the base language pairs. The results of BLEU score evaluation, subjective evaluations in terms of adequacy and fluency are shown in Table 2.

Table 2 Results of English-Indian language case studies

English–Malayalam MT system	Adequacy (%)	Fluency (%)
Rule-based	55.6	47
Statistical	77.23	87
English–Malayalam MT system	BLEU score	
Rule-based	20.8	
Statistical	39.90	
Malayalam–English MT system	Adequacy (%)	Fluency (%)
Rule-based	64.6	50.45
Statistical	74.89	84.34
Malayalam–English MT system	BLEU score	
Rule-based	29.9	
Statistical	37.90	

Table 3 Results of Indian-Indian language case studies

Marathi–Hindi MT system	Adequacy (%)	Fluency (%)
Rule-based	69.6	58
Statistical	79.8	88.4
Marathi–Hindi MT system	BLEU score	
Rule-based	23.3	
Statistical	51.60	
Hindi–Marathi MT system	Adequacy (%)	Fluency (%)
Rule-based	64.8	56.78
Statistical	75.89	85.14
Hindi–Marathi MT system	BLEU score	
Rule-based	17.9	
Statistical	43.30	

3.3 Indian to Indian Language Case Study Results

In order to perform the Indian-Indian language MT case study, we have used the Marathi–Hindi and the Hindi–Marathi system as the base pairs. The results of BLEU score evaluation, subjective evaluations in terms of adequacy and fluency are shown in Table 3.

3.4 SMT Versus RBMT Analysis

We have done a detailed error analysis on both RBMT and SMT systems. Table 4 shows the observations during the case study analysis. Further, we explain the observations of a detailed case study between English–Malayalam and Marathi–Hindi

Table 4 Performance comparison of SMT over RBMT

S. No.	Performance comparison of SMT over RBMT	
	SMT	RBMT
1	Being able to handle the rich morphology, can easily separate suffixes from inflected words with gender number person aspect and mood, leading to meaning transfer	Not able to split suffixes from inflected words with gender number person aspect and mood by itself and hence fails to handle rich morphology
2	Able to handle the verb phrases and the function words, since SMT follows memory-based training to learn phrases	Unable to effectively handle the appropriate translation and generation of function words, verb phrases, etc.
3	Rapid, easier to create, maintain, and improve upon; in short cost-effective development	Robust, high development and customization cost
4	Can handle ambiguity since it records phrase translations with its frequency of occurrence which acts as more natural word sense disambiguation	Fails to handle ambiguity due to poor quality WSD approaches
5	Good fluency and adequacy due to plentiful evidences of good quality phrase pairs recorded in phrase table	Lack of fluency
6	The language model used helped in generating more natural translations	Morph analyzers process word by word and hence fail to generate natural translations
7	Data-driven and hence domain-specific	Knowledge-driven and hence can work for out of domain data also

Table 5 Performance comparison of English–Malayalam SMT over Malayalam–English SMT

S. No.	English–Malayalam SMT	Malayalam–English SMT
1	English equivalents for the Malayalam agglutinative suffixes are in the form of prepositions. While translating from English to Malayalam, English word can easily get aligned to the agglutinated words in Malayalam, since it is a single word	While aligning from Malayalam–English, the agglutinated Malayalam word may map only to root words. Since it is separate words and there is a chance to miss out the preposition mapping in English
2	Require morphology generation for Malayalam	Require morphology analysis for Malayalam
3	Rapid, easier to create, maintain, and improve upon; in short cost-effective development	Rapid, easier to create, maintain, and improve upon; in short cost-effective development
4	Good fluency and adequacy, since there is more probability to get mapped to the inflected Malayalam word from English word	Less fluency, since multiple words have to get mapped from a single inflected form during translation is more erroneous
5	The language model used helped in generating more natural translations	The language model used helped in generating more natural translations

Table 6 Performance comparison of Malayalam–English RBMT over English–Malayalam RBMT systems

S. No.	Malayalam–English RBMT	English–Malayalam RBMT
1	English equivalents for the agglutinated Malayalam suffixes are in the form of prepositions	While translating from English to Malayalam, word by word processing of analysis and generation steps may not help to generate the correct agglutinative Malayalam word formation
2	Require morphology analysis for Malayalam	Require morphology generation for Malayalam
3	During morphology analysis from a single inflected word, agglutinated suffixes are getting separated and equivalent group words are translated during lexical transfer	During morphology generation from a group of English words, all words may not get properly formed. There is higher chances to get error in proper generation of the inflected form
4	Generating pre-positioned English words is easy	Generating rich morphological Malayalam agglutinative suffixed words is difficult
5	Fluency and adequacy will be more	Fluency and adequacy will be less

Table 7 Performance comparison of Marathi–Hindi SMT over Hindi–Marathi SMT

S. No.	Marathi–Hindi SMT	Hindi–Marathi SMT
1	Hindi equivalents for the Marathi agglutinative suffixes are in the form of post-positions. Since Marathi and Hindi have the same SOV order, it can easily map the inflections to a great level	While aligning form Hindi to Marathi, there is a probability that the agglutinative words may miss out from the post-position mapping of Hindi, since it is separate words in many cases compared to Marathi
2	Require morphology analysis for Marathi	Require morphology generation for Marathi
3	Rapid, easier to create, maintain, and improve upon; in short cost-effective development	Rapid, easier to create, maintain, and improve upon; in short cost-effective development
4	Good fluency and adequacy, since it is easy to map from the Marathi word to the Hindi equivalent form	Less fluency, since a single inflected word has to map from multiple words during the translation is more erroneous
5	The language model used helped in generating more natural translations	The language model used helped in generating more natural translations

language pairs with the SMT and RBMT experiments. Tables 5, 6, 7, and 8 show the performance comparison analysis of various aspects of SMT and RBMT approaches over Indian-Indian language and English-Indian language MT systems.

Figures 3 and 4 show SMT versus RBMT evaluation graphs for the English-Indian and Indian-Indian Language case study results.

Table 8 Performance comparison of Marathi–Hindi RBMT over Hindi–Marathi RBMT

S. No.	Marathi–Hindi RBMT	Hindi–Marathi RBMT
1	Hindi equivalents for the Marathi suffixes are in the form of post-positions. So group of Hindi words have to generate during the analysis from Marathi to Hindi	From group of Hindi words, agglutinative Marathi inflected form has to generate
2	Require morphology analysis for Marathi and generation for Hindi	Require morphology analysis for Hindi and morphology generation for Marathi
3	During morphology analysis from a single inflected word, agglutinated suffixes are getting separated and the equivalent Hindi words are translated during the lexical transfer	During morphology generation from word by word, all words may not get properly formed to generate the correct Marathi word. There is higher chance to get error in proper generation of the inflected form
4	Generating post-positioned Hindi words is easy	Generating the rich morphological Marathi agglutinative suffix word is difficult. Morph analyzers process word by word and hence fails to generate the natural Marathi translations
5	Fluency and adequacy will be more	Fluency and adequacy will be less

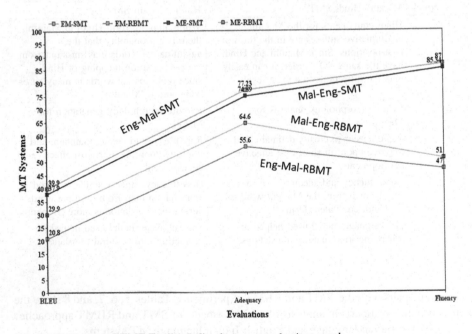

Fig. 3 SMT versus RBMT English-Indian language evaluations graph

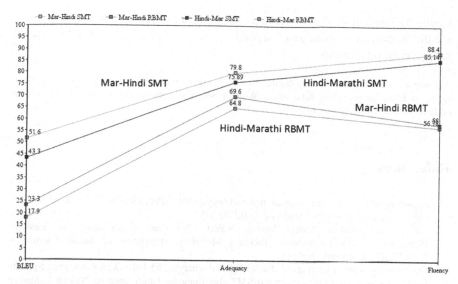

Fig. 4 SMT versus RBMT Indian-Indian language evaluations graph

4 Conclusions

In this paper, we have mainly focused on the comparative performance of SMT and RBMT systems on Indian to Indian language perspective and English to Indian language perspective. Our major observations are

1. Translation quality of SMT is relatively high as compared to the RBMT systems, considering that the efforts required to build RBMT systems are huge.
2. SMT performs better for English to Malayalam systems comparing to the Malayalam to English systems.
3. RBMT performs better for Malayalam to English compared to the English to Malayalam.
4. SMT system performs better for Marathi–Hindi compared to the Hindi–Marathi.
5. RBMT performs better for Marathi–Hindi compared to the Hindi–Marathi.
6. For English-Indian language scenario: SMT performs better for morphologically low language to rich language, and on the other hand RBMT performs better for morphologically rich language to low language.
7. Indian to Indian language MT performs better than English to Indian language MT in terms of SMT.
8. English to Indian language MT performs better than Indian to Indian language MT in terms of RBMT.

We observed that translation quality of statistical machine translation is relatively high than the rule-based system, since the efforts required to build RBMT systems are huge. Also, SMT which cannot split suffixes by itself was unable to handle the translation of inflected suffix words in some cases. RBMT, being able to use the morph analyzer can easily separate the suffixes from the inflected words and can generate translations.

Acknowledgements The authors would like to thank Department of Science & Technology, Govt. of India for providing fund under Woman Scientist Scheme (WOS-A) with the project code-SR/WOS-A/ET/1075/2014.

References

1. Anoop Kunchukuttan and Pushpak Bhattacharyya. 2012. *Partially modelling word reordering as a sequence labeling problem, COLING 2012.*
2. Anoop Kunchukuttan Abhijit Mishra, Rajen Chatterjee, Ritesh Shah and Pushpak Bhattacharyya, Shata-Anuvadak: Tackling Multiway Translation of Indian Languages, LREC 2014, Rekjyavik, Iceland.
3. Sreelekha. S., Piyush Dungarwal, Pushpak Bhattacharyya, Malathi D., Solving Data Sparsity by Morphology Injection in Factored SMT, International Conference on Natural Language Processing, ICON 2015.
4. Sreelekha, Pushpak Bhattacharyya, Malathi D. *Lexical Resources for Hindi—Marathi MT*, WIDRE Proceedings, LREC 2014.
5. Sreelekha, Pushpak Bhattacharyya. *Lexical Resources to enrich English-Malayalam Machine Translation*, LREC—International Conference on Lexical Resources and Evaluation, Slovenia, 2016.
6. Sreelekha S., Pushpak Bhattacharyya, Malathi D., "A Case study on English-Malayalam Machine Translation", iDravidian Proceedings, International Journal of Engineering Sciences, 2015.
7. Sreelekha, Raj Dabre, Pushpak Bhattacharyya 2013. *Comparison of SMT and RBMT, The Requirement of Hybridization for Marathi—Hindi MT* ICON, 10th International conference on NLP, December 2013.
8. Shachi Dave, Jignashu Parikh and Pushpak Bhattacharyya. 2002. *Interlingua based English-Hindi Machine Translation and Language Divergence*, JMT 2002.
9. Arafat Ahsan, Prasanth Kolachina, Sudheer Kolachina, Dipti Misra Sharma and Rajeev Sangal. 2010. *Coupling Statistical Machine Translation with Rule-based Transfer and Generation*. amta2010.amtaweb.org.
10. Latha R. Nair, David Peter S., Renjith Ravindran. 2012. *Design and Development of a Malayalam to English Translator—A Transfer based Approach*, International Journal of Computational Linguistics, Volume (3): Issue (1), 2012.
11. Franz Josef Och and Hermann Ney. *A Systematic Comparison of Various Statistical Alignment Models*. Computational Linguistics, 2003.
12. Franz Josef Och and Hermann Ney. 2001. *Statistical Multi Source Translation*. MT Summit 2001.
13. Peter E. Brown, Stephen A. Della Pietra. Vincent J. Della Pietra, and Robert L. Mercer*. The Mathematics of Statistical Machine Translation: Parameter Estimationn. *ACL 1993*.
14. Bonnie J. Dorr. 1994. *Machine Translation Divergences: A Formal Description and Proposed Solution*. Computational Linguistics, 1994.
15. Kevin Knight. 1999. *Decoding complexity in word-replacement translation models*, Computational Linguistics, 1999.
16. Ananthakrishnan Ramananthan, Pushpak Bhattacharyya, Karthik Visweswariah, Kushal Ladha, and Ankur Gandhe. 2011. *Clause-Based Reordering Constraints to Improve Statistical Machine Translation*. IJCNLP, 2011.

17. Philipp Koehn, Hieu Hoang, Alexandra Birch, Chris Callison-Burch, Marcello Federico, Nicola Bertoldi, Brooke Cowan, Wade Shen, Christine Moran, Richard Zens, Chris Dyer, Ondrej Bojar, Alexandra Constantin, Evan Herbst. 2007. *Moses: Open Source Toolkit for Statistical Machine Translation*, Annual Meeting of the ACL, demonstration session, Prague, Czech Republic, June 2007.
18. Kishore Papineni, Salim Roukos, Todd Ward and Wei-Jing Zhu. 2002. *BLEU: a Method for Automatic Evaluation of Machine Translation*, Proceedings of the 40th Annual Meeting of the Association for Computational Linguistics, Philadelphia, July 2002, pp. 311–318.
19. Ganesh Bhosale, Subodh Kembhavi, Archana Amberkar, Supriya Mhatre, Lata Popale and Pushpak Bhattacharyya. 2011. *Processing of Participle (Krudanta) in Marathi*. ICON 2011, Chennai, December, 2011.
20. Antony P. J. 2013. *Machine Translation Approaches and Survey for Indian Languages,* The Association for Computational Linguistics and Chinese Language Processing, Vol. 18, No. 1, March 2013, pp. 47–78.

Author Index

© Springer Nature Singapore Pte Ltd. 2018
S.S. Dash et al. (eds.), *International Conference on Intelligent Computing and Applications*, Advances in Intelligent Systems and Computing 632,
https://doi.org/10.1007/978-981-10-5520-1

677

Printed in the United States
By Bookmasters